T0178595

Fundamental Principles of Nuclear Engineering

Jiyang Yu

Fundamental Principles of Nuclear Engineering

Jiyang Yu
Tsinghua University
Beijing, China

ISBN 978-981-16-0841-4   ISBN 978-981-16-0839-1   (eBook)
https://doi.org/10.1007/978-981-16-0839-1

Jointly published with Tsinghua University Press
The print edition is not for sale in China (Mainland). Customers from China (Mainland) please order the
print book from: Tsinghua University Press.
ISBN of the Co-Publisher's edition: 978-730-24-9087-6

This Springer imprint is published by the registered company Springer Nature Singapore Pte Ltd.
The registered company address is: 152 Beach Road, #21-01/04 Gateway East, Singapore 189721,
Singapore

Contents

Chapter 1
Fundamentals of Mathematics and Physics

In engineering field, some engineering problems can be solved by using arithmetic and algebra. However, some other problems may not be adequately solved using arithmetic and algebra. Advanced mathematic tools such as calculus and integral are needed to understand physical process used in nuclear engineering [1].

1.1 Calculus

Arithmetic deals with some numbers with fixed values. While algebra deals with both literal and pure numbers. Although the value of number in algebraic can be changed from one case to the other, it also keeps a fixed value in a given problem.

Let's look at some very simple physical problems. When a solid object falls and allows it to fall freely to the ground, the speed of the object changes constantly. Another example is that the current in the alternating current (AC) circuit also changes over time. These two quantities have different values at a specific point in time. Physical problems involving values that change over time are often referred to as dynamic systems. The solution to dynamic system problems usually involves very different mathematical techniques than those described in arithmetic or algebra. Calculus involves not only all the techniques involved in arithmetic or algebra, such as addition, subtraction, multiplication and division, equations, and functions, but also more advanced techniques that are not difficult to understand [2].

There are plenty of dynamic systems encountered in nuclear engineering field. For example, the decay of radioactive materials, the startup of a nuclear reactor, and a power change of a steam turbine generator, all these examples involve quantities, which are functions of time. To analyze these examples, one has to use calculus. Although an operator of a nuclear power plant or general nuclear facility does not require a professional training on calculus, it will be helpful if he/she can understand some certain level of the basic ideas and terminology used in calculus.

© Tsinghua University Press 2022
J. Yu, *Fundamental Principles of Nuclear Engineering*,
https://doi.org/10.1007/978-981-16-0839-1_1

A brief introduction to these basic ideas and terminology of mathematics that helpful to understand a dynamic system will be discussed in this chapter.

1.1.1 Differential and Derivative

In mathematics, differential is a tool to describe the local characteristic of a function using linear techniques. Suppose a function is defined in a region. x_0 and $x_0 + \Delta x$ are two point (value) in this region. Then the incremental change of the function can be expressed as [3]:

$$\Delta y = f(x_0 + \Delta x) - f(x_0) \tag{1.1}$$

Using local linear technique, it can be expressed as:

$$\Delta y = A \cdot \Delta x + o(\Delta x) \tag{1.2}$$

where, A is a constant number independent with Δx, $o(\Delta x)$ is a higher order infinite small of Δx. We call the function $y = f(x)$ is derivable near the point of x_0 and $A \cdot \Delta x$ is called as the differential of the function $y = f(x)$ at point x_0 corresponding to Δx (the incremental change of argument x). It is denoted as dy. The incremental change of argument x is the differential of x. It is denoted as dx. So, we get:

$$dy = Adx \tag{1.3}$$

Here we use an example in physics to explain the concept of differential. One common example of dynamic system is the relationship between the position and the time for a moving solid ball or object. Figure 1.1 shows an object that is moving with a straight line from point P_1 to point P_2. The distance to P_1 along the line of travel from a fixed reference point O is represented by S_1; and by the same way, the distance to P_2 from point O by S_2.

If the time is measured by the clock, when the object is at position P_1 and the time of t_1, when the object is at position P_2, the time is indicated as t_2, and the average

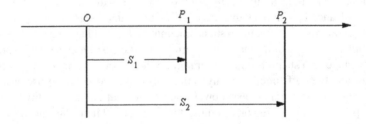

Fig. 1.1 Motion between two points

speed between points P_1 and P_2 is equal to the distance taken, divided by the time elapsed by the Eq. (1.4).

$$v_{av} = \frac{S_2 - S_1}{t_2 - t_1} \tag{1.4}$$

The traveled distance and the elapsed time are very small if the points P_1 and P_2 are very close together. A symbol Δ is used to denote the changes in these quantities. By using the symbol Δ, the average velocity between points P_1 and P_2 can be written as Eq. (1.5).

$$v_{av} = \frac{\Delta S}{\Delta t} = \frac{S_2 - S_1}{t_2 - t_1} \tag{1.5}$$

Although average velocity is one of the most important quantities, in many cases it is necessary to know the instantaneous velocity of a given time. Instantaneous velocity is usually different from average velocity unless the instantaneous velocity remains constant. Using the displacement graph in Fig. 1.2, S, versus time, t, in Fig. 1.2, the concept of derivatives is described here.

Using Eq. (1.4), you can see that the average velocity from S_1 to S_2 is $(S_2-S_1)/(t_2-t_1)$. If you connect points S_1 and S_2 in a straight line, you can see that it does not accurately reflect the slope of the curved line through all points between S_1 and S_2. Similarly, if you look at the average speed between t_2 and t_3 (very small periods), he will see that the lines connecting S_2 and S_3 follow the curved lines more closely. Assuming that the time between t_2 and t_3 is less than the time between t_2 and t_3, the line connecting S_3 and S_4 is very close to the curved line between S_3 and S_4.

When we further reduce the time interval between consecutive points, it means that the slope of the displacement curve will be closer to $\Delta S/\Delta t$. When the $\Delta t \to 0$, $\Delta S/\Delta t$ will equto to the instantaneous velocity. The expression of a derivative (in this case, the slope of the displacement curve) can be written as an Eq. (1.6). In words, this expression would be "the derivative of S with respect to time (t) is the limit of $\Delta S/\Delta t$ as Δt approaches 0."

Fig. 1.2 Displacement verses time

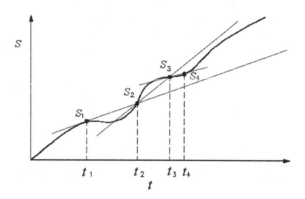

$$v = \lim_{\Delta t \to 0} \frac{\Delta S}{\Delta t} \qquad (1.6)$$

The symbols dS and dt are not the products of d and S, or of d and t, as in algebra. They are pronounced "dee-ess" and "dee-tee", respectively. And you also have to notice that the letter 'd' is written in normal style while the letter 'S' and 't' are all written in italic style. These kinds of expressions and quantities they represent are usually called as differentials.

$$v = \frac{\mathrm{d}S}{\mathrm{d}t} = \lim_{\Delta t \to 0} \frac{\Delta S}{\Delta t} \qquad (1.7)$$

where, dS is the differential of displacement S and dt is the differential of time t. These expressions are incremental changes, where dS is an incremental change in distance of S, and dt is an incremental change of time t.

The combined expression dS/dt is usually called as a derivative; it is the derivative of displacement S with respect to time t. In some simplest terms, a derivative is the rate of change of one quantity with respect to another one. So, dS/dt is the change rate of distance with the change of time.

According to Fig. 1.2, dS/dt is the instantaneous velocity at any point. This value of it is equal to the slope of the curve at that point.

While the equation of instantaneous velocity, $v = \mathrm{d}S/\mathrm{d}t$, may looks like a complicated expression, it will be familiar is one take it an example. For example, instantaneous velocity is the value shown by the speedometer of a moving car. The speedometer gives the value of the rate of change of displacement with respect to time; it provides the derivative of displacement S with respect to time t; i.e. it shows the value of dS/dt.

The terminology of differentials and derivatives are very important to understand dynamic systems. They are not only used to express relationships among distance moved and time which is velocity, but also can be used to express relationships among many other physical quantities. One of the most useful parts of understanding these ideas is getting a physical interpretation of their meaning. When a relationship is written using a derivative such as neutron power change with time, the physical meaning of rates of change should be readily known.

In some times, expressions are usually written as deltas, they can be understood as changes. Thus, the expression ΔT, where T could be the symbol for temperature, represents a step change in temperature. As previously discussed, a letter in lower case, written as d, is used to represent very small changes. So, dT is used to represent a very small change of temperature. The fractional change in a physical quantity is defined as the change divided by the value of that quantity. Thus, dT is an incremental change of temperature, and dT/T is a fractional change of temperature. When expressions are written using derivatives, they are in terms of rates of change. Thus, the dT/dt is the rate of change of temperature T with respect to time t.

Examples 1.1 Try to interpret the expression $\Delta v/v$, and try to write it in terms of a differential.

Solution: the fractional change of velocity is $\Delta v/v$. It is the step change of velocity divided by that velocity. In the case of Δv is taken as an incremental change, it can be written as a differential.

In terms of a differential, $\Delta v/v$ may be written as dv/v.

Examples 1.2 Try to express the physical meaning of the following equation relating the work done that is noted as W when a force F on a body moves through a distance of x.

Solution: the expression is $dW = Fdx$. This expression includes differentials dW and dx that can be interpreted as terms of incremental changes. The work done incremental equals to the incremental distance multiplied by the force.

Examples 1.3 Try to express the Eq. (1.8) with physical interpretation of relating the force F applied to an object, and its instantaneous velocity v, its mass m and time t.

$$F = m\frac{dv}{dt} \tag{1.8}$$

Solution: This equation has a derivative of velocity with time that is noted as dv/dt. It is the rate of velocity change with time. The mass of the object multiplied by the rate of change of velocity with time equals to the force applied to the object.

Equation (1.8) is the Newton's second law. It can be expressed as the second derivative of distance.

$$F = m\frac{dv}{dt} = m\frac{d^2 S}{dt^2} \tag{1.9}$$

To extend the concept of derivative to multi arguments function, partial derivative is introduced. For a multi arguments function f, it's partial derivative defined as Eq. (1.10).

$$\frac{\partial f}{\partial x}\bigg|_{\substack{x=x_0\\y=y_0\\z=z_0}} = \lim_{\Delta x \to 0} \frac{f(x_0 + \Delta x, y_0, z_0) - f(x_0, y_0, z_0)}{\Delta x} \tag{1.10}$$

1.1.2 Integral

When small change of one quantity is considered, differentials and derivatives will be used to illustrate a physical system. For example, the relationship of displacement and time for a moving object is defined as instantaneous velocity which is written as dS/dt. In some physical systems, when the rates of change can be measured directly,

Fig. 1.3 Graph of velocity
verses time

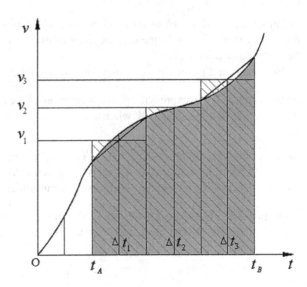

and the integral or summation value is concerned, it involves another aspect of the mathematics named as integral and summations.

Figure 1.3 shows the instantaneous velocity of an object as a time function. This is a typical graph that can be generated if the data of the car speedometer is recorded as a function of time.

At any given moment, the speed of an object can be determined by reference to Fig. 1.3. However, if you want to determine the distance within a certain interval, you must use some new terms [4].

Let's take a look at the velocity change between time t_A and t_B. The first method is to divide the past time into three intervals (Δt_1, Δt_2, Δt_3) and assume that the velocity remains constant during these intervals. During Δt_1, the velocity is assumed to be constant at the mean value v_1; During Δt_2, the velocity is also assumed to be a constant of another mean value v_2; During the interval of Δt_3, the velocity remains a constant value of the average velocity v_3. Then, the entire travel distance is the sum of the velocity products, and the past time of approximately three intervals. Equations (1.11) are roughly equivalent to the distance taken between t_A and t_B. During this interval, it is an approximate area under the curve.

$$S = v_1 \Delta t_1 + v_2 \Delta t_2 + v_3 \Delta t_3 = \sum_{i=1}^{3} v_i \Delta t_i \qquad (1.11)$$

Equation (1.11) is a summation equation. A summation indicates the sum of some quantities. The Greek letter in upper case $\sum$, is used to express a summation. Generalized subscripts are usually used to simplify writing summations. The summation given in Eq. (1.11) can be written in the form of the Eq. (1.11):

$$S = \sum_{i=1}^{\infty} v_i \Delta t_i \qquad (1.12)$$

The number below the summary symbol represents the value of i in the first term of sum; And the above number indicates the value of i in its last term.

As shown in Fig. 1.3, the time is divided into three smaller intervals, and the sum is only approximate to the distance. However, if the interval is divided into very small intervals, you can get the exact answer. When this task is done, the distance is written as a summary of an unlimited number of terms.

$$S = \sum_{i=1}^{\infty} v_i \Delta t_i = \int_{t_B}^{t_A} v \, dt \qquad (1.13)$$

Examples 1.4 Try to express the physical interpretation of the following equations related to the work done, W, when a force moves a body from position x_1 to x_2.

$$W = \int_{x_B}^{x_A} F \, dx \qquad (1.14)$$

Solution: The physical meaning of this equation is a summation. The total amount of work done is the integral of $F \, dx$ from $x = x_1$ to $x = x_2$. This can be visualized as taking the product of the instantaneous force and the incremental change in position dx between x_1 and x_2, and summing all of these products.

Examples 1.5 Try to express the physical interpretation of the following equation relating the amount of radioactive material as a function of the elapsed time, t, and the decay constant, λ.

$$\int_{N_0}^{N_1} \frac{dN}{N} = -\lambda t \qquad (1.15)$$

Solution: The physical meaning of this equation a summation. The negative of the product of the decay constant and the elapsed time equals the integral of dN/N from $N = N_0$ to $N = N_1$. This integral can be expressed as taking the quotient of the incremental change in N, divided by the value of N at points between N_0 and N_1, and summing all of these quotients.

1.1.3 Laplace Operator

The Laplace operator [5] is useful in nuclear engineering to express conservation of neutron, mass, momentum or energy. For n dimensional space, the Laplace operator is a two order differential operator. It is the divergence of gradient of a function. In rectangular plane coordinate system shown as Fig. 1.4, the Laplace operator has expression as shown in Eq. (1.16)

$$\nabla^2 u = \nabla \cdot (\nabla u) = \frac{\partial u}{\partial x^2} + \frac{\partial u}{\partial y^2} + \frac{\partial u}{\partial z^2} \tag{1.16}$$

where the gradient operator ∇ is defined as:

$$\nabla = \frac{\partial}{\partial x}\,\vec{i} + \frac{\partial}{\partial y}\,\vec{j} + \frac{\partial}{\partial z}\,\vec{k} \tag{1.17}$$

In cylindrical coordinate system shown as Fig. 1.5, the transform of coordinates are:

Fig. 1.4 Rectangular plane coordinate system

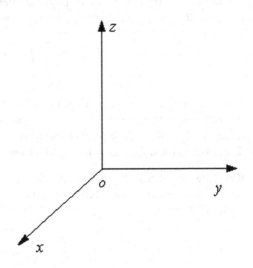

Fig. 1.5 Cylindrical coordinate system

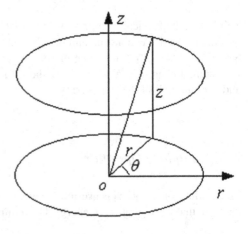

$$r = \sqrt{x^2 + y^2}, \quad \theta = \arctan\left(\frac{y}{x}\right), \quad z = z \tag{1.18}$$

Do partial derivative of coordinates, we get:

$$\frac{\partial r}{\partial x} = \frac{x}{r} = \cos\theta \tag{1.19a}$$

$$\frac{\partial r}{\partial y} = \frac{y}{r} = \sin\theta \tag{1.19b}$$

$$\frac{\partial \theta}{\partial x} = -\frac{\sin\theta}{r} \tag{1.19c}$$

$$\frac{\partial \theta}{\partial y} = \frac{\cos\theta}{r} \tag{1.19d}$$

Thus, we have:

$$\frac{\partial u}{\partial x} = \frac{\partial u}{\partial r}\frac{\partial r}{\partial x} + \frac{\partial u}{\partial \theta}\frac{\partial \theta}{\partial x} = \cos\theta\frac{\partial u}{\partial r} - \frac{\sin\theta}{r}\frac{\partial u}{\partial \theta} \tag{1.20a}$$

$$\frac{\partial u}{\partial y} = \frac{\partial u}{\partial r}\frac{\partial r}{\partial y} + \frac{\partial u}{\partial \theta}\frac{\partial \theta}{\partial y} = \sin\theta\frac{\partial u}{\partial r} + \frac{\cos\theta}{r}\frac{\partial u}{\partial \theta} \tag{1.20b}$$

Finally, we get:

$$\begin{aligned}
\frac{\partial^2 u}{\partial x^2} &= \left(\cos\theta\frac{\partial}{\partial r} - \frac{\sin\theta}{r}\frac{\partial}{\partial \theta}\right)\left(\cos\theta\frac{\partial u}{\partial r} - \frac{\sin\theta}{r}\frac{\partial u}{\partial \theta}\right) \\
&= \cos^2\theta\frac{\partial^2 u}{\partial r^2} + \frac{\sin^2\theta}{r}\frac{\partial u}{\partial r} \\
&\quad - \frac{2}{r}\sin\theta\cos\theta\frac{\partial^2 u}{\partial r\partial \theta} + \frac{\sin^2\theta}{r^2}\frac{\partial^2 u}{\partial \theta^2} + \frac{2\sin\theta\cos\theta}{r^2}\frac{\partial u}{\partial \theta}
\end{aligned} \tag{1.21}$$

$$\begin{aligned}
\frac{\partial^2 u}{\partial y^2} &= \left(\sin\theta\frac{\partial}{\partial r} + \frac{\cos\theta}{r}\frac{\partial}{\partial \theta}\right)\left(\sin\theta\frac{\partial u}{\partial r} + \frac{\cos\theta}{r}\frac{\partial u}{\partial \theta}\right) \\
&= \sin^2\theta\frac{\partial^2 u}{\partial r^2} + \frac{\cos^2\theta}{r}\frac{\partial u}{\partial r} + \frac{2}{r}\sin\theta\cos\theta\frac{\partial^2 u}{\partial r\partial \theta} \\
&\quad + \frac{\cos^2\theta}{r^2}\frac{\partial^2 u}{\partial \theta^2} - \frac{2\sin\theta\cos\theta}{r^2}\frac{\partial u}{\partial \theta}
\end{aligned} \tag{1.22}$$

$$\frac{\partial^2 u}{\partial z^2} = \frac{\partial^2 u}{\partial z^2} \tag{1.23}$$

Make an arrangement, it becomes:

$$\frac{\partial^2 u}{\partial x^2} + \frac{\partial^2 u}{\partial y^2} + \frac{\partial^2 u}{\partial z^2} = \frac{\partial^2 u}{\partial r^2} + \frac{1}{r}\frac{\partial u}{\partial r} + \frac{\sin^2 \theta}{r^2}\frac{\partial^2 u}{\partial \theta^2} + \frac{\partial^2 u}{\partial z^2} \tag{1.24}$$

Thus, the Laplace operator in cylindrical coordinate system is expressed as Eq. (1.25).

$$\nabla^2 = \frac{1}{r}\frac{\partial}{\partial r}\left(r\frac{\partial}{\partial r}\right) + \frac{1}{r^2}\frac{\partial}{\partial \theta^2} + \frac{\partial}{\partial z^2} \tag{1.25}$$

For spherical coordinate system shown as Fig. 1.6, one can get the expression of Eq. (1.26). We leave it as a homework for you to derive.

$$\nabla^2 = \frac{1}{r^2}\frac{\partial}{\partial r}\left(r^2\frac{\partial}{\partial r}\right) + \frac{1}{r^2 \sin\theta}\frac{\partial}{\partial \theta}\left(\sin\theta\frac{\partial}{\partial \theta}\right)$$
$$+ \frac{1}{r^2 \sin\theta}\frac{\partial}{\partial \phi^2} \tag{1.26}$$

Fig. 1.6 Spherical coordinate system

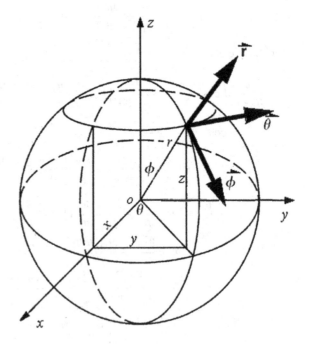

1.2 Units

Numbers alone are not enough to describe physical quantities. For example, "The pipeline must be 4 long to install" does not make sense unless the length unit of the data is also specified. By adding units to the numbers, it is clear that "the pipe must be 4 m long to be installed."

Units define the size of the measurement. If we measure length, the unit of length may be one meter or kilometer, all of which describe different length sizes. Emphasize the importance of giving units measurements to numbers that describe physical quantities when it is noted that the same physical quantity can be measured using different units. For example, length can be measured in meters, kilometers, inches, miles, or various other units.

A number of unit systems have been established to use each of the basic dimensions mentioned earlier. The unit systems used today are described in the following sections and provide some examples of the units used in each system.

1.2.1 Unit Systems

There are currently two unit systems in the field of nuclear engineering, the English units and the International System of Units (SI). In some countries, the English system is currently used. The unit system consists of different units of each basic size or measurement. The basic units of the system of International System of Units (SI) are shown in the Table 1.1. It is also known as an MKS system [6].

Other quantities can be expressed as the basic units. They are called as derived quantities and some of them are shown in Table 1.2.

MKS systems are much simpler than English systems because they use decimal-based systems where prefixes are used to represent the power of ten. For example, one kilometer is 1000 m and one centimeter is one hundredth of a meter. The English

Table 1.1 The basic units of the international system of units

Quantity	Symbol of quantity	Name	Symbol of unit
Length	L	Meter	m
Mass	m	Kilogram	kg
Time	t	Second	s
Current	I	Ampere	A
Temperature	T	Kelvin	K
Quantity of mass	$n(v)$	Mole	mol
Luminous intensity	$I\ (Iv)$	Candela	cd

Table 1.2 Some of derived quantities used in nuclear engineering

Quantity	Symbol of quantity	Symbol of unit	Relationship with the basic units
Energy	E	J	$\text{Kg m}^2 \text{ s}^{-2}$
Force	F	N	kg m s^{-2}
Power	P	W	$\text{kg m}^2 \text{ s}^{-3}$
Charge	C	C	A s
Voltage	V	V	$\text{kg m}^2 \text{ s}^{-3} \text{ A}^{-1}$
Resistance	R	Ω	$\text{kg m}^2 \text{ s}^{-3} \text{ A}^{-2}$
Capacity	C	F	$\text{kg}^{-1} \text{ m}^{-2} \text{ s}^4 \text{ A}^2$
Inductance	L	H	$\text{kg m}^2 \text{ s}^{-2} \text{ A}^{-2}$
Frequency	f	Hz	s^{-1}
Magnetic Flux	Φ	Wb	$\text{kg m}^2 \text{ s}^{-2} \text{ A}^{-1}$
Magnetic Flux Density	B	T	$\text{m}^{-1} \text{ A}$

Table 1.3 Prefixes of MKS system

Symbol	Prefixes	Powers of ten
y	yocto-	10^{-24}
z	zepto-	10^{-21}
a	atto-	10^{-18}
f	femto-	10^{-15}
p	pico-	10^{-12}
n	nano-	10^{-9}
u	micro-	10^{-6}
m	milli-	10^{-3}
k	kilo-	10^{3}
M	mega-	10^{6}
G	giga-	10^{9}
T	tera-	10^{12}
P	peta-	10^{15}
E	exa-	10^{18}
Z	zetta-	10^{21}
Y	yotta-	10^{24}

system has strange conversion constants. For example, a mile is 5,280 feet, and an inch is twelfth of a foot. The prefixes used in MKS systems are listed in Table 1.3.

Table 1.4 Relationship to convert units	Length	1 inch = 25.4 mm 1 foot = 12 inches = 0.3048 m 1 yard = 3 feet = 0.9144 m 1 mile = 1760 yards = 1.609 km 1 nautical mile = 1852 m
	Area	1 square inch = 6.45 square centimeter 1 square foot = 144 square inch = 9.29 square decimeter 1 square yard = 9 square foot = 0.836 square meter 1 acre = 4840 square yard = 0.405 hectare 1 square mile = 640 acre = 259 hectare
	Volume	1 cubic inch = 16.4 cubic centimeter 1 cubic foot = 1728 cubic inch = 0.0283 cubic meter 1 cubic yard = 27 cubic foot = 0.765 cubic meter
	Mass	1 pound = 16 oz = 0.4536 kg

1.2.2 Conversion of Units

To convert from one measurement unit to another measurement unit (for example, to convert 5 feet to meters), one can use the appropriate equivalent relationship from the conversion Table 1.4 [7].

1.2.3 Graphics of Physical Quantity

Graphics are used widely to illustrate quantity in the science [8]. It could be two dimensional, three dimensional or even higher order dimensional. Here we focus on two-dimensional graphics.

A two dimensional graph has two axes to express two quantities, for example, x and y. The graph or curve in the frame of x and y expresses the relationship of the two quantities.

Let us look at Fig. 1.7 [9]. In this graph, the horizontal axis is temperature. While the vertical axis is thermal conductivity of material (here is UO_2) [10]. Both axes are scaled by corresponding unit. For example, the point of (600, 5) in the graph means the conductivity of UO_2 at temperature of 600 °C is 5 W/(m·°C). A number in graph is a pure number without unit but quantities are measured by unit. That means a thermal conductivity of 5 W/(m·°C) divided by the unit of W/(m·°C), one gets a pure number of 5. That is the reason to mark the horizontal axis as $t/°C$ because a quantity of temperature divided by unit of °C, one gets a pure number in the graph.

There are some elements in a standard two-dimensional graph:

(1) Figure number and figure title. Here "Fig. 1.7"is a figure number and "Thermal Conductivity of UO_2 vs Temperature" is the title of the figure. The title of a figure is put at the bottom of the figure usually.

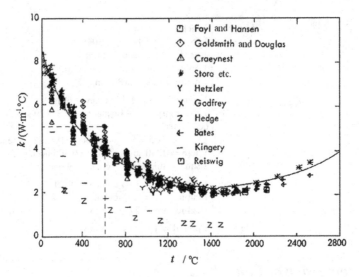

Fig. 1.7 Thermal conductivity of UO₂ verses temperature

(2) Framework of axes. The framework of Fig. 1.7 is a rectangle box. Top line and right line could be hidden sometimes.

(3) Scale of axis. In Fig. 1.7, we scale the y axis as "0, 2, 4, 6, 8, 10". A number in scale is a pure number without unit. To express the physical meaning of a quantity, it should be combined with the name of axis.

(4) Name of axis. In Fig. 1.7, the name of y axis is $k/(\text{W m}^{-1\cdot\circ}\text{C}^{-1})$. The italic letter k is a quantity means thermal conductivity. The normal letters of $\text{W m}^{-1\cdot\circ}\text{C}^{-1}$ is the unit of this quantity. A quantity with unit divided by a unit, one can get a pure number. For example, Goldsmith and Douglas measured a thermal conductivity of $k = 5.0$ W m$^{-1\circ}$C^{-1}, at temperature of 620°C. This is a point in Fig. 1.7 as (620, 5.0). The quantity t is a measurable physical quantity with unit of °C, $t = 620$°C. Thus $t/$°C $= 620$°C/°C $= 620$, we get a pure number. For $k = 5.0$ W m$^{-1\circ}$C^{-1}, in the same way, we get $k/(\text{W m}^{-1\circ}\text{C}^{-1}) = 5.0$. The combination of two pure number such as (620, 5.0) can be expressed as a point in two dimensional graph. It could be confused if the name of axis is texted as "t (°C)", "t [°C]" or "t °C".

(5) Legend. Legend is used to express different data from different authors in Fig. 1.7.

The x-axis of Figs. 1.7 and 1.8 are same because of the expression of "$t/$°C $= 1200$" is exactly same as "$t/(10^2$°C$) = 12$". It is quite different with Fig. 1.9. In Fig. 1.9, the x-axis is named as "$t(10^2$°C$)$". It could be confused with "$t(10^2$°C$) = 12$", one gets $t = 0.12$°C.

Exercises

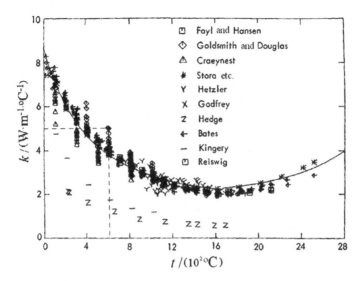

Fig. 1.8   Thermal conductivity of UO₂ verses temperature (same as Fig. 1.7)

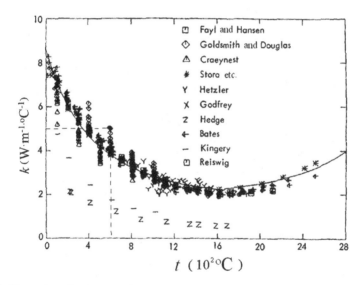

Fig. 1.9   Thermal conductivity of UO₂ verses temperature (lack of standardization)

1. A solid ball is dropped into a quiet pool, and waves will move in circles from the location of the splash outward at a constant velocity of 0.2 m/s. Determine the increasing rate of area of circle when the radius is 1 m.

2. Explain the physical interpretation of the equation bellow relating the amount of radioactive material as a function of elapsed time, t, and the decay constant, λ.

$$\int_{N_0}^{N_1} \frac{dN}{N} = -\lambda t$$

3. The unit of pressure is psi in English system, which means pound-force per square inch. 1 psi means a force of one pound-force exerted on an area of one square inch. One pound-force is the gravity force of one-pound mass in the gravity of the Earth. Determine the pressure in unit of Pa corresponding to 1 psi.

4. Derive the Laplace operator in spherical system.

$$\nabla^2 = \frac{1}{r^2}\frac{\partial}{\partial r}\left(r^2\frac{\partial}{\partial r}\right) + \frac{1}{r^2\sin\theta}\frac{\partial}{\partial \theta}\left(\sin\theta\frac{\partial}{\partial \theta}\right) + \frac{1}{r^2\sin\theta}\frac{\partial}{\partial \phi^2}$$

5. List the errors in the following figure.

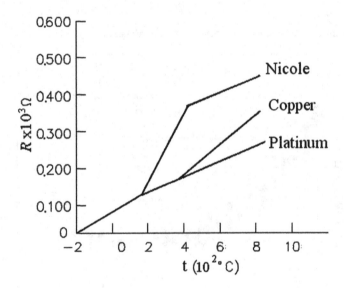

References

1. Grigsby LL. The electric power engineering handbook. Boca Raton: CRC Press; 2000.
2. Doe Fundamentals Handbook. Mathematics Volume 1. DOE-HDBK-1014/1-92. 1992.
3. Rudin W. Principles of mathematical analysis, 3rd ed. McGraw-Hill Education; 1976.
4. Godement Z. Mathematical analysis. Springer; 2004.
5. Ranjan P. Discrete laplace operator: theory and applications. Ohio: The Ohio State University; 2012.
6. Halliday D, Resnick R, Walker J. Fundamentals of physics, Extended, 7th Edn. Wiley Press; 2004.

7. Horvath AL. Conversion tables of units in science & engineering. London: Macmillan; 1986
8. Xin H, Fengho Y. Writing modern science and technology articles. Hefei: Anhui People's Press; 1997.
9. Callister WD. Fundamentals of materials science and engineering, 5th edn. John Wiley & Sons, Inc.
10. Chandramouli D. Thermalhydraulic and safety analysis of Uranium dioxide-Beryllium oxide composite fuel. Purdue: Purdue University; 2013.

Chapter 2
Thermodynamics

Thermodynamics [1] is a section of physics, which studies the dynamic behavior, and law of movement of microscopic molecules using macroscopic ways. Thermodynamics deals with thermal theory at the point of view of energy balance and transform. Energy can be transformed from one form to another but the total quantity of energy is conserved.

Some properties are used to analyze thermal state of material such as pressure, temperature, specific volume, and enthalpy etc. Based on observation and experiment, there are some relationships between these status parameters of matter. We will introduce the thermodynamic properties of matter, the relationship of energy and work, the first law of thermodynamics, and the second law of thermodynamics in this chapter. The efficiency of thermodynamic cycle is also discussed in this chapter.

2.1 Thermodynamic Properties

Thermodynamic properties describe the state or measurable characteristics of substance. Understanding these properties is essential to the understanding of thermodynamics [2].

1. **Pressure**

Pressure is defined as:

$$p = \frac{F_n}{A} \tag{2.1}$$

where F_n is the force normal to face, N; A is the area of face, m^2. The unit of pressure is Pa and 1 Pa $= 1$ N/m^2. One Newton force is equal to the force needed to accelerate 1 kg mass at 1 m/s^2.

© Tsinghua University Press 2022
J. Yu, *Fundamental Principles of Nuclear Engineering*,
https://doi.org/10.1007/978-981-16-0839-1_2

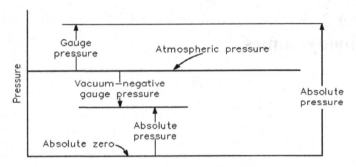

Fig. 2.1 Pressure relationships

When the pressure is measured by the relative to a perfect vacuum that is zero Pa, it is called an absolute pressure [3]; when it is measured relative to a normal atmospheric pressure which is around 10^5 Pa, it is called as a gauge pressure [4]. The gauge pressure scale was developed because most pressure gauges register zero when it is opend to the surrounding of atmosphere. Therefore, pressure gauges measure the difference between the pressure of the fluid to which they are connected and that of the surrounding space.

If the pressure is lower than that of the sapce, it is designated as a vacuum [5]. A perfect vacuum would correspond to zero Pa. All values of absolute pressure are positive, because a negative value would indicate tension that could happen in a solid, but which is considered impossible in any fluid in nature. Gauge pressures are negative if they are below atmospheric pressure and positive if they are above atmospheric pressure. Figure 2.1 shows the relationships between gauge, absolute, vacuum, and atmospheric pressures [6].

As we already learned in the first chapter, in the English unit system, pressure is measured as psi, which is pound-force per square inch. Some units of pressure can be converted each other as follows.

$$1 \text{ bar} = 10^5 \text{ Pa} \tag{2.2}$$

$$1 \text{ atm} = 1.01325 \times 10^5 \text{ Pa} \tag{2.3}$$

$$1\text{at} = 1 \text{ kgf/cm}^2 = 98066.5 \text{ Pa} \tag{2.4}$$

$$1 \text{ atm} = 14.7 \text{ psia} \tag{2.5}$$

Example 2.1 How deep can a diver descend down in ocean water (density $= 1000 \text{ kg/m}^3$) without damaging his watch, which will withstand an absolute pressure of 6 atm?

Solution: $p_{\text{abs}} = 1 \text{ atm} + p_{\text{g}}$

$$6 \times 1.01325 \times 10^5 \text{ Pa} = 1.01325 \times 10^5 \text{ Pa} + p_g$$
$$p_g = 5 \times 1.01325 \times 10^5 \text{ Pa} = 1000 \times 9.81 \times H$$
Get : $H = 51.54$ m

2.   Temperature

Temperature is quantity used to measure of the molecular activity in substance. The greater the molecules activity in micro space, the higher the temperature. It is a relative measure of how "hot" or "cold" a substance is and it can be used to predict the direction of heat transfer.

There are two temperature scales normally employed in the world [7]. They are Fahrenheit (F) and Celsius (°C) scales.

The Fahrenheit scale is based on a specification of 180 increments between the freezing point and boiling point of water at normal atmospheric pressure. The Celsius scale is based on a specification of 100 increments between the freezing point and boiling point of water at normal atmospheric pressure.

The zero points on the scales were arbitrary set in history. The temperature at which water boils was set at 100 on the Celsius scale and the body temperature of human body in fever was set to 100 on the Fahrenheit scale. The freezing point of water was selected as the zero point of the Celsius one while the coldest temperature achievable with a mixture of ice and salt water was selected as the zero point of the Fahrenheit one.

The relationship between these two scales is represented by Eq. (2.6) [8].

$$\frac{t_F}{°F} = \frac{9}{5} \frac{t}{°C} + 32 = \frac{9}{5} \frac{t_K}{K} - 459.67 \tag{2.6}$$

Example 2.2 Suppose one measures his/her body temperature and gets 100°F using Fahrenheit thermometer, what is his/her body temperature in Celsius scale?

Solution: according to Eq. (2.6), you can get:

$$t = \frac{\frac{t_F}{°F} - 32}{1.8} = 37.8 \text{ °C}$$

It is useful to define an absolute temperature scale that has only positive values for all temperature. The absolute temperature scale that corresponds to the Celsius scale is Kelvin (K) scale, and the absolute scale that corresponds to the Fahrenheit scale is Rankine (R) scale. The zero points on both absolute scales represent the same physical state that there is no molecular motion of individual atoms. The relationships between them are shown in Eq. (2.7).

$$\frac{t}{°C} = \frac{t_K}{K} - 273.15 \tag{2.7}$$

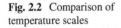

Fig. 2.2 Comparison of
temperature scales

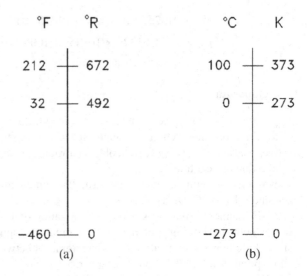

The relationship between absolute temperature and relative temperature can be expressed as Eqs. (2.8) and (2.9). The small part of decimal numbers is usually neglected in engineering field.

$$R = F + 460 \tag{2.8}$$

$$K = C + 273 \tag{2.9}$$

Figure 2.2 showes the comparison of different temperature scales.

3. Heat

Heat is energy in transit like work [9]. The transfer of energy as heat occurs at the molecular level because of a temperature difference. Heat is noted as a symbol of Q. In engineering applications, the unit of heat is J in the SI system or British thermal unit (Btu) in the English unit system.

Same as work, the amount of heat transferred depends upon the path of process, not simply on the initial and final conditions of the system. In addition, same as work, it is also important to distinguish the heat added to a system with that the heat removed from a system. In the case of heat is added to the system, a positive value of heat is used. This is in contrast to work, which is negative when work is transferred to the system and positive when work is transferred from the system.

A lower-case symbol q is sometimes used to indicate the heat added to or removed from a system per unit mass. It is the total heat (Q) divided by the total mass (m) of system. The name of "specific heat" is not used for q because specific heat is already used for another physical parameter which means the heat required to increase the temperature of unit mass in one degree. The quantity represented by q is called as *the heat transferred per unit mass*.

In order to quantify heat, the best definition method is to examine the relationship between the change of heat in the system (addition or removal) and the change of temperature of the system itself. You may be familiar with a physical phenomenon that when a substance is heated, its temperature will rise. Conversely, when it cools, its temperature decreases. If the heat added or removed from a substance changes its temperature, this heat is usually called sensible heat.

Another kind of heat is called latent heat. The so-called latent heat is the change of heat in a substance (added or removed from a substance) when the temperature remains unchanged, and the latent heat will produce a phase change. When latent heat is added to a substance, the temperature of the substance does not change. Latent heat can be divided into two different types. The first is the latent heat of polymerization. This is the heat that needs to be added or removed due to the phase transition between solid and liquid. The second type of latent heat is the latent heat of vaporization. This is the heat that needs to be added or removed to change the phase transition between liquid and steam. The latent heat of vaporization is sometimes referred to as the latent heat of condensation.

Different substances have different degrees of influence on temperature due to the increase of heat. When the same amount of heat is added to different substances, their temperature will increase by different amounts. The ratio of heat (Q, added or removed from the substance) to the change in temperature (ΔT) is called the heat capacity of the substance. The heat capacity per unit mass of the substance is called the specific heat capacity (CP) of the substance. The subscript p indicates that heat is added or removed under constant pressure.

Both work and heat are transient phenomena. The system itself does not have heat or work, but when the energy state of the system changes, heat or work transfer may occur. Both heat and work occur at the boundary of the system, because both are observed experimentally at the boundary of the system. Both represent energy across system boundaries.

4.   **Specific Volume**

Specific volume (v) is the total volume of the substance divided by the total mass of the substance (m) (volume per unit mass) [10]. The unit is m³/kg. Density (ρ) is the total mass of the substance (m) divided by the total volume (V) of the substance (mass per unit volume). The unit is kg / m³. Density is the reciprocal of its specific volume as followes.

$$\rho = 1/v \qquad (2.10)$$

5.   **Specific Gravity**

Specific gravity is a measure of the relative density of a substance compared to the density of water at a specific temperature [11]. Physicists use 39.2°f (4 °C) as the standard, but engineers usually use 60°F. In the SI system of units, the water density at standard temperature is 1000 kg/m³. Therefore, the specific gravity (dimensionless) of the liquid is the same as the density, in kg/m³. Since the density of the fluid varies with temperature, the specific gravity must be determined and specified at a specific temperature.

6. Humidity

Humidity in the air is the amount of moisture (water vapor) in the air. It can be divided into absolute humidity or relative humidity. Absolute humidity is the mass of water vapor per unit volume of air (g water / cm^3 air). Relative humidity is the ratio of the amount of water vapor present in the air to the maximum amount that the air may contain at this temperature. Relative humidity is usually expressed as a percentage. If the water vapor in the air is saturated, the relative humidity is 100%. If there is no water vapor in the air, the relative humidity is 0%.

2.2 Energy

Generalized energy is defined as the ability of a system to change (in thermodynamics, it refers to the change of work or heat).

1. Potential Energy

Potential energy (E_p) is defined as the energy of position [12]. It is defined by Eq. 2.11.

$$E_p = mgH \tag{2.11}$$

2. Kinetic Energy

Kinetic energy (E_k) is the energy of motion. It is defined as half of mass multiplied by square of velocity.

3. Specific Internal Energy

The potential energy and kinetic energy in mechanical energy are the macroscopic manifestations of energy. They can be quantified according to the position and speed of the object. In addition to these macroscopic forms of potential energy and kinetic energy, matter also has several microscopic forms of energy. Micro form energy includes energy generated by rotation, vibration, translation and interaction between material molecules. All these energies in the form of onlookers cannot be measured directly, but corresponding technologies have been developed to evaluate the change of the total energy of all these micro forms. These microscopic forms of energy are called the internal energy of matter and are usually represented by the symbol U. In engineering applications, the unit of internal energy is Joule, which is the same as that of heat.

The specific internal energy (u) of a substance is the internal energy of a substance per unit mass. Its value is equal to the total internal energy (U) divided by the total mass (m). It can be calculated as a function of temperature and specific volume as shown in the Eq. (2.12).

$$u = f(T, v) \tag{2.12}$$

The unit of the specific internal energy is J/kg.

4. **Specific p-V Energy**

In addition to the above internal energy (U), there is another form of energy, which is important to understand the energy transfer process. This form of energy is called p-V energy, which is numerically equal to the product of p and V, that is, the product of pressure and volume. It is produced by the pressure (p) and volume (V) of the fluid. Because energy is the ability of a system to make changes, a system that allows pressure and volume expansion, pressurized fluid has the ability to do work. In engineering applications, the unit of p-V energy, also known as flow energy, is the unit of pressure multiplied by volume.

The specific p-V energy of a substance is the p-V energy per unit mass. It equals the total p-V energy divided by the total mass, or the product of the pressure p and the specific volume v, and is written as pv.

5. **Specific Enthalpy**

Specific enthalpy (h) is defined as $h = u + pv$, where u is the specific internal energy of the system being studied, p is the pressure of the system and v is the specific volume of the system [13].

$$h = u + pv \tag{2.13}$$

The physical quantity enthalpy is usually used to describe the "open" system problem in thermodynamics. Enthalpy, like pressure, temperature and volume, is a property of matter. But it cannot be measured directly. Usually, the enthalpy of a substance is given relative to a reference value. For example, at 0.01 °C and a normal atmospheric pressure, the specific enthalpy of water is defined as the reference value of zero, and then the specific enthalpy of water or steam at other temperatures and pressures is determined. However, the fact that the absolute value of specific enthalpy cannot be measured is not a problem, because engineering is concerned about the change of specific enthalpy (Δh) rather than the absolute value. Usually, the enthalpy value will be included in the steam table as part of the list information.

2.2.1 Heat and Work

Physical quantities, kinetic energy, potential energy, internal energy and p-V energy are all different forms of energy and the properties of the system itself. Work is also a form of energy, but it is the transmission of energy, not an property of the system itself. Work is a process done by or to the system, and the system itself does not contain work. Distinguishing the difference between the energy form of the system itself and the energy form transmitted from the system is very important for understanding the energy transmission system.

For mechanical systems, work is defined as the force acting on an object through a certain distance. It is equal to the product of force (F) and displacement (S).

$$W = F S \qquad (2.14)$$

where: W is work, J; F is force, N; S is displacement, m.

When dealing with the work related to the energy transmission system, it is important to distinguish the work done by the system to its surrounding environment from the work done by the surrounding environment to the system. When the system is used to rotate the turbine to generate electricity in the turbine generator, the system does external work. When a pump is used to move fluid from one position to another, it does work to the system. The positive value of work indicates that the system does work to its surrounding environment; A negative value indicates that the surrounding environment does work to the system.

Heat and work are the same. They are the transmission of energy. Due to the temperature difference, energy is transferred at the molecular level in the form of heat. Like work, heat transfer depends on the path of transfer, not only on the initial and final state conditions of the system, which we call process quantity. In addition, like work, it is important to distinguish between adding heat to the system and discharging heat from the system to the surrounding environment. The positive value of heat indicates that the system absorbs heat from the surrounding environment, that is, the system energy is increased. This is just the opposite of work. The value of work transferred from the system is positive and transferred in is negative.

2.2.2 Energy and Power

Various forms of energy involved in the process of energy transfer, such as potential energy, kinetic energy, internal energy, p-V energy, work and heat, can be measured in many basic units. In engineering, there are usually three forms of units for measuring energy:

(1) Mechanical units, such as Newton meters (N-m);
(2) A unit of heat, such as Joule (J);
(3) Electrical energy unit, such as watt second (W-S).

Some important scientific experiments in history were carried out by J.P. Joule in 1843. He proved quantitatively that there is a direct correspondence between mechanical energy and thermal energy. These experiments show that a thousand calories is equal to 4186 J [14]. These experiments determine the equivalence of mechanical energy and thermal energy. Other subsequent experiments also confirmed the equivalence of electrical energy, mechanical energy and thermal energy.

In the MKS system, the mechanical units of energy are Joule (J) and erg, the thermal units are kilocalorie (kCal) and calorie (Cal), and the electrical energy units are watt second (W-s) and erg. It should be clear that although the units of various forms of energy are different, they are equivalent.

2.3 System and Process

System is a description of the analysis object in the science of thermodynamics. A system in thermodynamics is defined as a set of substances being studied. For example, the system can be water on one side of the heat exchanger, fluid in a section of pipe, or the whole lubricating oil system of diesel engine. Determining the boundary to solve the thermodynamic problem of the system will depend on the known system information and how to put forward the problem to the system.

Everything outside the system is called thermodynamic environment. The system is separated from the surrounding environment through the defined system boundary. These boundaries can be fixed or mobile. In many cases, thermodynamic analysis must be carried out for equipment (such as heat exchanger), which involves the inflow and outflow of fluid mass. The method followed in this analysis is to specify control surfaces, such as heat exchanger tube walls. Mass, heat and work (and momentum) may flow through the control surface.

According to the mass and energy across the system boundary, the system in thermodynamics can be divided into isolated, closed or open. An isolated system is a system that is not affected by the surrounding environment. This means that no energy in the form of heat or work can cross the boundary of the system. In addition, no mass can cross the system boundary.

Thermodynamic closed system is defined as a system with fixed mass. The closed system has no mass transfer with its surrounding environment, but may have energy transfer (heat or work) with its surrounding environment. An open system is a system that can transfer mass and energy with the surrounding environment.

1. **Control Volume**

The control volume is a fixed area in space, which is used to study the mass and energy balance of the flow system. The boundary of the control volume can be a real envelope or an imaginary envelope. The control surface is the boundary of the control volume.

2. **Thermodynamic Equilibrium**

When a system is in equilibrium with all possible state changes, the system is in thermodynamic equilibrium [15]. For example, if the gases that make up the system are in thermal equilibrium, the temperature of the whole system will be the same.

3. **Thermodynamic Process**

Whenever one or more properties of the system change, the state of the system changes. The continuous state path through which the system passes is called a thermodynamic process. An example of a thermodynamic process is to increase the temperature of the fluid while maintaining a constant pressure. Another example is to increase the pressure of the closed gas while maintaining a constant temperature.

4.   Cyclic Process

When a system in a given initial state experiences many different state changes (various processes) and finally returns to its initial value, the system experiences a cyclic process or cycle for short. Therefore, at the end of the loop, all properties have the same value as at the beginning. The steam (water) circulating through the closed cooling circuit goes through a cycle.

5.   Reversible Process

The reversible process of a system is defined as a process which will not cause any change to the system or the surrounding environment. In other words, the system and environment can be restored to the state before the process. In fact, there is no real reversible process; However, for analytical purposes, reversible processes can be used to simplify the analysis of some practical processes and determine the maximum theoretical efficiency. Therefore, reversible process is an appropriate assumption in engineering research.

Although the reversible process can be approached continuously, it can never be consistent with the actual process. One way to approximate a real process to a reversible process is to perform the process in a series of small or infinitesimal steps. For example, if heat transfer occurs due to a small temperature difference between the system and its surrounding environment, heat transfer can be considered reversible. For example, heat transfer through a temperature difference of 0.00001 °C "seems" to be more reversible than heat transfer through a temperature difference of 100 °C. Therefore, by cooling or heating the system in infinitesimal steps, we can approximate this process as a reversible process. Although this method is not practical for practical processes, it is conducive to the study of thermodynamic systems because the rate of process occurrence is not important.

6.   Irreversible Process

Irreversible process refers to the process that the system and surrounding environment cannot be restored to their original state. That is, if the process is reversed, the system and the surrounding environment will not return to their original state. For example, when the car engine taxis downhill, it will not return the fuel required for uphill.

Many factors make a process irreversible. The four most common causes of irreversibility are friction, unrestricted expansion of fluid, heat transfer with limited temperature difference, and mixing of two different substances. These factors exist in real and irreversible processes and prevent the reversibility of these processes.

7.   Isentropic Process

Isentropic process refers to the process in which the entropy of fluid remains unchanged. If the process experienced by the system is reversible and adiabatic, this will be true. Isentropic process can also be called constant entropy process.

8. Adiabatic Process

Adiabatic process refers to the process in which no heat enters or flows out of the system. The system can be considered completely insulated.

9. Throttling process

Throttling process is defined as the process in which the enthalpy does not change from state 1 to state 2, $h_1 = h_2$; No work, $W = 0$; This process is adiabatic, $Q = 0$.

In order to better understand the theory of ideal throttling process, let's take an example to compare the observed situation with the above theoretical assumptions. An example of a throttling process is the ideal gas flowing through the valve in the pipe. According to experience, we can observe: $p_{in} > p_{out}$, $v_{in} < v_{out}$ (where $p =$ pressure and $v =$ velocity).

These observations confirm the theory of $h_{in} = h_{out}$. Remember that $h = u + pv$ ($v =$ specific volume), so if the pressure decreases and if the enthalpy remains constant (assuming u is constant), the specific volume must increase. Because the mass flow is constant, the change of specific volume is observed with the increase of gas velocity, which is confirmed by our observations.

The theory also shows that $W = 0$. Our observation confirms this again, because the throttling process obviously does not do any "work". Finally, the theory points out that the ideal throttling process is adiabatic. This cannot be proved by observation, because the "real" throttling process is not an ideal process, and there will inevitably be some heat transfer.

10. Polytropic Process

When a gas undergoes a reversible process in which there is heat transfer, the process often occurs in such a way that the curve of logarithm pressure and logarithm volume is a straight line. Alternatively, it is stated in the form of Eq. (2.15).

$$pV^n = C \tag{2.15}$$

This type of process is called polytropic process. An example of a polytropic process is the expansion of the combustion gas in the cylinder of a water-cooled reciprocating engine.

2.4 Phase Change

The phase change of materials in the system is very important for thermodynamic analysis. The system can be designed to improve the performance of the system by using the phase transition between solid and liquid or between liquid and steam.

The properties of matter can be divided into intensive or extensive. If a property has nothing to do with the existing quality quantity, the property is intensive; If it is related to the amount of quality that exists, the property is a extensive property. The characteristics of pressure, temperature and density are intensive, while the volume and mass are extensive. By dividing a specific property by the total mass, you can make extensive properties to intensive. For example, total volume (V) is an extensive property that can be changed to specific volume by dividing by the mass of the system, which is an intensive property, $v = V/m$. Any specific property (specific volume, specific enthalpy, specific entropy) is an intensive property.

The use of intensive and extensive properties is demonstrated in the following discussion. As a system, 1 kg water is included in the piston cylinder arrangement, as shown in Fig. 2.3. Assume that the piston and weight maintain a pressure of 1 atm in the cylinder and the initial temperature is 20 °C, part (a) of Fig. 2.3. When heat is transferred to water, the temperature rises. The specific volume increases slightly and the pressure remains unchanged. As shown in part (b), when the temperature reaches 100 °C, additional heat transfer will lead to phase transition (boiling). That is, some liquids become steam, the temperature and pressure remain unchanged, but the specific volume increases significantly. When the last drop of liquid evaporates, further heat transfer will lead to an increase in the temperature and specific volume of the steam, as shown in part (c). In this example, temperature and pressure are intensive and therefore do not depend on the quality of existence. By examining the specific volume of water in the piston (intensive property) rather than the volume (extensive property), we can examine how any part of the water in the piston changes. The volume itself does not tell us the water in the piston. However, by knowing the specific volume of water, we can judge whether water is liquid or steam.

1. Saturation

The term saturation defines a special condition under which a mixture of steam and liquid can exist simultaneously at a given temperature and pressure. The temperature at which a liquid begins to vaporize (boil) at a given pressure is called the saturation temperature or boiling point. The pressure at which vaporization (boiling) begins at

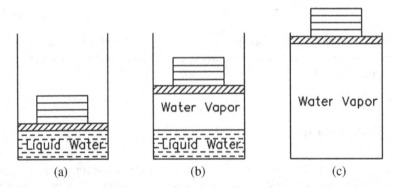

Fig. 2.3 Piston-cylinder arrangement

Fig. 2.4 Vapor pressure
curve

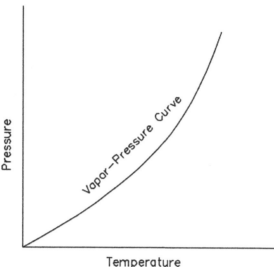

a given temperature is called saturation pressure. For example, for 100 °C water, the saturation pressure is 1 atm (14.7 psia) and for 1 atm water, the saturation temperature is 100 °C. For pure materials, there is a certain relationship between saturation pressure and saturation temperature. The higher the pressure, the higher the saturation temperature. The graph of the relationship between temperature and pressure under saturated conditions is called the vapor pressure curve. Typical steam pressure curve is shown in Fig. 2.4. When the pressure and temperature conditions drop to the curve, the steam / liquid mixture is saturated.

2.   Saturated and Subcooled Liquids

Saturated liquid means that a substance exists as a liquid at saturated temperature and pressure.

   If the liquid temperature is lower than the saturation temperature at the existing pressure, it is called supercooled liquid (meaning that the temperature is lower than the saturation temperature at a given pressure) or compressed liquid (meaning that the pressure is greater than the saturation pressure at a given temperature). The terms supercooled liquid and compressed liquid have the same meaning, so either term can be used.

3.   Quality

When a substance exists as part liquid and part vapor at saturation conditions, its quality (x) is defined as the ratio of the mass of the vapor to the total mass of both vapor and liquid [16]. Thus,

$$x = \frac{m_v}{m_v + m_l} \tag{2.16}$$

Fig. 2.5 T-V diagram
showing the saturation region

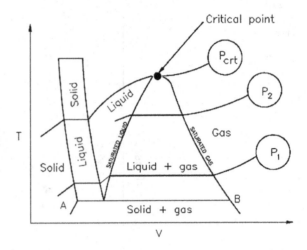

Example 2.3 If the mass of vapor is 0.2 kg and the mass of the liquid is 0.8 kg, the quality is 0.2 or 20%.

Quality is an intensive property. Quality has meaning when the substance is in a saturated state only, at saturation pressure and temperature. The area under the bell-shaped curve on Fig. 2.5 shows the region in which quality is important.

4. **Moisture Content**

The moisture content of a substance is opposite to its quality. Moisture (M) is defined as the ratio of the mass of liquid to the total mass of liquid and steam. The moisture content of the mixture in the previous paragraph is 0.8 or 80%. The following equation shows how to calculate the moisture of the mixture and the relationship between quality and moisture.

$$M = \frac{m_l}{m_v + m_l} \tag{2.17}$$

5. **Saturated and Superheated Vapors**

If there is a substance that exists completely in the form of steam at saturation temperature, it is called saturated steam. Sometimes, in engineering, we use the term dry saturated steam to emphasize that the mass is 100%. When the steam temperature is higher than the saturation temperature, it can be called superheated steam. The pressure and temperature of superheated steam are independent properties, because the temperature may rise when the pressure remains unchanged. In fact, what we usually call gas is highly superheated steam.

6. **Critical Point**

At a pressure of about 22 MPa (3200 psia) [17], there is no constant-temperature vaporization process. This is called a critical point. At the critical point, the state of liquid and steam is the same. The temperature, pressure, and specific volume of a critical point are called critical temperature, critical pressure, and critical volume.

The constant pressure process greater than the critical pressure does not have a significant phase change from the liquid phase to the steam phase, nor does it have a significant phase change from the liquid phase to the steam phase. For pressures greater than critical pressure, when the temperature is below critical temperature, the substance is often referred to as a liquid; when the temperature is above critical temperature, the substance is often called steam.

7.   **Fusion**

Consider one further experiment with the piston-cylinder arrangement of Fig. 2.3. Suppose the cylinder contained 1 kg of ice at 0 °C, 1 atm. As the heat transfers to the ice, the pressure remains the same, the specific volume increases slightly, and the temperature increases until the temperature reaches 0 °C, the ice melts, and the temperature remains the same. In this state, ice is called saturated solids. For most substances, a specific volume increases during melting, but for water, the specific volume of the liquid is smaller than the solid. This causes the ice to float on the surface of the water. When all the ice melts, any further heat transfer will cause the liquid temperature to rise. The melting process is also known as fusion. The heat added by melting ice into liquids is called latent fusion heat.

8.   **Sublimation**

For example, if the initial pressure of ice is 3.4 kPa at −18 °C, heat transfer to ice will cause the temperature to rise to −6.7 °C. However, at this time, ice enters the gas phase directly from the solid phase during sublimation. Sublimation is a special term used to describe the case where the transition between solid and gas phases occurs directly without passing through the liquid phase. Further heat transfer will cause the steam to overheat.

9.   **Triple Point**

Considering that the initial pressure of ice is 6 kPa, the temperature will rise until 0 °C due to heat transfer. However, at this point, further heat transfer may cause some ice to become steam and some ice to become liquid, because the three phases may be in equilibrium. This is called a three-phase point and is defined as a state in which all three phases are in equilibrium.

Figure 2.6 is the pressure temperature diagram of water, showing how the solid phase, liquid phase and vapor phase coexist in equilibrium. Along the sublimation line, the solid and vapor phases are in equilibrium, and along the melting line, the solid and liquid phases are in equilibrium; Along the vaporization line, the liquid and vapor phases are in equilibrium. The only possible point when the three phases are balanced is the three-phase point. The temperature and pressure at the three-phase point of water are 0.01 °C and 6 kPa respectively. The vaporization line ends at the critical point because there is no significant change from liquid phase to vapor phase above the critical point.

10.   **Condensation**

Fig. 2.6
Pressure–temperature
diagram

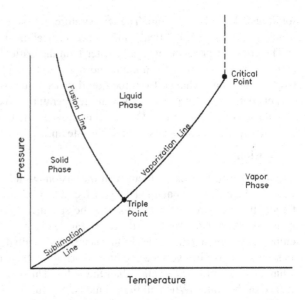

All the processes discussed earlier (evaporation, sublimation and melting) take place in the process of adding heat to matter. If heat is removed from the substance, the reverse of the process occurs.

As mentioned earlier, adding heat to a saturated liquid at constant pressure will cause the liquid to evaporate (from liquid to steam). If heat is removed from the saturated steam at constant pressure, condensation will occur and the steam will change into a liquid. Therefore, the evaporation and condensation processes are completely opposite.

Similarly, freezing is the opposite of melting. Sublimation also has an opposite process, gas directly into the solid, but this process usually has no unique term in English.

2.5   Property Diagrams

The relationship between the phase of a substance and its properties is most often shown on a property graph. A number of different properties are defined, and there are some dependencies between them. For example, water is steam, not liquid, when the standard atmospheric pressure and temperature exceeds 100 °C: it is present as a liquid at temperatures between 0 and 100 °C; Moreover, it exists as ice at temperatures below 0 °C. In addition, the properties of ice, water and steam are also related. Saturated steam has a specific volume of 1.695959407 m³/kg at 100 °C and standard atmospheric pressure. Saturated steam has different specific volumes at any other temperature and pressure. For example, at pressures of 200 °C and 10 MPa, the specific volume is 0.001148176 m³/kg.

Five basic properties are pressure (p), temperature (T), specific volume (v), specific entropy (h), and specific entropy (s). When it comes to a mixture of two phases, water and steam, a sixth property, quality (x), is also used.

There are six different types of property diagrams. They are: Pressure–Temperature (p-T) diagrams, Pressure-Specific Volume p-v diagrams, Pressure-Enthalpy (p-h) diagrams, Enthalpy-Temperature (h-T) diagrams, Temperature-entropy (T-s) diagrams, and Enthalpy-Entropy (h-s) or Mollier diagrams.

2.5.1 Pressure-Temperature (p-T) Diagram

A p-T diagram is one of the most common way to show the phases of a substance [18]. Figure 2.6 is the p-T diagram for pure water. The line separating the solid and liquid phases is called the fusion line. The line separating the solid and vapor phases is called the sublimation line. The line separating liquid phase and vapor phase is called evaporation line. The end point of the vaporization line is called the critical point. The point where the three lines intersect is called the three-phase point. The three-phase point is the only point where all three-phase balances exist. Under the temperature and pressure above the critical point, no matter how much pressure is applied, any substance can not exist in the form of liquid. A p-T diagram can be constructed for any pure substance.

2.5.2 Pressure-Specific Volume (p-v) Diagram

A p-v diagram is another common type of property diagram. A p-v diagram is different from a p-T diagram in one particularly important way. Figure 2.7 is the p-v diagram for pure water. There are regions on a p-v diagram in which two phases exist together. For example, at point A, water with a specific volume (v_f), given by point B, exists together with steam with a specific volume (v_g), given by point C. So, in the liquid–vapor region in Fig. 2.7, water and steam exist together. The dotted lines on Fig. 2.7 are lines of constant temperature. The quality of the mixture at any point in the liquid–vapor region can be found because the specific volumes of water, steam, and the mixture are all known. The quality can be found using the following relationship. A p-v diagram can be constructed for any pure substance.

$$x = \frac{v - v_f}{v_g - v_f} = \frac{v - v_f}{v_{fg}} \tag{2.18}$$

$$v = x v_g + (1 - x) v_f \tag{2.19}$$

Fig. 2.7 *p-v* diagram for water

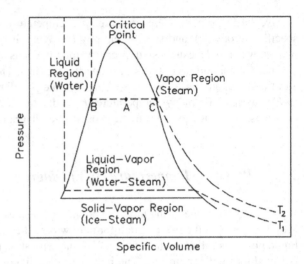

2.5.3 *Pressure-Enthalpy* (p-h) *Diagram*

A *p-h* diagram has the same functionality as a *p-v* diagram. Figure 2.8 is the *p-h* diagram for pure water. Like the *p-v* diagram, there are regions on a *p-h* diagram in which two phases exist together. In the liquid vapor area in Fig. 2.8, water and steam coexist. For example, at point A, water with an enthalpy (h_f), given by point B, exists together with steam with an enthalpy (h_g), given by point C. A *p-h* diagram can be constructed for any pure substance.

The quality of the mixture at any point in the liquid–vapor region can be found using the following relationship [19].

Fig. 2.8 *p-h* diagram for water

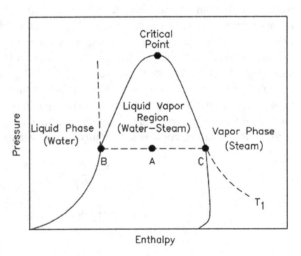

$$x = \frac{h - h_f}{h_{fg}} \tag{2.20}$$

$$h = xh_g + (1 - x)h_f \tag{2.21}$$

2.5.4 Enthalpy-Temperature (h-T) Diagram

The h-T diagram has the same functionality as the previous property map. Figure 2.9 is the h-T diagram of pure water. As with previous property diagrams, there are two phases of the area on the h-T diagram. The area between the saturated liquid line and the saturated steam line represents the area of the two phases that exist simultaneously. The vertical distance between the two saturation lines represents the potential heat of steaming. If pure water is present at a point on the saturated liquid line and the added heat is equal to the potential heat of evaporation, the water will change from saturated liquid to saturated steam (point B) while maintaining a constant temperature. As shown in Fig. 2.9, out-of-line operation can result in liquids or superheated steam that are below electricity. An h-T diagrams can be constructed for any pure substance.

Fig. 2.9 h-T diagram for water

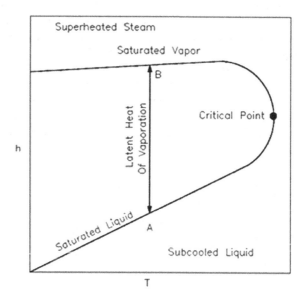

Fig. 2.10 *T-s* diagram for water

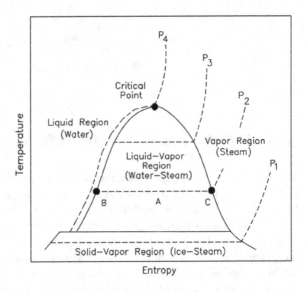

2.5.5 Temperature-Entropy (T-s) Diagram

A *T-s* diagram is the most commonly used diagram types for analyzing energy transfer system cycles. This is because the work done by or on the system, as well as the heat added to or removed from the system, can be visualized on the *T-s* diagram. According to the definition of entropy, the heat entering and exiting the system is equal to the area under the process *T-s* curve.

Figure 2.10 is the *T-s* diagram for pure water. A *T-s* diagram can be constructed for any pure substance. It exhibits the same features as *p-h* diagrams. In the liquid vapor area in Fig. 2.10, water and steam coexist. For example, at point A, water with an entropy (s_f) given by point B, exists together with steam with an entropy (s_g) given by point C. The quality of the mixture can be found at any point in the liquid steam area using the following relationships.

$$s = x s_g + (1 - x) s_f \qquad (2.22)$$

$$x = \frac{s - s_f}{s_{fg}} \qquad (2.23)$$

2.5.6 Enthalpy-Entropy (h-s) or Mollier Diagram

The Molire diagram [20] shown in Fig. 2.11 is a chart that plots enthalpy (h) and entropy (s). It is sometimes referred to as an *h-s* diagram and is completely different

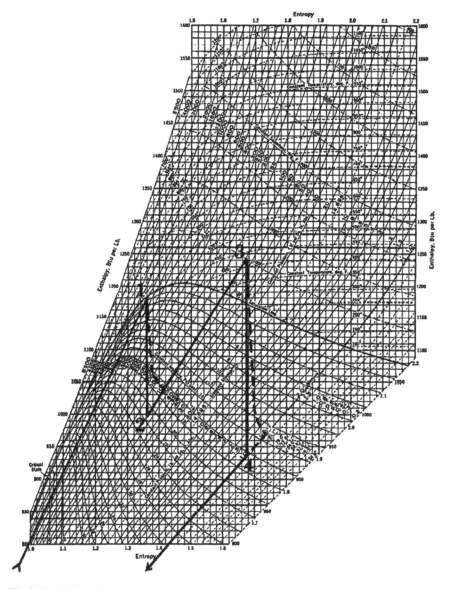

Fig. 2.11 Mollier diagram

in shape from a *T-s* diagram. The chart contains a series of constant temperature, a series of constant pressure lines, a series of constant humidity or quality lines. Molire diagrams are used only if the quality is greater than 50% or the steam is overheated.

2.6   The First Law of Thermodynamics

The first law of thermodynamics states [21]:

Energy can neither be created nor destroyed, but can only be changed in form.

For any system, energy transfer is related to the mass and energy passing through the control boundary, the external work and / or heat passing through the boundary, and the change of energy stored in the control volume. The mass flow of fluid is related to the kinetic energy, potential energy, internal energy and flow energy that affect the overall energy balance of the system. The exchange of external work and / or heat completes the energy balance.

The first law of thermodynamics is called the principle of energy conservation, which means that when studying the fluid in the control volume, energy can neither be created nor destroyed, but can be transformed into various forms. The energy balance described here is maintained within the system under study. A system is a spatial area (control volume) through which a fluid pass. When the fluid passes through the boundary of the system, various energy related to the fluid will be observed to achieve balance.

As described in the previous section of this chapter, the system can be one of three types: isolated, closed or open. Open system is the most common of the three systems. It shows that mass, heat and work can pass through the control boundary. Balance is expressed in words: all energy entering the system is equal to all energy leaving the system plus changes in energy storage in the system. Recall that the energy in a thermodynamic system is made up of kinetic energy (E_k), potential energy (E_p), internal energy (U), and flow energy (pv); as well as heat and work processes.

$$\sum E_{in} = \sum E_{out} + \Delta E_{storage} \tag{2.24}$$

Heat and/or work can flow directly into or out of the control volume. However, for convenience, as a standard practice, net energy exchange is proposed here, in which net heat exchange is assumed to enter the system and net work is assumed to leave the system. If no mass passes through the boundary, but work and / or heat passes through the boundary, the system is called a "closed" system. If the mass, work and heat do not cross the boundary (that is, only energy exchange occurs in the system), the system is called an isolation system. Isolated and closed systems are just special cases of open systems. In this book, the open system method of the first law of thermodynamics will be emphasized because it is more general. In addition, almost all practical applications of the first law require an open system analysis.

When analyzing thermodynamic problems or establishing energy balance, it is essential to understand the concept of control volume. There are two basic methods to study Thermodynamics: control mass method and control volume method. The former is called Lagrange method and the latter is called Euclidean method. In the concept of controlled mass, the "clump" of fluid and its related energy are studied. The analyzer "rides" anywhere with the clump to maintain the balance of all energy affecting the clump.

Fig. 2.12 Open system control volumes

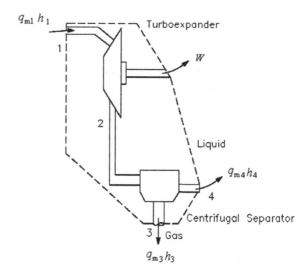

The control volume method is a method to establish a fixed area with a specified control boundary in space, as shown in Fig. 2.12. Then, the energy passing through the control volume boundary, including the energy of mass passing through the boundary, is studied and balanced. At present, the control volume method is often used to analyze thermodynamic systems. It is more convenient and requires less work in tracking energy balance.

$$W = q_{m1}h_1 - q_{m3}h_3 - q_{m4}h_4 \tag{2.25}$$

The energy forms that may pass through the boundary of CV include those related to the mass (m) passing through the boundary. The mass in motion has potential energy (E_p), kinetic energy (E_k) and internal energy (U). In addition, since the fluid is usually provided by some driving force, such as a pump, the fluid pressure generates another form of energy. This form of energy is called flow energy (pv-work). Therefore, the thermodynamic term represents various forms of energy passing through the control boundary is $m (u + pv + E_k + E_p)$.

In an open system analysis, the terms u and pv appear so frequently that another characteristic enthalpy is defined as $h = u + pv$ which leads to the above expression being written as $m (h + E_k + E_p)$. In addition to mass and energy, applied work (W) is usually called axial work, which is another form of energy that may cross the boundary of the system. In order to complete and satisfy the energy conservation relationship, energy generated neither by mass nor by shaft work is classified as thermal energy (Q). Then we can describe this relationship in the form of equation, as shown below.

$$q_m(h_{in} + E_{p,in} + E_{k,in}) + P_Q = q_m(h_{out} + E_{p,out} + E_{k,out}) + W \tag{2.26}$$

where:

q_m mass flow rate of working fluid, kg/s;
h_{in} specific enthalpy of the working fluid entering the system, J/kg;
h_{out} specific enthalpy of the working fluid leaving the system, J/kg;
$E_{p,in}$ specific potential energy of working fluid entering the system, J/kg;
$E_{p,out}$ specific potential energy of working fluid leaving the system, J/kg;
$E_{k,in}$ specific kinetic energy of working fluid entering the system, J/kg;
$E_{k,out}$ specific kinetic energy of working fluid leaving the system, J/kg;
W rate of work done by the system, W;
P_Q heat rate into the system, W.

Example 2.4 The enthalpies of steam entering and leaving a steam turbine are 3135 kJ/kg and 2556 kJ/kg, respectively. The estimated heat loss is 10 kJ/kg of steam. The flow enters the turbine at 50 m/s at a point 20 m above the discharge and leaves the turbine at 80 m/s. Determine the work of the turbine.
     Solution:

$$q_m(h_{in} + E_{p,in} + E_{k,in}) + P_Q = q_m(h_{out} + E_{p,out} + E_{k,out}) + W$$

Divide by mass flow rate,

$$(h_{in} + E_{p,in} + E_{k,in}) + q = (h_{out} + E_{p,out} + E_{k,out}) + w$$

where

q heat added to the system per pound (J/kg),

w work done by the system per pound.

     Then

$$3135 \text{ kJ/kg} + 9.8 \times 20 \text{ J/kg} + 0.5 \times 50^2 \text{ J/kg} - 10 \text{ kJ/kg}$$
$$= 2556 \text{ kJ/kg} + 0.5 \times 80^2 \text{ J/kg} + w$$

     Get

$$w = 567.25 \text{ kJ/kg}$$

This example demonstrates that potential and kinetic energy terms are insignificant for a turbine, since the ΔE_p and ΔE_k values are less than 0.196 and 1.95 kJ/kg.

2.6.1 Rankine Cycle

A thermodynamic process is happened whenever a property of a system changes. The change could be caused by work to or from the system, heat transfer to or from the system or mass transfer happened on boundary surface. One example of a thermodynamic process as introduced in later section is increasing the temperature of a fluid while maintaining a constant pressure. When a system in a given initial state goes through a number of different changes in state and finally returns to its initial values, the system has undergone a cyclic process or cycle. Rankine cycle is a kind of thermodynamic cycle used in nuclear engineering.

In a nuclear power plant, a typical Rankine cycle is composed with a pump, a steam generator, a turbine, a condenser and piping network as shown in Fig. 2.13 [22]. The feedwater is pressurized in the pump and flows into the steam generator. Heat transfer happens in the steam generator and the water is vaporized to steam. Steam flows into the turbine and expands in the turbine. With the expansion, work is done to turn the axis of turbine in a specified speed (1500 rpm) to generate electricity in 50 Hz. After the expansion, the vapour is condensed in the condenser and the condensation heat is transferred to seawater or atmosphere. Finally, the condensed water if feed back to the steam generator by the pump. In this cycle, the water in the system have four processes.

1-2: Adiabatic Isentropic Process. There is no heat transfer into or out of the system. The system can be considered to be perfectly insulated. Enthalpy is not changed during this process.

2-3: Constant Pressure Reversible Cooling Process. In this process, vapour is condensed to water.

3-4: Adiabatic Pressurized Process. Water is pressurized and kept flow in the system. Because the heat loss in the pump is ignorable, this process is also an adiabatic isentropic process.

Fig. 2.13 A typical rankine cycle

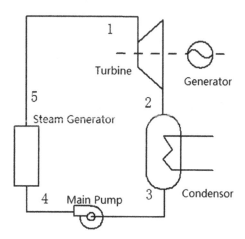

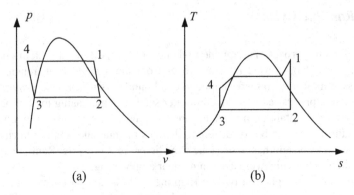

Fig. 2.14 Rankine cycle in *p-v* and *T-s* diagram

4-1: Boiling Process. During this process, the water in system is vaporized by the heat from the primary side of steam generator.

These four processes close the whole cycle and become a Rankine cycle. In *p-v* and *T-s* diagram, Rankine cycle is shown in Fig. 2.14.

In steam generator (or boiler), the heat transferred to 1 kg of water is shown as Eq. (2.27).

$$q_1 = h_1 - h_4 \tag{2.27}$$

The work done by 1 kg of steam in turbine is:

$$W_T = h_1 - h_2 \tag{2.28}$$

The heat transferred from 1 kg of vapour in condenser is:

$$q_2 = h_2 - h_3 \tag{2.29}$$

The work done to 1 kg of water is:

$$W_p = h_4 - h_3 \tag{2.30}$$

Then we get the efficiency of Rankine cycle as:

$$\eta = \frac{W_T - W_p}{q_1} = \frac{(h_1 - h_2) - (h_4 - h_3)}{(h_1 - h_4)} \tag{2.31}$$

2.6.2 Utilization of the First Law of Thermodynamics in Nuclear Power Plant

The first law of thermodynamics is useful in nuclear engineering to analyze components or systems in nuclear power plant. Here we take some examples to explain the utilization of the first law of thermodynamics in nuclear power plant.

Example 2.5 Components analysis of steam system.

In a pressurized water-cooled reactor, primary side coolant flows into steam generator at temperature of 300 °C and leaves at temperature 280 °C at flow rate of 5.0×10^7 kg/h. For simplification, suppose the heat capacity of water is 4.0 kJ/kg °C, find the heat transfer rate of the steam generator.

Solution: Kinetic energy and potential energy can be neglected because they are very small compared with the heat transferred. There is no shaft work done in steam generator.

$$q_m h_{in} + P_Q = q_m h_{out} P_Q = q_m (h_{out} - h_{in}) = q_m C_p \Delta T$$
$$= 5.0 \times 10^7 \text{ kg/h} (4.0 \text{ kJ/kgC}) (280 \text{ °C} - 300 \text{ °C})$$

Get:

$$P_Q = -4.0 \times 10^9 \text{ kJ/h} = -1.4 \times 10^9 \text{ W}$$

The solution is negative because heat is transferred out from primary side. The heat transferred in a heat exchanger can be expressed as Eqs. (2.32a) or (2.32b).

$$P_Q = q_m (h_{out} - h_{in}) \tag{2.32a}$$

Or

$$P_Q = q_m C_p \Delta T \tag{2.32b}$$

Equation (2.32b) is not suitable for phase change condition.

Pump also can be analyzed by the first law of thermodynamics.

Example 2.6 In a pressurized water reactor system, pump is used to drive coolant flow through reactor core. Suppose the pressure of coolant is 15 MPa, flow rate is 1.5×10^7 kg/h, inlet temperature is 280 °C (specific volume is 0.0013096 m³/kg). Pressure will be raised 6.5 atm after coolant passed the pump. Neglect heat loss, potential energy and kinetic energy. Fine the work done by the pump.

Solution:

k,in

Neglect heat loss, potential energy and kinetic energy, one gets:

$$q_m h_{in} = q_m h_{out} + W$$

Since

$$h_{in} = u_{in} + p_{in} v_{in}$$
$$h_{out} = u_{out} + p_{out} v_{out}$$

Internal energy does not change because heat loss is neglected. The specific volume of water also does not change because fluid is almost incompressible. Then,

$$(h_{in} - h_{out}) = v(p_{in} - p_{out})$$

Put values into this equation, one gets:

$$W = 1.5 \times 10^7 \text{ kg/h} \times 0.0013096 \text{ m}^3/\text{kg} \times (-6.5 \text{ atm})$$
$$W = 1.5 \times 10^7 \text{ kg/h} \times 0.0013096 \text{ m}^3/\text{kg} \times (-6.5 \times 1.01325 \times 10^5 \text{ Pa})$$
$$W = -3.59 \text{ MW} = -2640 \text{ hp}$$

Note: negative sign means work is done to fluid. "hp" means horsepower:

$$1 \text{hp} = 0.7354987 \text{ kW} \tag{2.33}$$

Example 2.7 Energy balance in nuclear reactor core. In a nuclear power system, core outlet temperature of 320 °C, and the temperature of inlet is 282 °C. Coolant flow rate is 6.5×10^7 kg/h, fluid average heat capacity C_p is 4.0 kJ/ (kg °C). Determine the power of heat source.

Solution: assuming that the gravitational potential energy and kinetic energy are small compared to other quantities, can be ignored, and obtained

$$P_Q = q_m(h_{out} - h_{in})$$
$$P_Q = q_m C_p(T_{out} - T_{in})$$
$$P_Q = 6.5 \times 10^7 \text{ kg/h} \times 4.0 \text{ kJ/(kg} \cdot {}^\circ\text{C)}$$
$$\qquad \times (320 \,^\circ\text{C} - 282 \,^\circ\text{C})$$
$$P_Q = 2744 \text{ MW}$$

In this example, if the specific enthalpy at inlet and outlet are given, can also be calculated using the specific enthalpy difference.

Example 2.8 Overall thermodynamic equilibrium. A nuclear reactor (primary side) can be regarded as a complete system. Heat source is 2744 MW, heat exchanger (steam generator) to take away 2749 MW heat, determine power of pump to maintain stable temperature.

Solution: assuming that the gravitational potential energy and kinetic energy are small compared to other quantities can be ignored, according to the conservation of energy, one get:

$$q_m h_{in} + W_p + P_{Q,C} = P_{Q,SG} + q_m h_{out}$$

For a closed system, the mass flow rate of entering and leaving the system is 0, and the energy entering and leaving the system is 0.

$$W_p + P_{Q,C} = P_Q$$
$$W_p = P_{Q,SG} - P_{Q,C} = 2754 - 2749 = 5MW$$

Now, for example, analyzing the two side of the heat exchanger will help to understand the importance of the heat exchanger in the energy conversion process.

Example 2.9 Heat exchanger secondary side heat transfer calculation. Steam through the condenser flow rate of 2.0×10^6 kg/h, 40 °C saturated steam enters ($h = 2574$ kJ/kg), and to the same pressure left at 30 °C liquid ($h = 125.8$ kJ/kg), cooling water temperature of 18 °C ($h = 75.6$ kJ/kg) and environmental requirements of export is restricted by the temperature of 25 °C ($h = 104.9$ kJ/kg). Determine the flow rate of the cooling water required.

Solution: By thermodynamic equilibrium

$$P_{Q,SG} = -P_{Q,CW} q_{m,SG} (h_{out} - h_{in})_{stm} = -q_{m,CW} (h_{out} - h_{in})_{CW}$$
$$q_{m,CW} = -q_{m,SG} (h_{out} - h_{in})_{SG} / (h_{out} - h_{in})_{CW}$$
$$= 2.0 \times 10^6 kg/h (125.8 - 2574 \text{ kJ/kg})/(104.9 - 75.6 \text{ kJ/kg})$$
$$q_{m,CW} = 1.67 \times 10^8 kg/h$$

In this example, we use the Eq. (2.32a), since the phase transition occurs when steam is condensed into liquid water, and hence the Eq. (2.32b) does not apply.

2.7 The Second Law of Thermodynamics

One of the earliest statements of the second law of thermodynamics was put forward by R. Clausius in 1850. He said the following statement [23].

It is impossible to build a device that operates circularly and has no effect, except that heat is discharged from the object at one temperature and absorbed by the object at a higher temperature.

The second law of thermodynamics is used to determine the maximum efficiency of any process. A comparison can then be made between the maximum possible efficiency and the actual efficiency obtained.

According to the second law of thermodynamics, the limitations on any process can be studied to determine the maximum possible efficiency of the process, and then the maximum possible efficiency can be compared with the actual realized efficiency. One of the application areas of the second law is the study of energy conversion system. For example, it is not possible to convert all the energy obtained from nuclear reactors into electricity. There must be a loss during the conversion process. Taking these losses into account, the second law can be used to express the maximum possible energy conversion efficiency. Therefore, the second law denies the possibility that all the heat provided to the system in a single cycle can be fully converted into work, no matter how perfect the system design. Max Planck described the concept of the second law as:

It is impossible to produce an engine that can work in a complete cycle and will not produce any effect other than increasing weight and cooling the accumulator.

The second law of thermodynamics is needed because the first law of thermodynamics cannot fully define the energy conversion process. The first law is used to correlate and evaluate the various energies involved in the process. However, information about the direction of the process cannot be obtained by applying the first law. In the early development of thermodynamic science, researchers pointed out that although work can be completely transformed into heat, work in the cycle can never be completely transformed into heat. It is also observed that some natural processes always proceed in a certain direction (for example, heat transfer from hot objects to cold objects). The second law is used to explain these natural phenomena.

2.7.1 Entropy

One result of the second law is the development of the physical properties of matter, called entropy. The introduction of entropy helps to explain the second law of thermodynamics. Changes in this property determine the direction in which a given process continues. Entropy can also be interpreted as a measure of the thermal inability to work during cycles. This is related to the second law, which predicts that not all heat supplied to the cycle can be converted to the same amount of work, so some cooling must occur. The change in entropy is defined as the ratio of heat transfer in a reversible process to the absolute temperature of the system.

$$\Delta S = \Delta Q / t_K \tag{2.34}$$

where

ΔS the change in entropy of a system during some process (J/K)
ΔQ the amount of heat added to the system during the process (J)

t_K the absolute temperature at which the heat was transferred (K)

The second law of thermodynamics can also be expressed as: for a complete thermodynamic cycle, $\Delta S \geq 0$. In other words, for a cyclic system, entropy must increase or remain unchanged and cannot decrease.

Entropy is the property of a system. It is an extensive quantity, just like total internal energy or total enthalpy. The entropy of the system can be obtained by calculating the sum of the entropy of each specific unit mass. For pure matter, the specific entropy can be obtained by looking up the table according to the specific enthalpy, specific volume and other relevant thermodynamic properties.

Because entropy is a property of matter, it can be better used as a coordinate when using a graph to represent a reversible process. In the temperature entropy diagram, the area under a reversible process curve represents the amount of heat conversion in this process.

In order to help understand the second law of thermodynamics, irreversible thermodynamic problems, processes and cycles are often replaced by reversible processes. Because only reversible processes can be represented by diagrams (such as enthalpy entropy diagram and temperature entropy diagram) for analysis, this substitution is very effective. Because the actual process and irreversible process cannot be continuous under equilibrium conditions, they cannot be drawn on the diagram. We can only know the beginning and end of the irreversible process. Therefore, some thermodynamic literatures use dotted lines to represent irreversible processes on charts.

2.7.2 Carnot's Principle

By trying to use reversible processes, in 1824, the French engineer Sadi Carnot published a theorem containing the following propositions, which further developed the second law of thermodynamics [24].

1. Among the heat engines operating between the same high-temperature heat source and low-temperature heat sink, no other heat engine can be more efficient than a reversible heat engine.
2. All reversible heat engines operating between the same high-temperature heat source and low-temperature heat sink have the same efficiency.
3. The efficiency of reversible heat engine only depends on the temperature of high-temperature heat source and low-temperature heat sink.

Carnot's theorem can be illustrated by a simple cycle (Fig. 2.15) and an example of heat into work. This cycle consists of following reversible processes.

1-2: adiabatic compression by doing work to the working medium, and the temperature changes from T_C to T_H;
2-3: endothermic isothermal expansion of working medium at temperature T_H;

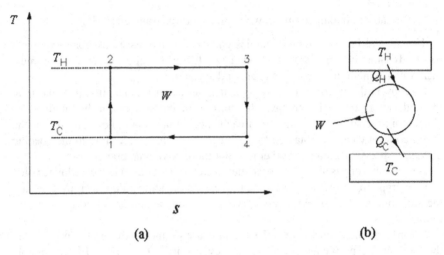

Fig. 2.15 Carnot cycle

3-4: make the working medium adiabatic expand to do external work, and the temperature drops from T_H to T_C;
4-1: exothermic isothermal compression of working medium at temperature T_C.

This cycle is called Carnot cycle. The heat (Q_H) entering the Carnot cycle is graphically shown in Fig. 2.15 as an area below line 2-3. The released heat (Q_C) is graphically displayed in the figure as the area below lines 1-4. The difference between the incoming heat and the released heat is the net work (sum of work), which is represented as rectangle 1-2-3-4 in the figure.

The efficiency (η) of the cycle is the ratio of net work to the heat entering the cycle and this ratio can be expressed by the following equation.

$$\eta = (Q_H - Q_C)/Q_H = (T_H - T_C)/T_H \qquad (2.35)$$

where:

η cycle efficiency
T_C designates the low-temperature reservoir (K)
T_H designates the high-temperature reservoir (K)

When T_H is at its maximum possible value and T_C is at its minimum possible value, there is the maximum possible efficiency. All actual systems and processes are truly irreversible. The efficiency given above marks the upper limit of the efficiency of any given system working at two temperatures. The maximum possible efficiency of the system is Carnot efficiency. However, because Carnot efficiency exists only in reversible process, the actual system will not reach this efficiency value. Therefore, Carnot efficiency is regarded as the upper limit of efficiency that can not be reached

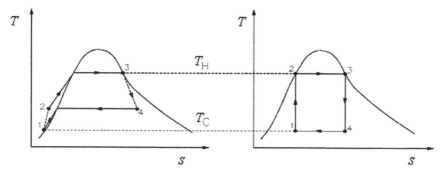

Fig. 2.16 Real process cycle compared to carnot cycle

by any real system efficiency. Figure 2.16 shows the comparison between real cycle and Carnot cycle.

The following example demonstrates the above principles.

Example 2.10 Carnot Efficiency

Is the claim is a valid claim if an inventor claims to have an engine operating between a source at 50 °C and a receiver at 0 °C that receives 100 kJ of heat and produces 25 kJ of useful work.

Solution: $T_H = 50\ K + 273\ K = 323\ K$

$$T_C = 0\ K + 273\ K = 273\ K$$
$$\eta = (323 - -273)/323 = 15.5\% < 25\%$$

Therefore, this statement is unreasonable.

For engineering practice, the most important use of the second law of thermodynamics is to determine the maximum possible efficiency we can obtain from a dynamic system. The actual efficiency will always be lower than this maximum. The loss of energy (such as friction) and the system are not truly reversible, so we can not obtain the maximum efficiency. The difference between ideal and actual efficiency is further illustrated by the following example.

Example 2.11 Actual versus Ideal Efficiency

The facility operates from a steam source at 170 °C and rejects heat to atmosphere at 15 °C. The actual efficiency of a steam cycle is 18.0%. Compare the Carnot efficiency to the actual efficiency.

Solution: $\eta = (170 - 15)/443 = 35\%$ as compared to 18.0% actual efficiency.

In the previous, we used the first law of thermodynamics to analyze open systems. The second law of thermodynamics is roughly the same in treatment. That is, an isolated, closed or open system is used for analysis according to the energy type passing through the boundary. As the same as the first law of thermodynamics, since

Fig. 2.17 Control volume

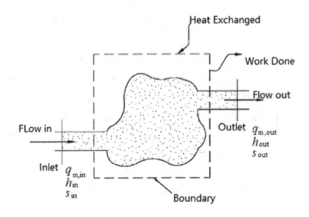

closed systems and isolated systems can be regarded as special cases of open systems, the second law of thermodynamics is suitable for more general cases. The solution to the problem of the second law of thermodynamics is very similar to the process used in the analysis of the first law of thermodynamics.

Figure 2.17 illustrates the control volume from the perspective of the second law of thermodynamics. In Fig. 2.17, the working medium moves through the control volume from the inlet to the outlet. Let's assume that the boundary of the CV (control volume) is ambient temperature, where all heat exchange (Q) occurs. We note that entropy is an extensive quantity, so it can be transported through the flow of working medium in and out of the control volume, just like enthalpy and internal energy.

Therefore, since the entropy flow generated by large-scale transportation enters the control volume is $q_{m,in}s_{in}$, and the entropy flow that flows out of the control volume is $q_{m,out}s_{out}$, assuming that the properties in the incoming and outgoing parts are uniform. Entropy may also be added to the control volume because of the heat transfer of the control volume boundary.

A simple demonstration of the use of this form of system in second law analysis will give us a better understanding of its use.

Example 2.12 Open System Second Law

Steam enters the nozzle of a steam turbine with a velocity of 3 m/s at a pressure of 7 atm and temperature of 280 °C at the nozzle discharge. The pressure and temperature are 1 atm at 150 °C. What is the increase in entropy for the system if the mass flow rate is 5000 kg/h?

Solution:

$$s_{in} = 7.225 \text{ kJ/kg}$$
$$s_{out} = 7.615 \text{ kJ/kg}$$
$$\Delta S = 5000 \times (7.615 - 7.225) = 1950 \text{ kJ/kg}$$

2.8 Power Plant Components

In order to analyze a complete steam power cycle of power plant, first analyze the components that make up a cycle (see Fig. 2.18), including steam turbine, pump and heat exchanger. We have obtained the ideal efficiency of the system when the heat source and heat sink temperature are known. The efficiency of each independent component of the system can also be calculated in this way.

The efficiency of each component can be calculated by comparing the real work done by these components with the work produced by the ideal component under the same conditions.

The function of a steam turbine is to extract energy from working medium (steam) and do work externally in the form of rotating steam turbine shaft. The working medium does work by expanding in the steam turbine. The work done by the turbine shaft is converted into electric energy by the generator. Applying the first law of thermodynamics and the energy conservation of steam turbine under steady-state conditions, we find that the decrease value of enthalpy $H_{in} - H_{out}$ of working medium is equal to the work value (W_t) of working medium in steam turbine, i.e.

$$H_{in} - H_{out} = W_t \qquad (2.36)$$

$$q_m(h_{in} - h_{out}) = P_t \qquad (2.37)$$

where

H_{in} working fluid enthalpy entering the turbine (J).
H_{out} enthalpy of the working fluid leaving the turbine (J).

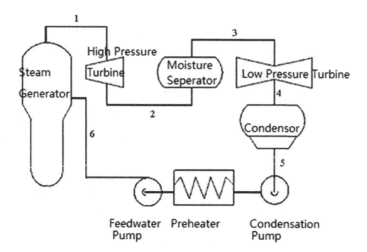

Fig. 2.18 Steam cycle

W_t work done by the turbine (N-m).
q_m mass flow rate of the working fluid (kg/s).
h_{in} specific enthalpy of the working fluid entering the turbine (J/kg).
h_{out} specific enthalpy of the working fluid leaving the turbine (J/kg).
P_t power of turbine (W).

2.8.1 Turbine Efficiency

Equations (2.36) and (2.37) are applicable to the change of kinetic energy and potential energy and the negligible heat loss of working medium in steam turbine. For most practical applications, these assumptions are reasonable. However, in order to use these relations, it is also necessary to assume that the work of the working medium is reversible through isentropic expansion in the ideal case. This kind of steam turbine is called ideal steam turbine. In an ideal steam turbine, the entropy of the working medium entering the steam turbine is equal to that of the working medium flowing out of the steam turbine.

An ideal steam turbine is defined to provide an analyzing basis for steam turbine performance. An ideal steam turbine can do the maximum possible work in theory.

Because of blade friction, blade leakage or small-scale mechanical friction, an actual steam turbine will do less work than ideal. Steam turbine efficiency is defined as the ratio of the work actual done by the steam turbine $W_{t,\,actual}$ to the work $W_{t,ideal}$ that the ideal steam turbine can do.

$$\eta_t = \frac{W_{t,actual}}{W_{t,ideal}} \tag{2.38}$$

$$\eta_t = \frac{(h_{in} - h_{out})_{actual}}{(h_{in} - h_{out})_{ideal}} \tag{2.39}$$

where,

η_t is turbine efficiency (no units).
$W_{t,actual}$ is the actual work done by a turbine (N-m).
$W_{t,ideal}$ is the work done by an ideal turbine (N-m).
$(h_{in} - h_{out})_{actual}$ is the actual change of enthalpy of the working fluid (J/kg).
$(h_{in} - h_{out})_{ideal}$ is the actual change of enthalpy of the working fluid in an ideal turbine (J/kg).

In many cases, turbine efficiency is determined independently. This allows the actual work to be calculated by directly multiplying the turbine efficiency by the ideal turbine efficiency under the same conditions. For small turbines, the turbine efficiency is approximately 60–80%; For large steam turbines, the value is approximately 90%.

The actual and ideal performance of a steam turbine can also be easily compared by temperature entropy diagram. Figure 2.19 shows such a comparison [25]. Ideally,

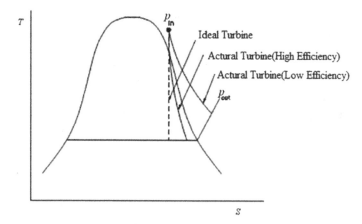

Fig. 2.19 Comparison of ideal and actual turbine performances

the entropy is constant. This is represented by a vertical line on the temperature entropy diagram. Actual turbine entropy increases. The smaller the entropy increase, the closer the turbine efficiency is to 100%.

2.8.2 Pump Efficiency

The function of the pump is to promote the flow of the working medium by doing work [26]. For a pump under steady flow conditions, according to the first law thermodynamics, we can find that the enthalpy increase of the working medium H_{out} − H_{in} is equal to the work done by the pump W_p, i.e.

$$H_{in} - H_{out} = W_p \tag{2.40}$$

$$q_m(h_{in} - h_{out}) = P_p \tag{2.41}$$

where:

H_{out}	enthalpy of the working fluid leaving the pump (J)
H_{in}	enthalpy of the working fluid entering the pump (J)
W_p	work done by the pump on the working fluid (N-m)
q_m	mass flow rate of the working fluid (kg/s)
h_{out}	specific enthalpy of the working fluid leaving the pump (J/kg)
h_{in}	specific enthalpy of the working fluid entering the pump (J/kg)
P_p	power of pump (W)

Equations (2.40) and (2.41) are applicable when the kinetic energy and potential energy change and the heat loss of working medium in the pump can be ignored. For most practical applications, these assumptions are reasonable. It is also assumed that the working medium is incompressible. Ideally, the work done by the pump is equal to the enthalpy changed by the working medium passing through the ideal pump.

Similar to a steam turbine, we can also define an ideal pump. Due to the inevitable loss of friction and working medium turbulence, the actual pump will need more work than the ideal pump.

Pump efficiency is defined as the ratio of the work required by a pump if it is an ideal pump $W_{p,\,ideal}$ to the work required by the actual pump $W_{p,\,actual}$.

$$\eta_p = \frac{W_{p,actual}}{W_{p,ideal}} \tag{2.42}$$

Example 2.13 A pump has an inlet specific enthalpy of 400 kJ/kg and operating at 75% efficiency. What is the exit specific enthalpy of the actual pump if the exit specific enthalpy of the ideal pump is 1200 kJ/kg?
Solution:

$$\eta_p = \frac{W_{p,actual}}{W_{p,ideal}}$$

$$W_{p,actual} = \frac{W_{p,ideal}}{\eta_p}$$

$$h_{in,actual} - h_{out,actual} = \frac{h_{in,ideal} - h_{out,ideal}}{\eta_p}$$

$$h_{out,actual} = \frac{h_{in,ideal} - h_{out,ideal}}{\eta_p} + h_{in,actual}$$

$$h_{out,actual} = (1200 - 400)/0.75 + 400 = 1467 \text{ kJ/kg}$$

2.8.3 Ideal and Real Cycle

Since the efficiency of Carnot cycle is only determined by the temperature of heat source and heat sink, in order to improve the cycle efficiency, we must increase the temperature of heat source and reduce the temperature of heat sink. In the real world, this approach is limited by the following constraints.

1. For a real cycle, the heat sink is limited to ambient temperature. Therefore, its temperature is limited to about 15 °C.
2. The temperature of the heat source is also limited to the combustion temperature of the fuel or the upper temperature limit of the structural material (such as graphite ball, cladding, etc.). In the case of fossil fuels, the upper temperature

limit is about 1670 °C. However, due to the limitations of metallurgy on boilers, even this temperature is impossible to reach at present. Therefore, at present, the maximum achievable upper limit of heat source temperature is only about 800 °C. These limitations make the maximum achievable efficiency of Carnot cycle 73%.

Therefore, in the Carnot cycle using ideal components, due to the limitations of the real world, 3/4 of the heat can be converted into work. However, as has been explained, this ideal efficiency is much higher than that of any real system at present.

In order to understand why 73% efficiency is impossible, we must first analyze the Carnot cycle, and then compare the real and ideal components recycled. We will make this comparative analysis by observing the temperature entropy diagram of Carnot cycle using real and ideal components.

In the Carnot isothermal expansion process, the increased energy q_s of the working medium is provided by the heat source, and part of q_r must be released to the environment, so not all the heat can be used by heat. The heat released to the environment is:

$$q_r = T_0 \Delta s \tag{2.43}$$

where T_o is the average heat sink temperature of 15 °C. The available energy [27] (A.E.) for the Carnot cycle may be given as:

$$\text{A.E.} = q_s - q_r \tag{2.44}$$

Substituting Eq. 2.43 for q_r gives:

$$\text{A.E.} = q_s - T_0 \Delta s \tag{2.45}$$

This is equal to the shaded area between 288 k (15 °C) and 1073 k (800 °C) marked with available energy given in Fig. 2.20. From Fig. 2.20, we can see that the efficiency of any working at a temperature lower than 1073 k will be lower. If materials capable

Fig. 2.20 The available energy of Carnot cycle

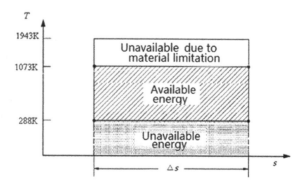

of withstanding the stresses of the high temperature can be improved to more than 1073 k, we can increase the available energy of the system.

From Eq. (2.43), we can find out why entropy can be used to calculate unavailable energy. If the temperature of the heat sink is known, then the change in entropy does correspond to a measure of the heat rejected by the engine.

Figure 2.21 shows the thermal cycle diagram of a typical fossil fuel power plant. Water is a working medium. If the working pressure is required to be less than 13 MPa, it is impossible to obtain a temperature of 1073 k. In fact, the nature and heat transfer process of water require that water be heated under a certain range of pressure. For this reason, the average temperature we heat is much lower than the maximum temperature allowed by the material.

The actual temperature that can be reached is far less than half of the temperature that can be reached by the ideal cycle. The efficiency of typical fossil energy cycle is about 40%, while that of nuclear power plant cycle is about 31%. Note that the efficiency of these cycles is only about half of the ideal cycle efficiency (73%).

Figure 2.22 shows the temperature entropy diagram of Carnot cycle. As can be seen from the figure, there are several factors that limit it from becoming a real cycle. Firstly, it is difficult for the pump to compress the gas–liquid mixture. The isentropic compression process from 1 to 2 will lead to cavitation of the system. In addition, the condenser manufacturing gas–liquid mixture may encounter technical problems at point 1.

Fig. 2.21 Carnot cycle versus typical power cycle available energy

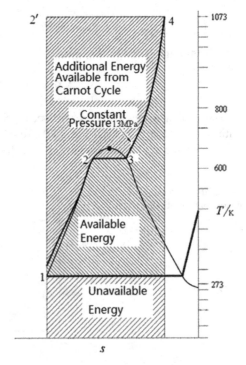

Fig. 2.22 Ideal carnot cycle

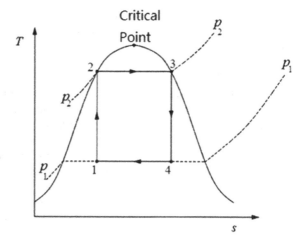

The early development of thermodynamics mainly focused on improving the performance of steam engine. It is extremely exciting to build a reversible cycle of Carnot cycle as close to steam as possible. Therefore, the Rankine cycle was found.

The characteristics of Rankine cycle are shown in Fig. 2.23. It only limits the isentropic compression process from 1 to liquid phase 2. In order to minimize the work, it is necessary to reach the operating pressure and avoid two-phase mixture. The compression process between 1 and 2 in the figure is deliberately exaggerated for display. In fact, when water is compressed from 1 to 70 atm, the temperature rises only about 1 °C.

Fig. 2.23 Rankine cycle

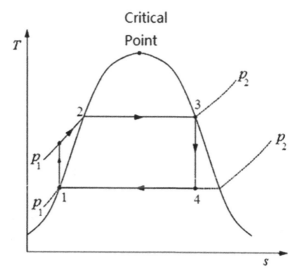

On the Rankine cycle temperature entropy diagram, the available or unavailable energy is shown on the curve. The greater the unavailable energy, the lower the efficiency of the cycle.

From the temperature entropy diagram in Fig. 2.24, it can be seen that if the ideal state is replaced by the non-ideal state, the cycle efficiency will be reduced. Because the non-ideal steam turbine leads to the increase of entropy, it is shown in the figure as the increase of curve area in the temperature entropy diagram. However, the area where the energy can be obtained is smaller than the area where the energy cannot be obtained.

The energy loss of Rankine cycle can also be found by comparing the two cycles, as shown in Fig. 2.25. By comparison, it can be seen that the unusable energy of one cycle can be compared with the other to determine which period is more efficient.

An enthalpy entropy diagrams can also be used to compare different systems and determine efficiency. Like the temperature entropy diagram, the enthalpy entropy

Fig. 2.24 Rankine cycle with real versus ideal

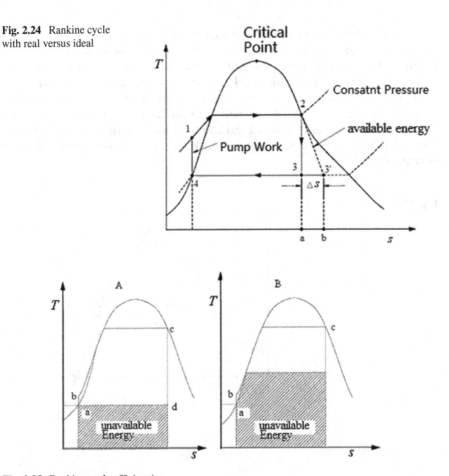

Fig. 2.25 Rankine cycle efficiencies

Fig. 2.26 h-s diagram of typical steam cycle

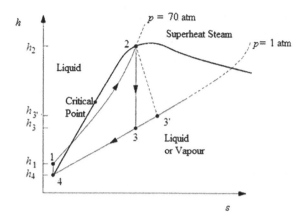

diagram (Fig. 2.26) shows that the ideal process is replaced by the non-ideal process in the cycle, which will reduce the cycle efficiency. This is because working, heating or cooling the non-ideal cycle will lead to the change of enthalpy. It can be clearly seen from the enthalpy entropy diagram that the deviation from 3 points to 3' points reduces the efficiency.

In the thermal cycle diagram of the nuclear power plant shown in Fig. 2.18, the main process of the cycle is:

1-2: under the condition of constant entropy, the saturated steam in the steam generator expands in the high-pressure cylinder to provide output work.

2-3: the wet steam from the high-pressure cylinder is dried and steam water separated in the steam water separation reheater.

3-4: under the condition of constant entropy, the gas separated from the steam water separator reheater expands in the low-pressure cylinder to provide output work.

4-5: under constant pressure, the exhaust steam from the steam turbine condenses in the condenser and transfers heat to the cooling water.

5-6: the inlet water is compressed in the condenser feed pump and preheated by the preheated feed water heater.

6-1: under the condition of constant pressure, the working medium is heated in the steam generator.

This cycle can also be represented by the temperature entropy diagram, as shown in Fig. 2.27. The points in the cycle correspond to the points in Fig. 2.18.

It must be noted that the cycle shown in Fig. 2.27 is an ideal cycle and does not represent the actual process. The steam turbine and pump in the ideal cycle are ideal, so there is no entropy increase. There will be entropy increase in real steam engines and pumps.

Figure 2.28 is a temperature entropy diagram closer to the actual cycle. The steam turbine and pump in this picture are closer to the actual situation, and there is entropy increase. In addition, in this cycle, as shown in point 5 in the figure, there is an obvious

Fig. 2.27 Steam cycle (ideal)

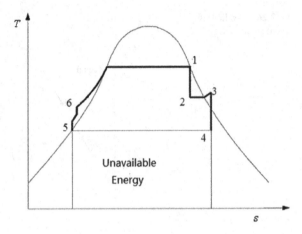

Fig. 2.28 Steam cycle (real)

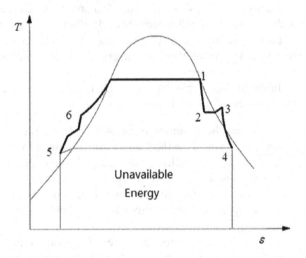

temperature drop in the condenser, which is caused by excessive cooling. Excessive cooling is necessary to prevent cavitation, so the efficiency of the whole cycle is reduced. Therefore, it can be seen that the increase of heat transfer temperature difference in the condenser will reduce the efficiency of the cycle. By controlling the temperature and flow rate of cooling water into the condenser, the operator can affect the efficiency of the whole cycle.

2.9 Ideal Gas Law

The ideal gas law is the equation that assumes the ideal gas state. It is a good approximation of the behavior of many gases under many conditions, although it has several limitations. In 1834, Émile Clapeyron first described it as a combination of the empirical Boyle's law, Charles' law and Avogadro's Law. The ideal gas law is usually written as [28]:

The gas pressure of isothermal expansion is negatively correlated with volume.

$$p_1 V_1 = p_2 V_2 = p_3 V_3 = \text{constant} \tag{2.46}$$

Charles also concluded from the test that under the condition of constant volume, the pressure and temperature of gas are positively correlated; In the case of constant pressure, the volume of gas is positively correlated with temperature.

By combining Boyle and Charles' experimental conclusions, there is the following relationship:

$$pV = RT \tag{2.47}$$

The constant R in the above formula is called the ideal gas constant. The gas constants of general gases are shown in Table 2.1.

The gas constant of a particular gas can be obtained by dividing the general gas constant by the molecular weight of the gas.

The real gas does not obey the equation of state of the ideal gas. When the temperature approaches the boiling point of the gas, the increase of pressure will cause liquefaction and the volume will decrease sharply. At very high pressure, the force between gas molecules becomes significant. However, when the pressure and temperature are higher than the boiling point, most gases are approximately the same.

Engineers use the ideal gas equation to solve the gas problem because it is easy to use and similar to the behavior of real gas. In most cases, the physical conditions of gas accord with the approximation of ideal gas.

The most common gas behavior studied by engineers is to approximate the compression process with an ideal gas. The compression process can be constant temperature, constant pressure, or adiabatic (no heat exchange). As pointed out in the first law of thermodynamics, the work done by any process depends on the process

Table 2.1 Ideal gas constant values [29]

Gas	Chemical symbol	Molecular weight	Gas constant	Specific heat		Specific heat ratio
		$M/(\text{g/mol})$	R	$C_p/(\text{kJ/kg}°\text{C})$	$C_V/(\text{J/kg}°\text{C})$	k
Air		28.95	53.35	0.860	1.200	1.40
Carbon dioxide	CO_2	44.00	35.13	0.800	1.025	1.28
Hydrogen	H_2	2.016	766.80	12.195	17.094	1.40
Nitrogen	N_2	28.02	55.16	0.880	1.235	1.40
Oxygen	O_2	32.0	48.31	0.775	1.085	1.40
Steam	H_2O	18.016	85.81	1.799	2.299	1.28

Fig. 2.29 Pressure–volume
diagram

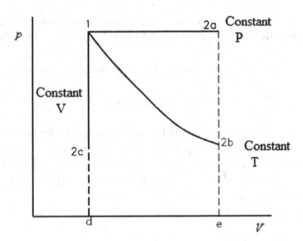

itself. The work done by the system is essentially the area under the *p-V* curve. It can be seen from Fig. 2.29 that different ideal gas processes, such as isothermal or isobaric processes, obtain different amounts of work.

In order to determine the process of isobaric work done, use the following equation:

$$W_{1-2} = p\Delta V \qquad (2.47)$$

Equation of constant volume process is not complicated. Constant volume process the work done is the product of volume and pressure changes.

$$W_{1-2} = \Delta p V \qquad (2.48)$$

Exercises

1. What is the absolute pressure at the bottom of a swimming pool which holds water of 2.2 m deep?
2. A volume of water steam has 12 pound mass and 23,000 Btu internal energy. Find the specific internal energy of steam at J/kg.
3. Energy can be expressed what kind of forms? What is the difference and similarly of heat and work?
4. What is thermal equilibrium?
5. A volume of water and steam, at temperature 200 °C and pressure 10 MPa. Suppose the density of mixture is 500 kg/m³, find the equilibrium quality.
6. A thermos holds 2.5 kg water at temperature 20 °C. A heater of 1000 W power heats the water in thermos. How long it takes to boil the water.
7. A volume of super heat steam at Pressure 5 MPa, temperature 360 °C, expands to pressure of 1 MPa. What is the change of specific enthalpy?

8.   A volume of mixture of water and steam at pressure 2.7 MPa, quality is 90%. Find the specific volume, specific enthalpy, and specific entropy of the mixture.
9.   The specific enthalpy of steam enters and leaves a steam turbine are 2915 kJ/kg and 2355 kJ/kg. Heat loss of the turbine is about 8 kJ/kg. Kinetic energy and potential energy can be neglected. Find the word done by 1 kg of steam.
10. The thermal power of a nuclear reactor core is 3000 MW. The power of steam generator is 3014 MW. Find the power of pump to maintain the temperature of coolant in primary loop. (The efficiency of pump is 95%.)
11. A nuclear power system has maximum coolant temperature of 750 °C. Find the maximum thermal efficiency of the system.
12. A thermo meter in a room shows temperature 35 °C and humidity 75%. At what temperature steam will condense at this room.
13. Find the enthalpy of steam at temperature 300 °C and pressure 2 MPa using steam table or graph.
14. Find the maximum efficiency of heat source 300 °C, and heat sink 15 °C.
15. Describe the main processes of Rankine cycle.
16. A heat exchanger works at power of 2.0×10^6 W, temperature of cold water enters the heat exchanger is 15 °C as secondary side coolant. Fine the minimum flow rate of coolant if the maximum outlet temperature is limited to 25 °C.

References

1. Borgnakke C, Sonntag RE. Fundamentals of thermodynamics. New Jersey: Wiley; 2014.
2. Wang Z. Thermodynamics. 2nd ed. Beijing: Peking University Press; 2014.
3. Yu J, Jia B. Reactor thermogeomenics. 2nd ed. Beijing: Tsinghua University Press; 2011.
4. Du S. Pressure measurement technology and instrumentation Beijing: Machinery Industry Press; 2005.
5. Jousten K. Handbook of vacuum technology. Weinheim: Wiley; 2008.
6. U.S. Department of Energy. Thermodynamics, heat transfer, and fluid flow. Washington, D.C.: Lulu.com; 2016.
7. Evolution of the International practical temperature scale of 1968. Philadelphia: American Society for Testing and Materials; 1974.
8. Jiyang Yu, Jia B. Reactor thermogeomenics. Beijing: Tsinghua University Press; 2003.
9. Rogers GFC, Mayhew YR. Engineering thermodynamics work and heat transfer. 4th ed. New Jersey: Prentice-Hall Inc.; 1996.
10. Sheikholeslami MK, Davood DG. Hydrothermal analysis in engineering using control volume finite element method. London: Academic Press; 2015.
11. Bikya R. Evaluation of methods for measuring aggregate specific gravity. Virginia: West University; 2012.
12. Wang K. Potential energy landscape of particulate matter. New York: City University of New York; 2012.
13. Dixon JR. Thermodynamics: an introduction to energy. New Jersey: Prentice-Hall; 1975.
14. James J. On the mechanical equivalent of heat. The British Association meeting in Cambridge; 1845.
15. Xu Y. Geometric analysis of thermodynamic equilibrium processes. Pittsburgh: Carnegie Mellon University; 2008.

16. Anderson PW. Basic notions of condensed matter physics. New York: Perseus Publishing; 1997.
17. Fisher ME. The renormalization group in the theory of critical behavior. Rev Mod Phys. 1974;46:597–616.
18. Kamisetty VK. Pressure-temperature phase diagram of tris (hydroxymethyl) aminomethane and phase diagram determination. Reno: University of Nevada; 2010.
19. Wagner W. International steam tables: properties of water and steam based on the industry. Berlin, Heidelberg: Springer, Berlin Heidelberg; 2008.
20. Rajpu RK. Engineering thermodynamics. 3rd ed. Boston: Jones & Bartlett Learning; 2009.
21. Rosenberg RM. From Joule to Caratheodory and Born: a conceptual evolution of the first law of thermodynamics. J Chem Edu. 2010;87:691–3.
22. Canada S, Cohen G, Cable R, etc. Parabolic trough organic rankine cycle solar power plant. Colorado: US Department of Energy NREL; 2009.
23. Adkins CJ. Equilibrium thermodynamics. 3rd ed. Cambridge UK: Cambridge University Press; 1983.
24. Tipler PA, MOSCA G. Physics for scientists and engineers. 6th ed. New York: W. H. Freeman; 2007.
25. Stodola A. Steam and gas turbines. New York: McGraw-Hill; 1927.
26. National Renewable Energy Laboratory. Improving pumping system performance: a sourcebook for industry. 2nd ed. Washington, D.C: Scholar's Choice; 2006.
27. Alekseev GN. Energy and entropy. Moscow: Mir Publishers; 1986.
28. Clausius R. Ueber die art der bewegung, welche wir wärme nennen. Annalen der Physik und Chemie (in German). 1857;176(3):353–79.
29. Moran MJ, Shapiro HN. Fundamentals of engineering thermodynamics. 8th ed. New Jersey: Wiley; 2014.

Chapter 3
Heat Transfer

Heat transfer is the science of study of heat transfer laws. Mainly the law of heat transfer caused by temperature difference.

Heat transfer is formed as a discipline in the nineteenth century. Estimating the temperature of a mathematical expression in a red-hot iron bar, British scientists Newton presented Newton's law of cooling in terms of thermal convection in 1701. However, it did not reveal the mechanism of convective heat transfer.

Real development of convective heat transfer is in the late nineteenth century. In 1904, German physicist Prandtl presented the boundary layer theory. In 1915, Nusselt presented the dimensionless analysis. From the theoretical and experimental understanding correctly and quantitative research laid the foundation for convective heat transfer. In 1929, Schmidt pointed out the resemblance of mass transfer and heat transfer.

In terms of thermal conductivity, thermal conductivity of the flat wall French physicist Biot results obtained in 1804 was the first expression of the law of heat conduction. Later, in France using mathematical Fourier method, more accurate to describe thermal conductivity, as a differential form became known as Fourier's Law.

Theoretical aspects of thermal radiation is more complex. In 1860, Kirchhoff analog absolute blackbody through artificial cavities, demonstrated at the same temperature blackbody radiation was the largest, and noted that the emissivity of the object with the same absorption rate of the object at a temperature equal to, after being called Kirchhoff's laws.

In 1878, Stefan found the fact that the radiation rate is proportional to the absolute temperature of the fourth power. In 1884, Boltzmann demonstrated in theory, known as the Stefan-Boltzmann law, commonly known as the fourth power of laws. In 1900, Planck Research cavity blackbody radiation, Planck obtained heat radiation law. This law not only describes the relationship between the temperature and the frequencies of blackbody radiation, but also demonstrates the displacement law of Wien about the blackbody energy distribution.

3.1 Heat Transfer Terminology

After briefly reviewing the history, we first introduce some terms frequently used in the field of heat transfer.

As long as there will be a temperature difference, there will be heat transfer. Heat transfer in three ways, namely, heat conduction, convection and radiation [1].

1. **Heat Conduction**

Need not occur relative movement between the parts of the object, or direct contact with different objects, the heat depend on the thermal motion of molecules, atoms and free electrons and other microscopic particles arising from the transfer referred to thermal conductivity.

Conduction involves the transfer of heat by the interactions of atoms or molecules of a material through which the heat is being transferred.

2. **Convection**

Relative displacement occurs between the portions of the fluid. Heat transfer happens due to the mixing of hot and cold fluids and move on each other. Convection is a process that uniforms the temperature between hotter and cooler parts of liquid or gas, with the way of flow to make temperature uniform. Convection is a unique way of liquids and gases heat transfer. The convection of gas usually stronger than liquid significantly. Convection can be divided into natural convection and forced convection. Natural convection is naturally occurs due to the uneven temperature caused. Forced convection is due to a pump to drive fluid forced convection. Increase the flow velocity of the fluid can accelerate convective heat transfer.

Convection involves the transfer of heat by the mixing and motion of macroscopic portions of a fluid.

3. **Thermal Radiation**

Objects have the ability to emit energy due to their own temperature. This way of heat transfer is called thermal radiation. Although thermal radiation is also a way of heat transfer, it is different from heat conduction and convection. It can transfer heat directly from one system to another without relying on medium. Thermal radiation emits energy in the form of electromagnetic radiation. The higher the temperature, the stronger the radiation. The wavelength distribution of radiation also changes with temperature. For example, when the temperature is low, it is mainly radiated with invisible infrared light. When the temperature is 500 °C or higher, it will emit visible light or even ultraviolet light. Thermal radiation is the main way of long-distance heat transfer. For example, the heat of the sun is transmitted to the earth through space in the form of thermal radiation.

4. **Heat Transfer Rate**

The heat transferred per unit time is defined as heat transfer rate. A common unit for heat transfer rate is J/s or W. Because the power and the heat transfer rate have the same dimension, sometimes it is called heat power. Its physical symbol is P_Q.

5. Heat Flux

Sometimes it is important to determine the heat transfer rate per unit area, or heat flux. Units for heat flux are W/m^2. The heat flux can be determined by dividing the heat transfer rate by the area through which the heat is being transferred.

6. Thermal conductivity

The heat transfer characteristics of a solid material are measured by a property called the thermal conductivity (k) measured in $W/(m\ ^\circ C)$. It is a measure of a substance's ability to transfer heat through a solid by conduction. The thermal conductivity of most liquids and solids varies with temperature. For vapors, it depends upon pressure.

Thermal conductivity is a property of material. Different material has different thermal conductivity usually. It also depends on the structure, density, moisture, temperature, pressure etc.

Thermal conductivity is generally refers to only the presence of thermally conductive heat transfer. When there are other forms of heat transfer, such as heat radiation transfer or convection, the conductivity is often referred to as the apparent thermal conductivity or effective thermal conductivity. In addition, the thermal conductivity for the purposes of homogeneous material. In engineering practice, there is also porous, multi-layer, multi-structure or anisotropic material, the thermal conductivity of these materials is actually an integrated thermal conductivity performance, also known as the average thermal conductivity.

Thermal conductivity is a property of the material itself, the thermal conductivity of different materials vary. The thermal conductivity of the material relates with its structure, density, humidity, temperature, pressure and other factors. Low moisture content of the same material, at lower temperatures, the thermal conductivity is small. In general, the thermal conductivity of solid is larger than that of liquid, and that of liquid larger than that of gas. This is largely due to the difference forces between molecules of different state.

Usually low thermal conductivity material called insulation material. According to the Chinese national standard, when the average temperature does not exceed 350 °C, the thermal conductivity of not more than 0.12 $W/(m\ K)$ material called insulation material and the thermal conductivity of 0.05 $W/(m\ K)$ or less material called effective thermal insulation materials.

High thermal conductivity of the material has excellent thermal conductivity. In the same heat flux density and thickness, the temperature difference of the surface sidewall temperature between the cold and the hot decreases if the thermal conductivity increases. For example, when no furring, due to the high thermal conductivity of material of the boiler heat transfer tubes, the wall temperature difference is small. However, when the heat transfer surface furring due to scale, the thermal conductivity of scale is very small, the temperature difference increases with the thickness of scale, thereby rapidly raise the temperature of the wall. When the thickness of the scale is quite large (about 1 ~ 3 mm), the tube wall temperature will exceed the allowable value, the resulting tube overheating damage.

Typically, the thermal conductivity of material can be obtained theoretically and experimentally in two ways. Now the thermal conductivity values of engineering calculations are measured out by specialized test. In theory, starting from the material microstructure, quantum mechanics and statistical mechanics, based on the material by heat conduction mechanism of establishing thermal physical model, through a complex mathematical analysis and calculation of the thermal conductivity can be obtained. However, due to the applicability of the theoretical approach to certain restrictions, but with the rapid increase in new materials, it is still yet to find a date and accurate enough for a wide range of theoretical equation, so the thermal conductivity experimental test method, is still the main source of the data of the thermal conductivity of material.

In Chap. 8, materials science, which, we will detail the internal lattice structure of the solid. According to materials science and the solid by free electrons of atoms, and atoms are bound in the crystal lattice of regularly arranged. Accordingly, the heat transfer is achieved by the two effects: the migration of free electrons and lattice vibrational waves. When regarded as quasi-particle phenomenon, lattice vibrations called phonons. Pure metals, the electronic contribution to the thermal conductivity of large; and in the non-conductor, the phonon contribution to play a major role. In all solids, the metal is the best conductor of heat. The thermal conductivity of pure metals generally decreases with increasing temperature. The effects of purity of the metal on the thermal conductivity is large, such as carbon-containing 1% of ordinary carbon steel thermal conductivity of 45 W/(m K), the thermal conductivity of the trace elements added after the stainless steel down to 16 W/(m K).

Liquid can be divided into two types of liquid metal and non- metallic liquid. The former high thermal conductivity and lower the latter. Non-metallic liquid, water, thermal conductivity maximum. Removal of water and glycerin, the majority of the thermal conductivity of the liquid decreases slightly with increasing temperature. In general, the thermal conductivity of the solution is lower than the thermal conductivity of pure liquid.

The thermal conductivity of the gas increases with increasing temperature. Pressure within the normal range, the thermal conductivity change with pressure is small, only when the pressure is greater than 200 MPa, or a pressure less than about 3 kPa, the thermal conductivity was increased with increasing pressure. Therefore, engineering calculations can often ignore the impact of the pressure on the gas thermal conductivity. The thermal conductivity of the gas is very small, so the thermal detrimental but beneficial for insulation.

7. Log Mean Temperature Difference

In heat exchanger applications, the inlet and outlet temperatures are usually specified based on the fluid in the tubes. The temperature change that takes place across the heat exchanger from the entrance to the exit is not linear. A precise temperature change between two fluids across the heat exchanger is best represented by the log mean temperature difference (LMTD or Δt_{ln}), defined in Eq. 3.1.

$$\Delta t_{\text{ln}} = \frac{\Delta t_2 - \Delta t_1}{\ln(\Delta t_2 / \Delta t_1)} \tag{3.1}$$

8. Convective Heat Transfer Coefficient

Convective heat transfer coefficient is a measure of the heat transfer capacity between fluid and solid surface, with the unit of W/(m² °C). The value of convective heat transfer coefficient is closely related to the physical properties of the fluid, the shape and position of the heat exchange surface, the temperature difference between the surface and the fluid and the flow velocity of the fluid. The greater the velocity of the fluid near the surface of the object, the greater the surface convective heat transfer coefficient. If people stand in the environment with high wind speed in winter, they will feel colder because of the large convective heat transfer coefficient on the skin surface and the large heat dissipation.

9. Overall Heat Transfer Coefficient

In the heat transfer analysis of the heat exchanger, in order to consider the convective heat transfer at the primary and secondary sides and the heat conduction of the heat transfer tube, an overall heat transfer coefficient is usually adopted, so that the total heat exchange power (or heat flow) can be calculated by the following formula:

$$P_Q = U_o A_o \Delta t_o \qquad (3.2)$$

where, U_o is the overall heat transfer coefficient, W/(m² K); Δt_o is the overall temperature difference, K; A_o is the overall area for heat transfer, m².

10. Bulk Temperature

The temperature of the fluid near the wall varies with the spatial position, and the mainstream temperature represents the temperature of the fluid away from the wall. Far away means that the distance from the wall is much larger than the thickness of the thermal boundary layer.

3.2 Heat Conduction

The temperature difference in an object or system is a necessary condition for heat conduction. In other words, heat transfer can occur only when there is a temperature difference in or between the media. However, the temperature difference is not a sufficient condition, because if the object with temperature difference is divided by vacuum, there is no heat conduction. In addition to temperature difference, heat conduction also requires medium.

The phenomenon of heat transfer from one part of a system to another or from one system to another is called heat transfer. Heat conduction is the main way of heat transfer in solids. Heat conduction also exists in the non-flowing liquid or gas layer. In the case of flow, it often occurs simultaneously with convection. The heat conduction rate depends on the distribution of the temperature field in the object. In conduction heat transfer, the most common means of correlation is through Fourier's Law of Conduction.

Objects or temperature difference within the system is a necessary condition for heat conduction. Alternatively, that as long as temperature difference exists, heat transfer can occur. However, the temperature difference is not a sufficient condition, because of the temperature difference between objects vacuum division, there is no heat conduction. In addition to the temperature difference, but also the media is necessary for heat conduction.

Heat from a part of the system to another or from one part of the system spread to another phenomenon called heat transfer. Solid heat conduction is the main mode of heat transfer. In the boundary layer of liquid or gas, thermal conduction also exists in the flow and often occurs simultaneously with convection. Heat transfer rate is dependent on the distribution of temperature field inside the object.

In gas, the thermal conductivity is the result of gas molecules collide with each other at irregular thermal motion. The higher the gas temperature, the greater the kinetic energy of molecular motion results molecules of different energy levels of collision of the heat transferred. In the conductive solid, a substantial number of free electrons in the lattice and heat is transmitted through the interaction of free electrons. In non-conductive solid, heat transfer through the vibration of the lattice structure, i.e. atoms, molecules near the equilibrium position of the vibration to achieve. As for the mechanism of conduction in liquid is not yet available with unified understanding. A view similar to that of the gas molecules collide with each other, the liquid heat transfer happens. Because the small distance between the liquid molecules, the impact force between the molecules is greater than the gas molecules. Another view is that the reason the thermal fluid similar to non-conductive solid, rely mainly on the elastic wave.

Therefore, heat conduction in essence, leaving the energy from the high temperature portion of the object in a large number of substances by diffusion thermal motion of molecules interact to a low temperature portion, low temperature process or object passed by a hot object. In solids, the kinetic energy of the crystal particle vibration at high temperature is large. In the low temperature portion, the kinetic energy of the particles vibration is less. Due to vibration interaction particles, and hence diffusion of heat inside the crystal part by the kinetic energy to the kinetic energy of the smaller parts. Heat transfer process and its essence is a process of energy diffusion.

In the conductor, due to the presence of a large number of free electrons, constantly make random thermal motion, energy transfer by lattice vibration is relatively small. The free electrons in the metal crystals on the heat conduction plays a major role. Therefore, the general electrical conductors is also a good conductor of heat. In the liquid heat transfer as follows: liquid molecules in areas of high temperature thermal motion of relatively strong due to the presence of the interaction energy of thermal motion will gradually transfer to the surrounding layers between the molecules of the liquid, causing the heat transfer phenomenon. Due to the small thermal conductivity of the fluid, the heat transfer is slow. Spacing between gas molecules is relatively large, the collision gas molecules rely on random thermal motion and molecules in the gas diffusion internally generated energy to form macroscopic heat transfer.

Reactor design must deal with the various components within the reactor core at steady state and transient heat conduction problem conditions, namely by solving the heat conduction equation to determine the temperature distribution within the various components, to meet the appropriate safety requirements.

Heat conduction inside the reactor core has the following main features:

(1) Heat source, such as the huge heat release rate in core and the uneven spatial distribution;
(2) Reactor thermal properties, such as thermal conductivity under conditions of nuclear radiation variability;
(3) The geometric complexity of reactor internals, the shape and boundary conditions.

Therefore, when addressing the issue within the thermal reactor, often they need to be reasonable under the specific circumstances and simplify engineering experience, and the introduction of certain methods and analytical model processing.

3.2.1 Fourier's Law of Conduction

The law, in its equation form of one dimensional, is expressed as,

$$q = k\frac{dt}{dx} \tag{3.3}$$

where, q is heat flux, W/m^2; k is thermal conductivity, $W/(m\ °C)$.

3.2.2 Rectangular

The use of Fourier's law of conduction in determining the amount of heat transferred by conduction is demonstrated in the following example.

Example 3.1 1000 W is conducted through a section of insulating material shown in Fig. 3.1 that measures 1 m^2 in cross-sectional area. The thickness is 1 cm. and the thermal conductivity is 0.12 W/ (m °C). Compute the temperature difference across the material.

Solution: using Eq. (3.3),

$$P_Q = qA = k\frac{\Delta t}{\Delta x}A = 1000W$$

Solving for $\Delta t = 1000/0.12 \times 0.01 = 83(°C)$.

Fig. 3.1 Thermal insulation
problem

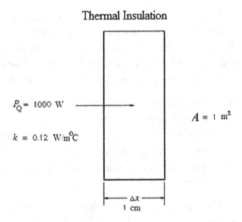

Thermal Insulation

$P_Q = 1000$ W

$k = 0.12$ W/m°C

$A = 1$ m²

Δx
1 cm

3.2.3   Equivalent Resistance

Thermal resistance is a comprehensive quantity reflecting the ability to prevent heat transfer. In the engineering application of heat transfer, in order to meet the requirements of production technology, sometimes the heat transfer is strengthened by reducing the thermal resistance; sometimes, the heat transfer is restrained by increasing the thermal resistance.

Thermal resistance is a reflection of the amount of the comprehensive ability to prevent heat transfer. In heat transfer engineering applications, in order to meet the requirements of the production process, sometimes to enhance heat transfer by reducing the thermal resistance; and sometimes by increasing the resistance to inhibit heat transfer.

When the heat transfers inside a body, the thermal resistance is the resistance of conduction. For the heat flow through the same cross-sectional area of the plate, thermal resistance is $\Delta x/(kA)$. Where Δx is the thickness of plate, A is the cross-sectional area of the plate perpendicular to the direction of heat flow, k is the thermal conductivity of the plate material.

In the convective heat transfer process, the thermal resistance between the fluid and solid wall called convection thermal resistance, $1/(hA)$. Where h is the convective heat transfer coefficient, A is the heat transfer area.

To objects with different temperature, will radiation to each other. If two objects are in bold, and ignored the gas between the two objects absorb heat, the radiation resistance is $1/(A_1 F_{1-2})$ or $1/(A_2 F_{2-1})$. Wherein A_1 and A_2 is the surface area of two objects mutually radiation, F_{1-2} and F_{2-1} are the radiation angle factors, it will be described in detail later.

As heat flows through the interface of two solid contact with each other, the interface itself will exhibit significant heat resistance. This resistance is called the thermal contact resistance. The main reason is that the contact resistance, the actual area of the appearance of any opinion good contact of two objects in direct contact only pay part of the interface, and the rest are gaps. Heat rely on thermal conduction and thermal radiation within the gap gas is passed, and their heat transfer capacity far less than generally solid material. So that the contact resistance when heat flows through the interface, there is a greater change in temperature along the direction of heat flow, which is the engineering applications need to avoid. The measures to reduce the contact resistance are: ① increase the pressure on the contact surface of the two objects, the protruding portion of the object boundary surface between each extrusion, thereby reducing the gap, increase the contact surface. ② jelly objects in the interface between the two objects painted with higher thermal capacity—thermal grease. For example, between the computer's CPU and heat sink often you need to add a layer of thermal grease.

In summary, the thermal resistance of the heat resistance is encountered in the heat flow path, which reflects the size of the heat transfer capability of the medium or between media. Greater the resistance, the smaller the heat transfer capability. Thermal resistance indicates the size of the temperature rise caused by heat flow of 1 W, unit °C/W or K/W. Therefore, thermal power multiplied by the resistance, you can get the temperature difference of the heat transfer path. You can use a simple analogy to understand the significance of thermal resistance. It corresponds to the current strength of the heat transfer, the voltage corresponding to the temperature difference, the thermal resistance equivalent.

The electrical analogy may be used to solve complex problems involving both series and parallel thermal resistances. Figure 3.2 shows the equivalent resistance circuit.

Example 3.2 A composite protective wall is formed of a 1 cm copper plate, a 0.1 cm layer of asbestos, and a 2 cm layer of fiberglass. The thermal conductivities of the materials are as follows: $k_{Cu} = 400$ W/(m °C), $k_{asb} = 0.08$ W/(m °C) and $k_{fib} = 0.04$ W/(m °C). The overall temperature difference across the wall is 500 °C. Calculate the thermal resistance of each layer of the wall and the heat transfer rate per unit area (heat flux) through the composite structure.

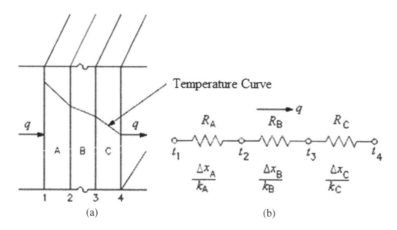

Fig. 3.2 Equivalent resistance

Solution:

$$q = \Delta t/(R_A + R_B + R_C) = 500/(0.01/400 + 0.001/0.08$$
$$+ 0.02/0.04) = 976\,\text{W/m}^2$$

3.2.4 Cylindrical

Generally, the heat conduction of cylindrical materials is the heat conduction along the radius direction, while along the z direction of the pipe axis, the temperature derivative dt/dz is generally small and ignored. For example, the heat conduction of the heat transfer tube wall in the heat exchanger. In this case, since the heat transfer area will change in the heat conduction direction, the logarithmic average area is usually used when calculating using the Fourier heat conduction law, which is defined as follows:

$$A_{ln} = \frac{A_o - A_i}{\ln(A_o/A_i)} \tag{3.4}$$

then

$$P_Q = k\frac{\Delta t}{\Delta r}A_{ln} \tag{3.5}$$

Since $A_o = 2\pi r_o L$, $A_i = 2\pi r_i L$, then (Fig. 3.3)

$$P_Q = \frac{2\pi k L \Delta t}{\ln(r_o/r_i)} \tag{3.6}$$

Fig. 3.3 Cross-sectional surface area of a cylindrical pipe

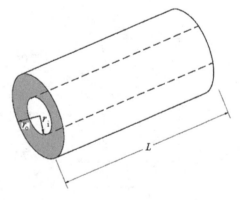

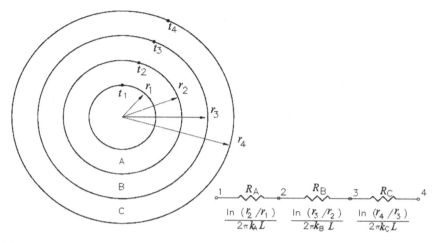

Fig. 3.4 Multi layers of cylindrical pipe wall

The heat resistance of cylindrical pipe wall is:

$$R_{th} = \frac{\ln(r_o/r_i)}{2\pi k L} \tag{3.7}$$

For multi layers of cylindrical pipe wall shown as Fig. 3.4, the heat resistance is:

$$P_Q = \frac{2\pi L (t_1 - t_4)}{\frac{\ln(r_2/r_1)}{k_A} + \frac{\ln(r_3/r_2)}{k_B} + \frac{\ln(r_4/r_3)}{k_C}} \tag{3.7}$$

3.3   Convective Heat Transfer

Convective heat transfer refers to the heat transfer between fluid and solid surface through fluid convection. Convective heat transfer is a heat transfer phenomenon in the process of fluid flow. It depends on the movement of fluid particles for heat transfer, which is closely related to the flow of fluid. When the fluid flows in laminar flow, the heat transfer in the direction perpendicular to the fluid flow is mainly carried out in the form of heat conduction (there is also weak natural convection). When the fluid flows in turbulence, convection is the main heat transfer mode.

If this movement and mixing is caused by the density change caused by the internal temperature difference of the fluid, the term *natural convection* is used. If this movement and mixing is caused by external forces (such as pumps), the term *forced convection* is used. Heat transfer from a hot water radiator to a room is an example of natural convective heat transfer. The heat transfer from the surface of the heat exchanger to most of the fluid pumped through the heat exchanger is an example of forced convection.

Convective heat transfer is more difficult to analyze than conductive heat transfer, because any single property of heat transfer medium, such as thermal conductivity, cannot be defined to describe its mechanism. Convective heat transfer varies from case to case (depending on fluid flow conditions) and is often coupled with fluid flow patterns. In practice, the analysis of convective heat transfer is empirical (through direct observation).

Convective heat transfer is handled empirically, because the factors affecting the thickness of stagnant film are:

- Fluid velocity
- Fluid viscosity
- Heat flux
- Surface roughness
- Type of flow (single-phase/two-phase)

Convection involves heat transfer between a surface at a given temperature and a fluid at an overall temperature. The exact definition of the overall temperature varies from case to case.

3.3.1 Convective Heat Transfer Coefficient

The basic relationship for heat transfer by convection is the Newton's cooling law:

$$q = h(t_w - t_b) \tag{3.8}$$

where, t_w is wall temperature, °C; t_b is bulk temperature of fluid, °C; h is convection heat transfer coefficient, W/(m^2 °C).

The convective heat transfer coefficient can be calculated by the empirical formula sorted out according to the experimental data, such as Dittus-Boelter correlation. The Dittus-Boelter correlation makes a distinction between the heating fluid and cooling fluid separately. In the case of heated fluid (e.g. in core channel and secondary side of steam generator) the correlation is:

$$Nu = 0.023 Re^{0.8} Pr^{0.4} \tag{3.9}$$

For cooling condition, such as primary coolant in steam generator, the correlation is:

$$Nu = 0.023 Re^{0.8} Pr^{0.3} \tag{3.10}$$

where Nu, Re, Pr are dimensionless number, defined as:

$$\text{Nu} \equiv \frac{hL}{k} \tag{3.11}$$

$$\text{Re} \equiv \frac{\rho V L}{\mu} \tag{3.12}$$

$$\text{Pr} \equiv \frac{\mu C_p}{k} \tag{3.13}$$

where k is thermal conductivity, μ is viscosity, L character length, V is bulk velocity, ρ is density of fluid.

After using Eq. (3.9) or (3.10) to get Nu, then according to Eq. (3.11) one can get the heat transfer coefficient h.

Example 3.3 A 6 m uninsulated steam line crosses a room. The outer diameter of the steam line is 45 cm and the outer surface temperature is 140 °C. The convective heat transfer coefficient for the air is 80 W/(m^2 °C). Calculate the heat transfer rate from the pipe into the room if the room temperature is 25 °C.

Solution:

$$P_Q = hA\Delta t = 80 \times 2\pi \times (0.45/2) \times 6 \times (140 - 25) = 7.8 \times 10^4 \text{(W)}$$

Many applications involving convective heat transfer occur in pipes, tubes or some similar cylindrical devices. In this case, the heat transfer surface area usually given in the convection equation changes with the heat passing through the cylinder. In addition, for the temperature difference inside and outside the pipeline and along the pipeline, some average temperature values need to be used to analyze the problem. This average temperature difference is called logarithmic mean temperature difference (LMTD), as described earlier.

It is the natural logarithm of the temperature difference at one end of the heat exchanger minus the temperature difference at the other end of the heat exchanger divided by the ratio of the two temperature differences. The above LMTD definition involves two important assumptions:

- the specific heat of fluid changes little with temperature, and
- in the whole heat exchanger, the convective heat transfer coefficient is relatively constant.

3.3.2   Overall Heat Transfer Coefficient

Many heat transfer processes encountered in nuclear facilities involve a combination of conduction and convection. For example, heat transfer in a steam generator involves convection from the reactor coolant body to the inner tube surface of the steam generator, conduction through the tube wall, and convection from the outer tube surface to the secondary side fluid.

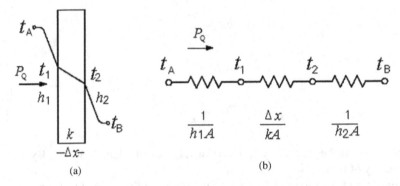

Fig. 3.5 Overall heat transfer coefficient

For the calculation of fluid heat transfer on both sides of solid heat exchange wall, the overall heat transfer coefficient or overall thermal resistance can be used. The thermal resistance of flat plate type is shown in Fig. 3.5.

Then the total rate of heat transfer is:

$$P_Q = \frac{t_A - t_B}{\frac{1}{h_1 A} + \frac{\Delta x}{kA} + \frac{1}{h_2 A}} \tag{3.14}$$

Recalling Eq. (3.2):

$$P_Q = U_o A \Delta t_o \tag{3.15}$$

where U_o is overall heat transfer coefficient and it is:

$$U_o = \frac{1}{\frac{1}{h_1} + \frac{\Delta x}{k} + \frac{1}{h_2}} \tag{3.16}$$

For cylindrical, an example is given to introduce the application of overall heat transfer coefficient. As shown in Fig. 3.6, for a circular tube, the temperature inside

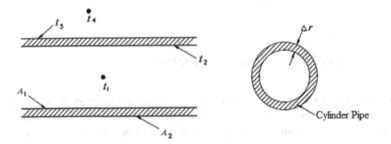

Fig. 3.6 Combined heat transfer

the tube is t_1, the temperature outside the tube is t_4, the inner wall temperature of the tube wall is t_2, and the outer wall temperature is T t_3, then there is convective heat transfer between 1 and 2, heat conduction between 2 and 3, and convective heat transfer between 3 and 4, the basic relationships for these three processes can be expressed as:

$$P_Q = h_1 A_1 (t_1 - t_2) \tag{3.17}$$

$$P_Q = (k/\Delta r) A_{\ln}(t_2 - t_3) \tag{3.18}$$

$$P_Q = h_2 A_2 (t_3 - t_4) \tag{3.19}$$

The total temperature difference can be expressed as the sum of the Δt of the three individual processes.

$$t_1 - t_4 = \Delta t_o = P_Q \left(\frac{1}{h_1 A_1} + \frac{\Delta r}{k A_{\ln}} + \frac{1}{h_2 A_2} \right) \tag{3.20}$$

This relationship can be modified by selecting a reference cross-sectional area A_o.

$$\Delta t_o = \frac{P_Q}{A_o} \left(\frac{A_o}{h_1 A_1} + \frac{A_o \Delta r}{k A_{\ln}} + \frac{A_o}{h_2 A_2} \right) \tag{3.21}$$

Recalling Eq. (3.15),

$$P_Q = U_o A_o \Delta t_o \tag{3.22}$$

One gets:

$$U_o = \frac{1}{\frac{A_o}{h_1 A_1} + \frac{A_o \Delta r}{k A_{\ln}} + \frac{A_o}{h_2 A_2}} \tag{3.23}$$

This is the overall heat transfer coefficient. If the pipe wall is very thin, the inner surface area is approximately equal to the outer surface area and the reference area, then Eq. (3.23) will degenerate into Eq. (3.16) of the overall heat transfer coefficient in the case of flat plate.

3.4 Radiant Heat Transfer

3.4.1 Thermal Radiation

Thermal radiation is the phenomenon that an object radiates electromagnetic waves due to its temperature. Thermal radiation is one of the three ways of heat transfer. The term thermal here means to distinguish it from other radiation (such as ionizing radiation). The transfer of heat from a fireplace across a room in the line of sight is a typical example of heat radiation.

All objects whose temperature is higher than absolute zero can produce thermal radiation. The higher the temperature, the greater the total energy radiated and the more short-wave components. The spectrum of thermal radiation is a continuous spectrum, and the wavelength coverage can theoretically range from 0 to ∞. General thermal radiation mainly depends on visible light and infrared light with long wavelength. Because the propagation of electromagnetic wave does not need any medium, thermal radiation is the only way of heat transfer that can be carried out in vacuum.

When the temperature is 300 °C, the strongest thermal radiation wavelengths in the infrared region. When the temperature of the object at more than 500–800 °C, the thermal radiation of the strongest components in the visible wavelength region. Objects at the same time external radiation, but also absorb radiation energy from other objects. Alternatively, radiation energy absorbed by the object and its temperature, surface area, blackness and other factors.

Thermal radiation has the following main features:

(1)   Any object, as long as the temperature is higher than 0 K, it will keep emit thermal radiation to the surrounding space.
(2)   Can travel through a vacuum and air.
(3)   With the change in the form of energy.
(4)   Has a strong directivity.
(5)   Temperature and wavelength are related.
(6)   Emits radiation depends on the temperature to the fourth power.

3.4.2 Black Body Radiation

A body that emits the maximum amount of heat for its absolute temperature is called a black body. Radiant heat transfer rate from a black body to its surroundings can be expressed by the following equation [2].

$$q = \sigma T^4 \tag{3.24}$$

where, σ is the Stefan–Boltzman constant, 5.67×10^{-8} W/(m^2 K^4).

Two black bodies that radiate toward each other have a net heat flux between them. The net flow rate of heat between them is given by an adaptation of Eq. 3.24.

$$q = \sigma\left(T_1^4 - T_2^4\right) \tag{3.25}$$

Blackbody radiation refers to the radiation released by the radiator. Release the maximum amount of radiation at a specific temperature and wavelength. At the same time, blackbody is an object that can absorb all incident radiation and will not reflect any radiation. However, a blackbody is not necessarily black. For example, the sun is a gas planet. It can be considered that the electromagnetic radiation emitted to the sun is difficult to be reflected back, so it is considered that the sun is a blackbody. In theory, black light radiates electromagnetic waves of all wavelengths. To describe the relationship between the peak wavelength of blackbody electromagnetic radiation energy flux density and its own temperature, we need to use Wien's displacement law, which will not be introduced here.

But there is no such ideal blackbody in the real world, so what can be used to describe this difference? For any wavelength, a parameter called emissivity is defined as the ratio of the radiation energy of the real object to the radiation energy of the blackbody at the same temperature within a small wavelength interval of the wavelength. Obviously, the emissivity is a positive number between 0 and 1. Generally, the emissivity depends on material properties, environmental factors and observation conditions. If the emissivity is independent of the wavelength, the object can be called a grey body, otherwise it is called a selective radiator. To take into account the fact that real objects are gray bodies, Eq. 3.24 is modified to be of the following form.

$$q = \varepsilon\sigma T^4 \tag{3.26}$$

where, ε is called emissivity of the gray body (dimensionless).

Emissivity is a factor by which we multiply the black body heat transfer to take into account that the black body is the ideal case. Emissivity is a dimensionless number and has a maximum value of 1.0.

3.4.3 Radiation Configuration Factor

The radiation angle coefficient is the percentage of the radiant energy emitted by one surface falling on another surface [3]. It is a coefficient reflecting the geometric shape and position relationship between different objects radiating each other. The angular coefficient of any two surfaces with areas of A_i and A_j (see Fig. 3.7) is a pure geometric factor, which has nothing to do with the temperature and emissivity of the two surfaces.

Fig. 3.7 The configuration
factor of any two surfaces

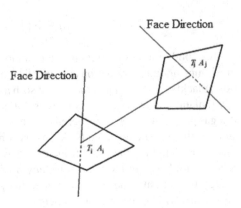

3.5 Heat Exchangers

Heat exchanger is a device that transfers part of the heat of hot fluid to cold fluid, also known as heat exchanger. Heat exchanger plays an important role in chemical industry, petroleum, power, food and many other industrial productions. In nuclear energy engineering, heat exchanger can be widely used as heater, cooler, condenser, steam generator, reboiler and so on.

Heat exchanger is a kind of equipment to transfer heat between two or more fluids at different temperatures. The heat exchanger transfers heat from the fluid with higher temperature to the fluid with lower temperature, so that the fluid temperature reaches the index specified in the process, so as to meet the needs of process conditions.

According to the structure of heat exchanger, it can be classified into floating head heat exchanger, fixed tubesheet heat exchanger, U-shaped tubesheet heat exchanger, plate heat exchanger, shell and tube heat exchanger, etc. Figure 3.8 is a schematic diagram of a typical shell and tube heat exchanger.

The direction of the fluid flow within the exchanger can be classified as parallel flow and counter flow, as shown in Fig. 3.9. The temperature distributions of these

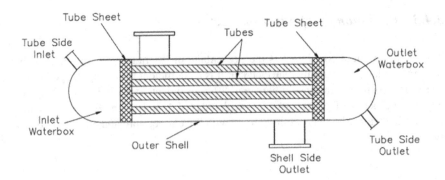

Fig. 3.8 Typical tube and shell heat exchanger

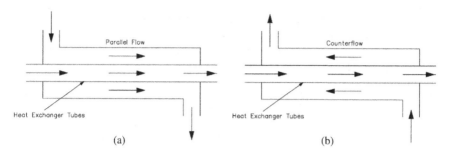

Fig. 3.9 Fluid flow direction

two types of heat exchangers are shown in Figs. 3.10 and 3.11. The advantage of parallel flow heat exchanger is that the temperature at the outlet of the primary and secondary sides of the heat exchanger is the closest. Compared with the parallel flow heat exchanger, the advantage of counter flow is that the temperature difference between the primary and secondary sides is relatively uniform, which is conducive to reducing the thermal stress, the heat transfer distribution is relatively uniform, and the temperature at the outlet of the cold side can be relatively high.

The calculation of heat exchanger generally adopts logarithmic average temperature difference, which is illustrated by an example below:

Fig. 3.10 Heat exchanger temperature profiles

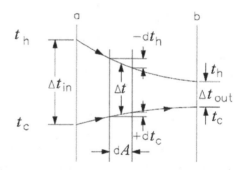

Fig. 3.11 Heat exchanger temperature profiles of counter-flow

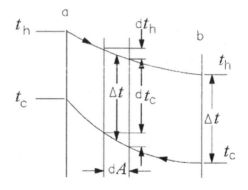

Example 3.4 A liquid-to-liquid counter flow heat exchanger is used as part of an auxiliary system at a nuclear facility. The heat exchanger is used to heat a cold fluid from 50 to 120 °C. Assuming that the hot fluid enters at 250 °C and leaves at 200 °C, calculate the LMTD for the exchanger.

Solution:

$$\Delta t_{\text{ln}} = (\Delta t_2 - \Delta t_1)/\ln(\Delta t_2/\Delta t_1) = 60°C$$

3.6 Boiling Heat Transfer

In nuclear facilities, convective heat transfer is used to remove heat from the heat transfer surface. The liquid used for cooling is usually in a compressed state (i.e. supercooled fluid) and the pressure is higher than the normal saturation pressure at a given temperature. Under certain conditions, some type of boiling (usually nucleate boiling) may occur. Therefore, when discussing convective heat transfer, it is best to study the boiling process, because it is suitable for the nuclear field.

More than one type of boiling may occur in a nuclear facility, especially in the event of a rapid loss of coolant pressure. Discussing boiling processes, especially local boiling and bulk boiling, will help students understand these processes and explain more clearly why bulk boiling (especially film boiling) should be avoided in the operation of nuclear facilities.

Boiling phenomenon refers to a liquid when heated above its saturation temperature, vaporization occurs in the liquid surface and the interior. It refers to boiling heat transfer of heat from the wall to pass liquid, vaporized liquid boiling convective heat transfer process.

Classification according to the space in which the liquid boiling can be divided into: ① pool boiling. Also known as boiling a large container. Liquid in a large space heating surface side, relying on the bubble disturbances and natural convection flows. For example, jacket-heating kettle boiling of the liquid. ② Flow boiling. Liquid flowing through a constant flow rate of heating tube, boiling phenomenon occurred. Then the generated bubbles are not free floating, but mixed with a liquid to form a vapor-iquid two-phase flow. The secondary side of steam generator heating tube boiling is flow boiling.

Closely related to the boiling heat transfer and bubble generation and detachment. Conditions bubble formation are overheating and must have vaporized liquid core. These conditions are the force balance and thermal equilibrium with the surrounding liquid bubble of the decision. On an absolutely smooth surface, it is unlikely to produce a bubble, there must be vaporized core. Gas or or scratches on the surface, can be used as the core of vaporization. After close to this core vaporization of the liquid, forming bubbles and growing up, and then out of the surface, followed by another new bubble formation. Causing the liquid to the wall when a strong impact

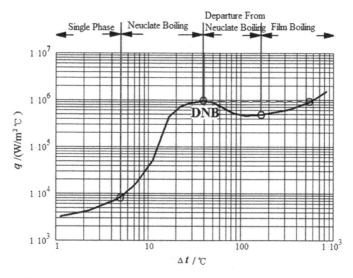

Fig. 3.12 Boiling curve of pool boiling

and disturbance bubble formed from the surface, so the same liquid, the boiling heat transfer coefficient is much greater than when there is no phase transition.

Nukiyama [4] got a boiling curve in 1934 as shown in Fig. 3.12. Where the vertical axis is heat flux and the horizontal axis is temperature difference between wall and fluid. The DNB point is departure of nucleate boiling which will be explained later.

3.6.1 Flow Boiling

Figure 3.13 shows flow patterns and heat transfer regions of a pipe with flow boiling [5]. Flow boiling is different with pool boiling.

The difference between flow boiling and pool boiling is that the fluid is heated in the process of flow. Fluid flow can be natural circulation or forced circulation driven by pump. The flow boiling is described below by taking the flow in the tube as an example. Figure 3.13 shows a vertically placed uniform heating channel. The under heated liquid flows upward from the bottom into the pipe. The flow pattern encountered and the corresponding heat transfer zone are shown in the figure. The changes of wall temperature and fluid temperature along the height are given on the left side of the figure.

Single phase liquid convection zone (zone A): when the fluid just enters the channel, it is a single-phase convection zone. The temperature of the liquid in this zone increases when heated, the fluid temperature is lower than the saturation temperature, and the wall temperature is also lower than the temperature necessary to produce bubbles.

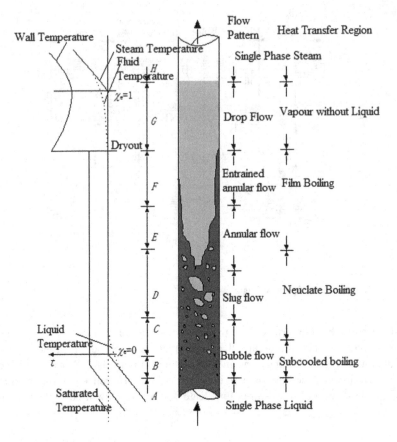

Fig. 3.13 Flow boiling in a pipe

Subcooled boiling zone (zone B): the characteristic of subcooled boiling is that water vapor bubbles on the heating surface are formed at the points conducive to the formation of bubbles. After these bubbles leave the wall, they are generally considered to be condensed in the subcooled liquid core.

Nucleate boiling zone (Zone C, D): nucleate boiling zone is characterized by that the mainstream temperature of the fluid reaches the saturation temperature, and the steam bubbles generated no longer disappear. The flow patterns of zones C and D are different, but their heat transfer zones are the same.

Liquid film forced convection zone (E, F zone): these zones are characterized by the formation of a liquid film on the wall. Through the forced convection of the liquid film, the energy from the wall is transmitted to the interface between the liquid film and the mainstream steam, and evaporation occurs at the interface.

Liquid deficient zone (Zone G): after the flow mass vapor content reaches a certain value, the liquid film is completely evaporated and turned to dry. The dividing point between zone F and zone G is the dryout point. Generally, the interruption or dryout of liquid film during annular flow is called critical (CHF), and sometimes this critical is called dryout critical. The section from the dryout point to all single-phase steam is called the liquid shortage area. At the dryout point, the wall temperature rises abruptly.

Single phase steam convection zone (zone H): this zone is characterized by that the fluid is single-phase superheated steam, the fluid temperature breaks away from the limit of saturation temperature, begins to increase rapidly, and the wall temperature increases accordingly.

3.6.2 Departure from Nucleate Boiling and Critical Heat Flux

The critical heat flux (CHF) [6] is the surface heat flux when nucleate boiling changes to film boiling. The point of transition from nucleate boiling to film boiling is called the point of departure from nucleate boiling, commonly written as DNB. It is a limiting quantity in the thermal hydraulic design of reactor, that is, the heat flux is not allowed to reach or too close to the critical heat flux, so as to prevent overheating or melting of fuel elements. The calculation of critical heat flux is a very important task for reactor thermal hydraulic analysis. The equations for calculating the critical heat flux can be divided into two categories, as shown in Fig. 3.14. One is to calculate the DNB critical at low steam quality, and the other is to calculate the dryout critical at high steam quality.

Critical heat flux is very important in design of nuclear reactor core. The heat flux in any region of core should be limited to critical heat flux otherwise the temperature of wall will jump and fuel rod would be damaged. Figure 3.14 shows two kind of critical heat flux: DNB and CHF. DNB happens [7] at low quality and high heat flux. CHF happens at high quality and relative low heat flux.

To determine the DNB heat flux, W-3 correlation is used [8]. For uniformly heat condition, the W-3 correlation is:

Fig. 3.14 DNB critical and CHF critical

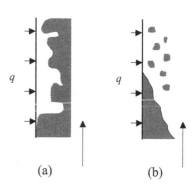

(a)   (b)

$$q_{\text{DNB,en}} = f(p, \chi_e, G, D_h, h_{in}) = \xi(p, \chi_e)\zeta(G, \chi_e)\psi(D_h, h_{in}) \qquad (3.27)$$

where,

$$\xi(p, \chi_e) = (2.022 - 0.06238p) + (0.1722 - 0.001427p)$$
$$\times \exp[(18.177 - 0.5987p)\chi_e]$$

$$\zeta(G, \chi_e) = [(0.1484 - 1.596\chi_e + 0.1729\chi_e|\chi_e|) \times 2.326G + 3271]$$
$$\times (1.157 - 0.869\chi_e)$$

$$\psi(D_e, h_{in}) = [0.2664 + 0.8357\exp(-124.1D_h)]$$
$$\times [0.8258 + 0.0003413(h_f - h_{in})]$$

where, q is heat flux in kW/m^2; p is pressure in MPa; G is mass flux in $\text{kg/(m}^2\text{ s)}$; h is specific enthalpy in kJ/kg; D_h is heated equivalence diameter in m; χ_e is equilibrium quality. The scope of the W-3 correlation is:

$$p = (6.895 \sim 16.55)\text{MPa}$$
$$G = (1.36 \sim 6.815) \times 10^3 \text{ kg/(m}^2\text{ s)}$$
$$L = (0.254 \sim .668)\text{m}$$
$$\chi_e = -0.15 \sim 0.15$$
$$D_h = (0.0051 \sim 0.0178)\text{m}$$
$$h_{in} \geq 930.4 \text{ kJ/kg}$$

Figure 3.15 shows the comparison of experimental data and W-3 correlation. 95% of points are located in the region of $\pm 23\%$. That means the limit of DNBR should be $1/(1 - 0.23) = 1.3$. Where DNBR is defined as:

$$\text{DNBR} = \frac{q_{\text{DNB}}}{q(z)} \qquad (3.28)$$

3.7 Heat Generation

The fission of nuclear fuel releases enormous energy. Although the fission energy of different nuclear fuel elements is different, it is generally believed that ^{235}U, ^{233}U or ^{239}Pu nucleus releases about 200 meV usable energy per fission. These energies can be roughly divided into three categories, and each category has its own characteristics, as shown in Table 3.1.

Fig. 3.15 The comparison
of experimental data and
W-3 correlation

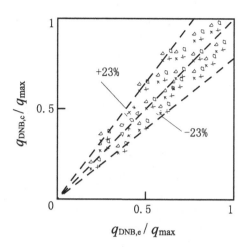

Table 3.1 Classification and distribution of fission energy

Types	From	Energy/MeV	Distance	Distribution
Instance	Kinetic Energy of Fission Debris	167	Very Short, <0.025 mm	Mostly in fuel pin
	Kinetic Energy of Released Neutrons	5	Medium	Mostly in Moderator
	γ Ray	5	Long	Around the Whole Core
Delayed	β Ray from Fission Debris	7	Short, <10 mm	Fuel Pin and Moderator
	γ Ray from Fission Debris	6	Long	Around the Whole Core
Redundant Neutrons (n, γ)	(n, γ) of Redundant Neutrons and β, γ Ray from Fission Debris	About 10	Short or Long	Around the Whole Core
Total		About 200		

The first class is instance release energy, which is released mostly in fuel pin. It
includes the kinetic energy of fission debris, neutrons, and γ ray. There is some kind
of energy released directly to moderator such as kinetic energy of neutron. In nuclear
engineering, it is assumed that 97.4% of fission energy is released in fuel pin while
other part is released outside of fuel pin.

3.7.1 Total Power of Reactor Core

The fission rate within a nuclear reactor is controlled by several factors. The density of the fuel, the neutron flux, and the type of fuel all affect the fission rate and, therefore, the heat generation rate. The following equation is presented here to show how the heat generation rate is related to these factors.

$$P_t = 1.6021 \times 10^{-13} E_f N_{235} \sigma_f \overline{\varphi} V_c \tag{3.29}$$

where, P_t is the total heat generation rate of reactor in W; V_c is the total volume of fuel pin in m^3; σ is micro cross section of fission in cm^2; $\overline{\varphi}$ is the neutron flux present in $n/(cm^2 s)$; N_{235} is atom density of ^{235}U in $1/m^3$; E_f is the energy released per fission in MeV. 1 meV $= 1.6021 \times 10^{-13}$ J.

The thermal power produced by a reactor is directly related to the mass flow rate of the reactor coolant and the temperature difference across the core. The relationship between power, mass flow rate, and enthalpy is given in Eq. 3.30.

$$P_t = q_m(h_{out} - h_{in}) \tag{3.30}$$

where, P_t is the total thermal power of reactor in W; q_m是 is the mass flow rate of coolant in kg/s; h_{out} is the enthalpy of outlet coolant in J/kg; h_{in} is the inlet coolant enthalpy in J/kg.

3.7.2 Flatten of Power

The total power of a reactor is limited by how much energy can safely be carried away by the coolant. The peak power density limits the total power of a reactor. To raise the total power as high as possible, the power distribution should be flatted. This is called as flatten of power.

There are many methods to flat power distribution such as arrangement of different enrichment fuels in different regions, arrangement of adjust rods, using poison materials, etc.

Figure 3.16 shows the impact of enrichment of uranium.

Reflector reflects neutrons back to core. It can raise neutron flux near the reflector as shown in Fig. 3.17.

Figures 3.18 and 3.19 show the effect of control rods or adjust rods in radial and vertical direction specifically.

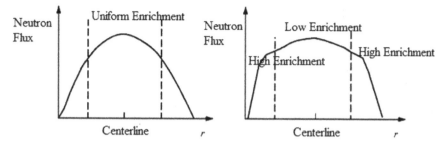

Fig. 3.16 Arrangement of different enrichment fuel

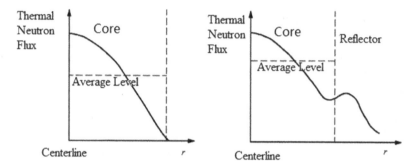

Fig. 3.17 The effect of reflector

Fig. 3.18 Effect of control rods or adjust rods in radial direction

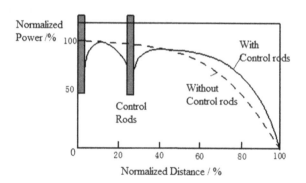

Fig. 3.19 Effect of control rods or adjust rods in vertical direction

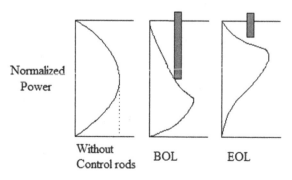

Table 3.2 Hot channel factors of PWR

Items	Symbol	1960's	1097's	1990's
Neutron flux factor	F_q^N	3.11	2.59	
Engineering flux factor	F_q^E	1.04	1.03	
Heat flux factor	F_q	3.24	2.67	2.35
Enthalpy raise factor of neutron flux	$F_{\Delta H}^N$	1.73	1.545	
Enthalpy raise factor of engineering	$F_{\Delta H}^E$	1.22	1.075	
Enthalpy raise factor	$F_{\Delta H}$	2.11	1.67	1.55

3.7.3 Hot Channel Factor

To measure the power peak factor, hot channel factor was introduced in history. Hot channel factor is defined as the ratio of peak power to overall average power of a channel.

Example 3.5 The total power of a reactor core is 3400 MW. Length of fuel rod is 12 inches. There are 264 fuel rods in an assembly and totally has 193 assemblies. If the hot channel factor is 1.83, find the peak linear power density.

Solution: Overall mean linear power density is,

$$\overline{q_l} = \frac{3400 \times 10^6}{12 \times 0.3048 \times 264 \times 193} = 18.24 \text{kW/m}$$

The peak linear power density is:

$$q_{l,\max} = 18.24 \times 1.83 = 33.38 \text{kW/m}$$

Table 3.2 shows some reasonable hot channel factors of pressurized water reactor [9].

3.7.4 Decay Heat

Residual heat in the core is the heat release in the core after reactor shutdown. It consists of two parts, one is residual fission heat, the other is decay heat. The decay heat production is a particular problem associated with nuclear reactors. The decay heat consists of two parts, including the radioactive decay heat of fission products and neutron capture reaction products. The power corresponding to residual heat release in the reactor after shutdown is called residual power.

After shutdown, the remaining neutrons continue to cause fission, resulting in continued heating of the reactor. Residual neutrons include prompt neutrons and delayed neutrons. The contribution of prompt neutrons usually decays very quickly with time, while delayed neutrons last a little longer. The calculation of residual power can be divided into two cases. One is that it is assumed to operate for an infinite time before shutdown (conservative calculation), and the other is that it only operates for a limited time before shutdown.

For light water reactors that have been operating at constant power for a long time, if the negative reactivity introduced during shutdown is large enough, for reactors that use ^{235}U as fuel during the period when decay heat plays an important role, the change of relative power with time can be approximately estimated by Eq. (3.31).

$$\frac{P(t)}{P_0} = 0.15e^{-0.1t} \tag{3.31}$$

where, t is time after shut down in second; P_0 is the power before shut down in W; $P(t)$ is the power after t seconds of shun down in W.

Reactors using ^{239}Pu as fuel have only about 1/3 of decay heat as ^{235}U. For heavy water reactor, Eq. 3.31 should be modified as Eq. 3.32.

$$\frac{P(t)}{P_0} = 0.15e^{-0.06t} \tag{3.32}$$

When a reactor has not been maintained full power for a long time, the decay heat can be calculated by Eq. 3.33 (suitable to light water reactor).

$$\frac{P(t)}{P_0} = 0.005Ae^{-0.06t}\left[t^{-a} - \left(t + t'\right)^{-a}\right] \tag{3.33}$$

where, P is decay power after t(s) of shut down; P_0 is the power before shun down that maintained t'(s) time; A, a is coefficients shown in Table 3.3.

The decay heat of (n, γ) can be evaluated by Eq. (3.34).

$$\frac{P(t)}{P_0} = 1.63 \times 10^{-3}e^{-4.91\times10^{-4}t} + 1.60 \times 10^{-3}e^{-3.41\times10^{-6}t} \tag{3.34}$$

Compared with decay heat from fission debris, decay heat from (n, γ) is relatively small but takes a longer time.

Glasstone correlation is used sometimes for design [10], it is:

Table 3.3 Coefficients of Eq. 3.33

Time/s	A	a	Maximum positive error	Maximum negative error
$10^{-1} \leq t < 10^1$	12.05	0.0639	4%(1 s)	3%(1 s)
$10^1 \leq t \leq 1.5 \times 10^2$	15.31	0.1807	3%(150 s)	1%(30 s)
$1.5 \times 10^2 < t < 4 \times 10^6$	26.02	0.2834	5%(150 s)	5%(3×10^3 s)
$4 \times 10^6 \leq t \leq 2 \times 10^8$	53.18	0.3350	8%(4×10^7 s)	9%(2×10^8 s)

$$\frac{P(t)}{P_0} = 0.1\{(t + 10)^{-0.2} - (t + t_0 + 10)^{-0.2}$$
$$+0.87(t + t_0 + 2 \times 10^7)^{-0.2} - 0.87(t + 2 \times 10^7)^{-0.2}\} \qquad (3.35)$$

where, t_0 is the seconds maintaining full power before shut down. t is the seconds after shut down.

Reactor decay heat may be a major problem. In the worst case, it may cause the reactor core to melt and/ or damage, such as the case of Three Mile Island. The degree of concern for decay heat varies depending on reactor type and design. Due to the decay heat of low power pool reactor, few people are concerned about the core temperature.

Each reactor has some limitations based on decay heat considerations during shutdown. These limits may vary due to steam generator pressure, core temperature, or any other parameters that may be related to the generation of decay heat. Even during refueling, heat removal from fuel rods is a control factor. For each established limit, some safety devices or protection functions are usually established.

Methods for removing decay heat from reactor core can be divided into two categories. One involves circulating fluid through the reactor core in a closed loop, using some type of heat exchanger to transfer heat out of the system. The other includes methods operating in open systems, sucking coolant from certain sources and discharging hotter liquids to certain storage areas or environments.

In most reactors, decay heat is usually removed by the same method used to remove heat from fission during reactor operation. In addition, many reactors are designed to allow natural circulation between the core and its normal or emergency heat exchangers to remove decay heat. These are examples of the first type of decay heat removal method.

If the reactor design requires the safe removal of decay heat from the core, but an accident may occur, resulting in the unavailability of the above closed-loop heat transfer method, some emergency cooling system will be included in the reactor design. Generally, the emergency cooling system consists of some reliable water sources, which are injected into the core at a relatively low temperature. This water will be heated by the decay heat of the core and leave the reactor through some paths in which the water will be stored in some structures or released to the environment. Using this type of system is almost always inferior to using the above closed-loop system. Students should study the systems, limitations, and protection characteristics applicable to their specific facilities.

Exercises

1.   A pipe with 3 m length, inner diameter is 15 cm and outer diameter is 18 cm, out surface temperature is 250 °C, inner surface temperature is 260 °C. The material of wall is carbon steel. Find the heat loss of this pipe.

2.   A pipe with inner diameter of 15 cm and outer diameter of 18 cm, inner surface temperature is 280 °C. The temperature of air around the pipe is 25 °C and

the heat transfer coefficient is 100 W/(m² °C). The material of wall is carbon steel. Find the outside wall temperature of the pipe.

3. A room has a floor with 4 m × 6 m, height in 3 m, the temperature of floor keeps at 40 °C, ceiling is −2 °C. Black body emission is assumed; find the radiant heat transfer rate between floor and ceiling.

4. A counter-flow heat exchanger with inlet fluid temperature at 120 °C and leaves at temperature 310 °C. Hot side inlet temperature is 500 °C, outlet temperature is 400 °C, find the LMTD.

5. The LMTD of a heat exchanger is supposed to be 23.2 °C. The outside diameter of heat transfer tube is 2 cm. The thickness of tube is 3 mm. The heat transfer coefficient of inner side of tube is 9000 W/(m² °C) and 6000 W/(m² °C) for outside. The thermal conductivity of tube wall is 20 W/ (m °C). Find the linear power density of the tube. (Power of 1 m tube).

6. Explain the physical meaning of critical heat flux.

7. Explain the reason to flat power distribution. What kind of method can be used to flat power distribution?

8. Fukushima nuclear power plant accident happened in 12th, March 2011. Evaluate the reactor power at present.

9. The maximum temperature difference in a heat transfer tube is 25 °C, LMTD is 16 °C, find the minimum temperature difference.

10. A heat exchanger with power of 100 MW, has an overall heat transfer coefficient of 2300 W/(m² °C). Suppose the LMTD is related with heat transfer area as following table. Find the area needed.

Cases	Area/m²	LMTD/°C
1	4000	15.4
2	5000	11.2
3	6000	8.1
4	7000	6.9

11. In a material test experiment, thickness of sample is 1 cm, temperature difference is 50 °C, and heat flux measured at this case is 100 W/m². Find the thermal conductivity of this sample.

12. The thickness of a flat plate is 2.54 cm, thermal conductivity of the plate is 2 W/(m °C), find heat resistance of the plate.

13. In an experiment of boiling heat transfer, pressure is 1 atm, wall temperature is measured as 110 °C, heating power is 15 kW, heat transfer area is 0.05 m², find the boiling heat transfer coefficient.

14. A body in space, 200 km diameter, and surface temperature is 200 K, find radiant power.

15. What is the limit of DNBR if one use a DNB correlation with 95% experimental points falls in the region between ±18% at 95% confidence?

16. The overall linear power density of a PWR core is 18 kW/m. there are 15,900 fuel rods arranged in this core, find the total power of this reactor core.

References

1. Holman JP. Heat transfer, 10th ed. Beijing: Machinery Industry Press; 2011.
2. Kogure T, Leung KC. The astrophysics of emission-line stars. New York: Springer; 2007.
3. Baughn J, Monroe G. Optical measurements of the radiation configuration factor. In: 3rd Thermophysics conference Los Angeles, USA; 1968.
4. Nukiyama S. Film boiling water on thin wires. Soc. Mech. Engng. 1934; 37.
5. Kutateladze SS. Hydromechanical model of the crisis of boiling under conditions of free convection. J Tech Phys USSR. 1950;20(11):1389–92.
6. Zuber N. Hydrodynamic aspects of boiling heat transfer. Los Angeles: California. Univ; 1959.
7. Incropera F. Fundamentals of heat and mass transfer. 6th ed. New Jersey: Wiley; 2011.
8. Tong LS. Thermal analysis of pressurized water reactors. 2nd ed. Illinois: American Nuclear Society; 1979.
9. Ziabletsev DN. PWR integrated safety analysis methodology using multi-level coupling algorithm. Pennsylvania: The Pennsylvania State University; 2002.
10. Glasstone S, Sesonske A. Nuclear reactor engineering. Princeton, New Jersey: D. Van Nostrand Co.; 1963.

Chapter 4
Fluid Flow

Fluid flow is very important in most industrial processes, especially those related to heat transfer. When heat needs to be transferred from one point to another, it often needs to be designed with fluid flow. For example, cooling water system in gasoline or diesel engine, air flow of air-cooled motor, coolant loop of nuclear reactor, etc. Fluid flow is also often used to provide lubrication.

The fluid flow in the field of nuclear engineering is sometimes extremely complex and can not always get good results through mathematical analysis [1]. Unlike solids, various parts of the fluid flowing in pipes or components can have different velocities and accelerations. Although it is very difficult to obtain detailed flow structure information, some simplified and basic concepts and methods are very helpful to solve practical engineering problems. Even though such simplification may not be enough for practical engineering design, it is still very valuable for understanding the operation of the system and predicting the dynamic response of the system.

The basic principles of fluid flow include three: momentum conservation, energy conservation and mass conservation. Let's first discuss the conservation of mass, that is, the continuity equation.

4.1 Continuity Equation

A fluid is a substance that can flow freely because there is no rigid link between molecules or atoms. According to Newton's definition of fluid, fluid refers to an object that will deform under the action of any force. Therefore, the fluid includes liquid and gas. Fluid is an object form corresponding to solid. Its basic feature is that it has no certain shape and has fluidity.

All fluids have certain compressibility. The compressibility of liquid is very small, while the compressibility of gas is large. When the shape of the fluid changes, there is also a certain motion resistance (i.e. viscosity) between the layers of the fluid. When the viscosity and compressibility of fluid are very small, it can be approximately

© Tsinghua University Press 2022
J. Yu, *Fundamental Principles of Nuclear Engineering*,
https://doi.org/10.1007/978-981-16-0839-1_4

regarded as an incompressible inviscid ideal fluid. It is an ideal model introduced by people to study the motion and state of fluid.

Mass conservation is described as the difference between the mass flow into the control body and the flow out of the control body, which is the change rate of the system mass, that is:

$$\frac{\Delta m}{\Delta t} = q_{m,in} - q_{m,out} \tag{4.1}$$

where, $\Delta m/\Delta t$ is the increase or decrease of the mass within the control volume over a specified time period. q_m is mass flow rate in kg/s. The mass flow rate (q_m) of a system is a measure of the mass of fluid passing a point in the system per unit time.

For steady-state flow, the mass in the control body does not change with time. This means that there is no accumulation of mass within any component in the system. The continuity equation is simply a mathematical expression of the principle of conservation of mass.

$$q_{m,in} = q_{m,out} \tag{4.2}$$

Or

$$(\rho A V)_{in} = (\rho A V)_{out} \tag{4.3}$$

For a control volume with multiple inlets and outlets, the principle of mass conservation requires that the sum of the mass flow entering the control volume is equal to the sum of the mass flow leaving the control volume. More generally, the continuity equation is represented by Eq. 4.4:

$$\sum_i (\rho A V)_{in} = \sum_j (\rho A V)_{out} \tag{4.4}$$

One of the simplest applications of the continuity equation is determining the change in fluid velocity due to an expansion or contraction in the diameter of a pipe.

Example 4.1 A piping system has a "Y" configuration for separating the flow as shown in Fig. 4.1. The diameter of the inlet leg is 12 cm, and the diameters of the outlet legs are 8 and 10 cm. The velocity in the 10 cm leg is 5 m/s. The flow through the main portion is 50 kg/s. The density of water is 1000 kg/m^3. What is the velocity out of the 8 cm pipe section?

Solution:

$$A_8 = 0.005024\,m^2,\ A_{10} = 0.00785\,m^2,\ A_{12} = 0.011304\,m^2.$$

$$\text{Then, } (\rho A V)_{12} = 50\,kg/s,\ (\rho A V)_{10} = 39.25\,kg/s$$

Fig. 4.1 "Y" configuration
for example problem

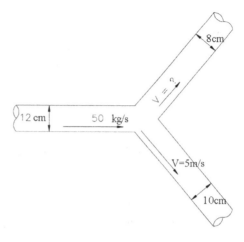

One gets, $(\rho A V)_8 = 50 - 39.25 = 10.75\,\text{kg/s}$

$$V_8 = 10.75/1000/0.005024 = 2.14\,\text{m/s}$$

4.2 Laminar and Turbulent Flow

Laminar flow is a flow state of fluid, which flows in layers. When the fluid flows at low speed in the pipe, it presents laminar flow, and its particles move smoothly and linearly along the direction parallel to the pipe axis [2]. For the laminar motion of viscous fluid, the trajectory of fluid micro cluster has no obvious irregular pulsation. There is only momentum exchange caused by molecular thermal motion between adjacent fluid layers. Laminar flow only occurs when the Reynolds number Re is small.

In nature, we often encounter turbulence. For example, river rapids, air flow, chimney smoke exhaust, etc. are turbulent. Turbulence occurs at large Reynolds numbers. When the Reynolds number is small, the influence of viscous force on the flow field is greater than inertial force, the disturbance of velocity in the flow field will be attenuated due to viscous force, the fluid flow is relatively stable, and the flow is laminar; On the contrary, if the Reynolds number is large, the influence of inertial force on the flow field is greater than viscous force, the fluid flow is more unstable, and the small change of velocity is easy to develop and enhance, forming disordered and irregular turbulence.

The basic characteristic of turbulence is the randomness of fluid micelle motion. Turbulent micro clusters not only have transverse pulsation, but also have reverse motion relative to the total motion of the fluid. Therefore, the trajectory of the fluid micro clusters is extremely disordered and changes rapidly with time. The most important phenomenon in turbulence is the transfer of momentum, heat and mass caused by this random motion, and its transfer rate is several orders of magnitude higher than that of laminar flow. Turbulence has both advantages and disadvantages. On the one hand, it can enhance heat transfer; On the other hand, it greatly increases the friction resistance and energy loss.

4.2.1   Reynolds Number and Hydraulic Diameter

The flow regime (laminar or turbulent) is determined by calculating the Reynolds number of the fluid. According to the study of Osborne Reynolds, Reynolds number is a dimensionless number composed of fluid physical characteristics. Equation 3.12 is used to calculate the Reynolds number of fluid flow [3].

$$\mathrm{Re} \equiv \frac{\rho V D_e}{\mu} \tag{3.12}$$

It is found that when Reynolds number exceeds a critical Reynolds number, the laminar flow begins to transition to irregular turbulence due to disturbance, and the motion resistance increases sharply. The critical Reynolds number mainly depends on the flow pattern. For a circular tube, it is around 2000, at this time, the characteristic velocity is the mainstream average velocity on the cross section of the tube, and the characteristic length is the hydraulic diameter of the tube.

In fact, if the Reynolds number is less than 2000, the flow is laminar. If greater than 3500, the flow is turbulent. The flow with Reynolds number between 2000 and 3500 is sometimes called transition flow [4].

In the Eq. (3.12), D_e is hydraulic diameter. It is defined as Eq. (4.5).

$$D_e \equiv \frac{4A}{P} \tag{4.5}$$

Example 4.2 Calculate the hydraulic diameter of a fuel bundle. Fuel rods are arranged in square form in a PWR core shown as Fig. 4.2. The diameter of fuel rod is 10 mm. The pitch of fuel rods is 12 mm.
   Solution:

$$D_e = \frac{4A}{P} = \frac{4 \times \left(12^2 - \frac{1}{4}\pi \times 10^2\right)}{\pi \times 10} = 8.335 \mathrm{mm}$$

Fig. 4.2 Channel of square
form fuel bundle

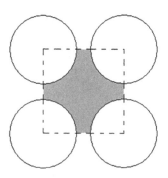

4.2.2 Flow Velocity Profiles

When a fluid flows in a pipe, not all fluid micro elements flow at the same speed. The
velocity of fluid near the wall is small, while that far away from the wall is large.
The distributions of velocity at different Re are shown in Fig. 4.3.

The velocity distribution diagram is closely related to the flow pattern. If the flow
is laminar, the velocity distribution is a parabolic distribution, and the maximum
velocity is twice the average velocity. For turbulence, the velocity distribution tends
to be flatter, as shown in Fig. 4.3. The velocity of the fluid close to the wall is zero.
It can also be seen from Fig. 4.3 that the velocity distribution diagram is also related
to the wall conditions. The velocity distribution will be flatter on the coarser wall.

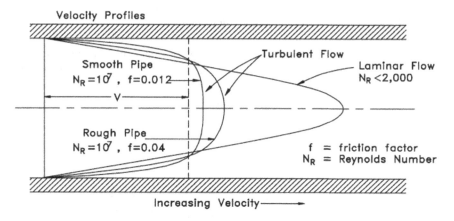

Fig. 4.3 Laminar and turbulent flow velocity profiles

4.2.3   Average (Bulk) Velocity

In most fluid flow analysis, we do not need to know the accurate distribution of cross section velocity, but only need to get the average velocity of cross section, that is

$$V_m = \frac{q_m}{\rho A} \tag{4.6}$$

That is, the ratio of the total volume flow to the area of the cross section. Because the cross-section velocity distribution is relatively flat in the case of turbulence, this average velocity is basically the velocity at the central axis. For laminar flow, the average velocity is half of that at the central axis.

4.2.4   Viscosity

Viscosity coefficient is a physical quantity that measures the viscosity of a fluid. Also known as viscosity coefficient, dynamic viscosity, proportional coefficient, viscous damping coefficient, etc. Newton's viscosity law points out that in pure shear flow, the shear stress (or viscous friction stress, which is the force per unit area) between two adjacent fluid layers is directly proportional to the normal velocity gradient in the vertical direction. The viscosity coefficient is numerically equal to the shear stress of the fluid under the unit velocity gradient. The velocity gradient also represents the angular deformation rate in fluid motion, so the viscosity coefficient also represents the ratio between shear stress and angular deformation rate.

According to the international system of units, the unit of viscosity is Pa·s. Viscosity is an attribute of a fluid, and the viscosity values of different fluids are different. The viscosity coefficient of the same fluid is significantly related to temperature, but almost independent of pressure. The viscosity coefficient of gas increases with the increase of temperature, while that of liquid decreases.

4.3   Bernoulli's Equation

Daniel Bernoulli [5] put forward the Bernoulli principle in 1726. This is the basic principle of hydraulics before the establishment of the continuum theoretical equation of fluid mechanics. Its essence is the conservation of mechanical energy of fluid. Namely:

$$KE + PE + PV = \text{constant} \tag{4.7}$$

Substituting appropriate expressions for the potential energy and kinetic energy, Eq. 4.7 can be rewritten as Eq. 4.8.

$$p + \frac{\rho v^2}{2} + \rho g H = \text{const} \tag{4.8}$$

where, p is pressure, v is average velocity, ρ is density, g is acceleration due to gravity, H is height above reference.

According to energy conservation, fluid flow system holds:

$$Q + \sum_i (U + E_p + E_k + pV)_{in} = W + \sum_j (U + E_p + E_k + pV)_{out}$$
$$+ (U + E_p + E_k + pV)_{stored} \tag{4.9}$$

where, Q is heat transferred to fluid, W is work done by fluid, U in internal energy, E_p is potential energy, E_k is kinetic energy, pV is flow energy.

If the flow is stable and there is no heat transfer and no work done by or to fluid. Between point 1 and point 2, Eq. 4.9 becomes:

$$(E_p + E_k + pV)_1 = (E_p + E_k + pV)_2 \tag{4.10}$$

Then,

$$\left(mgH + \frac{mv^2}{2} + pV \right)_1 = \left(mgH + \frac{mv^2}{2} + pV \right)_2 \tag{4.11a}$$

Divide V, one gets

$$\left(\rho g H + \frac{\rho v^2}{2} + p \right)_1 = \left(\rho g H + \frac{\rho v^2}{2} + p \right)_2 \tag{4.11b}$$

That means between any two points, $\rho g H + \rho v^2/2 + p$ keeps constant.

$$\left(H + \frac{v^2}{2g} + \frac{p}{\rho g} \right)_1 = \left(H + \frac{v^2}{2g} + \frac{p}{\rho g} \right)_2 \tag{4.12}$$

Example 4.3 A cone pipe with inlet diameter of 0.5 m, outlet diameter of 1.0 m. Inlet head is 5 m water, volume flow rate is 3 m³/s, find the inlet velocity, outlet velocity and outlet head.

Solution:

$$V_1 = 3/(3.14 \times 0.25^2) = 15.3 \, \text{m/s}$$
$$V_2 = 3/(3.14 \times 0.5^2) = 3.82 \, \text{m/s}$$

Fig. 4.4 Flow in a nozzle

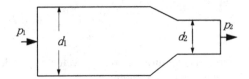

$$\left(\frac{p}{\rho g}\right)_2 = \left(\frac{V^2}{2g} + \frac{p}{\rho g}\right)_1 - \left(\frac{V^2}{2g}\right)_2$$
$$= 15.3^2/19.62 + 5 - 3.82^2/19.62 = 16.2(mH_2O)$$

For a nozzle shown as Fig. 4.4, suppose $z_1 = z_2$, according to Eq. 4.12, one gets:

$$\left(\frac{p}{\rho g} + H + \frac{V^2}{2g}\right)_1 = \left(\frac{p}{\rho g} + H + \frac{V^2}{2g}\right)_2 \qquad (4.13)$$

According to mass conservation, one gets:

$$\rho V_1 \frac{\pi d_1^2}{4} = \rho V_2 \frac{\pi d_2^2}{4} \qquad (4.14)$$

Then,

$$\frac{V_2^2}{2}\left[\left(\frac{d_2}{d_1}\right)^4 - 1\right] = \frac{p_2 - p_1}{\rho} \qquad (4.15)$$

or,

$$V_2 = \sqrt{\frac{2(p_1 - p_2)}{\rho\left[1 - (d_2/d_1)^4\right]}} \qquad (4.16)$$

4.3.1 Venturi Meter

Figure 4.5 shows a kind of flow meter called Venturi meter [6] using Bernoulli's Equation.

Example 4.4 $d_1 = 0.711$ m, $d_2 = 0.686$ m, $h_1 = 0.914$ m, $\rho = 1000$ kg/m³. The density of fluid in U figure pressure meter $\rho_1 = 13,550$ kg/m³. Find the mass flow rate in this pipe.

Solution: According to mass conservation, one can get,

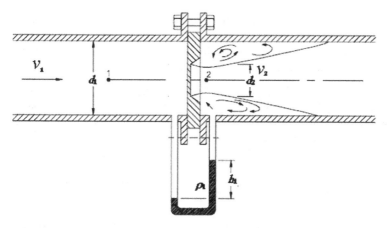

Fig. 4.5 Flow meter schematic

$$V_1 = V_2 \left(\frac{d_2}{d_1}\right)^2$$

Using Eq. (4.16), one gets,

$$V_2 = \sqrt{\frac{2(p_1 - p_2)}{\rho[1 - (d_2/d_1)^4]}} = \sqrt{\frac{2(\rho_1 - \rho)gh}{\rho[1 - (d_2/d_1)^4]}}$$

$$= \sqrt{\frac{2 \times (13550 - 1000) \times 9.81 \times 0.914}{1000 \times \left[\left(\frac{0.711}{0.686}\right)^4 - 1\right]}} = 41.07 \,(\text{m/s})$$

Then,

$$q_\text{m} = \rho V_2 A_2 = 1000 \times 41.07 \times \frac{\pi \times 0.686^2}{4} = 15179 \,(\text{kg/s})$$

4.3.2 Extended Bernoulli Equation

As mentioned earlier, Bernoulli equation is only applicable to ideal fluids with negligible viscosity and incompressible. That is, there are restrictions for using Bernoulli equation, which can be summarized as follows:

(1) Steady flow: in a flow system, the properties of a fluid at any point do not change with time.
(2) Incompressible flow: the density is constant. When the fluid is a gas, it is suitable for Mach number less than 0.3.

(3)   Frictionless flow: the friction effect can be ignored, and the viscous effect can
      be ignored.
(4)   Fluid flows along the streamline: fluid elements flow along the streamline, and
      the streamlines do not intersect each other.

In the real case, the pipe wall will produce resistance and head loss due to viscous
force. Moreover, there are usually pumps driving fluid flow in the pipeline system.
In this case, the extended Bernoulli equation needs to be used. Equation 4.17 is one
form of the Extended Bernoulli equation.

$$\left(\frac{p}{\rho g} + H + \frac{V^2}{2g} + H_p \right)_1 = \left(\frac{p}{\rho g} + H + \frac{V^2}{2g} + H_f \right)_2 \qquad (4.17)$$

where, H_p is pump head, H_f is risistance head including friction, local form, valves
and inlet and outlet.

The head loss caused by fluid friction represents the energy used to overcome
pipe wall friction. Although it represents energy loss from the perspective of fluid
flow, it usually does not represent a significant loss of total fluid energy. It also does
not violate the law of energy conservation, because the head loss caused by friction
will lead to the equivalent increase of fluid internal energy. These losses are greatest
when fluid flows through inlets, outlets, pumps, valves, fittings and any other pipes
with rough internal surfaces.

Most methods for calculating friction head loss are empirical. Let's introduce the
calculation of head loss.

4.4 Head Loss

Head loss, also known as pressure drop, refers to the reduction of total head when
fluid flows in the pipeline system. Total head includes potential head, dynamic head
and static head. When the real fluid flows in the pipeline, the head loss is inevitable.
This is because the real fluid is viscous, so there is friction between the fluid and the
pipe wall, and there is mutual friction between adjacent fluids with different speeds in
the fluid. In addition, there is turbulence effect, local head loss caused by the change
of channel shape at the inlet and outlet of the channel, and so on.

4.4.1 Frictional Loss

The head loss caused by wall friction is also called friction pressure drop, which
refers to the head loss caused by the viscosity of the wall when the fluid flows along
the straight pipe. The friction pressure drop is directly proportional to the length of
the pipe, the square of the flow velocity and the friction coefficient, but inversely

proportional to the diameter of the pipe

$$\Delta p_f = f \frac{L}{D} \frac{\rho V_m^2}{2} \tag{4.18}$$

where f is the friction factor, which is related to Re of flow and wall roughness. The measure of pipe wall roughness is the relative roughness, which is the ratio of absolute roughness to pipe diameter, i.e., ε/D.

The Moody Chart [7] can be used to determine the friction factor based on the Reynolds number and the relative roughness. The friction factor can be obtained by Moody Chart as shown in Fig. 4.6.

Example 4.5 Determine the friction factor (f) for fluid flow in a pipe that has a Reynolds number of 40,000 and a relative roughness of 0.01.

Solution: Using the Moody Chart, a Reynolds number of 40,000 intersects the curve corresponding to a relative roughness of 0.01 at a friction factor of 0.039.

The frictional head loss can be calculated using a mathematical relationship that is known as Darcy's equation for head loss. The equation takes two distinct forms. The first form of Darcy's equation determines the losses in the system associated with the length of the pipe.

$$H_f = f \frac{L}{D} \frac{V_m^2}{2g} \tag{4.19}$$

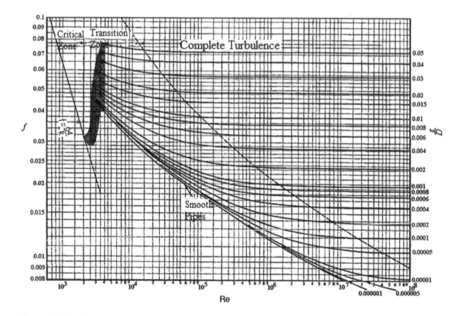

Fig. 4.6 Moody chart

where g is acceleration due to gravity, 9.81 m/s^2.

Example 4.6 A pipe 30 m long and 50 cm diameter contains water at 90 °C flowing at a mass flow rate of 300 kg/s. The water has a density of 1000 kg/m^3 and a viscosity of 314 μPa s. The relative roughness of the pipe is 0.00008. Calculate the head loss for the pipe.

Solution: First, it is necessary to determine the average velocity of the fluid, then calculate Re, then determine the friction factor, and finally obtain the friction pressure drop.

$$\text{Average velocity}: V_{\text{m}} = q_{\text{m}}/(rA) = 300/\left(1000 \times 3.14 \times 0.25^2\right) = 1.53 (\text{m/s})$$

$$\text{Re} = rV_{\text{m}}D/m = 1000 \times 1.53 \times 0.5/\left(314 \times 10^{-6}\right) = 2.4 \times 10^6$$

Use the Moody Chart for a Reynolds number of 2.4×10^6 and a relative roughness of 0.00008.

$$f = 0.012$$

$$H_{\text{f}} = f(L/D) \times V_{\text{m}}^2/(2g) = 0.012 \times 30/0.5 \times 1.53^2/19.62 = 0.086 (\text{mH}_2\text{O})$$

4.4.2 Minor Losses

Minor losses, also known as form resistance pressure drop, is caused by the change of channel geometry. Such as elbow, flange, valve, inlet and outlet, etc. The local pressure drop is described by the local resistance coefficient K. Since the local pressure drop is proportional to the square of the velocity, it is convenient to describe the local pressure drop as follows:

$$\Delta p_{\text{k}} = K \frac{\rho V_{\text{m}}^2}{2} \tag{4.20}$$

or

$$H_{\text{k}} = K \frac{V_{\text{m}}^2}{2g} \tag{4.21}$$

Minor losses is sometimes expressed by equivalent length, as shown in Table 4.1. The so-called equivalent length is the pipe length equivalent to the friction pressure drop. Make the friction pressure drop equal to the local pressure drop to obtain:

Table 4.1 Typical values of L_{eq}/D

Item	Types	L_{eq}/D
Globe valve	Conventional	400
	Y-pattern	160
Gate valve	Fully open	10
	75% open	35
	50% open	150
	25% open	900
Standard tee	Flow through run	10
	Flow through branch	60
90° standard elbow		30
45° standard elbow		16
Return bend		50

$$f\frac{L}{D}\frac{V_m^2}{2g} = K\frac{V_m^2}{2g} \tag{4.22}$$

This yields two relationships that are useful.

$$L_{eq} = K\frac{D}{f} \tag{4.23}$$

or

$$K = f\frac{L_{eq}}{D} \tag{4.24}$$

Typical values of L_{eq}/D for common piping system components are listed in Table 4.1. The equivalent length of piping that will cause the same head loss as a particular component can be determined by multiplying the value of L_{eq}/D for that component by the diameter of the pipe [8].

The higher the value of L_{eq}/D, the longer the equivalent length of pipe.

Example 4.7 A fully-open gate valve is in a pipe with a diameter of 25 cm. What equivalent length of pipe would cause the same head loss as the gate valve?
Solution: $L_{eq}/D = 10$, $L_{eq} = 2.5$ m.

4.5 Natural Circulation

Natural circulation is a kind of flow mode that only depends on the density difference between cold and hot fluids caused by different temperatures to drive the fluid circulation in a closed system. In the natural circulation system, the flow can be maintained automatically without mechanical driving device. Natural circulation system is widely used in many industrial fields, especially in nuclear energy utilization, boiler circulation, solar thermal utilization system and so on.

As the fluid flows in the pipeline, there is the head loss described above. Head loss includes friction pressure drop, local pressure drops and other losses. Therefore, in order to maintain flow, a mechanically driven equipment pump is usually required to offset head loss. The cycle driven by the pump is called forced cycle. If the head loss can be offset by the lifting head caused by density difference, a certain flow can be maintained without pump, which is natural circulation.

4.5.1 Thermal Driving Head

The thermal driving head is the power to cause natural circulation, which is caused by the different density of fluids at different temperatures.

Consider two pieces of the same fluid with the same volume, one with high temperature and the other with low temperature. Since the density of the fluid with high temperature is small, the gravity received by the fluid with high temperature is also smaller than that of the fluid with low temperature, so the fluid with high temperature floats upward and the fluid with low temperature settles downward.

This phenomenon can be seen in many places. For example, hot-air balloons rely on the high temperature and low density of the air in the balloon to make the balloon float. The heat added to the air in the balloon adds energy to the air molecules. The movement of air molecules increases and air molecules occupy more space. The air molecules inside the balloon occupy more space than the same number of air molecules outside the balloon. This means that the hot air is less dense and lighter than the surrounding air. Because the air density in the balloon is low, gravity has little effect on it. As a result, the balloon weighs less than the surrounding air. Gravity sucks cooler air down into the space occupied by the balloon. The downward movement of the cooler air forces the balloon out of the previously occupied space and the balloon rises.

4.5.2 Conditions Required for Natural Circulation

The occurrence of natural cycle is conditional. Even if the natural cycle has occurred and any of these conditions is lost, the natural cycle will stop. These conditions are:

(1)   There is a temperature difference (with heat source and heat sink);
(2)   The heat source is lower elevation than the heat sink;
(3)   The fluid between the heat source and the heat sink is continuous.

First, there must be a temperature difference. Therefore, there may be two parts of the fluid, one of which has a high temperature and the other has a low temperature. The temperature difference is necessary, because the temperature difference can cause the density difference, and the density difference is the driving force.

The temperature difference should be maintained continuously, otherwise the natural cycle will stop. Therefore, the heat source needs to continuously add heat to the fluid, and the heat sink needs to continuously discharge heat from the fluid. Otherwise, the temperature of the hot and cold fluid will soon converge and the cycle will stop.

Secondly, the heat source should be located at a lower position and the heat sink at a higher position, so that the hot fluid can float and the cold fluid can sink.

Of course, the hot and cold zones must be filled with fluid so that the fluid can flow.

Swimming pool reactor is a good reactor type that uses natural circulation to cool fuel rods.

The spent fuel pool of nuclear power plant also uses natural circulation to cool the decay waste heat of spent fuel. At this case, the heat source is the fuel rod and the heat sink are the surrounding pool. After the water at the bottom of the fuel rod absorbs the heat released from the fuel rod, the temperature increases and the density decreases. Driven by gravity, the surrounding cold water is pushed into the fuel rod area, and the hot fluid is squeezed out to form a natural circulation. After the hot fluid flows to the top of the fuel rod, it mixes with the surrounding cold water to reduce the temperature. In this way, the temperature of the pool will continue to rise. In order to maintain the natural cycle, other heat exchangers must be used to cool the water in the pool (heat sink).

Generally speaking, the greater the temperature difference between the cold source and the heat source, the greater the thermal drive head, so the greater the flow of natural circulation. However, it should be noted that boiling of hot fluid should be avoided, because boiling will cause phase change, which may interrupt the cycle. The natural circulation of two-phase flow may also occur, but it is difficult to maintain due to the possibility of separation between gas and liquid.

In nuclear power plants, some parameters can be used as the basis for judging whether there is a natural cycle. For example, for PWR nuclear power plants, the following parameters can be used to judge whether natural circulation exists:

(1) The temperature difference between the cold and hot legs of the coolant system should be 25~80% of the full power temperature difference, and the temperature difference should be basically stable or slightly reduced. This means that the decay heat of the core is being continuously removed, and the core temperature is maintained or reduced.
(2) The temperature difference between hot and cold legs needs to be maintained or decreased slightly.
(3) The pressure on the secondary side of the steam generator should follow the temperature of the primary circuit, which means that the steam generator can remove heat from the primary circuit.

In order to ensure that the primary circuit has sufficient natural circulation, it is necessary to maintain the water level of the pressurizer at more than 50%, keep the primary circuit with sufficient degree of superheat (for example, a nuclear power plant requires that the degree of superheat is greater than 8 °C), and keep the water level of the steam generator higher than the normal water level, so as to have enough heat sink to ensure that the heat of the primary circuit can be removed and avoid boiling of the primary loop. Because once boiling occurs, it is easy to interrupt the natural cycle (There was a lesson of the Three Mile Island accident in the United States).

4.6   Two-Phase Fluid Flow

The liquid will boil, and once it boils, it will enter the two-phase flow. The two-phase flow is very different from the previous introduction.

For frictional pressure drop, there are several ways to deal with two-phase flow. When the channel size is the same and the mass flow rate is the same, the friction pressure drops caused by two-phase flow is greater than that of single-phase flow. This is mainly caused by the large velocity of two-phase flow under the same mass flow rate. The frictional pressure drop of two-phase flow is also related to the specific flow pattern.

Several techniques can be used to predict the head loss caused by fluid friction in two-phase flow. For the same duct size and mass flow, the friction of two-phase flow is greater than that of single-phase flow. This difference seems to be a function of the flow type and is the result of an increase in flow rate. The two-phase friction loss is experimentally measured by measuring the pressure drop of different pipeline elements. The two-phase loss is usually related to the single-phase loss through the same element.

4.6.1   Two-Phase Friction Multiplier

A widely accepted method to deal with the friction pressure drop of two-phase flow is to use a two-phase pressure drop multiplication factor or multiplier, which is defined as follows

$$\phi^2 = \frac{H_{f,2p}}{H_{f,L}} \tag{4.25}$$

where the subscript 2p represents two-phase flow and L represents saturated liquid. Square is usually used to indicate that this number must be nonnegative.

The friction multiplier [9] of two-phase flow can be calculated using Fig. 4.7.

Fig. 4.7 Martinelli-Nelson two-phase friction multiplier

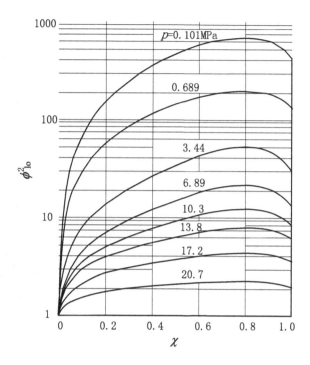

4.6.2 Flow Patterns

It is very important to distinguish the flow pattern of two-phase flow for the analysis of two-phase flow, which is no less important than whether laminar flow or turbulence should be distinguished first in the analysis of single-phase flow. For two-phase flow, generally, the density difference between vapor phase and liquid phase is very large, so the flow situation in vertical channel and horizontal channel is very different. When analyzing two-phase flow, it is usually necessary to distinguish horizontal channel and vertical channel.

For the vertical upward flow channel, the flow pattern is generally divided into bubble flow, slug flow, stirred flow and annular flow according to the distribution and flow of bubbles in the channel [10] (see Fig. 4.8).

1. Bubbly flow: there is dispersion of vapor bubbles in a continuum of liquid.
2. Slug flow: in bubbly flow, the bubbles grow by coalescence and ultimately become of the same order of diameter as the tube. This generates the typical bullet-shaped bubbles that are characteristic of the slug-flow regime.
3. Stir Flow: in stir flow, the bubbles broken sometimes and fluid is stirred by bubble.
4. Annular flow: the liquid is distributed between a liquid film flowing up the wall and a dispersion of droplets flowing in the vapor core of the flow.

Fig. 4.8 Flow patterns for vertical channel

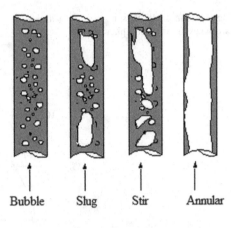

Bubble   Slug   Stir   Annular

Fig. 4.9 Flow patterns for horizontal channel

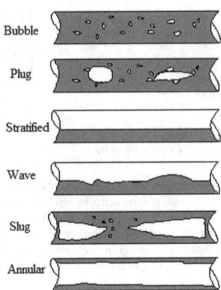

Bubble

Plug

Stratified

Wave

Slug

Annular

The flow patterns for horizontal channel shown as Fig. 4.9 are to be defined as bubble flow, plug flow, stratified flow, slug flow and annular flow [11].

4.6.3 Flow Instability

Flow instability [12] refers to the phenomenon of flow oscillation, drift or counter-current. There are many forms of flow oscillation, which may be caused by bubbles or unreasonable mechanical resistance caused by design or processing.

The bubbles in two-phase flow are prone to flow drift or flow oscillation after being slightly disturbed. The amplitude of oscillation can be constant or variable. The flow oscillation of one channel in the core will cause the flow redistribution of the surrounding channels, resulting in a wider range of flow instability.

Flow instability should be avoided as much as possible, because it may cause fatigue or thermal fatigue failure of components; it may also deteriorate the heat transfer performance of the system, resulting in the deterioration of heat transfer. The existing experience shows that flow instability can reduce the critical heat flux required to prevent departure from nucleate boiling by about 40%; it may also interfere with the control system, especially in reactors with coolant as moderator. Therefore, the study of flow instability is very important for reactor design and safety analysis. The existing experience shows that the flow instability is not prominent under the normal operation of PWR nuclear power plant. The flow instability needs to be considered only when the power rises to 150% of the full power under the rated flow. However, in the case of natural circulation, due to the small flow rate, the flow instability needs great attention.

Flow instability includes Ledinegg instability (or flow drift) and flow pattern instability. The characteristic of Ledinegg instability is that the flow in the system will drift aperiodically, that is, the flow will change from one value to another. This is because a certain area of the hydraulic characteristic curve of the system is multivalued (Fig. 4.10), that is, one driving head corresponds to several flow rates. The hydraulic characteristic curve with multivalued area can be transformed into a single value curve by adding throttling parts at the inlet of the system or increasing the system pressure, so as to eliminate the Ledinegg instability. In Fig. 4.10, the slope of point 2 is less than zero and is unstable. Curve I is a single value curve after changing the hydraulic characteristics of the pipeline, eliminating the unstable point with negative slope.

Transient flow instability includes density wave instability, pulsation, acoustic oscillation and thermal oscillation. In the heated boiling channel, if a thermal parameter is disturbed, for example, temporarily reducing the flow at the channel inlet

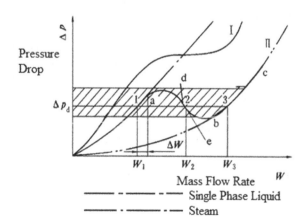

Fig. 4.10 The Ledinegg instability

will increase the specific enthalpy of the fluid, increase the average void fraction and decrease the average density. Then this disturbing potential must affect the flow pressure drop and heat transfer performance. In some cases, after multiple feedback actions, oscillations of flow, void fraction (or density) and pressure drop will be formed. It is called density wave (or cavitation wave) instability, which is a common transient flow instability. In boiling water reactor, the oscillation of density will also cause the change of reactivity. The feedback of nuclear effect makes the general hydrodynamic instability more complex. Experiments show that increasing the system pressure, increasing the flow resistance at the inlet of the channel, shortening the heating length of the channel and increasing the mass flow rate can increase the flow stability and reduce the possibility of density wave oscillation.

4.7   Some Specific Phenomenon

4.7.1   Pipe Whip

When a pipeline breaks, the fluid with high internal pressure and high mass flow rate will quickly spray out at the break, which will cause the phenomenon of pipe whipping. Pipe whipping will cause major damage to system components, instruments and equipment. Therefore, the reactor building shall provide protection against pipe whipping, water flow impact and flying objects that may occur in case of loss of coolant accident, so as to protect the safety-related equipment in the containment.

4.7.2   Water Hammer and Steam Hammer

Water hammer [13] refers to the phenomenon that the mass flow rate changes suddenly and the pressure fluctuates greatly due to a sudden opening or closing of the valve, a sudden stop and sudden opening and closing of the water pump during the transportation of water (or other liquids).

When a feed pump starts and stops, the water flow impacts the pipeline, which will produce a serious water hammer. Because inside the water pipe, the inner wall of the pipe is smooth and the water flows freely. When an opening valve is suddenly closed or a feed pump stops, the water flow will produce a pressure on the valve and pipe wall. Because the pipe wall is smooth, the subsequent water flow quickly reaches the maximum under the action of inertia, and produces destructive effect, which is the "water hammer effect" in hydraulics, that is, positive water hammer. On the contrary, after a closed valve is suddenly opened or a feed pump is started, it will also produce a water hammer, called negative water hammer, but it is not as big as the former.

According to the principle of momentum conservation, when water hammer is generated, the amplitude of pressure exceeding static pressure is:

$$\Delta p = \rho c \Delta v \qquad (4.26)$$

where c is the wave speed of pressure (sound speed).

Example 4.8 Water at a density of 1000 kg/m^3 and a pressure of 1 MPa is flowing through a pipe at 3 m/s. The speed of sound in the water is 1457 m/s. A check valve suddenly closed. What is the maximum pressure of the fluid in Pa?
Solution:

$$p_{max} = p_0 + \Delta p = 1 \times 10^6 + 1000 \times 1457 \times 3 = 5.371 \times 10^6 \text{Pa}$$

When the steam pipeline is suddenly closed, a steam hammer will occur. The steam hammer is not as serious as a water hammer because of the following three reasons:

(1) Gas is easily compressed;
(2) The propagation velocity of pressure wave in gas is small;
(3) The density of the gas is small.

Therefore, the pressure increase calculated according to (4.22) is much smaller than that in the case of water.

Water hammer and steam hammer are not rare in nuclear power plants, which should be paid enough attention. During operation, in order to prevent water hammer, ensure that the gas in the water system and the water in the gas system are drained when the system is started. As far as possible, start the pump with the valve closed, and then slowly open the valve. As far as possible, start the pump with small capacity first, and then increase the pump capacity. Use the preheating valve around the main steam shut-off valve as far as possible. Close the valve as slowly as possible and then stop the pump. The function of the steam or degassing device shall also be checked regularly.

Exercises

1. What is natural circulation? It is affected by what kind of factors. How to enhance a natural circulation?
2. A vertical heat channel with inlet temperature at 280 °C and outlet 320 °C, pressure 15.5 MPa, diameter 10.72 mm, height 3.89 mm, inner wall surface temperature 320 °C, inlet mass flux 1.138×10^7 kg/(m^2 h). Find frictional pressure drop.
3. Find the frictional coefficient of pipe with Re $= 3 \times 10^5$, $\varepsilon = 0.00005$.
4. A container is filled with water; there is a pipe with diameter of 5 cm near bottom. Water level is 5 m, find the flow rate when the valve is fully opened.

5. Density of fluid is 930 kg/m³, at pressure 15 MPa, with a velocity of 5 m/s, suppose the sound speed is 1457 m/s. Find the pressure caused by water hammer.
6. What is Ledinegg instability?
7. What kind of flow patterns exist in a horizontal flow channel?
8. Suppose the system pressure is 7.5 MPa, flow quality is 0.15. Find the two-phase friction multiplier.
9. A nozzle flow with outlet velocity of 10 m/s (density is 1000 kg/m³). Diameter of the nozzle is 2 cm shown as in Figure follows. Find (1) the force exerted on the plate when the plate is stable. (2) The force exerted on the plate when the plate is moving with a speed of 1 m/s.

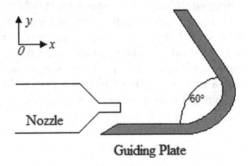

Guiding Plate

10. Density is 1000 kg/m³, mass flow rate is 1000 kg/s, find the inlet velocity

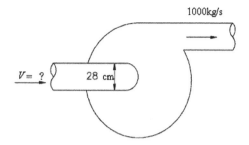

11. Calculate the hydraulic diameter of the following channel. The diameter of fuel rods is 9 mm with a pitch of 12 mm.

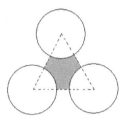

References

1. Institution of Mechanical Engineers. Heat and fluid flow in nuclear and process plant safety. London: Mechanical Engineering Publications; 1983.
2. Davidson PA. Turbulence: an introduction for scientists and engineers. Oxford UK: Oxford University Press; 2004.
3. Tennekes H, Lumley JL. A first course in turbulence. Massachusetts: MIT Press; 1972.
4. Fouz I. Fluid mechanics. Oxford UK: University of Oxford; 2001.
5. Mulley R. Flow of industrial fluids: theory and equations. Boca Raton: CRC Press; 2004.
6. de Sa DO. Instrumentation fundamentals for process control. Oxford UK: Taylor & Francis; 2001.
7. Moody LF. Friction factors for pipe flow. Trans. ASME. 1944;66(8):671–84.
8. Chanson H. Hydraulics of open channel flow: an introduction. Oxford UK: Butterworth-Heinemann; 2004.
9. Cengel Y, Ghajar A. Heat and mass transfer: fundamentals and applications. New York: McGraw-Hill; 2014.
10. Levy S. Two-phase flow in complex systems. New Jersey: Wiley; 1999.
11. Ghiaasiaan SM. Two-phase flow, boiling, and condensation: in conventional and miniature systems. Cambridge: Cambridge University Press; 2008.
12. Ruspini. Two-phase flow instabilities: a review. Int. J. Heat Mass Transf. 71; 2013.
13. Thorley ARD. Fluid transients in pipelines. 2nd ed. London: Professional Engineering Publishing; 2004.

Chapter 5
Electrical Science

Electrical theory and electrical systems are very important in the operation, maintenance and technical support of nuclear engineering, especially nuclear power plants. The contents of this chapter include electrical foundation, DC, alternator, generator, electric motor, transformer and so on.

What is electricity? Current refers to the flow of electrons within a material or device, or the force that drives the flow of electrons. Scientists believe that electric currents are generated by the movement of microscopic particle electrons and protons [1].These microparticles are smaller than atoms and are the microscopic particles that make up them [2]. In order to understand the electricity, you need to understand the structure of the atom first.

5.1 Basic Electrical Theory

5.1.1 The Atom

Atoms are the elementary particles that make up all matter and are the basic particles that chemical reactions can no longer be separated. Atoms are inseparable in chemical reactions, but they can be divided in physical reactions. Atoms consist of positively charged nucleus and negatively charged extranuclear electrons. The amount of positive and negative charges is equal, so the atoms are electrically neutral to the outside world. The nucleus consists of neutrons and protons [20], and the nuclear model of the atom is shown in Fig. 5.1.

Protons carry a positive charge (+) of one unit, electrons carry a negative charge (−) of one unit, and neutrons do not have any charge. Neutrons have a slightly larger mass than protons. The combination of different neutrons and protons forms a variety of atoms, such as the ^{4}He nucleus consisting of 2 neutrons and 2 protons, while the ^{3}He nucleus is made up of 2 protons and 1 neutron composition, they are isotopes. In a natural state, the number of protons in an atomic nucleus determines the number

© Tsinghua University Press 2022
J. Yu, *Fundamental Principles of Nuclear Engineering*,
https://doi.org/10.1007/978-981-16-0839-1_5

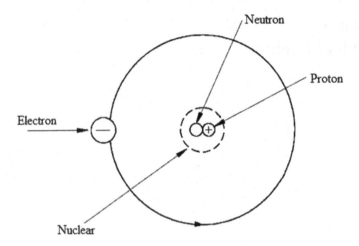

Fig. 5.1 The atom

of extranuclear electrons and therefore their chemical properties. We refer to nuclei with different numbers of protons as different elements, and the periodic table of elements is arranged according to the number of protons (see Chapter 7). Electrons with the same number of protons but with negative electricity move in a circular orbit (or shell) outside the nucleus, therefore, the two opposite charges cancel, and the atom is said to be electrically neutral, or in balance.

5.1.2 Electrostatic Force

In the nucleus, negatively charged electrons move around the positively charged nucleus, and the interaction force between them is an electrostatic force [3]. The electrostatic force between the nucleus and electrons is shown in Fig. 5.2. Without the electrostatic force, the high-speed electrons cannot move around the nucleus, it will escape from the nucleus.

The interaction between charges occurs through the electrostatic field. As long as there is an electric charge, there is an electrostatic field around the charge, the basic nature of the electrostatic field is that it puts a powerful effect on the charge placed in it, this force is called electrostatic field force.

The electrostatic field force to which the positive charge is subjected is in the tangent direction of the electrostatic field line, and the electrostatic field force to which the negative charge is received is in the opposite direction in the direction of the tangent of the electrostatic field line. The first law of electricity tells us that unlike charges (like electrons and protons) attract each other, and like charges repel each other. This is one of the basic laws of electricity.

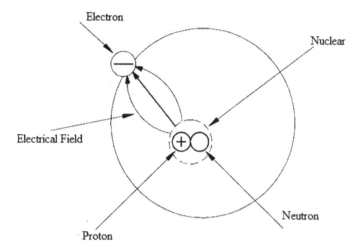

Fig. 5.2 Electrostatic force

Atoms can lose electrons or get electrons, and when the number of electrons is larger than the number of protons, the atoms are negatively electrical, and vice versa. Atoms are also likely to follow the unlike charges attract each other, and like charges repel each other. Electrons that can be freed from the nucleus are called free electrons, and the greater the number of these free electrons an object contains, the greater its negative electric charge, so they can be obtained by measuring the number of free electrons.

The interaction between charges occurs through the electrostatic field. "Lines of force" are typically used to represent electrostatic fields, as shown in Fig. 5.3. Lines that unlike charges attract with each other, and like charges repel with each other are

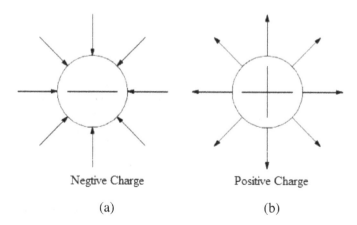

Fig. 5.3 Electrostatic field

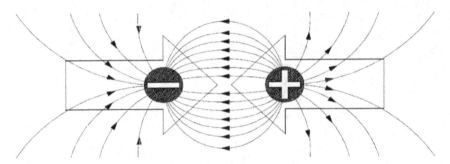

Fig. 5.4 Electrostatic field between two charges of opposite polarity

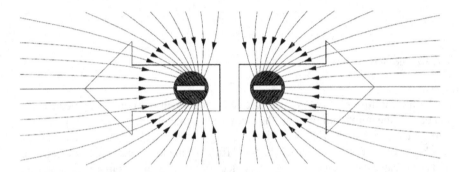

Fig. 5.5 Electrostatic field between two charges of like polarity

shown in Figs. 5.4 and 5.5. The direction of the small arrows shows the direction of the force as it would act upon an electron if it were released into the electric field.

When two objects of like charge are brought near one another, the lines of force repel each other, as shown in Fig. 5.5.

5.1.3 Coulomb's Law of Electrostatic Charges

The strength of the attraction or of the repulsion force depends upon two factors: (1) the amount of charge on each object, and (2) the distance between the objects. The greater the charge on the objects, the greater the electrostatic field. The greater the distance between the objects, the weaker the electrostatic field between them, and vice versa. Coulomb's law [4] is expressed as: the interaction force between two stationary point charges in a vacuum, proportional to the product of their charge, inversely proportional to the secondary side of their distance, the direction of the force in their connection [5]. Unlike charges attract each other, and like charges repel each other and the force is:

Fig. 5.6 Potential difference
between two charged objects

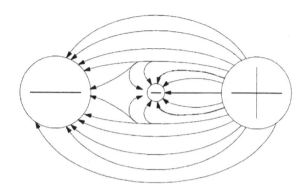

$$F = k\frac{q_1 q_2}{r^2} \tag{5.1}$$

where, F is force of electrostatic attraction or propulsion, N; r is distance between two particles, m; q is charge of particle, C; k is constant of proportionality, $k = 9.0 \times 10^9$ N·m²/C².

The law was proposed by french physicist Coulomb in his 1785 paper *The Law of Electricity*. Coulomb's law is the first quantitative law in the history of electrical development, and it is one of the basic laws of electrostatic field theory.

The magnitude of the electrostatic field force between two charges is usually expressed as a potential difference, and if a charged particle is placed between two particles with a potential difference, shown as in the Fig. 5.6, the charged particle moves in one direction. For example, the negative charge in the middle of Fig. 5.6 moves to the right.

Due to the force of the electrostatic field, these charges can do work by attracting and/or repelling another charged particle. This working ability is called "potential"; Therefore, if one charge is different from another, there is a potential difference between them. The sum of the potential differences of all charged particles in the electrostatic field is called electromotive force (EMF).

The basic unit for measuring potential difference is "volt". The sign of the potential difference is "V", indicating the ability to force electrons to move. Due to the use of volt units, the potential difference is also known as "voltage". The volt unit will be described in detail in the next chapter.

5.2 Electrical Terminology

Conductor
Conductors are substances with low resistivity and easy to current conduction. The presence of a large number of free-moving charged particles in a conductor is

called a carrier. Under the action of the external electrostatic field, the carrier makes directional motion, forming a current.

A metal wire is a common electrical conductor. Metals are the most common type of conductor. The valence electrons at the outermost layer of metal atoms easily become free electrons. Free electrons are very concentrated in metals, so metal conductors usually have a higher conductivity than other conductor materials. Atoms with only one valence electron, such as copper, silver, and gold, are examples of good conductors. Most metals are good conductors.

The solution of the electrolyte is also a conductor. Its carriers are positive and negative ions. Experiments have found that most pure liquids can also be dissociation, but the degree of dissociation is very small. For example, pure water has a resistivity of up to 10^4 ohms m, which is 10^{10}–10^{12} times greater than that of metals. However, if a little electrolyte is added to pure water, the ion concentration is greatly increased, so that the resistivity is greatly reduced and becomes a conductor. The resistivity of electrolytes is usually much larger than that of metals because the concentration of carriers in electrolytes is much smaller than the concentration of free electrons in metals, and the force of ions and surrounding media is greater, making their migration rate in the outer electrostatic field much smaller. Electrolytes are often used in electrochemical industries, such as electrolytic purification, electroplating and so on.

Ionized gases can also conduct electricity. The carriers are electrons and positive/negative ions. In general, gases are good insulators. If by external causes, such as heating or exposure to X-rays, γ-rays or ultraviolet light, the gas molecules can be dissociation, the ionized gas becomes a conductor. The conductivity of ionized gas has a great relationship with the supply voltage, and is often accompanied by physical processes such as sounding and luminescence. Ionizing gases are often used in the electric light source industry. The conductivity of a gas due to the action of an external ionizer is called the non-self-contained discharge of a gas. With the supply voltage increase, the current also increases, the voltage increases to a certain value when the non-self-holding discharge reaches saturation, continue to increase the voltage to a certain value after the current suddenly increases sharply, then even if the ionizer is removed, can still maintain conductivity, the gas from non-self-holding discharge transition to self-holding discharge. The characteristics of gas self-holding discharge depend on the type of gas, pressure, electrode material, electrode shape, electrode temperature, distance between the poles and many other factors. If the conditions are different, self-holding discharge has different forms, there are glow discharge, arc discharge and corona discharge.

Insulators
Insulators are substances also known as dielectrics, that are not good at conducting current. They have very high resistivity. Insulators and conductors, there is no absolute boundary. Insulators can be converted into conductors under certain conditions.

Insulators are characterized by the positive and negative charges in the molecule is bound very tightly, can move freely with very few charged particles, its resistivity is very large, so in general can ignore the free charge under the action of the external electrostatic field to form the macro current, and is considered to be non-conductive substances. Insulators can be classified into gaseous (e.g. hydrogen, oxygen, nitrogen and all gases in non-ionized states), liquid (e.g. pure water, oil, paint and organic acids, etc.) and solid states (e.g. glass, ceramics, rubber, paper, quartz, etc.). Solid-state insulators are classified into crystal and non-crystalline. The actual insulator is not completely non-conductive, under a strong electrostatic field, the positive and negative charge inside the insulator will break the bondage, and become a free charge, insulation performance is destroyed, this phenomenon is called dielectric breakdown. The maximum electrostatic field strength that a dielectric material can withstand is called a breakdown field strength. In the insulator, there is a binding charge, under the action of the external electrostatic field, this charge will be micro-displacement, thus producing a polarized charge, which is called the polarization of the dielectric. Dielectrics can be classified into anisotropic dielectrics and anisotropic dielectrics according to their physical properties. Insulators are used in engineering as electrical insulation materials, capacitors and special dielectric devices such as piezoelectric crystals.

Conductors such as Ge and Si are often referred to as semiconductors. The resistivity of these conductors is between metals and insulators and decreases rapidly as the temperature increases. There are a certain number of free electrons and cavities in this type of material, which can be thought of as carriers with positive charges. Unlike in the case of metals or electrolytes, the amount of impurities in a semiconductor and changes in external conditions (e.g. light, or changes in temperature, pressure, etc.) can cause significant changes in their conductivity.

Resistors
A resistor is a current-limiting element, typically two pins, that limits the amount of current passing through the branch it connects to. A resistor that cannot be changed is called a fixed resistor. A variable resistance value is called a potential or variable resistor. The ideal resistor is linear, i.e. the current passing through the resistor is proportional to the voltage. Some special resistors, such as thermistors, have a non-linear relationship between voltage and current. Resistors are represented by the letter R in ohms (Ω).

The resistance value of resistor is generally related to temperature, material, length, and cross-sectional area. The physical amount of resistance affected by temperature is the temperature coefficient, which is defined as the percentage of the resistance value that changes with temperature rises by 1 °C. The main physical characteristic of a resistor is that it can generate thermal energy, or it can be said that it is an energy-consuming element, and the current passing through it produces internal energy. Resistance usually acts as a divider and current limiter in the circuit.

Capacitor
A capacitor is a passive two-terminal electrical component used to store electrical energy temporarily in an electric field. Capacitor is represented by the letter C. As the name implies, it is a "capacitive container" and a device that holds the charge.

Capacitors are one of the electronic components widely used in electronic devices and are widely used in circuits such as straight-through, coupling, bypassing, filtering, tuning circuits, energy conversion, control and so on.

Actual capacitors come in many forms, but all contain at least two conductors (plates) separated by a dielectric (i.e., an insulator that can store energy by polarization). The conductor may be a thin film, foil or metal sintered bead or a conductive electrolyte, etc. The function of non-conductive dielectric is to increase the charging capacity of capacitor. Materials commonly used as dielectrics include glass, ceramics, plastic film, air, vacuum, paper, mica and oxide layer. As a part of the circuit, capacitors are widely used in many common electrical equipment. Unlike resistors, ideal capacitors do not dissipate energy. Instead, capacitors store energy in the form of an electrostatic field between plates.

An ideal capacitor is characterized by a single constant, that is, its capacitance. Capacitance is the ratio of the charge Q on each conductor to the potential difference V between them. The international system of units for capacitance is farad (f), equal to 1 C (1 C/V) per volt.

The larger the surface area of the "plate" (conductor), the narrower the gap between them, and the greater the capacitance. In fact, the dielectric between plates passes through a small amount of leakage current and has an electric field strength limit, that is, breakdown voltage. Conductors and leads introduce unwanted inductance and resistance.

Capacitors are widely used in electronic circuits to block DC and allow AC to pass through. In analog filter networks, they smooth the output of the power supply. In a resonant circuit, they tune the radio to a specific frequency. In power transmission systems, they stabilize voltage and power flow.

Inductor

An inductor is a component that can be stored by converting electrical energy into magnetic energy. The structure of the inductor is similar to that of a transformer, but with only one winding. An inductor resists change in electric current passing through it. An inductor will attempt to prevent the current from flowing through it when the circuit is switched on, and if the inductor is in a state where the current is passing through, it will attempt to maintain the current when the circuit is open. Inductors are also known as chokes, resistors, dynamic resistors, etc.

The characteristic of inductor is its inductance, that is, the ratio of voltage to current change rate, in Henry (H). The inductance depends mainly on the number of turns of the coil (number of turns), the winding mode, the material of the core, etc. In general, the more coils there are and the denser the coils are wound, the greater the inductance. Coils with cores have a greater inductor than coils without cores, and the greater the magnetic conductivity of the core, the greater the inductance. The physical symbol of inductor is L, and the basic unit is Henry (H).

The value of inductance is usually 1 μ H. Inductors are widely used in AC electronic equipment, especially radio equipment. They are used to block AC and allow DC to pass through; inductors designed for this purpose are called chokes. They are also used in electronic filters to separate signals of different frequencies and combine with capacitors to form tuning circuits for tuning radio and television receivers.

Electric Current

Electric Current or current for short, is the amount of electricity that flows per unit of time. Usually expressed in the letter I, its unit is ampere (A). Current is the directional movement of an electric charge in a conductor.

Ampere is the basic unit in the system of international units. In addition to A, commonly used units are kA, mA, μA. 1 A is 1000 mA or 100000 μ A. The direction of positive charge flow is the direction of current. Figure 5.7 shows that the current direction and the electron motion direction are opposite. In some countries (e.g. the United States) the direction of current is set in the direction of electron flow, which should be noted when reading the relevant literature.

The free charge in the conductor forms an electric current by making regular directional motion under the action of electrostatic field forces. The power supply produces electrostatic field force, under the action of electrostatic field force, the charge in the electrostatic field is directed to move, forming a current. One ampere of current is defined as the movement of one coulomb of charge past any given point of a conductor during one second of time.

If the copper wire is placed between two charged objects with potential difference, all negatively charged free electrons will feel a force to push them from negative charge to positive charge. This force is opposite to the conventional direction of the electrostatic force line, as shown in Fig. 5.7.

According to whether the current direction changes periodically over time, it is divided into AC and DC. AC is also known as "alternated current". Generally, refers to sine wave voltage or current that changes periodically over time in size and direction. Direct current (DC), direction do not change over time. Pulsating DC refers to the direction (positive and negative poles) unchanged, but the magnitude changes over

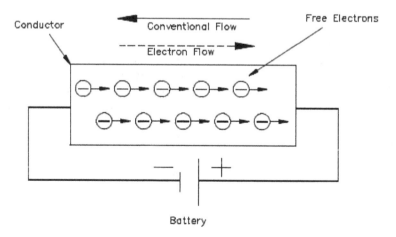

Fig. 5.7 Potential difference across a conductor causes a current to flow

time, for example: we put 50 Hz of alternating current through the diode rectifier to get a typical pulsating DC, they only after the filtered wave (inductive or capacitive) into smooth DC.

Electric Source

An electrical source is a device that converts other forms of energy into electrical energy. The ideal power supply is one power supply when the power loss of the power supply itself is negligible. Electrical source can be classified into voltage sources and current sources. Ideal sources are used for analytical purposes only since they cannot occur in nature.

An ideal voltage source is a two-ended circuit component that if its end voltage can be maintained at a given value in any case, regardless of the current passing through it. An ideal current source is if the current strength passed through it can in any case maintain a given value regardless of its end voltage. Voltage and current sources are collectively referred to as power supplies.

A real source is a real life current or voltage supply that has some losses associated with it.

5.3   Ohm's Law

In 1827, George Simon Ohm discovered that there was a definite relationship between voltage, current, and resistance in an electrical circuit [6]. Ohm's law states that in the same circuit, the current in a conductor is proportional to the voltage at both ends of the conductor and inversely proportional to the resistance of the conductor. namely

$$I = \frac{U}{R} \tag{5.2}$$

where, I is current, A; U is potential difference, V; R is resistance, Ω.

The work that the current does in a unit of time is called electrical power. It is the amount of physical energy consumed, expressed in P, in watts (W). As a physical amount that indicates how quickly the current is doing its work, the magnitude of an electrical power is numerically equal to the amount of power it consumes in 1 s.

$$P = U \cdot I \tag{5.3}$$

For pure resistance circuits, it is also useful to calculate electrical power according to Ohm's law:

$$P = I \cdot R^2 \tag{5.4}$$

or

$$P = \frac{U^2}{R} \qquad (5.5)$$

5.4 Methods of Producing Voltage (Electricity)

This is about the generation of electricity, not thermal power, nuclear power and other power generation methods. There are many methods, such as electrochemical methods, friction, electromagnetic induction, pressure, heating, photoelectric and so on. Here are some of the main methods of generating electricity.

5.4.1 Electrochemistry

Electrochemical is the study of the relationship between electricity and chemical reactions. Electrical and chemical interactions can be done by batteries. Figure 5.8 shows a schematic of a chemical battery.

An electrochemical [7] cell is a device capable of either generating electrical energy from chemical reactions or facilitating chemical reactions through the introduction of electrical energy.

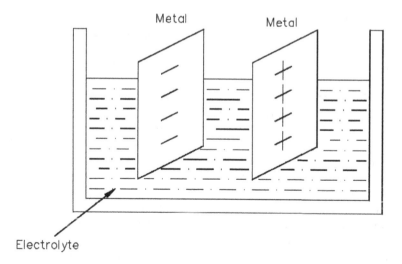

Fig. 5.8 Electrochemical cell

Electrochemical cell uses the difference of metal chemical activity between the two electrodes to produce potential difference, which makes the electron flow and generates the current. Also known as a non-chargeable battery, is a chemical battery, its electrochemical reaction cannot be reversed, that is, only the chemical energy can be converted into electrical energy, cannot be recharged. A primary battery is a device that converts chemical energy into electrical energy. Therefore, by definition, ordinary dry batteries, fuel cells can be called as electrochemical cell.

Basic conditions for the formation of electrochemical cell:

(1)   Two metals with different vibrancy (i.e. one is live metal and the other is non-active metal) or an inert electrode such as metal and graphite is inserted into an electrolyte solution.
(2)   After being connected with a wire, it is inserted into the electrolyte solution to form a closed circuit.
(3)   Spontaneous redox reactions are to occur.

The electrochemical cell is an oxidation reaction and reduction reaction that can be carried out spontaneously on the negative and positive poles of the battery, thus generating current in the outer circuit.

A common example of an electrochemical cell is a standard 1.5.volt cell meant for consumer use. This type of device is known as a single Galvanic cell. A battery consists of two or more cells, connected in either parallel or series pattern.

5.4.2 Static Electricity

If electrons are removed from the atoms in this body of matter, as happens due to friction when one rubs a glass rod with a silk cloth, it will become electrically positive as shown in Fig. 5.9.

Static electricity is a static charge, or non-flowing charge (the flowing charge forms an electric current). Static electricity is formed when the charge gathers on an object, and the charge is divided into positive and negative charges. That is, static electricity can also be divided into two kinds: positive static and negative static electricity. Positive static electricity is formed when the positive charge gathers on an object, and negative static electricity is formed when the negative charge gathers on an object. But whether it is positive static or negative static electricity, when an electrostatic object comes into contact with a zero-potential object (grounded object) or an object with which it has a potential difference, there is a charge transfer, which is what we see every day spark discharge phenomenon.

For example, the northern winter weather is dry, the human body is easy to bring static electricity, when contact with others or metal conductors will appear discharge phenomenon. People will have an electric shock, at night can see sparks, which is the clothing fibers and human body friction with static electricity. Rubber rod friction with fur, rubber rod with negative charges, rubber belt with positive charges.

Fig. 5.9 Static electricity

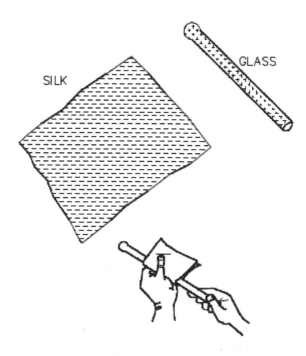

Any two objects of different materials can be separated after contact to produce static electricity. When two different objects come into contact with each other, an object loses some of its charge. The object that is transferred out by the electron is positively charged, while the other object gets some residual electrons with negative charge. If the charge is difficult to neutralize during separation, the charge accumulates to carry static electricity to the object.

5.4.3 Magnetic Induction

When a conductor in a closed circuit makes the movement of cutting magnetic lines in a magnetic field, an electric current is generated in the conductor, a phenomenon called electromagnetic induction, and the resulting current is called an induced current. The polarity of the voltage is given by the Faraday's law. The law states that the magnetic field that senses the current should hinder the change of the primary flux. For inductive voltage can also be used right-handed rules to determine the direction of the induced current, and then to determine the polarity of the induced voltage. Michael Faraday [8] is generally credited with the discovery of induction in 1831, and mathematically described it as Faraday's law of induction.

Fig. 5.10
Generator—electromagnetic
induction

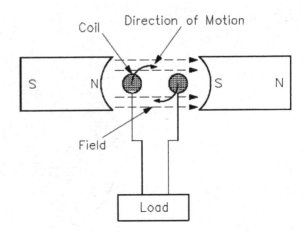

A generator is a machine that uses magnetic induction principles to convert mechanical energy into electrical energy. The magnetic induction is applied to generating voltage by rotating the coil through a stationary magnetic field, as shown in Fig. 5.10, or by rotating the magnetic field through the coil.

The voltage caused by Faraday's law of electromagnetic induction, caused by the relative motion of circuits and magnetic fields, is the basic principle of magnetic induction generators. Voltage is generated when a permanent magnet moves relative to a conductor (and vice versa). If the wire is then connected to an electrical load, the current flows, turning the energy of mechanical motion into electrical energy.

Lenz's law describes the direction of the induced field. Faraday's law was later extended to Maxwell's equation, which is an equation in James Clark Maxwell's electromagnetic field theory. Electromagnetic induction has many applications in technology, including electrical components such as inductors and transformers, as well as equipment such as motors and generators.

5.4.4 Piezoelectric Effect

By applying pressure to some crystals (e.g. quartz or Rochelle salts) or ceramics (e.g. barium titanate, an iron-electric material) [9], electrons can be driven out of orbit in the direction of force. It polarizes their interiors, with positive and negative opposite charges on their two opposite surfaces (see Fig. 5.11). When the pressure is released, the electrons return to orbit. Some materials respond to bending pressure, while others respond to twisting pressure. If external wires are connected in the presence of pressure and voltage, electrons will flow and current will be generated. If the pressure remains constant, the current will flow until the potential difference is equalized.

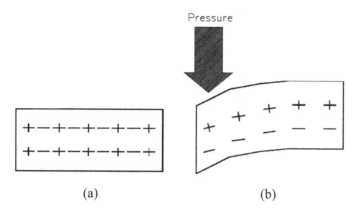

Fig. 5.11 Pressure applied to certain crystals produces an electric charge

This generation of voltage is called piezoelectric effect. When the force is removed, the material is decompressed and immediately causes an electric force in the opposite direction. As the direction of the force changes, so does the polarity of the charge. The power capacity of these materials is extremely small. However, these materials are very useful because of their extreme sensitivity to changes of mechanical force.

When an electrostatic field is applied in the direction of polarization of the dielectric, the dielectrics are deformed, and when the electrostatic field is removed, the deformation of the dielectric disappears, a phenomenon known as the anti-piezoelectric effect.

5.4.5 Thermoelectricity

The so-called thermoelectric effect is the electron separation phenomenon produced by heating on the contact surface of two different metals. Copper and zinc, shown in Fig. 5.12, tend to move towards the zinc side after the contact surface is heated, creating a voltage. When the heat source is removed, the voltage disappears. This is the basic principle of thermocouples, which are one of the basic sensors for measuring temperature in industry and can be used to measure relatively high temperatures.

The thermoelectric voltage [10] in the thermocouple depends on the heat energy applied at the junction of two different metals. Thermocouples are widely used to measure temperature and as thermal sensors in automatic temperature control equipment.

Compared with other power supplies, the power capacity of thermocouple is very small, but slightly larger than that of crystal. Generally speaking, a thermocouple can be subjected to higher temperature than that of ordinary mercury or alcohol thermometer.

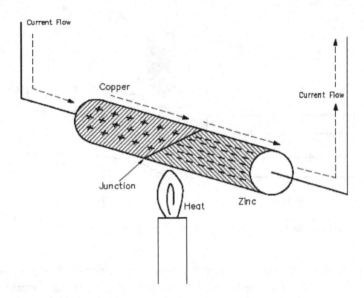

Fig. 5.12 Heat energy causes copper to give up electrons to zinc

5.4.6 Photoelectric Effect

Light is a form of energy. Many scientists believe that it is composed of small energy particles called photons. Under the exposure of light higher than a certain frequency, electrons inside certain substances are excited by photons to form an electric current, i.e. the photoelectric effect (see Fig. 5.13).

Photoelectric effects are divided into photoelectric, photoelectric and photoconductivity. These uses of photoelectric effects are described below.

Photovoltaic: The combined energy of light in one of two plates causes one plate to release electrons to another. Plates accumulate opposite charges, such as batteries (Fig. 5.13).
Photoemission: Photon energy from the beam can cause the surface to release electrons in a vacuum tube. Then, a plate will collect electrons.

Fig. 5.13 Producing electricity from light using a photovoltaic cell

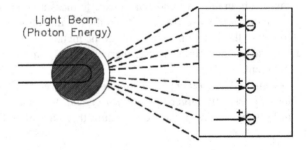

Photoconduction: Light energy is applied to materials that are usually less conductors, resulting in the production of free electrons in the material, making them better conductors.

According to the particle's theory of light, light is made up of a discontinuous photon whose energy can be absorbed by an electron in the metal when it hits a light-sensitive metal, such as selenium. When electrons absorb the energy of photons, kinetic energy increases immediately, and if kinetic energy increases enough to overcome the gravitational pull of atomic nucleation, it can escape the metal surface, become photo electrons, and form currents. In a unit of time, the more incident photons, the more photo electrons fly out, the stronger the light current.

Heinrich Hertz discovered the photoelectric effect in 1887, and Einstein was the first to successfully explain the photoelectric effect. Light wavelengths are less than a certain threshold when electrons, or limit wavelengths, can be emitted, and the corresponding frequency of light is called the limit frequency. The critical value depends on the metal material, and the energy emitting the electron depends on the wavelength of light and is independent of the intensity of the light.

The emission of typical metal conduction electrons usually requires several electron volts, corresponding to visible or ultraviolet light. The study of photoelectric effect is an important step for understanding the quantum properties of light and electron, and affects the formation of the concept of wave particle duality.

Photoelectric emission can occur on any material, but it is most easily observed on metal or other conductors, because this process will produce charge imbalance. If the current cannot eliminate the charge imbalance, the emission barrier will increase until the emission current stops. If the metal is exposed to oxygen, the thin oxide layer on the metal surface will usually increase the energy barrier of photoelectric emission. Therefore, most practical experiments and equipment based on photoelectric effect use clean metal surface in vacuum.

When photoelectrons are emitted into a solid rather than a vacuum, the term "internal photoemission" is usually used, while emission into a vacuum is distinguished as external photoemission.

5.4.7 Thermionic Emission

Thermionic emission [11] or thermal excited charge emission is the thermally induced flow of charge carriers from a surface or over a potential-energy barrier.

Thermal electron emission is the phenomenon of escaping by heating a large number of electrons in a metal to overcome surface potential. Similar to gas molecules, free electrons in metals exercise irregularly in thermal motion at a certain rate. On the metal surface there is a force that prevents electrons from escaping out, and electron escape needs to overcome resistance to do the work, called escape work. At room temperature, only a very small number of electrons have more kinetic energy than escape, and very little electrons escape from the metal surface.

Thermionic energy converter is a device composed of two electrodes, which are close to each other in vacuum. One electrode is usually called cathode or emitter, and the other is called anode or plate. Usually, electrons in the cathode are prevented from escaping from the surface by a potential energy barrier. When an electron begins to leave the surface, it generates a corresponding positive charge in the material, pulling it back to the surface. In order to escape, electrons must obtain enough energy in some way to overcome this energy barrier. At room temperature, almost no electron can get enough energy to escape. However, when the cathode is very hot, the thermal motion will greatly increase the electron energy. At a sufficiently high temperature, a considerable number of electrons can escape. The release of electrons from hot surfaces is called hot electron emission.

Generally, when the metal temperature rises above 1000 K, the number of electrons whose kinetic energy exceeds the escape work begins to increase, and a large number of electrons escape from the metal, which is thermal electron emission. Without an external electrostatic field, the escaping thermal electrons accumulate near the metal surface, becoming a space charge, which prevents the thermal electrons from continuing to emit. The classical example of thermionic emission is the emission of electrons from a hot cathode into a vacuum (also known as thermal electron emission or the Edison effect) in a vacuum tube. The hot cathode can be a metal filament, a coated metal filament, or a separate structure of metal or carbides or borides of transition metals.

Constantly emission forms an electric current. As the voltage increases, all electrons emitted from the cathode reach the anode, and the current is saturated. The cathode of many electro-vacuum devices is operated by thermal electron emission, because thermal electron emission depends on the escape work of the material and its temperature, the material with high melting point and low escape work should be used to make cathode.

Figure 5.14 shows a diode made from the principle of thermal electron emission which is a simplest example of a thermionic device called as a vacuum tube diode in which the only electrodes are the cathode and plate, or anode. The diode can be used to convert alternating current (AC) flow to a pulsating direct current (DC) flow.

5.5 Magnetism

Certain metals and metallic oxides have the ability to attract each other. This property is called magnetism, and the materials, which have this property, are called magnets.

The composition of magnets is iron, cobalt, nickel and other atoms, the internal structure of their atoms is relatively special, itself has magnetic moment. Magnets can be classified into "permanent magnets" and "non-permanent magnets". Permanent magnets can be natural products, also known as natural magnets, or they can be manufactured by hand. Non-permanent magnets, such as electromagnets, are magnetic only under certain conditions. Permanent magnets are usually made of cobalt steel.

Fig. 5.14 Vacuum tube diode

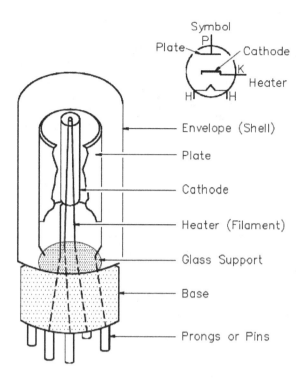

- Envelope (Shell)
- Plate
- Cathode
- Heater (Filament)
- Glass Support
- Base
- Prongs or Pins

Magnetism is the result of electrons rotating around the nucleus on their own axis [12] (Fig. 5.15). In magnetic materials, atoms have certain areas called domains. These domains are aligned so that their electrons tend to rotate in the same direction.

Figure 5.15 is a diagram of the magnetic field generated by electrons rotating and spinning around the nucleus, and modern magnetism believes that the root cause of magnetism is the movement of electrons. Objects that are polarized have magnetic

Fig. 5.15 Electron spinning around nucleus produces magnetic field

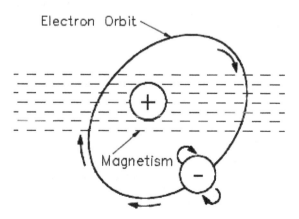

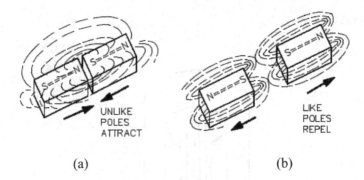

UNLIKE
POLES
ATTRACT

LIKE
POLES
REPEL

(a)   (b)

Fig. 5.16 The law of magnetic attraction and repulsion

poles called the South Pole (S) and the North Pole (N), respectively. The law of magnetism state that like magnetic poles repel and unlike magnetic poles attract one another (Fig. 5.16).

Because of the motion of electrons around the nucleus, all matter has some degree of magnetic effect. But in nature, iron, nickel, cobalt and other materials show strong magnetic properties, so magnetism is also known as ferromagnetic.

5.5.1 Magnetic Flux

The group of magnetic field lines emitted outward from the north pole of a magnet is called magnetic flux. The symbol for magnetic flux is Φ. The SI unit of magnetic flux is the weber (Wb). One weber is equal to 1×10^8 magnetic field lines.

Example 5.1 $\Phi = 5000$, the number of webers is 50 μ Wb.

Magnetic flux density is the amount of magnetic flux per unit area of a section, perpendicular to the direction of flux. Equation (5.6) is the mathematical representation of magnetic flux density.

$$B = \Phi/A \qquad\qquad (5.6)$$

where **B** is magnetic flux density in teslas, Φ is magnetic flux in webers; A is area in square meters.

Example 5.2 Find the flux density in teslas, when the flux is 800 μ Wb and the area is 0.004 m^2.

Solution: $\Phi = 5000$, the number of webers is 50 μ Wb, Area is 0.005 m^2, then $B = \Phi/A = 0.01 \text{Wb/m}^2$.

5.5.2 Electromagnetism

Electromagnetism is a branch of physics [13] that studies the interaction between electricity and magnetism and its laws and applications.

The electromagnetic force usually shows electromagnetic fields, such as electric fields, magnetic fields, and light. The electromagnetic force is one of the four fundamental interactions (commonly called forces) in nature. The other three fundamental interactions are the strong interaction, the weak interaction, and gravitation. Lightning is an electrostatic discharge that travels between two charged regions.

The relationship between magnetism and electrical current was discovered by a Danish scientist named Oersted in 1819. He found that if an electric current was caused to flow through a conductor, the conductor produced a magnetic field around that conductor (Fig. 5.17).

According to the view of modern physics, the phenomenon of magnetism is produced by the motion of charge, so it is necessary to contain the content of magnetism within the field of electricity. Therefore, the content of electromagnetism and electricity is difficult to divide, and "electricity" sometimes as "electromagnetics" shorthand. Electromagnetics has developed from two separate sciences (electricity, magnetism) into a complete branch of physics, based mainly on two important experimental findings, namely, the magnetic effect of current and the electrical effect of changing magnetic fields. These two experimental phenomena, together with Maxwell's hypothesis that changing the magnetic field generated by electrostatic fields, laid down the whole theoretical system of electromagnetics and developed electrical and electronic technologies that had a great impact on modern civilization.

Right-handed rules can be used to determine the direction of the magnetic field around the conductor: the direction of the magnetic field generated by the current in the conductor can be remembered with the direction of the right thumb and other fingers. That is, extend your right hand so that your thumb is in the same direction as the current, and the direction in which the remaining four fingers bend is the direction of the magnetic line, as shown in Fig. 5.18. It is important to note that in some countries the direction of electron motion is defined as the direction of current, in which case the left hand is required.

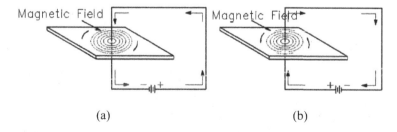

(a)   (b)

Fig. 5.17 The magnetic field produced by current in a conductor

Fig. 5.18 Figure 5.22
right-hand rule for current
carrying conductors

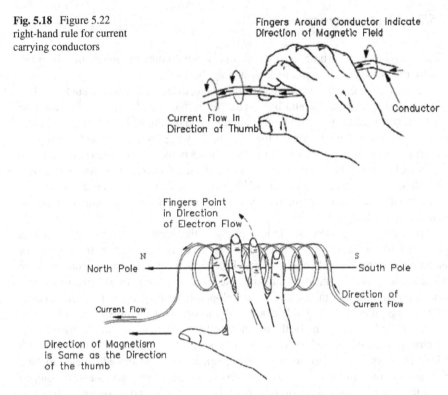

Fig. 5.19 Right hand rule for coils

In the case of coils, the direction of the magnetic pole can also be determined by the right-hand rule: the direction of the current can be followed by the right-hand finger other than the thumb, the direction of the thumb is the direction of the magnetic north pole, as shown in Fig. 5.19.

If the core is added to the coil, the magnetic field can be greatly enhanced to obtain an electromagnet-like magnet. Electromagnetic iron's magnetic north pole can be judged by the same rules as right-handed, as shown in Fig. 5.20.

5.5.3   Magnetomotive Force

Magnetomotive force (mmf) is the strength of a magnetic field in a coil of wire. It is proportional to the current strength in the coil, and also proportional to the number of turns in the coil. The current times the number of turns of the coil is expressed in units called "ampere-turns" (At), also known as mmf. Equation (5.7) is the mathematical representation for ampere-turns [14] (At).

Fig. 5.20 Right hand rule
for coils with a core

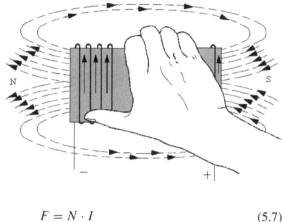

$$F = N \cdot I \qquad (5.7)$$

where, N—number of turns, I—current.

Example 5.3 Calculate the ampere-turns for a coil with 1000 turns and a 5 mA current.

Solution: $N = 1000$ turns, $I = 5$ mA, then $F = 5$At.

5.5.4 Magnetic Field Intensity

The longer the length of the coil, the smaller the magnetic field strength, which can be defined as:

$$H = \frac{F}{L} \qquad (5.8)$$

where, L—Length between poles of coil, m; H—Field intensity, At/m.

Example 5.4 Find field intensity of an 80 turn, 20 cm coil, with 6A of current.

Solution: 2400At/m.
If the same coil in Example 5.4 were to be stretched to 40 cm with wire length and current remaining the same, the new value of field intensity is 1200 At/m.
The 20 cm coil used in Example 5.4 with the same current is now wound around an iron core 40 cm in length. Then the field intensity is 2400 At/m.

5.5.5   *Permeability and Reluctance*

The permeability (μ) of a magnetic material is a physical quantity of magnetic properties that characterizes magnetic media. It is the ratio of B to H. Equation (5.10) is the mathematical representation for magnetic material permeability.

$$\mu = \frac{B}{H} \qquad\qquad (5.10)$$

where, H is magnetic field intensity, B is magnetic flux density. μ is permeability, T m/At or H/m.

The relative magnetic permeability of the magnetic medium (μ_r) is defined as the ratio of magnetic permeability μ to vacuum magnetic permeability (μ_0), i.e.

$$\mu_r = \frac{\mu}{\mu_0} \qquad\qquad (5.11)$$

where, $\mu_0 = 4\pi \times 10^{-7}$ H/m or 1.26×10^{-6} T·m/At.

Magnetic reluctance which corresponds to resistance is a parameter in the magnetic path due to the presence of leakage in the magnetic path. Magnetic reluctance is represented by the symbol R in At/wb, defined as follows:

$$R = \frac{L}{\mu A} \qquad\qquad (5.12)$$

where, L is length of coil, m; A is cross section area of coil, m^2.

Reluctance is inversely proportional to permeability (μ). Because iron core has a high magnetic permeability, the magnetic reluctance is small, while the air has a larger magnetic permeability. Generally, different types of materials have different values of reluctance (Fig. 5.21). Since air gap has a very high reluctance, the size of the air gap affects the value of reluctance: the larger the gap, the greater the reluctance. Air is nonmagnetic. Air can not concentrate magnetic lines. The larger air gap only provides space for the magnetic lines to spread out.

5.5.6   *Magnetic Circuits*

A closed circuit consisting of a strong magnetic material in which a magnetic field produces a certain strength is called a magnetic circuit. Magnetic paths generally contain magnetic components, such as permanent magnets, ferromagnetic materials, or electromagnets, but may also contain air gaps and other substances. Magnetic circuits are generally composed of coils that current to excite magnetic fields (and in

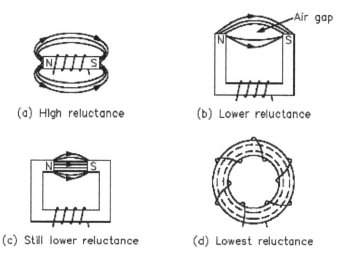

(a) HIgh reluctance   (b) Lower reluctance

(c) Still lower reluctance   (d) Lowest reluctance

Fig. 5.21 Different physical forms of electromagnets

some cases permanent magnets can also be used as an excitation source for magnetic fields), iron cores made of soft magnetic materials, and air gaps of the appropriate size.

The physical quantities involved in the magnetic circuit are flux, magnetic potential, magnetic resistance, magnetic position difference, etc. The magnetic circuit has some similarities to the circuit, as shown in Fig. 5.22, it can be compared with an electric current in which EMF, or voltage, produces a current flow. The ampere-turns (NI), or the magnetomotive force (Fm or mmf), will produce a magnetic flux F.

If a flux in the magnetic path passes through several sections, the total magnetomotive force (mmf) of each section is equal to the sum of the magnetomotive force

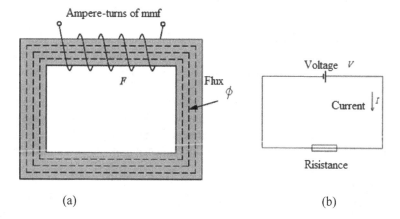

(a)   (b)

Fig. 5.22 Magnetic current with closed iron path

on each section of the magnetic path. The magnetomotive force of each section of the magnetic path is equal to the product of the magnetic reluctance and flux of that section, so that the total magnetic reluctance is equal to the sum of the magnetic reluctance of each section. This is equivalent to the sum of the resistances of each of the series circuits. Similarly, if there are multiple magnetic branch parallels in the magnetic path, then the two ends of each branch have the same magnetomotive force, the sum of the flux of each magnetic branch is equal to the total flux, so that the total magnetic conduction of these parallel branches is equal to the sum of the magnetic conductors of each branch. This is equivalent to the sum of the currents in parallel circuits.

The main purpose of magnetic path analysis is to determine the relationship between excitation flux and the flux it produces, which is necessary to understand the performance of the device and to design accordingly, such as determining the shape, size, size of excitation current, selecting suitable materials, etc.

According to Ohm's law, there is a relationship between flux, magnetomotive force, and magnetic resistance as follows:

$$\Phi = F/R \tag{5.13}$$

where Φ is magnetic flux, Wb; F is the magnetomotive force, At; R is magnetic resistance defined as Eq. (5.12), At/Wb.

Example 5.5 A coil has an mmf of 600 At, and a reluctance of 3×10^6 At/Wb. Find the total flux.

Solution: 200 μWb.

5.5.7   BH Magnetization Curve

Figure 5.23 shows the comparison of the B-H curve of the two ferromagnetic materials with the B-H curve of the air. These curves can be measured by experimental methods. There is a nonlinear relationship between B and H. Initially, as H increases, B increases almost in a straight line, with a slope of magnetic permeability defined by the Eq. (5.10). As H increases further, the increase in B becomes slower, and after the turn point, the H value increases almost no more, i.e. it reaches saturation. Different ferromagnetic materials have different magnetization curves, and their B saturation values are not the same. But for the same material, the saturation value of B is certain.

Air, which is nonmagnetic, has a very low B-H curve [15], as shown in Fig. 5.23.

When current in a coil reverses direction thousands of times per second, hysteresis can cause considerable loss of energy. The magnetic flux in an iron core lags behind the magnetizing force. Hysteresis is defined as "a lagging behind." The magnetization

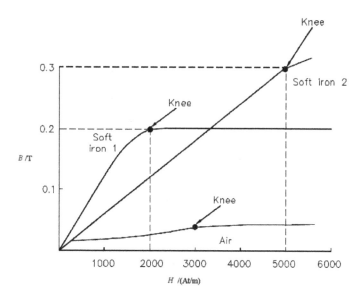

Fig. 5.23 Typical BH curve for two types of soft iron

of magnets is obviously irreversible, when ferromagnetic magnets are magnetized to saturation, if the magnetic field intensity (H) is gradually reduced from the maximum value, the magnetic flux density (B) is not returned in the original way, but decreased along a curve slightly higher than the original path. When H is 0, B is not equal to zero, i.e. the change in B in the magnet lags behind the change in H, which is called hysteresis, as shown in Fig. 5.24. The B_r in the Fig. is called the retentivity of that magnetic material, and H_c is the force that must be applied in the reverse direction to reduce flux density to zero, is called the coercive force of the material. The greater the area inside the hysteresis loop, the larger the hysteresis losses.

5.5.8 Magnetic Induction

Electromagnetic induction was discovered by Michael Faraday in 1831. Faraday found that if a conductor "cuts across" lines of magnetic force, or if magnetic lines of force cut across a conductor, a voltage, or EMF, is induced into the conductor. Electromagnetic induction is a phenomenon in which a conductor is placed in a magnetic field and moves to produce voltage. This voltage is called the inductive voltage. If this conductor is closed into a circuit, the voltage drives the electrons to flow, creating an induced current. The discovery of electromagnetic induction is one of the greatest achievements in the field of electromagnetics. It not only reveals the intrinsic relationship between electricity and magnetism, but also lays the foundation for the mutual transformation between electricity and magnet, and opens

Fig. 5.24 Hysteresis loop
for magnetic materials

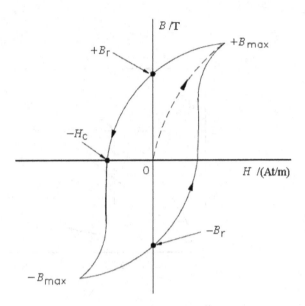

the way for mankind to obtain huge and cheap electricity, which is of great practical
significance. The discovery of electromagnetic induction marks the arrival of a major
industrial and technological revolution. It has been proved that the wide application
of electromagnetic induction in electrician, electronic technology, electrification and
automation has played an important role in promoting the development of social
productivity and science and technology.

As shown in Fig. 5.25, if the magnetic line is from the N-pole to the S-pole, i.e.
from top to bottom, conductor C moves from left to right, the conductor generates
voltage. If the circuit is closed, an electric current is generated, and the direction of
the current can be determined by the right-handed rule. Open your right hand, keep
your five fingers in a plane with your palms, and your thumb is perpendicular to
the rest of your four fingers, so that the magnetic line passes vertically through your

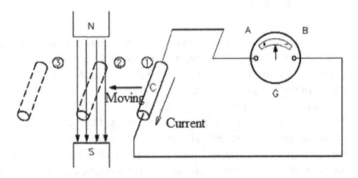

Fig. 5.25 The current of Induced EMF

palm and your thumb points in the direction of the conductor's motion. The direction of the remaining four fingers is the direction of the current.

5.5.9 Faraday's Law of Induced Voltage

Faraday's law of induction is a basic law of electromagnetism predicting how a magnetic field will interact with an electric circuit to produce an electromotive force (EMF), a phenomenon called electromagnetic induction. When a conductor in a closed circuit makes the movement of cutting magnetic lines in a magnetic field, an induced voltage is generated in the conductor. The area of the closed coil is unchanged, the magnetic field intensity is changed, the flux will also change, and electromagnetic induction will also occur. So, the exact meaning is that the induction voltage is generated by changes in flux.

The polarity of voltage is pointed out by the Lenz's law: the magnetic field that induces the current should hinder the change of the primary flux. The direction of the induced current can be determined by the right-handed rule, and then the polarity of the induced voltage can be determined.

The magnitude of the inductive voltage is determined by Faraday's law of electromagnetic induction:

$$V = -N\frac{d\Phi}{dt} \tag{5.14}$$

where, V is induced voltage, V; N is number of turns in a coil; Φ is Flux, Wb. The negative sign in Eq. (5.14) is an indication that the emf is in such a direction as to produce a current whose flux, if added to the original flux, would reduce the magnitude of the emf.

Example 5.6 Given: Flux $= 4$ Wb. The flux increases uniformly to 8 Wb in a period of 2 s. Find induced voltage in a coil that has 12 turns, if the coil is stationary in the magnetic field.

Solution: induced voltage is 24 V.

The polarity of inductive voltage is given by the Lenz's law.

Lenz's law is named after Russian scientist Heinrich Lenz in 1834, and it says:

If an induced current flows, its direction is always such that it will oppose the change which produced it.

The Lenz's law can also be expressed as: the effect of induced current always resists the factor that causes the induced current. For example, in Fig. 5.25, as the wire moves to the left, the downward magnetic line through the right circuit increases, so the magnetic line generated by the induced current generated by the conductor in the circuit should be up, and the current direction should be shown in the Figure.

5.6 DC Theory

Direct current is a current that does not change direction.

5.6.1 Dc Sources

When we talk about DC power, the first thing that comes to mind is probably all kinds of batteries. Chemical battery refers to a kind of device that converts the chemical energy of positive and negative active substances into electrical energy through electrochemical reaction. In addition to batteries, however, there are other devices that produce DC which are frequently used in modern technology.

In a chemical cell (Fig. 5.26), the main part is the electrolyte solution, the positive and negative electrodes immersed in the solution, and the wires connecting the electrodes. Positive and negative electrodes are made of different metals, and electrolytes provide negative and positive ions. The chemical reaction causes the metal on one of the electrodes to be ionized and continuously enter the electrolyte. The chemical reaction process is complex, and we just need to know that through this reaction, electrons are deposited in the cathode, generating voltage. If two electrodes are connected, the current from the anode to the cathode is obtained.

Chemical batteries can be classified into: original battery and rechargeable battery.

A good example of an original battery is a battery containing zinc and copper electrodes. Zinc electrode contains a large number of negatively charged atoms, and copper electrode contains a large number of positively charged atoms. When these electrodes are immersed in the electrolyte, the chemical action begins. The zinc electrode will accumulate a greater negative charge because it will dissolve into the electrolyte. The atoms leaving the zinc electrode are positively charged and attracted

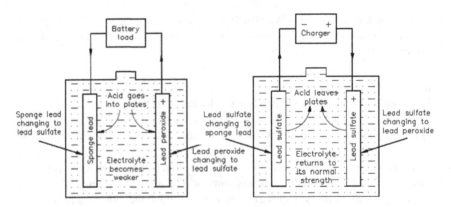

Fig. 5.26 Basic chemical battery

by negatively charged ions in the electrolyte; Atoms repel positively charged ions in the electrolyte to the copper electrode (Fig. 5.26).

A good example of rechargeable battery is lead-acid batteries. Lead-acid batteries use electrolytes of sulfate with a specific gravity of 1.835 to pure sulfuric acid. Because the electrolyte contains water, the electrolyte has a specific gravity of 1.000 to 1.835. Typically, the proportion of sulfate electrolytes in lead-acid batteries is 1.350. The change in the specific gravity measures how much charge is charged.

Specific gravity is defined as the ratio comparing the weight of any liquid to the weight of an equal volume of water. The specific gravity is measured with a hydrometer. One end of the hydrometer float is weighed and both ends are sealed. The balance calibrated at specific gravity is placed longitudinally along the float body. Place the float in the glass tube and suck the fluid to be measured into the glass tube. When the fluid is sucked into the pipe, the hydrometer float will sink to a certain level in the fluid. The extent to which the hydrometer float protrudes from the liquid level depends on the specific gravity of the liquid. The reading on the fluid surface buoy is the specific gravity of the fluid.

Another parameter describing the battery is ampere hour, defined as 1 A of current flowing through 1 h of the corresponding energy. Ampere hour is typically used to measure the storage capacity of the battery.

Figure 5.26 is a diagram of the charge and discharge of lead-acid batteries, electrodes are mainly made of lead and its oxides, and electrolytes are sulfuric acid solutions. In the discharge state, the main component of the positive pole is lead dioxide, the main component of the negative pole is lead, and in the charging state, the main component of the positive and negative pole is lead sulfate. The equations for charging and discharging are:

$$2PbSO_4 + 2H_2O \xrightarrow{\text{discharge}} PbO_2 + Pb + 2H_2SO_4 \tag{5.15}$$

$$PbO_2 + Pb + 2H_2SO_4 \xleftarrow{\text{charge}} 2PbSO_4 + 2H_2O \tag{5.16}$$

At discharge, H_2SO_4 is broken down into H_2 and SO_4^{2-} ions. H_2 is oxygenated into water released from the positive pole, which weakens the acidity of the electrolyte. SO_4^{2-} ions synthesize $PbSO_4$ with lead on cathodes and anodes as shown in Eq. (5.15).

At charging, SO_4^{2-} ions in $PbSO_4$ in cathodes and anodes are driven into the electrolyte to form H_2SO_4. PbO_2 is formed at the anode and Pb is formed at the cathode, a process that increases the acidity of the electrolyte as shown in Eq. (5.16).

When the lead-acid battery is nearly fully charged, hydrogen is released near the cathode and oxygen is released near the anode. This is because, towards the end of the charge, the inflow current is greater than the current required to break down the residual lead sulfate on the electrode, which electrolytes water into hydrogen and oxygen. Therefore, when charging lead-acid batteries, attention needs to be paid to the risk of hydrogen burning or exploding.

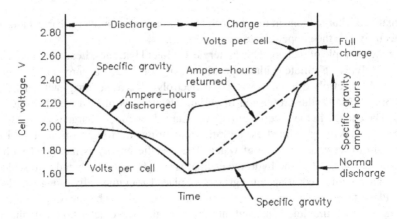

Fig. 5.27 Voltage and specific gravity during charge and discharge

Figure 5.27 shows how the parameters of the lead-acid battery change during charging and discharging. During discharge, the proportion of the electrolyte is increased with the release of amperes linear decrease, and the specific gravity will resume when charging.

Example 5.7 A lead-acid battery reads 1.175 specific gravity. Its average full charge specific gravity is 1.260 and has a normal gravity drop of 120 points (or.120) at an 8 h discharge rate.

Solution:
  Fully charged 1.260.
  Present charge 1.175.
  The battery is 85 points below its fully charged state. It is therefore about 85/120, or 71%, discharged.
  In addition to the battery that can provide DC power, DC generators can also generate DC electricity, and we'll talk about the principles of DC generators later.
  In addition to batteries and generators, we know that thermocouples can also generate DC voltages, as shown in Fig. 5.28, a diagram of how thermocouples generate DC voltages. A thermocouple is a device that converts heat from a heat source into electrical energy. When the connection points of the two metals are heated, the electrons of one metal gain enough energy to become free electrons to enter the other metal and generate voltage. The metal combinations available are iron and copper nickel alloys, copper and copper nickel alloys, palladium and palladium, chromium alloys and aluminum-nickel alloys. Thermocouples are generally used to measure temperature and the voltage obtained by measurement is marked as a reading of temperature.
  Most power stations produce alternating current. The main reason for the generation of AC is that it can be transmitted over a long distance with less loss than DC; However, many devices used today operate only on DC or more effectively on DC.

Fig. 5.28 Production of a
DC voltage using a
thermocouple

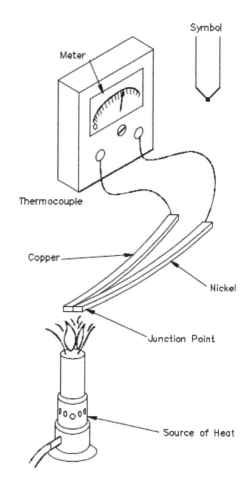

For example, transistors, tubes, and some electronic controls require direct current
to operate. If we want to operate these devices from ordinary AC sockets, they must
be equipped with rectifier devices to convert AC to DC.

Rectifiers convert alternators into direct current. Rectifiers typically use semiconductor diodes to convert AC into DC. Let's first look at how diodes work.

Semiconductor diodes are also known as crystal diodes, or diodes. It is an electronic device capable of conducting current in one direction. Inside the semiconductor
diode there is a PN junction (P for positive; N for negative), with terminals at both
ends. The diode has a one-way current conductivity according to the polarity of the
additional voltage.

In Fig. 5.29, the diode is on when the P-side is connected to the positive pole
of the power supply and the N-side is connected to the negative pole of the power
supply. This is because the voltage provided by the supply tends to push the holes

Fig. 5.29 Forward-biased
diode

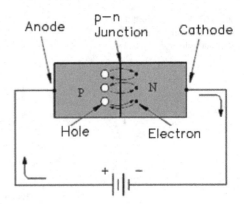

in the P-junction toward the N-junction, while the electrons in the N-junction are pushed to the P-junction to form a current.

The physical concept of the hole is introduced here. An electron-neutral atom whose number of positive electrons and negative electrons is equal. When a negative electron is missing, there is a positive electrical void we call it a hole. A hole, also known as an electron hole, loses an electron on a covalent bond in physics and ends up leaving a blank space on the covalent bond. That is, some valence electrons in the covalent bond get some energy, thus getting rid of the constraint of the covalent bond and becoming free electrons, while leaving a space on the covalent bond, which we call a hole.

If the P end of the diode is connected to the negative pole of the power supply, the N end is connected to the positive pole of the power supply, as shown in Fig. 5.30. As the cavities in the P-junction move to the left, the electrons in the N-junction move to the right and cannot form an electric current. The diode is in an off state at this time.

Figure 5.31 is a half-wave rectifier circuit diagram made of such semiconductor diodes.

Fig. 5.30 Reverse-biased
diode

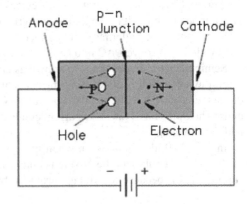

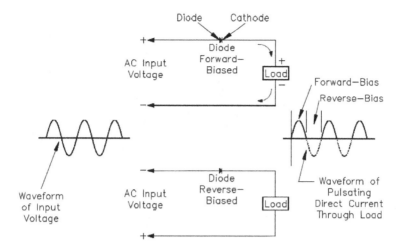

Fig. 5.31 Half-wave rectifier

When a diode is connected to a circuit with an AC input, only half of the AC wave can make the circuit on, and the other half of the wave makes the circuit in off state, so a half-wave DC voltage can be obtained at the output of the diode. This is a pulsed DC power supply. So, can you design a full-wave rectifier power supply? The answer is yes, as shown in Fig. 5.32 is a full-wave rectifier power supply using transformer coils.

Another full wave rectifier circuit is a full wave bridge rectifier. The circuit uses four diodes. The action of these diodes in each half cycle of applying AC input voltage is shown in Fig. 5.33. Then, the output of the circuit becomes pulsating DC, and all waves input to AC are transmitted. The output appears to be the same as the output obtained from the full wave rectifier (Fig. 5.32).

5.6.2 Resistance and Resistivity

The resistance described earlier is a current-limiting element. When the resistor is attached to the circuit, it limits the amount of current passing through the branch it connects to. A resistor that cannot be changed is called a fixed resistor. A variable resistance value is called a potential or variable resistor. The ideal resistor is linear, i.e. the current passing through the resistor is proportional to the added voltage.

Resistivity [16]is defined as a measure of the resistance of a material to the current. The resistance of the fixed conductor length depends on the resistance of the material, the length of the conductor, and the cross-sectional area of the conductor. While resistivity is independent of the length of the conductor, cross-sectional area and other factors, and is the electrical properties of the conductor material itself. The resistivity is determined by the material of the conductor and is temperature-related.

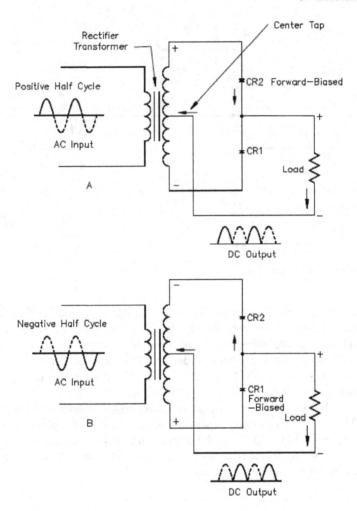

Fig. 5.32 Full-wave rectifier

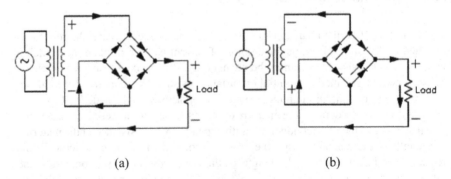

(a)   (b)

Fig. 5.33 Bridge rectifier circuit

Table 5.1 Properties of conducting materials

Material	Resistivity at 20 °C $\rho/(10^{-8} \cdot \Omega \text{ m})$	Temperature coefficient $\alpha /(\Omega/°C)$
Aluminum	2.86	0.004
Carbon	Carbon has 2500–7500 times the resistance of copper	−0.0003
Constantan	49.6	0
Copper	1.75	0.004
Gold	2.36	0.004
Iron	9.76	0.006
Nichrome	114	0.0002
Nickel	8.75	0.005
Silver	1.65	
Tungsten	5.69	

Knowing the resistivity of the material, the resistance value of the device can be calculated according to Eq. (5.17).

$$R = \rho \frac{L}{A} \qquad (5.17)$$

where, ρ is specific resistance or resistivity, $\Omega \cdot$m; L is length of conductor,, m; A is cross-sectional area of conductor, m^2.

The resistivity ρ (rho) allows different materials to be compared for resistance, according to their nature, without regard to length or area. Resistivity is also temperature-dependent. Table 5.1 gives resistivity values for metals having the standard wire size of one foot in length and a cross-sectional area of 1 cm.

Resistance temperature coefficient, α (alpha), is defined as the change of material resistance under a given temperature change. A positive value indicates that R increases with the increase of temperature; a negative value indicates that R decreases; and zero indicates that R is a constant. Typical values are listed in Table 5.1.

$$R_t = R_{20} + R_{20}\alpha \Delta t \qquad (5.18)$$

5.6.3 Kirchhoff's Law

Simple circuits can be solved with Ohm's law, but the actual circuit can be much more complex, and it is not easy to solve with Ohm's law. Through extensive experimental research, in 1857 German physicist Gustav Kirchhoff developed a method for solving complex circuits. Kirchhoff's main conclusions are two, today known as Kirchhoff's first law and Kirchhoff's second law [17].

Kirchhoff's law is the basic law of voltage and current in the circuit, and it is the basis for analyzing and calculating more complex circuits, which can be used for both DC circuit analysis and AC circuit analysis, as well as nonlinear circuits containing electronic components.

To analyze circuits with complex network, Kirchhoff introduces some of the following basic concepts:

1, branch:

(1)   Each component is a branch;
(2)   Series components as a branch;
(3)   The current is equal everywhere in a branch.

2, node:

(1)   the connection point between the branch and the branch;
(2)   the connection point of more than two branches;
(3)   Generalized node (any closed face).

3, loop:

(1)   Closed branch;
(2)   A collection of closed nodes.

4, mesh:

(1)   the inside does not contain any loops of the branch;
(2)   The mesh hole must be a loop, but the loop is not necessarily a mesh hole.

Law 1: The sum of the voltage drops around a closed loop is equal to the sum of the voltage sources of that loop (Kirchhoff's Voltage Law).

Law 2: The current arriving at any node in a circuit is equal to the current leaving that node (Kirchhoff's Current Law).

Kirchhoff's first law is the continuity of current in the circuit, the basic principle is the principle of charge conservation. Kirchhoff's first law is the law that determines the relationship between the currents of each branch at any node in the circuit, and is therefore also known as the law of node currents. Kirchhoff's first law states that the **sum of all currents entering a node is equal to the sum of all currents leaving the** node. Or describe it as: Assuming that the current entering a node is positive and the current leaving the node is negative, the algebraic sum of all currents involving the node is equal to zero. Figure 5.34 is a diagram of Kirchhoff's current law.

Kirchhoff's second law, also known as Kirchhoff's law of voltage, is the embodiment of the single-value of voltage in the circuit when the electrostatic field is a potential field, and its deep physical basis is the principle of energy conservation. Kirchhoff's second law is the law that determines the relationship between voltages at various points in any circuit, and is therefore also known as the law of circuit voltage. Kirchhoff's second law states: Algebra of the voltage difference and equal to zero at all ends of all components along the closed circuit.

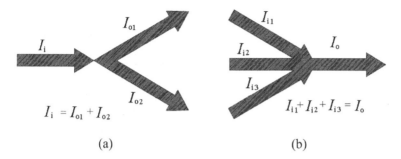

Fig. 5.34 Illustration of Kirchhoff's current law

Kirchhoff's second law can only be used for closed loops, one of which can become a closed loop, and two conditions must be met:

(1) Must have at least one power supply;
(2) There must be a complete connecting circuit, the current from any point, along the circuit can return to the starting point.

As shown in Fig. 5.35, a closed loop diagram. Let's use an example to illustrate the application of Kirchhoff's law.

Example 5.8 Using Kirchhoff's laws, it is possible to take a circuit with two loops and several power sources (Fig. 5.36) and determine loop equations, solve loop currents, and solve individual element currents.

Solution: The first step is to draw an assumed direction of current flow (Fig. 5.37). It does not matter whether the direction is correct. If it is wrong, the resulting value for current will be negative.

The second step in the analysis is to indicate the polarity of each device, as shown in Fig. 5.38. At this point, assume the current direction of the middle branch, assuming I_2-I_1, facing down.

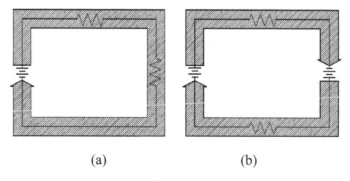

Fig. 5.35 Closed loop

Fig. 5.36 Example circuit
for loop equations

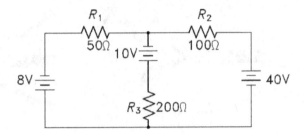

Fig. 5.37 Assumed
direction of current flow

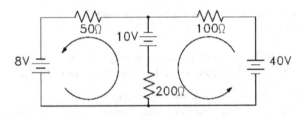

Fig. 5.38 The polarity of
each device

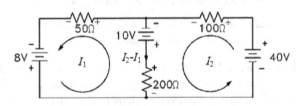

The third step is to analyze the two circuits separately, starting with the analysis of the first circuit, using Kirchhoff's law, there are:

$$8 + 200(I_2 - I_1) - 50I_1 - 10 = 0 \qquad (5.19)$$

Using the same procedure for Loop 2, the resulting equation is shown in Eq. (5.20)

$$10 - 200(I_2 - I_1) + 40 - 100I_2 = 0 \qquad (5.20)$$

Fourth, solve Eqs. (5.19) and (5.20) simultaneously,
$I_1 = 268.6\,\text{mA}, I_2 = 345.8\,\text{mA}$.

We can also apply Kirchhoff's law to establish an equivalent relationship between type Y circuits and triangular circuits. The diagram of the Y-type circuit and the triangular circuit is shown in Fig. 5.39.

Readers are invited to use Kirchhoff's law to prove that they have the following relationship:

$$R_a = \frac{R_1 R_3}{R_1 + R_2 + R_3} \qquad (5.21)$$

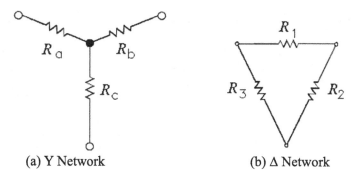

(a) Y Network   (b) Δ Network

Fig. 5.39 T or Y network and π (pi) or Δ (delta) network

$$R_b = \frac{R_1 R_2}{R_1 + R_2 + R_3} \tag{5.22}$$

$$R_c = \frac{R_2 R_3}{R_1 + R_2 + R_3} \tag{5.23}$$

or

$$R_1 = \frac{R_a R_b + R_b R_c + R_c R_a}{R_c} \tag{5.24a}$$

$$R_2 = \frac{R_a R_b + R_b R_c + R_c R_a}{R_a} \tag{5.24b}$$

$$R_3 = \frac{R_a R_b + R_b R_c + R_c R_a}{R_b} \tag{5.24c}$$

Example 5.9 Let us consider a bridge circuit (Fig. 5.40). Find R_t at terminals a and d.

Fig. 5.40 Bridge circuit

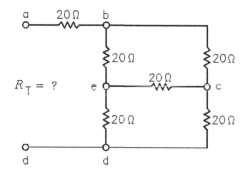

Solution: Convert the Y network (b-e, e-c, e-d) to the equivalent Δ network.

Using (5.24), $R_1 = 60\Omega$, $R_2 = 60\Omega$, $R_3 = 60\Omega$, Now, we can redraw the Y circuit as a Δ circuit and reconnect it to the original circuit (Fig. 5.41), Reduce and simplify the circuit. Note that the 20 and 60 W branches are in parallel in Fig. 5.41. Refer to Figs. 5.41 and 5.42 for redrawing the circuit in each step below.

Finally, $R_t = 40\Omega$.

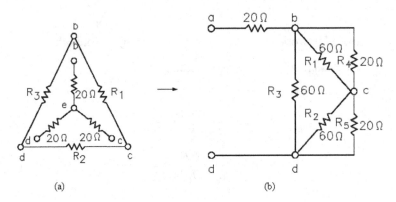

(a)   (b)

Fig. 5.41 Y—Δ Redrawn circuit

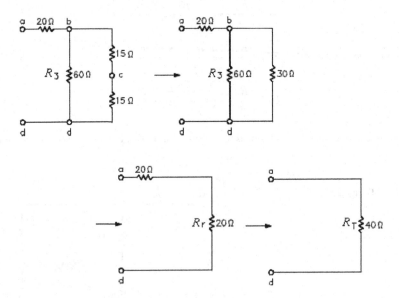

Fig. 5.42 Steps to simplify redrawn circuit

5.6.4 Inductors

Inductor is a component that can convert electrical energy into magnetic energy and store it. It is usually a wire wound around a core of permeable material. A magnetic field is produced when current flows through a wire. If the two circuits are arranged as shown in Fig. 5.43, a magnetic field will be generated around wire a, but electromotive force (EMF) will not be induced by wire B, because there is no relative movement between the magnetic field and wire B.

If we turn on the switch now, the current will stop flowing in wire A and the magnetic field will disappear. When the magnetic field collapses, it moves relative to wire B. When this happens, electromotive force is induced in wire B.

Inductor, also known as coil or reactor, is a passive double ended electrical component that can resist the change of current passing through it. It consists of conductors such as wires, usually wound into coils. As long as the current flows as shown in Fig. 5.43, energy will be stored in the magnetic field of the coil.

This is Faraday's law mentioned earlier, the area of the closed coil on the right is unchanged, changing the strength of the magnetic field, the flux will also change, electromagnetic induction will occur. The polarity of voltage is pointed out by the Lenz's law: the magnetic field that senses the current should hinder the change of the primary flux. The direction of the induced current can be determined by the right-handed rule, and then the polarity of the induced voltage can be determined. Therefore, inductive voltage is also known as induced EMF (Fig. 5.44).

To generate inductance voltage, three conditions are required:

(1)   has a conductor (coil);
(2)   has a magnetic field;
(3)   Conductors and magnetic fields have relative motion or relative changes.

The faster the relative movement or change, the greater the inductance voltage. The magnitude of the inductive voltage is also related to the number of turns in coil A and coil B, as shown in Fig. 5.45.

Self-induced EMF is another electromagnetic induction phenomenon. Inductor L, shown in Fig. 5.45, is in a circuit with an electric current in the starting moment and a magnetic field stored in the inductor. When switch S is turned on, the circuit current

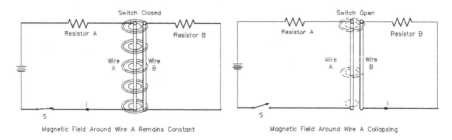

Fig. 5.43 Induced EMF

Fig. 5.44 Induced EMF in coils

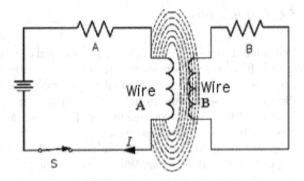

Fig. 5.45 Self-induced EMF

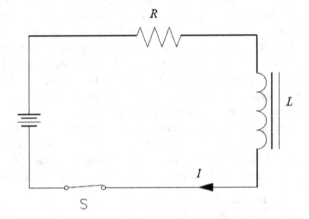

decreases and the magnetic field in the inductor decreases, creating a Self-induced EMF in the coil.

The magnitude of the inductive voltage is proportional to the derivative of the current over time, and the proportional constant is called inductor. Inductors are therefore used to describe the physical amount of inductor-generated inductive voltage. The magnitude of the induction voltage is shown in Eq. (5.25).

$$V_c = -L\frac{dI}{dt} \tag{5.25}$$

where, V_c is induced voltage (volts), V. The unit of L is V·s/A. The minus sign shows that the CEMF is opposite in polarity to the applied voltage.

Inductors in series are combined like resistors in series.

$$L_{eq} = L_1 + L_2 \tag{5.26}$$

Inductors in parallel are combined like resistors in parallel

$$\frac{1}{L_{eq}} = \frac{1}{L_1} + \frac{1}{L_2} \tag{5.27}$$

Inductors will store energy in the form of a magnetic field. Because inductors can store energy, Circuits containing inductors will behave differently from a simple resistance circuit. In inductor circuits, a drop or increase in current presents a pattern of exponential function changes, as shown in Figs. 5.46 and 5.47.

We define the time it takes for the current to rise to 63.2% of the maximum, or 63.2% of the drop maximum, as a time constant for the circuit, as shown in Fig. 5.48.

The time constant of the circuit is proportional to the inductor's inductor and inversely proportional to the resistance of the circuit, i.e.:

$$t_L = \frac{L}{R} \tag{5.27}$$

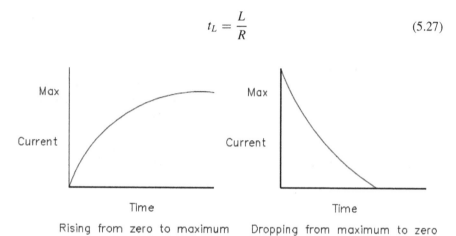

Rising from zero to maximum   Dropping from maximum to zero

Fig. 5.46 Current in an inductor circuit

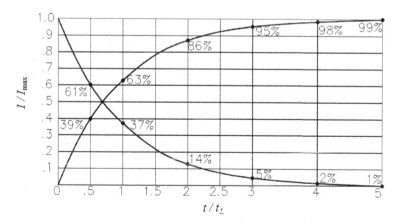

Fig. 5.47 Time constant

Fig. 5.48 Voltage applied to an inductor

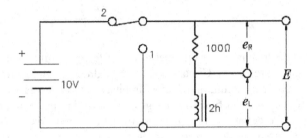

It is generally assumed that after 5 time constants, the current reaches a stable value.

Let's look at an example to analyze the circuits shown in Fig. 5.48 to see how the circuit parameters of the inductor change over time. At the beginning, assuming that the switch is in the position"1", so there is no current in the inductor, when we put the switch to the position "2", the 10 V power supply tries to establish a 10 V/100 Ω in the circuit but when the current increases, the induced EMF in the inductor attempts to prevent the current from increasing, and after 5 times of the time constant, a steady-state current is reached. When the switch is put back to the position "1" again, the magnetic field in the inductor decreases and the inductive voltage attempts to maintain the original current, which is reduced to zero after a 5 time constants. The current and voltage changes are shown in Fig. 5.49.

5.6.5   Capacitor

A capacitor is a device that holds an electric charge. Figure 5.50 is a diagram of the capacitor connected to a DC power supply charging, consisting of two metal plates and a medium (insulating material) between them.

When the power supply is connected, the electrons in the A-plate move to the positive pole of the power supply, and the positive charge in the B-plate moves to the negative pole of the power supply, leaving a positive charge in the A-plate, and a negative charge in the B-plate, in A, an electrostatic field is formed between the plates, storing electrical energy.

Since there is little charge that can cross between the plates, the capacitor will remain charged even if the battery is removed. Since the charges on the opposite plates attract each other, they will tend to oppose any change in charge. In this way, the capacitor will resist any voltage changes sensed on it.

If both ends of the charged capacitor are connected by a conductor, the discharge process begins, as shown in Fig. 5.51. After discharge, the two plates of the capacitor are returned to a neutral state.

In electricity, the capacitor's ability to store charges, given the voltage difference, is called capacitance. It is the ratio of the voltage V between the poles and the charge Q of the capacitor. The international unit of capacitors is Faraday (F). A farad is the

Fig. 5.49 The current and
voltage changes when a
voltage is applied to an
inductor

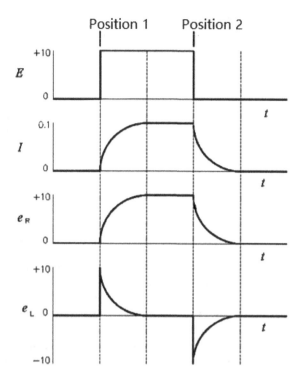

Fig. 5.50 Charging a
capacitor

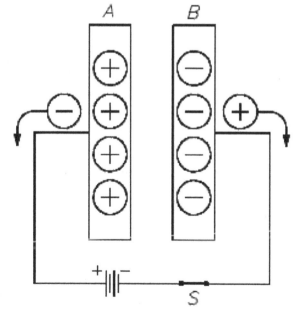

Fig. 5.51 Discharging a
capacitor

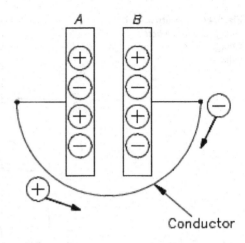

Conductor

capacitance that will store one coulomb of charge when one volt is applied across
the plates of the capacitor [20].

The capacitor's storage capacity is related to three factors: the area of the electrode
plate, the distance between the plates, and the material of the medium. The dielectric
constant (K) describes the ability of the dielectric to store electrical energy. Air is
used as a reference and is given a dielectric constant of 1. Therefore, the dielectric
constant is dimensionless. Some other dielectric materials are paper, Teflon, Bakelite,
mica, and ceramic.

$$K = \frac{\varepsilon}{\varepsilon_0} \tag{5.28}$$

where, ε_0 is vacuum dielectric constant, $\varepsilon_0 = 8.85 \times 10^{-12}$ F/m. Equation (5.29)
illustrates the formula to find the capacitance of a capacitor with two parallel plates.

$$C = K\frac{A}{d}\varepsilon_0 \tag{5.29}$$

where, A is area, d is distance between two plates.

Example 5.10 A capacitor at 4 V DC, 8C of electric charge is stored, find the
capacity.

Solution: 2F.

Example 5.11 What is the capacitance if the area of a two plate mica capacitor is
0.0050 m^2 and the separation between the plates is 0.04 m?

Solution: 7.74 pF. (1 pF $= 10^{-12}$F.)

Capacitors are usually distinguished by the type of medium, commonly used media are air, mica, ceramics, etc. Table 5.2 is an introduction to capacitors for several common media.

Capacitors can also be used in series or in parallel, and the equivalent capacitor calculation formula for two capacitors in series is:

$$\frac{1}{C_{eq}} = \frac{1}{C_1} + \frac{1}{C_2} \tag{5.30}$$

The equivalent capacitor calculation formula for paralleling the two capacitors is:

$$C_{eq} = C_1 + C_2 \tag{5.31}$$

Example 5.12 Find the total capacitance of Fig. 5.52.

Solution: Series of C_3 and C_4, is 2.44 pF, total capacitance is 27.44 pF.

When a capacitor is connected to a DC voltage source, it charges very rapidly. If no resistance were present in the charging circuit, the capacitor would become charged almost instantaneously. Resistance in a circuit will cause a delay in the time for charging a capacitor. The exact time required to charge a capacitor depends on the resistance (R) and the capacitance (C) in the charging circuit. Equation (5.32) illustrates this relationship.

$$t_c = RC \tag{5.32}$$

Here's a look at the concept of the charge time constant for capacitors. The charging time constant of the capacitor is the time required when the end voltage of the capacitor reaches 0.632 times of the maximum value. Figure 5.53 is the charging curve of the capacitor and is generally considered to be full after reaching a charging time constant of 5 times.

5.6.6 DC Generators

DC equipment is often used in today's technology. Before introducing the construction and operation of these devices, it is necessary to understand some common terms.

DC motor is an energy transfer device. These machines can be used as both motors and generators. DC motor and generator have the same basic structure, mainly different in energy conversion. In order to better understand the operation and structure of DC motor, several basic terms must be understood.

Figure 5.54 is a diagram of a DC generator. Because the voltage drawn by brush A through the commutator is always the voltage in the coil that cuts the N-pole magnetic line, so brush A always has positive polarity and the end of brush A can get

Table 5.2 Types of capacitors

Media	Manufacture method	Scope of capacity	Features
Mica	Use foil or deposited silver on the mica sheet to make electrode plates. The assembly is dipped in epoxy	10–5000pF	The medium loss is small, the insulation resistance is large, the temperature coefficient is small, and it is suitable for high frequency circuit
Paper	Use two pieces of foil to make electrodes, clamped in extremely thin capacitive paper, rolled into cylindrical or flat-column cores, and sealed in a metal shell or insulated material shell	0.001–1μF	Small size, capacity can be larger. However, the inherent inductors and losses are relatively large, and it is more suitable for low frequencies
Ceramic	Using ceramics as medium, deposited silver on both sides of the ceramic substrate and are then stacked together to make a capacitor	0.5–1600pF	Ceramic capacitors have a high dielectric constant and are available so that relatively high capacitance's can be obtained in a small physical size. and it is suitable for high frequency circuit
Electrolyte	The negatives are made of aluminum cylinders, which are filled with liquid electrolytes and inserted into a curved aluminum belt to make positive poles. DC voltage treatment is also required to form a film of oxidation on the positive plate as a medium	5–1000 μF	Electrolytic Capacitors are generally used when very large capacitance values are required
Tantalum or Niobium	Use metal palladium or vanadium as a positive pole, use liquid such as dilute sulfuric acid as a negative pole, and make a medium of oxidation film generated by the surface of tantalum or vanadium	0.01–300 μF	Small size, large capacity, stable performance, long life, large insulation resistance, good temperature characteristics which makes them suitable for use in blocking, by-passing, decoupling, filtering and timing applications

Fig. 5.52 Equivalence capacity

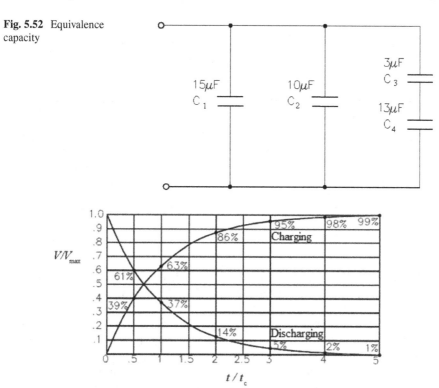

Fig. 5.53 Capacitive time constant for charging capacitor

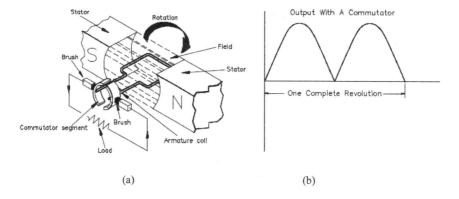

(a)   (b)

Fig. 5.54 Basic DC machine

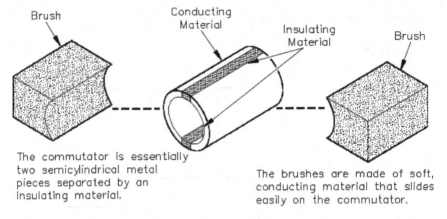

The commutator is essentially
two semicylindrical metal
pieces separated by an
insulating material.

The brushes are made of soft,
conducting material that slides
easily on the commutator.

Fig. 5.55 Commutator segments and brushes

a pulsating voltage with the same direction but a change in value. The commutator is made of conductor and insulator materials spaced between each other, as shown in Fig. 5.55.

The direction of current in the DC generator rotor can be determined by the right-hand rule shown in Fig. 5.56. The main magnetic field in the generator is usually generated electromagnetically, so there are excitation windings in the stator that need to provide current to the excitation windings. Excitation current is usually DC, which can be supplied by other DC power supplies or by self-generated DC. A self-generated DC is called a self-excitation generator, and most modern DC generators use self-excitation. Where self-excitation is not sufficient, such as when the generator is required to respond quickly with the control signal, or when the generator voltage needs to operate over a wide range, the excitation method of external DC power supply will be used.

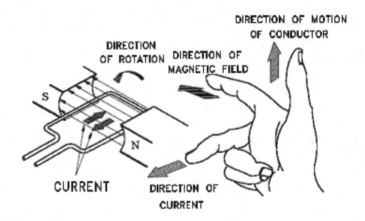

Fig. 5.56 Right-hand rule for generators

Fig. 5.57 Varying generator terminal voltage

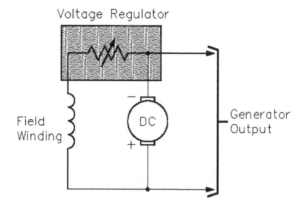

When the magnetic field and the conductor move relative in the magnetic field, and the rotation direction is that the conductor cuts off the magnetic flux line, electromotive force will be generated in the conductor. The magnitude of the induced electromotive force depends on the magnetic field strength and the flux line cutting rate, as shown in Eq. (5.33). In a given time period, the stronger the magnetic field or the more flux line cutting, the greater the induced electromotive force.

$$V_G = K_G \cdot \Phi \cdot N \qquad (5.33)$$

where, K_G is fixed constant, Φ is magnetic flux strength, N is speed in RPM, V_G is generated voltage.

How can one adjust the output voltage of a DC generator? The output voltage of the DC generator is related to three factors: (1) the number of coils in the armature, (2) the speed of the armature, and (3) the strength of the main magnetic field. The first two are not easy to adjust in the operation of the generator, so the main magnetic field intensity is generally adjusted. Figure 5.57 is a schematic for output voltage regulation for self-excitation DC motors, usually adding an adjustable resistor to regulate the current of the excitation coil to adjust the strength of the main magnetic field generated by the excitation coil.

The so-called armature is a key and pivot component in the process of converting mechanical energy and electrical energy from motor to each other. The armature consists of an armature core and an armature winding, which is the circuit part of a DC generator that produces inductive voltage for electromechanical energy conversion. The armature core is both a part of the main magnetic circuit and a supporting component of the armature winding, which is embedded in the groove of the armature core. A DC generator may be constructed in a variety of ways depending upon the relationship and location of each of the fields. Each type of construction contains certain advantages.

When the excitation winding of the generator is connected in parallel with the generator armature, the generator is called a parallel generator (Fig. 5.58a). The

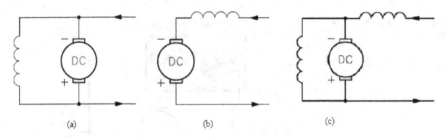

Fig. 5.58 Shunt-wound, series-wound and compounded DC generator

excitation current in the shunt generator depends on the output voltage and excitation resistance. Generally, the excitation is maintained between 0.5 and 5% of the total current output of the generator.

Parallel generators called as shunt-wound generators operate at constant speed under different load conditions, and their voltage output is much more stable than series generators. The output voltage does change. The reason for this change is that as the load current increases, the voltage drop (I_aR_a) on the armature coil increases, resulting in a decrease in the output voltage. As a result, the current passing through the magnetic field decreases, the magnetic field decreases, resulting in a further decrease in the voltage. If the load current is much higher than the design value of the generator, the output voltage drops seriously. For the load current within the design range of the generator, the output voltage drop is the smallest (Fig. 5.59).

When the excitation winding of a DC generator is connected in series with the armature, the generator is called a series-wound generator (Fig. 5.58b). For a serial excitation generator, the relationship between its output voltage and the load current is shown in Fig. 5.60.

For a series-wound excitation generator, the excitation winding and armature have the same current, so when the load current is small, the excitation current is small and the output voltage is small. The magnitude of the output voltage is positively correlated with the load current. Due to this feature of the serial excitation generator, it is generally not used in situations where the load varies greatly.

Fig. 5.59 Output voltage-verses-load current for shunt-wound DC generator

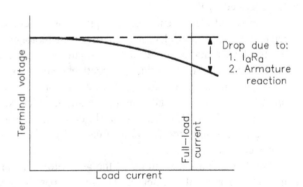

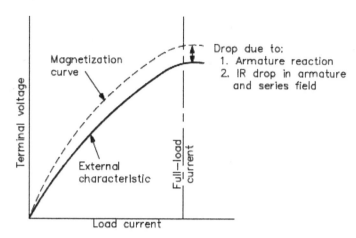

Fig. 5.60 Output voltage-verses-load current for series-wound DC generator

On the contrary, if the load consumes large current, the excitation current is also high. Therefore, the magnetic field of the series magnetic field winding is very strong and the generated voltage is very high. As shown in Fig. 5.60, in a series generator, the change of load current will seriously affect the generator output voltage. The voltage regulation of the series generator is poor, so the series generator is not used for fluctuating load. Like parallel generators, series generators also have some losses due to winding resistance and armature reaction. These losses cause the terminal voltage to be lower than the terminal voltage of the ideal magnetization curve.

The disadvantage of series and parallel generators is that the change of load current will cause the change of generator output voltage. Many applications using generators require more stable output voltages. One way to provide a stable output voltage is to use a compound generator. The compound generator has a field winding in parallel with the generator armature (the same as the parallel winding generator) and a field winding in series with the generator armature (the same as the series winding generator) (Fig. 5.58c).

The two windings of the compound generator shall be manufactured to ensure that their magnetic fields are aid or oppose to each other. If the two magnetic fields are wound so that their flux fields are opposite to each other, the generator is called differential compounded. Due to the nature of this type of generator, it is only used in special cases, which will not be discussed further in this text. If the two magnetic fields of a compounded generator are wound together to make their magnetic fields aid each other, it is called a cumulatively compounded. With the increase of load current, the current through the series field winding increases, which increases the overall magnetic field strength and leads to the increase of generator output voltage. Through proper design, the increase of series winding magnetic field strength will compensate for the decrease of parallel magnetic field strength. Therefore, the overall strength of the combined magnetic field remains almost unchanged, so the output voltage will remain unchanged. In reality, these two magnetic fields can not completely

Fig. 5.61
Voltage-verses-current for a
compounded DC generator

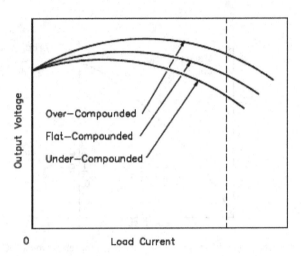

compensate each other for their magnetic field strength. The output voltage will
change from no-load state to full load state.

In an actual compounded generator, the change of output voltage from no-load to
full load is less than 5%. The generator with this characteristic is called flat compound
generator (Fig. 5.61).

In some applications, series windings are wound so that they can over compensate for changes in shunt magnetic field. Within the normal operating range of the
machine, the output increases gradually with the increase of load current. This type
of generator is called over-compounded generator.

The series winding can also be wound to make it insufficient to compensate for the
change of shunt magnetic field strength. With the increase of load current, the output
voltage decreases gradually. This type of generator is called under-compounded
generator.

5.6.7 DC Motors

DC motor is an electromechanical device that converts DC electrical energy into
mechanical energy, and is widely used in engineering because of its good speed
control performance.

Conductors (referred to as electrified conductors) that have current flowing
through the electromagnetic field are acted upon by force, which is perpendicular to
the direction of the magnetic field. This is the basic theoretical basis of DC motor,
to understand DC motor, first need to understand this principle.

Every current-carrying conductor has a magnetic field around it. The direction
of this magnetic field may be found by using the left-hand rule for current-carrying

Fig. 5.62 Left-hand rule for current-carrying conductors

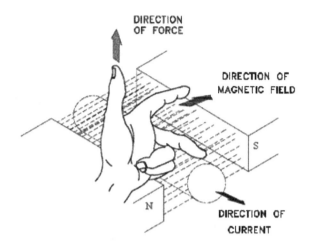

conductors. When the thumb points in the direction of current flow, the fingers will point in the direction of the magnetic field produced, as shown in Fig. 5.62.

To judge the direction of the current one uses the right hand, to judge the direction of force one uses the left hand. Once right hand, once left hand, it's extremely easy to confuse. We can also determine the force direction of the electrified conductor in another way that is easier to remember. When the electrified conductor is in a magnetic field, the magnetic field generated by the conductor itself interacts with the external magnetic field to form a composite magnetic field, as shown in Fig. 5.63. The direction of current flow through the conductor is indicated with an "x" or a ".". The "x" indicates the current flow is away from the reader, or into the page. The "." indicates the current flow is towards the reader, or out of the page.

In the upper region of Fig. 5.63a, the magnetic field generated by the conductor is in the opposite direction to the external magnetic field, weakening the external

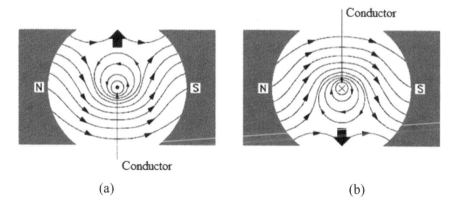

(a)   (b)

Fig. 5.63 Current-carrying conductor in a magnetic field

Fig. 5.64 Armature current
in a basic DC motor

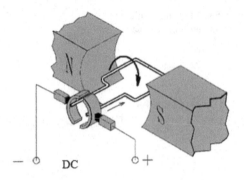

magnetic field and strengthening the external magnetic field in the lower area. The force direction of the conductor points in the direction (upward) of the weakening magnetic field. The Figure on the right, on the other hand, is forced downward.

The force that causes and sustains rotation is called torque, and the torque provided by the DC motor is used to drive the load around. If the direction of the torque changes after the rotor in Fig. 5.64 has turned 90°, the direction of the torque will change, and a current changer is required to maintain the torque in the original clockwise direction, as shown in Fig. 5.64.

When the DC motor is connected to the DC power supply, the current flows not only through the rotor coils (armatures) shown in Fig. 5.64, but also through the stator coils used to generate the magnetic field to produce the desired magnetic field.

How much torque does the coil in Fig. 5.64 produce? Torque is defined as the product of the force and rotation radius of the conductor in N m. The torque of the DC motor is calculated in the following form:

$$T = K_M \cdot \Phi \cdot I \tag{5.34}$$

where, K_M is a constant depending on physical size of motor, Φ is field flux, number of lines of force per pole, I is armature current.

When the armature is energized and rotated, an induced voltage is generated as the conductor cuts the magnetic line. The polarity of this induced voltage and the polarity of the input voltage are the opposite, so called the induced EMF. The relationship of the induced EMF, the strength of the magnetic field and the speed of the armature are:

$$V_C = K_C \cdot \Phi \cdot N \tag{5.35}$$

where, N is speed of the armature, K_C is a constant.

According to Ohm's law, the actual current in the armature is:

$$I_a = \frac{V - V_C}{R_a} \tag{5.36}$$

where, V is terminal voltage, R_a is armature resistance.

The field of a DC motor is varied using external devices, usually field resistors. For a constant applied voltage to the field (V), as the resistance of the field (R_f) is lowered, the amount of current flow through the field (I_f) increases as shown by Ohm's law in Eq. (5.37).

$$I_f = \frac{V}{R_f} \tag{5.37}$$

If the excitation resistance is reduced, the excitation current increases, thus increasing the magnetic flux, the induced EMF E_c increases according to (5.35), and the armature current decreases according to (5.36) and the torque (5.34). This results in a reduction in the speed of the motor. The speed decrease will make the reverse voltage decrease, the armature current will increase, the torque will increase, the speed will increase. This process is constantly fed back until a stable constant speed is reached.

Figure 5.65 shows schematically the different methods of connecting the field and armature circuits in a DC motor. The circular symbol represents the armature circuit, and the squares at the side of the circle represent the brush commutator system. The direction of the arrows indicates the direction of the magnetic fields.

Figure 5.65a shows the external excited DC motor. The construction of this type of DC motor prevents the field from connecting to the armature. This type of DC motor is not typically used. Figure 5.65b shows a shunt DC motor. The motor is called a " shunt " motor because the field is parallel to the armature. Figure 5.65c shows a series of DC motors. The motor field windings for a series motor are in series with the armature. Figures 5.65d and e show compounded DC motors. Build a compounded DC motor that contains both shunt fields and series fields. Figure 5.65d is called a " cumulatively-compounded " DC motor because shunts and series fields

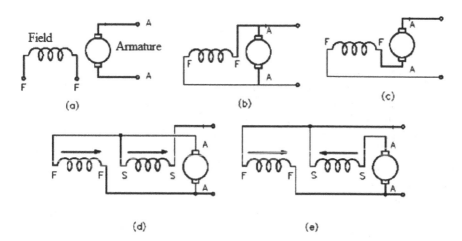

Fig. 5.65 DC motor connections

Fig. 5.66
Torque-verses-speed for a
shunt-wound DC motor

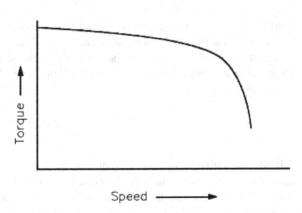

help each other. Figure 5.65e is referred to as the " differentially-compounded " DC motor because the shunts and series fields oppose to each other.

Figure 5.66 is the characteristic curve of the speed and torque of the motor. The torque of the motor decreases with the speed, which is due to the increase in the speed causing the induced EMF in the armature to increase and thus the current to decrease, depending on the type of torque as Eq. (5.34). When the speed reaches approximately 2.5 times the rated speed, the curve suddenly turns down until the torque is zero and the motor stops turning.

The characteristics of shunt-wound motor make it have good speed regulation performance. Even if the speed decreases slightly with the increase of load, it is still classified as constant speed motor. Shunt motors are used in industrial and automotive applications that require precise control of speed and torque.

Since the armature and field in a series-wound motor are connected in series, the armature and field currents become the same, and the torque can be expressed as shown in Eq. (5.38).

$$T = K \cdot I^2 \tag{5.38}$$

Figure 5.67 shows the torque and speed characteristics of series motor with constant voltage source. With the decrease of speed, the torque of series-wound motor increases sharply. When the load is removed from the series motor, the speed will increase sharply. For these reasons, series wound motors must be connected to the load to prevent damage at high speed.

The advantage of series-wound motor is that it can produce large torque and run at low speed. This is a very suitable motor for starting heavy load; It is usually used in industrial cranes and winches, where very heavy loads must move slowly, while lighter loads move faster.

The compounded motor combines the characteristics of series-wound and shunt-wound, and is more widely used in presses, shearers, reciprocating motion and other mechanical equipment.

Fig. 5.67
Torque-verses-speed for a
series-wound motor

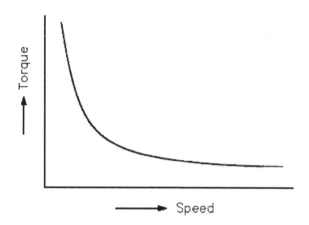

When the DC motor is first started, the armature current is larger because the induced EMF has not yet formed in the armature. The magnitude of the armature current depends on the armature resistance and is determined by (5.36). To reduce the starting current, a starting resistor is usually externally attached. Let's give an example: if a 10-horsepower DC motor has a armature resistance of 0.4Ω and a DC power supply is 260 V, the starting current reaches 650A. This current value is approximately 12 times the current value at the rated speed. Such a large current will undoubtedly cause damage to the brush, the commutator and the windings. The upper limit of the starting current is typically between 125 and 200% of the rated current. This requires the addition of a starting resistor to be possible. The magnitude of the starting resistor is calculated by:

$$R_s = \frac{V}{I_s} - R_a \tag{5.39}$$

where
Rs = starting resistance.
Et = terminal voltage.
Is = desired armature starting current.
Ra = armature resistance.

Example 5.13 If the full load current of the motor mentioned previously is 50 amps, and it is desired to limit starting current to 125% of this value, find the required resistance that must be added in series with the armature.

Solution:

$$R_s = \frac{V}{I_s} - R_a = \frac{260}{125\% \times 50} - 0.4 = 3.76\Omega$$

Starting resistors are usually designed in the starting circuit of DC motors and are commonly used as adjustable resistors. The resistance value can be controlled manually or automatically, and the moment of initial starting is set at the position where the resistance value is greatest. Slowly lower the resistance as the speed increases until the starting resistance is separated when the rated speed is reached.

5.7   Alternating Current

Alternating current is referred to simply as AC. Generally, refers to the magnitude and direction of the voltage or current with periodic changes over time, its most basic form is sine alternating current. Sine AC is the most widely used, non-sine AC can generally be transformed by Fourier to become a sine AC. The AC discussed in this text refers to sine AC.

Understanding how AC electricity is generated is necessary to understand and analyze AC circuits. When the closed coil rotates at a uniform speed around an axis perpendicular to the magnetic field in a well-stabilized magnetic field, the coil produces sine alternated current that periodically changes in size and direction, as shown in Fig. 5.68.

5.7.1   Development of a Sine-Wave Output

At the moment when the coil is in the vertical position (Fig. 5.69 0°), the coil side moves parallel to the magnetic field without cutting off the magnetic line of force. At this moment, there is no induced voltage in the circuit. When the coil rotates counterclockwise, the side of the coil will cut off the magnetic lines of force in the

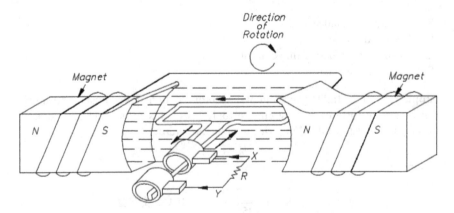

Fig. 5.68 Simple AC generator

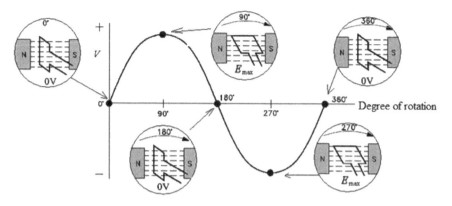

Fig. 5.69 Sine-wave output

opposite direction. The direction of the induced voltage depends on the direction of movement of the coil.

The induced voltage is connected in series so that slip ring X is positive (+) and slip ring Y is negative (−). The potential across resistor R will cause current to flow from Y to X through the resistor. When the coil is horizontal with the magnetic line of force, the current will increase until the maximum value is reached (Fig. 5.69, 90°). The horizontal coil moves perpendicular to the magnetic field and cuts the maximum number of magnetic lines of force. When the coil continues to rotate, the induced voltage and current decrease until it reaches zero, and the coil is in the vertical position again (Fig. 5.69 180°). In the other half of the rotation, the generated voltage is equal, but the polarity is opposite (Fig. 5.69 270°, 360°). Now the current through R goes from X to Y (Fig. 5.68).

As shown in Fig. 5.69, periodic polarity reversal will generate voltage. The coil rotates 360° to produce AC sine wave output.

According to Eq. (5.14), the potential difference is proportional to the derivative of Φ on time.

$$V = -n\frac{d\Phi}{dt} \tag{5.14}$$

The resulting inductive voltage can be represented by a function as follows:

$$e(t) = V_{max}\sin(\omega t) \tag{5.40}$$

where, ω is angular velocity (radians/sec)

$$\omega = 2\pi f \tag{5.41}$$

Theoretically, the magnitude of the sine AC current can be described in I_{max}, and the magnitude of the AC voltage can be described in V_{max}, i.e. peak current and

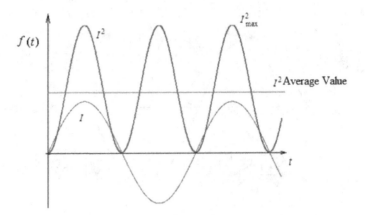

Fig. 5.70 Effective value of current

peak voltage. Or double them to represent the amplitude between the highest and lowest values. However, effective currents and voltages are usually used in practical engineering. Described by the magnitude of the DC current and DC voltage that flow through the same resistor and generate the same heat as DC, we call it the effective current and the effective voltage.

The effective value of AC can be calculated by squaring all sine wave amplitudes in a cycle, taking the average of these values, and then taking the square root. The effective value is the root of the average square of the current, which is called the root mean square or RMS value. To understand the meaning of the effective current applied to the sine wave, refer to Fig. 5.70. The values of I are plotted on the upper curve, and the corresponding values of I^2 are plotted on the lower curve. The I^2 curve has twice the frequency of I and varies above and below a new axis. The new axis is the average of the I^2 values, and the square root of that value is the RMS, or effective value, of current. The average value is $\frac{1}{2} I_{max}^2$.

The RMS value is then

$$I_{eff} = \sqrt{\frac{I_{max}^2}{2}} = \frac{\sqrt{2}}{2} I_{max} = 0.707 I_{max} \tag{5.42}$$

Example 5.14 The peak value of voltage in an AC circuit is 200 V. What is the RMS value of the voltage?

Solution: 141.4 V.

Example 5.15 The peak current in an AC circuit is 10 amps. What is the average value of current in the circuit?

Solution: 311V.

Another useful value is the average value of the amplitude during the positive half of the cycle. Equation (5.43a) is the mathematical relationship between Iav, Imax, and I.

$$I_{av} = 0.637I_{max} = 0.90I \tag{5.43a}$$

Equation (5.43b) is the mathematical relationship between V_{av}, V_{max}, and V.

$$V_{av} = 0.637V_{max} = 0.90V \tag{5.43b}$$

Phase angle is a fraction of cycle that has passed since voltage or current passed through a given value. The given value is usually zero. Refers back to Fig. 5.69, with point 1 as the starting point or zero phase. The phase at point 2 is 30°, point 3 is 60°, point 4 is 90°, and so on, until point 13 is 360° or 0. The more common term is phase difference. It can be used to describe different voltages with the same frequency, passing through zero values in the same direction at different times.

Another physical quantity that describes AC is the phase angle. Phase angles are angles at which voltage or current passes through a particular value, which is usually selected as 0. In Fig. 5.71, the inductance voltage at point 1 is 0, so the phase angle of point 1 is 0° and the point 2 is 30°, point 3 is 60°, point 4 is 90°, and until point 13 is 360°, or back to 0°.

With the concept of phase angles, you can define more useful phase differences. The so-called phase difference is the difference in phase angle between two AC voltages of the same frequency. The phase difference between the two AC voltages shown in Fig. 5.72 is 60°. If the phase difference between two currents, two voltages, or a voltage and a current is 0°, they are said to be "in-phase." If the phase difference is an amount other than zero, they are said to be "out-of-phase."

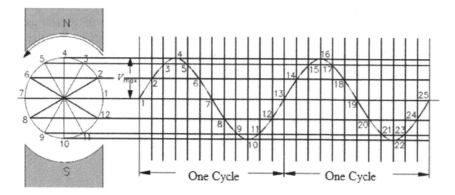

Fig. 5.71 Voltage sine wave

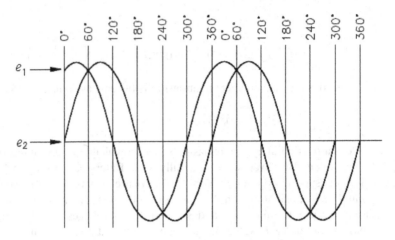

Fig. 5.72 Phase angle

5.7.2 Basic AC Reactive Components

With the basic knowledge of AC, let's discuss the response of circuit devices such as resistors, capacitors, inductors, etc. in AC circuits.

Let's start with the inductor. Any device that relies on magnetic or magnetic field operation is an inductor form. Motors, generators, transformers and coils are inductors. The use of inductors in circuits can cause current and voltage to become out-of-phase and inefficient unless corrected.

When alternated electricity passes through the circuit of the inductive coil, a self-induced EMF is generated in the circuit, which hinders the change of current and forms a sense of resistance. The greater the self-induced coefficient, the greater the self-induced EMF, and the greater the resistance. If the AC frequency is large then the current change rate is also large, then the self-induced EMF is also bound to be large, so the resistance also increases with the frequency of AC.

The opposition of the inductance to the flow of an alternating current is called *inductive reactance* (XL). Equation (5.45) is the mathematical representation of the current flowing in a circuit that contains only inductive reactance.

$$I = \frac{V}{X} \tag{5.45}$$

where
I = effective current (A).
X_L = inductive reactance (Ω).
V = effective voltage across the reactance (V).

The value of XL in any circuit is dependent on the inductance of the circuit and on the rate at which the current is changing through the circuit. This rate of change depends on the frequency of the applied voltage. Equation (5.46) is the mathematical representation for XL.

$$X_L = 2\pi f L \tag{5.46}$$

where.

$\pi = \sim 3.14$.

f = frequency (Hertz).

L = inductance (Henries).

The magnitude of an induced EMF in a circuit depends on how fast the flux that links the circuit is changing. In the case of self-induced EMF (such as in a coil), a counter EMF is induced in the coil due to a change in current and flux in the coil. This CEMF opposes any change in current, and its value at any time will depend on the rate at which the current and flux are changing at that time. In a purely inductive circuit, the resistance is negligible in comparison to the inductive reactance. The voltage applied to the circuit must always be equal and opposite to the EMF of self-induction.

As mentioned earlier, any change (rise or fall) in the current in the coil will result in a corresponding change in the magnetic flux around the coil. Since the current changes at its maximum rate when passing through zero at 90° and 270°, the change of magnetic flux is also the largest at these times. Therefore, the self-induced electromotive force in the coil is at its maximum (or minimum) value at these points. Since the current does not change when passing through the 0°, 180° and 360° peaks, the magnetic flux changes to zero at this time. Therefore, the self-induced electromotive force in the coil is zero at these points.

In an ideal pure inductor circuit, the internal resistance of the inductor is negligible. The induced EMF generated by the inductor and the voltage of the additional input are equal at all times in the opposite direction, as shown in Fig. 5.73a.

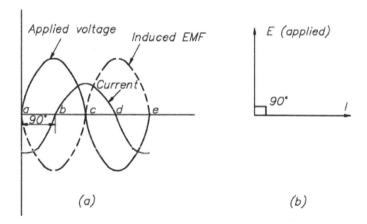

Fig. 5.73 Current, self-induced EMF, and applied voltage in an inductive circuit

According to the Lenz's Law, the induced EMF is always the opposite of the direction of the current change. So, at point a, the current is the maximum negative, the induced EMF is 0, and it starts to drop. The induced EMF is negative when the current changes in a positive direction (point a to point c). The point at which the induced EMF is zero is the point at which the current is zero over time (points a, c, e). The point with the largest current change rate is the point at which the induced EMF is greatest (b, d). The phase angle of the current is followed by a voltage input of 90°, while the induced EMF is 90° behind the current.

If the agreed input voltage is in a vertical upward direction and rotates counterclockwise (in the direction where the angle increases), the phase angle schematic of 90° behind the current is shown in Fig. 5.73b. Diagrams of this type are referred to as phasor diagrams.

Example 5.16 A 0.4H coil with negligible internal resistance, connected to an AC power supply of 220 V, 50 Hz, calculates resistance and current, and draws a phase diagram.

Solution:

$$X_L = 2\pi f L = 2 \times 3.14 \times 50 \times 0.4 = 125.6\Omega$$

$$I = \frac{V}{X_L} = \frac{220}{125.6} = 1.75(A)$$

The phase diagram is shown in Fig. 5.74b.

There are many reasons for using capacitors in AC circuits, such as transmission lines, fluorescent lamps and computer displays. Typically, these are counteracted by the inductors discussed earlier. However, when the number of capacitors far exceeds that of induction devices, we must calculate the increase or decrease of capacitance in AC circuit by artificial means.

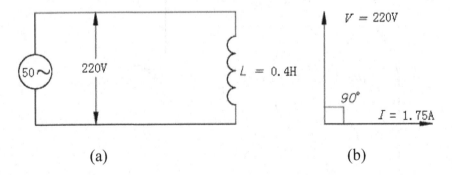

(a)   (b)

Fig. 5.74 Example of inductor

Fig. 5.75 Voltage, charge, and current in a capacitor

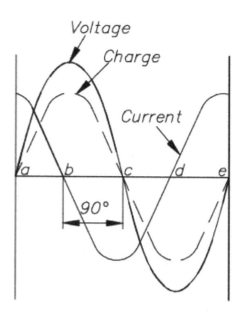

Let's look at the capacitor. The relationship between charging of a capacitor, the current flowing through, and the AC voltage entered in an AC circuit is shown in Fig. 5.75.

The current flow in Fig. 5.57 is greatest at a, c, e points, and the voltage change rate at these points is the largest. Between point a and point b, the AC voltage and charging are increasing, charging the capacitor, so the current is greater than zero, but decreasing in value. When point b is reached, the capacitor is fully charged, and the current is zero, then discharges and the current direction opposite to the voltage.

The characteristics of the capacitor are described as capacitive reactance, X_C and the unit is also Ω, it is the opposition by a capacitor or a capacitive circuit to the flow of current and the mathematical expression is as follows:

$$X_C = \frac{1}{2\pi f C} \tag{5.47}$$

where.

f = frequency (Hz).

$\pi = \sim 3.14$.

C = capacitance (F).

Example 5.17 A $10\mu F$ capacitor is connected to a 120 V, 60 Hz power source (see Fig. 5.76). Find the capacitive reactance and the current flowing in the circuit. Draw the phasor diagram.

Solution: $X_c = 1/(2 \times 3.14 \times 50 \times 10 \times 10^{-6}) = 318.5(\Omega)$

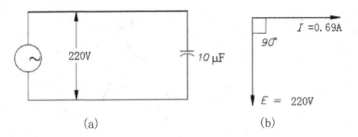

(a)   (b)

Fig. 5.76 Circuit and Phasor diagram

$I = \frac{V}{X_C} = \frac{220}{318.5} = 0.69(A)$

Phasor diagram showing current leading voltage by 90° is drawn in Fig. 5.76b.

Whenever inductive and capacitive components are used in an AC circuit, the calculation of their effects on the flow of current is important. No circuit is without some resistance, whether desired or not. Resistive and reactive components in an AC circuit oppose current flow. The total opposition to current flow in a circuit depends on its resistance, its reactance, and the phase relationships between them. Impedance is defined as the total opposition to current flow in a circuit. The impedance of a circuit is defined as the Eq. (5.48).

$$Z = \sqrt{R^2 + X^2} \qquad\qquad (5.48)$$

where

Z = impedance (Ω).

R = resistance (Ω).

X = net reactance (Ω).

The relationship between resistance, reactance, and impedance is shown in Fig. 5.77.

As shown in Fig. 5.77, the phase angle of the pure resistance is always the same as the input voltage, and we put it in the position of 0°. The current in the inductor lags by 90°, pointing in the vertical upward direction, and the current in the capacitor is

Fig. 5.77 Relationship between resistance, reactance, and impedance

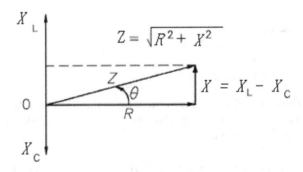

Fig. 5.78 Simple *R-L* circuit

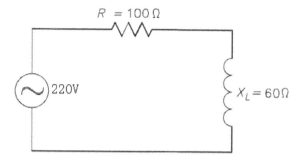

$R = 100\,\Omega$

$220V$

$X_L = 60\,\Omega$

90° ahead pointing in the vertical downward direction. The net reactance of inductive and capacitive is the difference between then.

Impedance is the vector sum of R and X in Fig. 5.77. Let's give a few examples to familiarize with impedance calculations.

Example 5.18 If a 100 Ω resistor and a 60 Ω X_L are in series with a 220 V applied voltage (Fig. 5.78), what is the circuit impedance?

Solution: $Z = \sqrt{R^2 + X_L^2} = \sqrt{100^2 + 60^2} = 116.6(\Omega)$.

Example 5.19 A 50 Ω X_C and a 60 Ω resistance are in series across a 220 V source (Fig. 5.79). Calculate the impedance.

Solution: $Z = \sqrt{R^2 + X_C^2} = \sqrt{60^2 + 50^2} = 78.1(\Omega)$.

Example 5.20 Find the impedance of a series *R–C-L* circuit, when $R = 6\Omega$, $X_L = 20\Omega$, and $X_C = 10\ \Omega$ (Fig. 5.80).

Solution: $Z = \sqrt{R^2 + (X_L - X_C)^2} = \sqrt{6^2 + (20 - 10)^2} = 11.66(\Omega)$.

With the concept of impedance, it can be used to analyze complex circuits. For R-L-C parallel circuits, the total current is:

$$I_T = \sqrt{I_R^2 + (I_C - I_L)^2} \tag{5.49}$$

Fig. 5.79 Simple *R–C* circuit

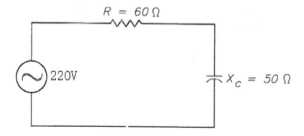

$R = 60\,\Omega$

$220V$

$X_C = 50\ \Omega$

Fig. 5.80 Simple R–C-L
circuit

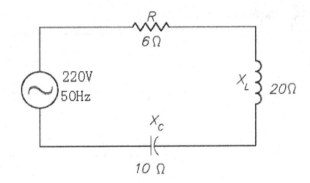

Fig. 5.81 Simple parallel
R–C-L circuit

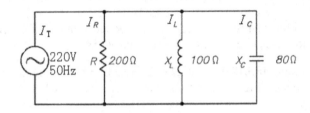

Example 5.21 A 200 Ω resistor, a 100 Ω X_L, and an 80 Ω X_C are placed in parallel across a 220 V AC source (Fig. 5.81). Find: (1) the branch currents, (2) the total current, and (3) the impedance.

Solution:

Branch currents:
$$I_R = \frac{V_T}{R} = \frac{220}{200} = 1.1(A), \ I_L = \frac{V_T}{R_L} = \frac{220}{100} = 2.2(A), \ I_C = \frac{V_T}{R_C} = \frac{220}{80} = 2.75(A).$$

Total current: $I_T = \sqrt{I_R^2 + (I_C - I_L)^2} = \sqrt{1.1^2 + (2.75 - 2.2)^2} = 1.23(A).$

Impedance: $Z = \frac{V_T}{I_T} = 178.8(\Omega).$

5.7.3   AC Power

Direct current only has one form of power, while alternating current has three different forms of power, which are related to each other in a unique relationship. In this chapter, you will learn that the power calculation method in AC circuit is different from that in DC circuit.

When discussing the power of the DC circuit, according to the Eq. (5.3), the power is:

$$P = U \cdot I \qquad\qquad (5.3)$$

Fig. 5.82 Power triangles

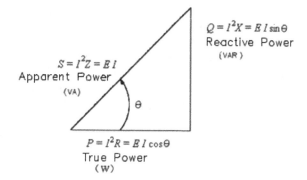

However, for AC circuits, the actual power consumed by the circuit is no longer a product of the active voltage and current because of the potential phase difference between voltage and current. We define the product of the voltage and the current as apparent power. The power triangle, as shown in Fig. 5.82, equates AC power with DC power by showing the relationship between generator output (apparent power-S) (in VA), available power (real power-P) (in W) and wasted or stored power (reactive power-Q) (in VAR). Phase angle (θ) indicates the inefficiency of the AC circuit and corresponds to the total reactive impedance (Z) to the current in the circuit.

Because there are both energy-consuming elements such as resistors and energy storage elements such as inductors and capacitors in the AC circuit, in addition to providing the available power (real power) required for their normal operation, there must also be a portion of the energy stored in inductors, capacitors and other components, called reactive power. This is why the apparent power must be greater than the real power so that the circuit or device can function properly.

The power triangle represents the ratio that can be directly used to find the efficiency level of generated power and usable power, expressed as power factor (discussed later). Apparent power, reactive power and real power or true power can be calculated by using the DC equivalent value (RMS value) of AC voltage and current components along with the power factor.

Apparent power (S) is the power delivered to an electrical circuit. The measurement of apparent power is in volt-amperes (VA).

True power (P) is the power consumed by the resistive loads in an electrical circuit. The measurement of true power is in watts.

Reactive power (Q) is the power consumed in an AC circuit because of the expansion and collapse of magnetic (inductive) and electrostatic (capacitive) fields. Reactive power is expressed in volt-amperes-reactive (VAR).

Unlike true power, reactive power is not useful power because it is stored in the circuit. This energy is stored by inductors, which expand and contract the magnetic field to keep the current constant, while capacitors charge and discharge to keep the voltage constant. Circuit inductance and capacitance consume and give back reactive power. Reactive power is a function of system current. When the magnetic field expands, the power transmitted to the inductor is stored in the magnetic field.

When the magnetic field collapses, the power transmitted to the inductor returns to the source. When the capacitor is charged, the power transmitted to the capacitor is stored in the electrostatic field. When the capacitor discharges, the power transmitted to the capacitor returns to the power supply. All the power from power supply to the capacitor or inductor is not consumed. It will return all of it to the source. Therefore, the actual power, that is, the power consumed, is zero. We know that alternating current is constantly changing; Therefore, the expansion and collapse cycles of magnetic field and electrostatic field continue to occur.

The total power provided by the power supply is the apparent power. Part of this apparent power is called true power, which is dissipated by circuit resistance in the form of heat. The remaining apparent power is returned to the power supply through the circuit inductance and capacitance.

Power factor (pf) is the ratio between true power and apparent power. The $\cos\theta$ is called the power factor (pf) of an AC circuit. It is the ratio of true power to apparent power, where θ is the phase angle between the applied voltage and current sine waves and between P and S on a power triangle. A mathematical representation of power factor is

$$\cos\theta = P/S$$

where, $\cos\theta$ = power factor (pf), P = true power (watts), S = apparent power (VA).

The power factor also determines which part of the apparent power is real power. It can range from 1, when the phase angle is 0°, to 0, when the phase angle is 90°. In inductive circuits, the current lags voltage is said to have a lagging power factor. In capacitive circuits, current leads voltage, which is said to have a leading power factor,

Example 5.22 A 200 Ω resistor and a 50 Ω X_L are placed in series with a voltage source, and the total current flow is 2 A, as shown in Fig. 5.83. Find: 1. Power factor; 2. applied voltage, V; 3. P; 4. Q; 5. S.

Solution: $\theta = \arctan\left(\frac{X_L}{R}\right) = \arctan\left(\frac{50}{200}\right) = 14°$.
Then $\cos\theta = 0.97$

Fig. 5.83 Series R-L circuit

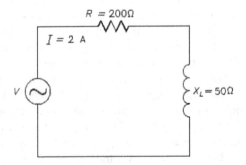

Fig. 5.84 Parallel R-L circuit

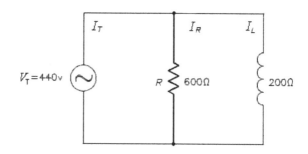

$$Z = \sqrt{R^2 + X_L^2} = \sqrt{200^2 + 50^2} = 206.16(\Omega)$$

$V = ZI = 412.3(V)$.

True power: $P = VI \cos\theta = 412.3 \times 2 \times 0.97 = 799.86(W)$.

Reactive power: $Q = VI \sin\theta = 412.3 \times 2 \times 0.242 = 199.6(VAR)$.

Apparent power: $S = VI = 412.3 \times 2 = 824.6(V \cdot A)$.

Example 5.23 A 600 Ω resistor and 200 Ω X_L are in parallel with a 440 V source, as shown in Fig. 5.84. Find: 1. I_T; 2. Power factor; 3. P; 4. Q; 5. S

Solution: $I_R = \frac{V_T}{R} = \frac{440}{600} = 0.73(A)$, $I_L = \frac{V_T}{X_L} = \frac{440}{200} = 2.2(A)$.

Then, $I_T = \sqrt{I_R^2 + I_L^2} = \sqrt{0.73^2 + 2.2^2} = 2.3(A)$

$$\theta = \arctan\left(-\frac{I_L}{I_R}\right) = \arctan\left(-\frac{2.2}{0.73}\right) = -71.5°$$

Power factor: $\cos\theta = \cos(-71.5) = 0.32$.
True power: $P = VI \cos\theta = 440 \times 2.3 \times 0.32 = 323.84(W)$.
Reactive power: $Q = VI \sin\theta = 440 \times 2.3 \times 0.948 = 959.4(VAR)$.
Apparent power: $S = VI = 440 \times 2.3 = 1012(VA)$.

Example 5.24 An 80 Ω X_c and a 60 Ω resistance are in series with a 120 V source, as shown in Fig. 5.85. Find: 1. Z; 2. I_T; 3. Power factor; 4. P; 5. Q; 6. S.

Solution: $Z = \sqrt{R^2 + X_C^2} = \sqrt{60^2 + 80^2} = 100(\Omega)$.
Then, $I_T = \frac{V_T}{Z} = \frac{120}{100} = 1.2(A)$
$\theta = \arctan\left(-\frac{X_C}{R}\right) = \arctan\left(-\frac{80}{60}\right) = -53°$
Power factor: $\cos\theta = \cos(-71.5) = 0.32$.
True power: $P = VI \cos\theta = 120 \times 1.2 \times 0.6 = 86.4(W)$.
Reactive power: $Q = VI \sin\theta = 120 \times 1.2 \times 0.798 = 114.9(VAR)$.
Apparent power: $S = VI = 120 \times 1.2 = 144(VA)$.

Fig. 5.85 Series R–C circuit

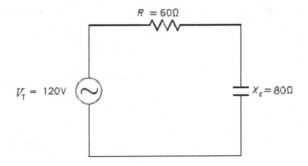

5.7.4   Three-Phase Circuits

The design of three-phase AC circuits lends itself to a more efficient method of producing and utilizing an AC voltage.

A three-phase (3ϕ) system is a combination of three single-phase systems. In a 3ϕ balanced system, power comes from a 3ϕ AC generator that produces three separate and equal voltages, each of which is 120° out of phase with the other voltages [18] (Fig. 5.86).

The power supply of the three-phase circuit is a three-phase generator, which consists of a fixer and a rotor. The inner grooves of the stator cores are symmetrically arranged with three windings, which are spaced 120° apart from each other. The rotor is a rotating electromagnet with an excitation winding around its core. When there is current in the excitation winding, a magnetic flux density is generated in the gap between the rotor and the stator, distributed along the circumference according to the sine wave.

The weight of three-phase equipment (motor, transformer, etc.) is less than that of single-phase equipment with the same rated power. They have a wide voltage range

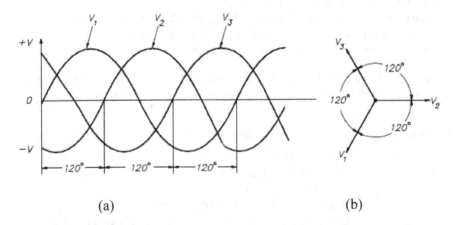

Fig. 5.86 Three-phase AC

and can be used for single-phase loads. Compared with single-phase equipment, three-phase equipment has the advantages of small volume, light weight and high efficiency.

Three phase systems can be connected in two different ways. If the three common ends of each phase are connected at a common point and the other three ends are connected to a 3ϕ line, it is called Y-connection (Fig. 5.87). If three phases are connected in series to form a closed loop, it is called triangular or Δ-connection.

Balanced loads, in a 3ϕ system, have identical impedance in each secondary winding (Fig. 5.88). The impedance of each winding in a delta load is shown as Z_Δ (Fig. 5.88a), and the impedance in a wye load is shown as Z_y (Fig. 5.88b). For either the delta or the wye connection, the lines A, B, and C supply a 3ϕ system of voltages.

In a balanced delta load, the line voltage (V_L) is equal to the phase voltage (V_ϕ), and the line current (I_L) is equal to the square root of three times the phase current ($\sqrt{3}\,I$). While in a balanced wye load, the line voltage (V_L) is equal to the square root of three times phase voltage ($\sqrt{3}\,V_\phi$), and line current (I_L) is equal to the phase current (I_ϕ).

Fig. 5.87 3ϕ AC power connections

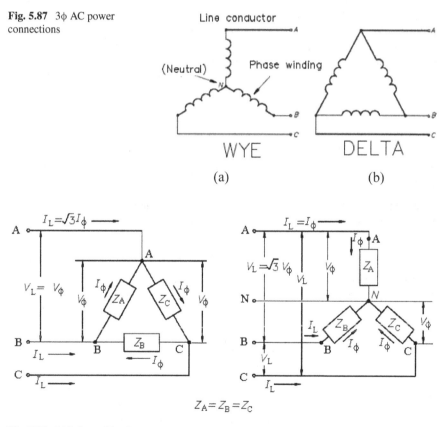

Fig. 5.88 3ϕ balanced loads

Phase power accounts for one-third of total power because the impedance of each phase of the balanced delta or wye load has an equal current.

$$P_\phi = V_\phi I_\phi \cos\theta \tag{5.50}$$

Total power

$$P_T = 3V_\phi I_\phi \cos\theta \tag{5.51}$$

In a Y-connected load, $I_L = I_\phi$, $V_\phi = \sqrt{3}/3\, V_L$, so:

$$P_T = \sqrt{3}V_L I_L \cos\theta \tag{5.52}$$

In a delta-connected load, $V_L = V_\phi$, $I_\phi = \sqrt{3}/3\, I_L$, so:

$$P_T = \sqrt{3}V_L I_L \cos\theta \tag{5.53}$$

As you can see, the total power formulas for delta- and wye-connected loads are identical. Total apparent power (S_T) in volt-amperes and total reactive power (Q_T) in volt-amperes-reactive are related to total real power (P_T) in watts. A balanced three-phase load has the real, apparent, and reactive powers given by:

$$P_T = \sqrt{3}V_L I_L \cos\theta \tag{5.53}$$

$$S_T = \sqrt{3}V_L I_L \tag{5.54}$$

$$Q_T = \sqrt{3}V_L I_L \sin\theta \tag{5.55}$$

Example 5.25 Each phase of a delta-connected 3ϕ AC generator supplies a full load current of 200 A at 440 V with a 0.6 lagging power factor, as shown in Fig. 5.89. Find: 1. V_L; 2. I_L; 3. P_T; 4. Q_T; 5. S_T.

Fig. 5.89 Three-phase delta generator

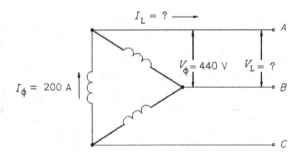

Fig. 5.90 Three-phase wye generator

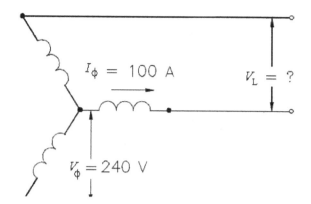

Solution: $V_L = V_\phi = 440$ V.

$I_L = \sqrt{3}\, I_\phi = 1.73 \times 200 = 346$ A.

$P_T = \sqrt{3}\, V_L I_L \cos\theta = 1.73 \times 440 \times 346 \times 0.6 = 158.2$ kW.

$Q_T = \sqrt{3}\, V_L I_L \sin\theta = 1.73 \times 440 \times 346 \times 0.8 = 210.7$ kVAR.

$S_T = \sqrt{3}\, V_L I_L = 1.73 \times 440 \times 346 = 263.4$ kVA.

Example 5.26 Each phase of a wye—connected 3ϕ AC generator supplies a 100 A current at a phase voltage of 240 V and a power factor of 0.9 lagging, as shown in Fig. 5.90. Find: 1. V_L;2. P_T;3. Q_T;4. S_T.

Solution: $V_L = \sqrt{3}\, V_\phi = 1.73 \times 240 = 415.2$ V.

$P_T = \sqrt{3}\, V_L I_L \cos\theta = 1.73 \times 415.2 \times 100 \times 0.9 = 64.6$ kW.

$Q_T = \sqrt{3}\, V_L I_L \sin\theta = 1.73 \times 415.2 \times 100 \times 0.436 = 31.3$ kVAR.

$S_T = \sqrt{3}\, V_L I_L = 1.73 \times 415.2 \times 100 = 71.8$ kVA.

An important characteristic of three-phase balance system is that the phasor sum of three-phase line or phase voltage is zero, and the phasor sum of three-phase line or phase current is zero. When the three load impedances are not equal, the phasor and neutral current (I_n) are not zero, so the load is unbalanced. When the load is open or short circuited, imbalance will occur.

If the three-phase system has unbalanced load and unbalanced power supply, the method of fixing the system is very complex. Therefore, we only consider unbalanced loads with balanced power supply.

Example: A 3ϕ balanced system, as shown in Fig. 5.91a, contains a wye load. The line-to-line voltage is 240 V, and the resistance is 40 Ω in each branch. Find line current and neutral current for the following load conditions.

1. balanced load
2. open circuit phase A (Fig. 5.91b)
3. short circuit in phase A (Fig. 5.91c)

Solution:

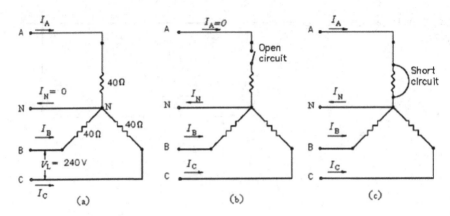

Fig. 5.91 3ϕ unbalanced load

(a) balanced load:
$$I_L = I_\phi, I_\phi = V_\phi/R_\phi, V_\phi = V_L/\sqrt{3}.$$
$$I_L = (V_L/\sqrt{3})/R_\phi = (240/1.73)/40 = 3.5 \text{ A}.$$
$$I_N = 0.$$

(b) Current flow in lines B and C becomes the resultant of the loads in B and C connected in series. $I_B = I_C$.
$$I_B = V_L/(R_B + R_C) = 240/(40 + 40) = 3 \text{ A}.$$
$$I_N = I_B + I_C = 6 \text{ A}.$$

(c) short circuit in phase A, $I_B = I_C, I_A = I_N$.
$$I_B = V_L/R_B = 240/40 = 6 \text{ A}.$$
$$I_C = I_B = 6 \text{ A}.$$
$$I_N = \sqrt{3} I_B = 1.73 \times 6 = 10.4 \text{ A}.$$

Under a fault conditions, the neutral connection in the Y-connected load will carry more current than the phase under the balanced load. An unbalanced three-phase circuit is indicated by an abnormally high current in one or more phases. If the imbalance is allowed to continue, this may cause equipment damage.

5.7.5 AC Generator

AC generators are widely used to generate AC power. To understand how these generators work, you must first understand the function of each component of the generator.

The magnetic field in the AC generator consists of coil conductors in the generator. The coil receives voltage from the power supply (called excitation) and generates magnetic flux. The magnetic flux in the magnetic field cuts off the armature to generate voltage. This voltage is ultimately the output voltage of the AC generator.

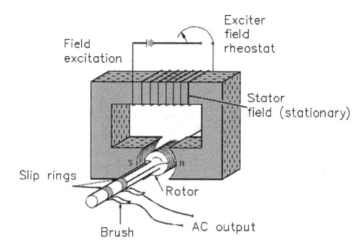

Fig. 5.92 Basic AC generator—stationary field, rotating armature

The armature is the part of the generator that generates voltage. The component consists of a number of coils large enough to carry the full load current of the generator.

The prime mover is the component that is used to drive the AC generator. The prime mover may be any type of rotating machine, such as a diesel engine, a steam turbine, or a motor.

The rotor is the rotating part of the generator, as shown in Fig. 5.92. The rotor is driven by the prime mover of the generator, which can be steam turbine, gas turbine or diesel engine. Depending on the type of generator, the component may be an armature or a magnetic field. If a voltage output is generated there, the rotor will be an armature; If magnetic field excitation is applied to the rotor, the rotor will be the magnetic field.

The stator is a stationary part. Like the rotor, this component may be an armature or a magnetic field, depending on the type of generator. If a voltage output is generated there, the stator will be an armature; If a magnetic field excitation is applied to the stator, the stator will become a magnetic field.

The slip rings are electrical connectors used to transfer power to and from the rotor. The slip ring consists of a circular conductive material connected to the rotor winding and insulated from the shaft. When the rotor rotates, the brush rides on the slip ring. The electrical connection with the rotor is made by the connection with the brush.

Slip rings are used in AC generators because the desired output of the generator is a sine wave. In a DC generator, a commutator is used to provide an output in which current always flows in a positive direction. This is not necessary for AC generator.

Therefore, an AC generator may use slip rings, which will allow the output current and voltage to oscillate between positive and negative values. The oscillation of voltage and current is in the shape of sine wave.

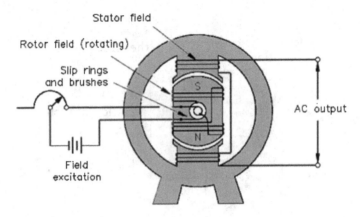

Stator field

Rotor field (rotating)

Slip rings
and brushes

AC output

Field
excitation

Fig. 5.93 Simple AC generator—rotating field, stationary armature

Figure 5.93 is an AC generator schematic in which the excitation coil is in the rotor. The coil in the rotor generates the main magnetic field when it receives an excitation current.

The strong magnetic field is generated by the current flowing through the rotor field coil. The field coils in the rotor receive excitation by using slip rings and brushes. The two brushes are spring-held and in contact with the slip ring to provide a continuous connection between the field coil and the external excitation circuit. The armature is contained in the stator windings and connected to the output. Each time the rotor rotates a full turn, a complete AC cycle is generated. The generator has many wires wound in the rotor slot.

The AC voltage generated depends on the magnetic field strength and speed of the rotor. Most generators operate at constant speed; Therefore, the voltage generated depends on the magnetic field excitation or intensity. The frequency at which the voltage is generated depends on the number of poles and the operating speed of the generator, as shown in Eq. (5.56).

$$f = nN_p/120 \tag{5.56}$$

where
f = frequency (Hz).
N_p = total number of poles.
n = rotor speed (rpm).
120 = conversion from minutes to seconds and from poles to pole pairs.
The 120 in Eq. (5.56) is derived by multiplying the following conversion factors. (60 s/1 min × 2 poles/pole pair).
Typical nameplate data of AC generator (Fig. 5.94) include: (1) Manufacturer; (2) Serial number and type number; (3) Speed (RPM), number of poles, output frequency, number of phases and maximum power supply voltage; (4) Rated capacity (kVA and kW) at specified power factor and maximum output voltage; (5) Armature and field

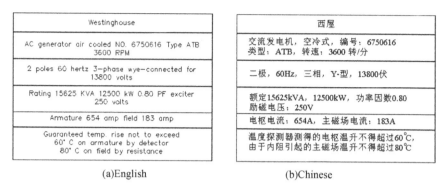

Westinghouse
AC generator air cooled NO. 6750616 Type ATB 3600 RPM
2 poles 60 hertz 3-phase wye-connected for 13800 volts
Rating 15625 KVA 12500 kW 0.80 PF exciter 250 volts
Armature 654 amp field 183 amp
Guaranteed temp. rise not to exceed 60° C on armature by detector 80° C on field by resistance

西屋
交流发电机，空冷式，编号：6750616 类型：ATB，转速：3600 转/分
二极，60Hz，三相，Y-型，13800伏
额定15625kVA，12500kW，功率因数0.80 励磁电压：250V
电枢电流：654A，主磁场电流：183A
温度探测器测得的电枢温升不得超过60℃， 由于内阻引起的主磁场温升不得超过80℃

(a)English   (b)Chinese

Fig. 5.94 AC generator nameplate ratings

current per phase; And (6) maximum temperature rise. The power (kW) rating of the AC generator is based on the ability of the prime mover to overcome generator losses and the ability of the machine to dissipate internally generated heat. The rated current of the AC generator is based on the insulation rating of the machine [19].

Most power grids and distribution systems have multiple AC generators operating simultaneously. Typically, two or more generators operate in parallel to increase the available power. Three conditions must be met before paralleling (or synchronizing) AC generators.

Firstly, their terminal voltages must be equal. If the voltages of the two generators are not equal, one of the generators can be picked up to the other as a reactive load. This will result in high current exchange between the two machines, which may damage the generator or distribution system.

Secondary, their frequencies must be equal. The frequency mismatch between the two alternators will cause the generator with the lower frequency to act as the load of the other generator (this situation is called "motoring"). This may cause overloading of the generator and distribution system.

Finnaly, their output voltage must be in phase. Phase mismatch will result in large opposing voltage. The worst-case mismatch is 180° out of phase, resulting in a opposing voltage between the two generators twice the output voltage. This high voltage will damage the generator and distribution system due to high current.

In parallel operation, a voltmeter shall be used to indicate the voltage of two parallel generators. Use the output frequency meter to accomplish frequency matching. Phase matching is accomplished by synchroscope which is a device that senses two frequencies and gives phase difference indication and relative comparison of frequency difference.

The principle of three-phase generator is basically the same as that of single-phase generator, except that there are three equidistant windings and three output voltages, with a phase difference of 120° from each other. Physically adjacent loops (Fig. 5.95) are separated by 60° rotation; However, the loops shall be connected to the slip rings in such a way that there is a current of 120° between phases.

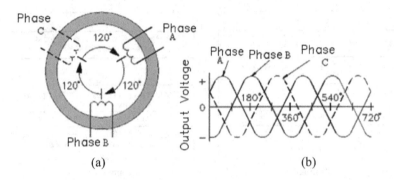

Fig. 5.95 Stationary armature 3φ generator

Individual coils of each winding are combined and represented as a single coil. The significance of Fig. 5.95 is that it shows that the three-phase generator has three independent armature windings with a phase difference of 120°.

As shown in Fig. 5.95, the armature of the three-phase generator has six leads, and the output is connected to the external load. In fact, the windings are connected together, and only three leads are led out and connected to the external load.

There are two ways to connect the three armature windings. In one connection type, the windings are connected in series or delta (Δ). In a delta connected generator, the voltage between any two phases (called line voltage) is the same as the voltage generated in any one phase. The three-phase voltage is the same as the three-wire voltage. The current in any line is three times the phase current. You can see that the delta connected generator provides an increase in current, but the voltage does not increase.

One advantage of delta connected AC generator is that if one phase is damaged or disconnected, the other two phases can still provide three-phase power supply. During three-phase operation, the capacity of the generator is reduced to 57.7%.

In another connection type, one of the leads of each winding is connected and the remaining three leads are connected to the external load. This is called a Y-connection (y).

The voltage and current characteristics of Y-connected AC generator are opposite to those of triangular connection. The voltage between any two lines in the Y-connected generator is 1.73 times the voltage of any phase, and the line current is equal to the phase current. Y-connected generator can increase voltage but not current.

One advantage of Y-type generator is that each phase only needs to carry 57.7% of line voltage, so it can be used for high-voltage power generation.

5.7.6 AC Motor

AC motors are widely used in mechanical drives in various applications. To understand the working principle of these motors, we must understand the basic working principle of AC motors.

The operating principle of all AC motors depends on the interaction between the rotating magnetic field generated by the AC current in the stator and the revolving magnetic field induced on the rotor or provided by a separate DC current source. The usable torque generated by the resulting interaction can be easily coupled to the required load of the entire facility.

Before discussing the specific types of AC motors, some common terms and principles must be introduced. Before discussing how the rotating magnetic field causes the motor rotor to rotate, we must first understand how the rotating magnetic field is generated. Figure 5.96 shows the three-phase stator to which the three-phase AC current is supplied.

The windings are connected in a Y-shape. Two windings in each phase are wound in the same direction. At any time, the magnetic field generated by a particular phase will depend on the current passing through that phase. If the current is the maximum value, the resulting field is the maximum value. If the current through this phase is zero, the generated magnetic field is zero. Since the current in the three windings is 120° out of phase, the resulting magnetic field will also be 120° out of phase. The three magnetic fields will be combined to produce a magnetic field that acts on the rotor.

In an AC induction motor, the polarity of the magnetic field induced in the rotor is opposite to that in the stator. Therefore, when the magnetic field rotates in the stator, the rotor also rotates to keep it aligned with the stator magnetic field. The rest of this chapter discusses AC induction motors [20].

Figure 5.96 is a winding diagram for a three-phase AC motor that is connected in a Y-type manner. The two coils of each phase are arranged relatively and winding in the same direction. The magnetic field intensity and current generated by the coil are proportional. When the current is zero, the magnetic field intensity is zero, and when the current reaches its peak, the magnetic field intensity also peaks.

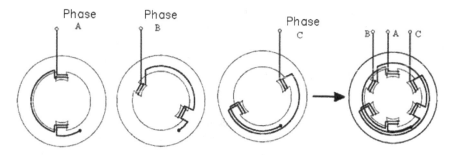

Fig. 5.96 Three-phase stator

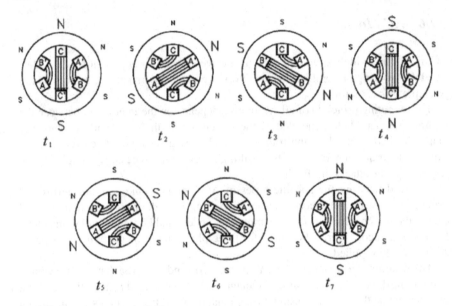

Fig. 5.97 Rotating magnetic field

From one moment to the next, the magnetic fields of each phase are combined to produce a magnetic field, and its position moves by a certain angle. At the end of an AC cycle, the magnetic field will move 360° or one revolution (Fig. 5.97). Because there is an opposing magnetic field induced on the rotor, it will also rotate through one revolution.

For purpose of explanation, rotation of the magnetic field is developed in Fig. 5.97 by "stopping" the field at six selected positions, or instances. These instances are marked at 60-degree intervals on sine waves representing the current flow in the three stages of A, B, and C. For the following discussion, when the current flow of a stage is positive, the magnetic field forms the North Pole at the poles marked A, B, and C. When the phase current is negative, the magnetic field forms the North Pole at the poles marked A, B and C'.

In Fig. 5.97, t_1 is the moment with the largest C-phase current and therefore the strongest magnetic field (the magnetic field intensity is shown in Fig. 5.98). Together with the magnetic fields generated by phases A and B, it forms a mixed magnetic field at the t_1 moment. In the Fig., we use a large font size of the letters to indicate the strength of the magnetic field, and the N with the largest font size represents the strongest North Pole. At the t_1 moment, the current in phase C is at its maximum positive value. In the same time, the current in phases A and B is half the maximum negative value. The generated magnetic field is established vertically downward, forming the maximum field strength on phase C between pole C (North Pole) and pole C' (south pole). The magnetic field is aided by the weak magnetic field developed across phase A and phase B with A 'and B' being North Pole and A and B being South Pole.

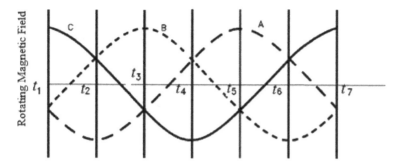

Fig. 5.98 Curves of the rotating magnetic field

At the t_2 moment the phase A has the greatest magnetic field intensity, and we can see that the mixed magnetic field rotates 60° clockwise. It was not until the t_7 moment that the magnetic field of the t_1 moment was restored. So, the mixed magnetic field rotates for a whole turn (360°) over a period of time.

When the stator winding of AC induction motor is connected with AC, a rotating magnetic field will be generated. The rotating magnetic field cuts the bars of the rotor and induces current in it due to the action of the generator. The direction of the current can be found by using the left-hand rule of the generator. This induced current will generate a magnetic field around the rotor conductor with opposite polarity to the stator magnetic field, which will attempt to align with the stator magnetic field. Since the stator magnetic field rotates continuously, the rotor cannot be aligned or locked with the stator magnetic field, so it must follow the stator magnetic field (Fig. 5.99).

The rotor of an AC induction motor can hardly rotate at the same speed as the rotating magnetic field. If the speed of the rotor is the same as that of the stator, there will be no relative motion between them, and there will be no induced electromotive force in the rotor. Without this induced electromotive force, there would be no field interaction to produce motion. Therefore, if there is relative motion between the rotor and the stator, the rotor must rotate at a speed lower than the stator.

The percentage difference between the rotor speed and the rotating magnetic field speed is called slip. The smaller the percentage, the closer the rotor speed is to the rotating magnetic field speed. The slip ratio can be calculated by Eq. (5.57).

$$s = \frac{n_S - n_R}{n_S} \times 100\% \tag{5.57}$$

where
n_S = synchronous speed (rpm).
n_R = rotor speed (rpm).
The speed of the rotating magnetic field or synchronous speed of a motor can be found by using Eq. (5.58).

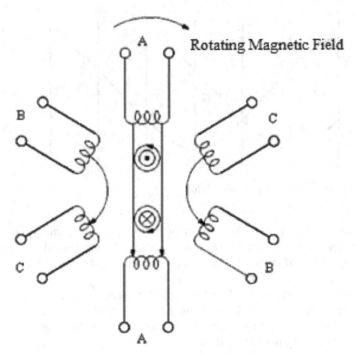

Rotating Magnetic Field

Fig. 5.99 Induction motor

$$n_s = 120f/n_p \tag{5.58}$$

Example 5.28 A two pole, 60 Hz AC induction motor has a full load speed of 3554 rpm. What is the percent slip at full load?

Solution: $n_s = 120f/N_p = 120 \times 60/2 = 3600$ rpm.
   Then $s = (n_s - n_R)/n_s \times 100\% = (3600 - 3554)/3600 \times 100\% = 1.3\%$
   The torque of an AC induction motor is dependent upon the strength of the interacting rotor and stator fields and the phase relationship between them. Torque can be calculated by using Eq. (5.59).

$$T = K\Phi I_R \cos\theta_R \tag{5.59}$$

where
   T = torque (N m).
   K = constant.
   Φ = stator magnetic flux.
   IR = rotor current (A).
   $\cos\theta R$ = power factor of rotor.

Fig. 5.100 Torque verses
slip

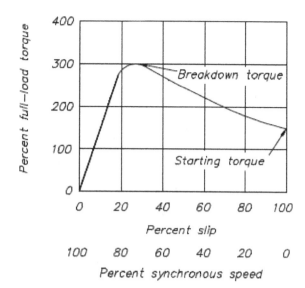

During normal operation, K, Φ, and $\cos\theta_R$ are constant in all cases, so the torque is directly proportional to the rotor current. The increase of rotor current is almost proportional to the slip. The torque variation associated with respect to slip (Fig. 5.100) shows that the torque increases linearly when slip increases from 0 to −10%. When the load and slip increase more than the full load torque, the torque will reach the maximum at about 25% slip. The maximum torque is called the breakdown torque of the motor. If the load increases beyond this point, the motor will stall and stop quickly. The breakdown torque of a typical induction motor is 200% to 300% of the full load torque. The starting torque is the torque value at 100% slip, usually 150% to 200% of the full load torque. When the rotor accelerates, the torque will increase to the breakdown torque and then decrease to the value required to carry the motor load at a constant speed (usually between 0–10%).

Various types of AC motors are used for specific applications [21]. By matching the type of motor to the appropriate application, increased equipment performance can be obtained.

The previous explanation of AC motor operation relates to induction motors. Induction motor is the most commonly used AC motor in industrial applications because of its simple structure, rugged construction and relatively low manufacturing cost. The reason why induction motors have these characteristics is that the rotor is a self-contained unit without external connection. The name of this type of motor comes from the fact that AC current is induced into the rotor through a rotating magnetic field.

The induction motor rotor (Fig. 5.101) is made of a laminated cylinder with slots on the surface. The winding in the slot is one of two types. The most commonly used is the "squirrel cage" rotor. The rotor is made of heavy copper bars and each end is connected by metal rings made of copper or brass. Due to the induction of

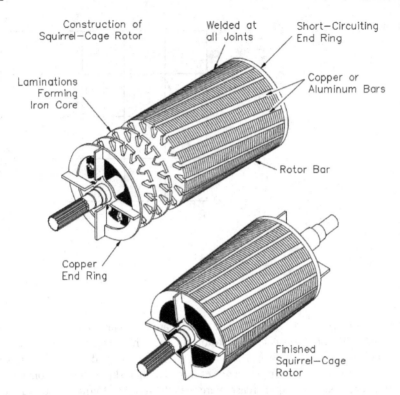

Fig. 5.101 Squirrel-cage induction rotor

low voltage in the rotor rod, there is no need for insulation between the core and the rod. The size of the air gap between the rotor rod and the stator winding required to obtain the maximum magnetic field strength is small.

If two stator windings with unequal impedance are 90° apart and connected in parallel with a single-phase power supply, the generated magnetic field will rotate. This is called phase splitting.

In split phase motors, starting windings are used. This winding has higher resistance and lower reactance than the main winding (Fig. 5.102). When the same voltage V_T is applied to the starting winding and the main winding, the current in the main winding (I_M) lags behind the current is of the starting winding I_S (Fig. 5.102). The angle between the two windings has sufficient phase difference to provide a rotating magnetic field to produce starting torque. When the motor reaches 70 to 80% of the synchronous speed, the centrifugal switch on the motor shaft opens and disconnects the starting winding.

Single phase motors are used in very small commercial applications, such as household appliances and buffers.

Synchronous motors are similar to induction motors because they all have stator windings that produce rotating magnetic fields. Unlike induction motors,

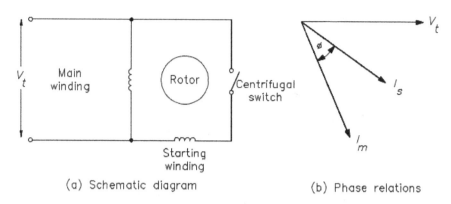

(a) Schematic diagram

(b) Phase relations

Fig. 5.102 Split-phase motor

synchronous motors are excited by external DC power supply, so slip rings and brushes are required to provide current to the rotor.

In a synchronous motor, the rotor is synchronized with the rotating magnetic field and rotates at a synchronous speed. If the synchronous motor is loaded to a position where the rotor is not synchronized with the rotating magnetic field, no torque will be generated and the motor will stop. Synchronous motors are not self-starting motors because torque is generated only when operating at synchronous speed; Therefore, the motor needs some type of device to make the rotor reach synchronous speed.

Synchronous motors use wound rotors. This type of rotor consists of coils placed in the rotor slot. Slip rings and brushes are used to supply current to the rotor. (Fig. 5.103).

The synchronous motor can be started by a DC motor on a common shaft. When the motor reaches synchronous speed, AC current is applied to the stator winding. The DC motor now acts as a DC generator to provide DC magnetic field excitation to the rotor of the synchronous motor.

The load can now be placed on the synchronous motor. Synchronous motors are usually started by squirrel cage windings embedded in the surface of the rotor poles. Then, the motor starts as an induction motor and reaches 95% of the synchronous speed. At this time, DC is applied and the motor begins to pull into the synchronous state. The torque required to pull the motor into synchronization is called pull-in torque.

We already know that the rotor of synchronous motor is synchronized with the rotating magnetic field and must continue to operate at synchronous speed under all loads. Under no-load condition, the center lines of rotating magnetic field pole and DC magnetic field pole coincide. When the load is applied to the motor, the rotor pole moves backward relative to the stator pole. The speed has not changed. The angle between the rotor and stator poles is called the torque angle (α).

If the mechanical load on the motor increases to the extent that the rotor is out of synchronization ($\alpha = 90°$), the motor will stop. The maximum torque generated by the motor without losing synchronization is called pullout torque.

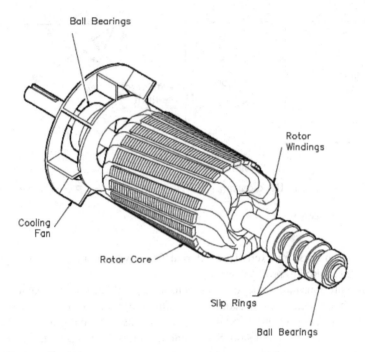

Ball Bearings

Rotor
Windings

Cooling
Fan

Rotor Core

Slip Rings

Ball Bearings

Fig. 5.103 Wound rotor

For constant load, the power factor of synchronous motor can be changed from leading value to lagging value by adjusting DC excitation. The excitation can be adjusted to make pf = 1. When the motor load is constant, when the excitation increases, the counter EMF (V_G) increases. The result is a change in phase between stator current (I) and terminal voltage (V_t), so the motor operates at a leading power factor. If we reduce the excitation, the motor will operate at a lagging power factor. Note that the torque angle, α, will also varies field excitation is adjusted to change power factor.

In large industrial complex, in order to adapt to large load and improve the power factor of transformer, synchronous motor is adopted.

5.7.7 Transformer

Transformers are widely used in AC power transmission and various control and indication circuits. Understanding the basic theory of how these components operate is necessary to understand the role of transformers in today's nuclear facilities [22].

If the flux line of magnetic field from one coil cuts off the winding of another nearby coil, a voltage will be generated in the coil. The induction of electromotive force in the coil by the magnetic flux line generated in another coil is called mutual

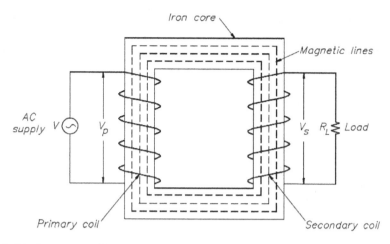

Fig. 5.104 Core-type transformer

induction. The magnitude of the induced electromotive force (EMF) depends on the relative position of the two coils.

Each winding of the transformer contains a certain number turns of wire. The turns ratio is defined as the ratio of the number of wires turns in the primary winding to the number of wires turns in the secondary winding.

The transformer coil energized from the AC power supply is called the primary winding (coil), and the coil that transmits AC to the load is called the secondary winding (coil) (Fig. 5.104).

The working principle of the transformer is that by changing the magnetic flux, energy can be transferred from one group of coils to another group of coils through magnetic induction. Magnetic flux is generated by AC power supply.

In Fig. 5.104, the primary and secondary coils are shown on the independent legs of the magnetic circuit, so we can easily understand the working principle of the transformer [20]. In fact, half of the primary and secondary coils are wound on ach of the two legs, and there is sufficient insulation between the two coils and the iron core to properly insulate the windings from each other and the iron core. Due to magnetic flux leakage, the working efficiency of transformer winding will be greatly reduced. Magnetic flux leakage is a part of the magnetic flux passing through one coil, but not through both. The greater the distance between the primary winding and the secondary winding, the longer the magnetic circuit and the greater the leakage.

When an AC voltage is applied to the primary winding, the AC current will flow first in one direction and then in the other direction, thereby magnetizing the magnetic core. This alternating magnetic flux flowing along the full length of the magnetic circuit will generate voltage in both the primary and secondary windings. Since the two windings are connected by the same magnetic flux, the induced voltage per turn of the primary winding and the secondary winding must be the same and in the same direction. This voltage is opposite to the voltage applied to the primary winding and is called counter electromotive force (CEMF).

The voltage of the transformer winding is directly proportional to the number of turns of the coil. This relationship is expressed by Eq. (5.14).

$$\frac{V_p}{V_s} = \frac{N_p}{N_s} \tag{5.14}$$

where
VP = voltage on primary coil.
VS = voltage on secondary coil.
NP = number of turns on the primary coil.
NS = number of turns on the secondary coil.

The symbol of the transformer does not indicate the phase of the secondary side voltage. The phase of this voltage depends on the direction of the windings around the core. In order to solve this problem, polarity points are used to represent the phases of primary and secondary signals. The voltage is in phase with the primary voltage (Fig. 5.105a) or 180° out of phase (Fig. 5.105b).

All transformers have copper and core losses. Copper loss is the power loss caused by the ohmic resistance of the winding in the primary and secondary windings of the transformer. A transformer without loss is called an ideal transformer. For an ideal transformer, the input power is equal to the output power, as shown in Eq. (5.60).

$$V_p I_p = V_s I_s \tag{5.60}$$

The subscript p means primary side while s means secondary side.

Efficiency of an actual transformer is the ratio of the power output to the power input, as illustrated by Eq. (5.61).

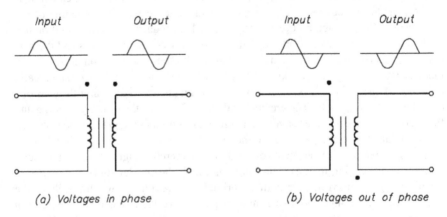

(a) Voltages in phase   (b) Voltages out of phase

Fig. 5.105 Polarity of transformer coils

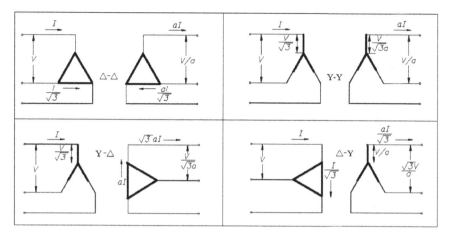

Fig. 5.106 3φ Transformer Connections

$$\eta = \frac{P_s}{P_p} \times 100\% \tag{5.61}$$

A three-phase transformer may have three separate but identical single-phase (1φ) transformers or a single 3φ unit containing three-phase windings. The transformer windings may be connected to form a 3φ bank in any of four different ways (Fig. 5.106). Figure 5.106 shows the voltages and currents in terms of applied line voltage (V) and line current (I), where the turns ratio (a) is equal to one. Voltage and current ratings of the individual transformers depend on the connections.

Example 5.29 If line voltage is 380 V to a 3φ transformer bank, find the voltage across each primary winding for all four types of transformer connections.

Δ-Δ: 380V
Y-Y: 380/1.73 = 219.7V
Y-Δ: 380/1.73 = 219.7V
Δ-Y: 440V

Example 5.30 Find the primary phase current if line current is 10.4 A in a 3 phase transformer connection.

Δ-Δ: 10.4/1.73 = 6A
Y-Y: 10.4A
Y-Δ: 10.4A
Δ-Y: 10.4/1.73 = 6A

Example 5.31 If primary line current is 20 amps, and the turns ratio is 4:1, find the secondary line current and phase current for each type of transformer connection.

Δ-Δ: 4×20 = 80A
Y-Y: 4×20 = 80A
Y-Δ: 1.73×4×20 = 138.4A
Δ-Y: 4×20/1.73 = 46.2A

All transformers have copper and core losses. Copper loss is the power lost in the main and secondary windings of transformers due to winding Omic resistors.

The copper loss, in watts, can be found using Eq. (5.62).

$$P_{\text{loss,Cu}} = I_p^2 R_p + I_s^2 R_s \tag{5.62}$$

where

I_P = primary current.
I_S = secondary current.
R_P = primary winding resistance.
R_S = secondary winding resistance.

Core losses are caused by two factors: hysteresis and eddy current losses. Hysteresis loss is that energy lost by reversing the magnetic field in the core as the magnetizing AC rises and falls and reverses direction. Eddy current loss is a result of induced currents circulating in the core.

The efficiency of a transformer can be calculated using Eqs. (5.63), (5.64), and (13.10).

$$\eta = \frac{P_s}{P_p} \times 100\% = \frac{P_s}{P_s + P_{\text{loss,Cu}} + P_{\text{loss,Te}}} \times 100\% \tag{5.63}$$

$$\eta = \frac{P_s}{V_s I_s \cos\theta + P_{\text{loss,Cu}} + P_{\text{loss,Te}}} \times 100\% \tag{5.64}$$

Example 5.32 A 5:1 step-down transformer has a full-load secondary current of 20 amps. A short circuit test for copper loss at full load gives a wattmeter reading of 100 W. If RP = 0.3 W, find RS and power loss in the secondary.

Solution: $P_{\text{loss,Cu}} = I_p^2 R_p + I_s^2 R_s = 100$ W.
    To find IP: $I_p = 20/5 = 4$A,
    To find $R_s = \frac{100 - I_p^2 R_p}{I_s^2} = \frac{100 - 4^2 \times 0.3}{20^2} = 0.24\ \Omega$,
    Power loss in secondary $= I_s^2 R_s = 20^2 \times 0.24 = 96$ W,

Example 5.33 An open circuit test for core losses in a 10 kVA transformer [Example (5.32)] gives a reading of 70 W. If the PF of the load is 90%, find efficiency at full load.

$$\eta = \frac{P_s}{V_s I_s \cos\theta + P_{\text{loss,Cu}} + P_{\text{loss,Te}}} \times 100\%$$

$$= \frac{10000 \times 0.9}{10000 \times 0.9 + 100 + 70} = 98.2\%$$

Exercises

1.   Calculate the ampere-turns for a coil with 2000 turns and a 12 mA current.
2.   A lead-acid battery reads 1.210 specific gravity. Its average full charge specific
     gravity is 1.265 and has a normal gravity drop of 133 points at full discharge.
     What is the status of the battery?
3.   Find the equivalent resistance of the following circuit.

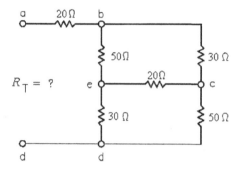

4.   Find the equivalent capacitance of the following circuit.

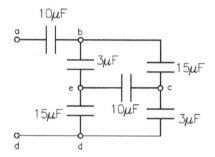

5.   What is the value of effective voltage if the maximum voltage of an AC is
     380 V?
6.   Find the total current of the following R-L-C circuit.

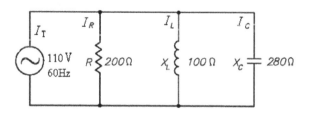

7.   AC generator supplies a 220 V as shown in the following Fig.. Find: 1. Power factor; 2. P; 3. Q; 4. S.

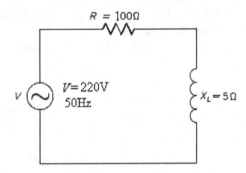

8.   Each phase of a delta-connected 3ϕ AC generator supplies a phase voltage of 240 V and a power factor of 0.8, as shown in the following Figure. Find: 1. V_L; 2. P_T; 3. Q_T; 4. S_T.

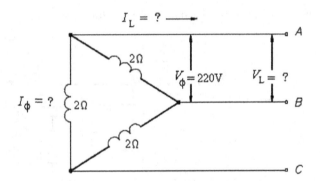

9.   Find the core loss of a transformer with apparent power of 100kVA, copper loss of 1 kW, power factor 85%, and efficiency of 98%.

10. The ampere-turns of a coil is 200At, magnetic resistance is 5×10^6 At/Wb, find the magnetic flux.

11. Calculate the impedance of the following R–C–L circuit.

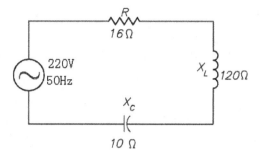

12. If line current is 10 A in a 3φ Y-△ type transformer connection with turn ratio of 4:1, find the secondary phase current.
13. A four poles, 50 Hz AC induction motor has a full load speed of 1450 rpm. What is the percent slip at full load?
14. What is the effective voltage of an AC current with maximum voltage of 380 V?
15. A plate capacitor with area of $0.8 \times 10^{-5} m^2$, the distance between two plates is 0.01 m, the dielectric constant of insulation material is 8.5, Find the capacitance.

References

1. Jones DA. Electrical engineering: the backbone of society. Proc IEE: Sci Measure Technol. 1991;138(1):1–10.
2. Purcell EM. Berkeley Physics Tutorial (SI Edition) Volume 2: Electromagnetics (Photocopy, Original2nd Edition). Machinery Industry Press; 2014.
3. Berkson W. Fields of force: the development of a world view from Faraday to Einstein. London: Routledge; 2014.
4. National Research Council. Physics through the 1990s. New York: The National Academies Press; 1998.
5. DOE Fundamentals Handbook Electrical Science. U. S. Department of Energy. Washington, D.C. DOE - HDBK - 1011/3 - 92. 1992.
6. Keithley JF. The story of electrical and magnetic measurements: from 500 BC to the 1940s. New Jersey: John Wiley & Sons; 1999.
7. Mosel PT. Electrochemical energy storage for renewable sources and grid balancing. Amsterdam: Elsevier; 2015.
8. Faraday M. The 1911 Encyclopaedia britannica a dictionary of arts, sciences, literature and general information. 11th ed. London: Encyclopedia Britannica; 1911.
9. Curie J, Curie P. Development, via compression, of electric polarization in hemihedral crystals with inclined faces. Bulletin de la Société minérologique de France. 1880;3:90–3.
10. Rowe DM. Thermoelectrics handbook: macro to nano. Oxford: Taylor & Francis; 2006.
11. Guthrie F. On a relation between heat and static electricity. London, Edinburgh Dublin Philosoph Mag J Sci. 1873;46:257–66.
12. Furlani EP. Permanent magnet and electromechanical devices: materials, analysis and applications. London: Academic Press; 2001.
13. Tipler P. Physics for scientists and engineers: electricity, magnetism, light, and elementary modern physics. 5th ed. New York: W. H. Freeman; 2004.
14. Smith RJ. Circuits, devices and systems. New York: Wiley; 1966.

15. Choudhury DR. Modern control engineering. New Delhi: Prentice-Hall of India; 2005.
16. Xiangjun H. Circuit analysis (Version 2). Higher Education Press; 2007.
17. Oldham KTS. The doctrine of description: Gustav Kirchhoff, classical physics, and the "purpose of all science" in 19th-century Germany. California: University of California, Berkeley; 2008.
18. Thomas JD. Rebhun's diseases of dairy cattle. Amsterdam: Elsevier Health Sciences; 2008.
19. Klempner G, Kerszenbaum I. Handbook of large turbo-generator operation and maintenance. New Jersey: John Wiley & Sons; 2011.
20. U. S. Department of Energy. Electrical Science. Washington, D.C.: Lulu.com; 2016.
21. Gilmour R. Canadian electrical code part I. Canadian Standards Association. C22.1–02 Safety standard for electrical installations. 19th ed. Toronto, Ontario Canada; 2002.
22. Leslie A. Geddes handbook of electrical hazards and accidents. Boca Raton: CRC Press; 1995.

Chapter 6
Instrumentation and Control

Instrument and control refers to the automatic control of the controlled variable. It will measure the signal with a given value, and then the deviation signal will be determined using control method. The output signal will be adjusted by the deviation. This chapter mainly introduces the temperature, pressure, water level, flow, location, radiation measurement principle and the system control principle [1].

6.1 Temperature Detect

The hotness or coldness of a piece of plastic, wood, metal, or other material depends upon the molecular activity of the material. Kinetic energy is a measure of the activity of the atoms, which make up the molecules of any material. Therefore, temperature is a measure of the kinetic energy of the material in question [2].

Temperature measurement based on the zeroth law of Thermodynamics: if two thermodynamic systems are with the three thermodynamic systems in thermal equilibrium, they each other must also be in thermal equilibrium. The importance of the zeroth law of thermodynamics is it gives the temperature of the definition and measurement method. The law of thermodynamic system is composed of a large number of molecules and atoms of a body or system.

The law reflects the same thermal equilibrium state of all the thermodynamic system that has common macro characteristics. This feature is by the mutual heat balance system is decided by a value equal to the numerical value of the state function and the state function is defined as the temperature and equal temperature is the necessary condition for thermal equilibrium.

Temperature can only be measured indirectly through the object with some characteristics of temperature change. The scale used to measure the value of temperature is called temperature scale. It regulates the temperature readings starting point (zero

© Tsinghua University Press 2022
J. Yu, *Fundamental Principles of Nuclear Engineering*,
https://doi.org/10.1007/978-981-16-0839-1_6

point) and measure the temperature of the basic unit. Unit of international of temperature for thermodynamic temperature scale is the Kelvin scale (K). At present the with more other scale Fahrenheit (°F), Celsius scale (°C), etc.

The temperature measurement includes contact temperature measurement and non-contact temperature measurement method. The characteristics of contact temperature measuring element method is direct contact with the object, carries out the sufficient heat exchange between the two, and finally reached a thermal equilibrium, then the physical parameters of the temperature sensing element value represents the temperature of the object. The advantage of this method is intuitive and reliable, the disadvantage is the distribution of the temperature-sensing element may affect the temperature field; the measurement error can be caused by poor contact. In addition, high temperature and corrosive medium on the temperature-sensing element will adversely affect the performance and life. The characteristics of non-contact temperature measurement is the temperature sensing element is not in contact with the object to be measured, but through the radiation heat exchange, so it can avoid contact thermometry shortcomings. The upper limit value of contactless temperature measurement method is high. In addition, the contactless temperature measurement method of thermal inertia is small to be convenient for temperature measurement of moving objects and the rapid changes in temperature. Due to the emission rate of object, the object to be measured to the influence of other mediators of the distance between the gauges, dust and water vapor, measuring error of this method is difficult to control well.

6.1.1 Resistance Temperature Detector

Whether you want to meature the temperature of the surrounding air, the water that cools car engines, or the components of nuclear facilities, you must have some ways to measure the kinetic energy of materials. Most temperature measuring devices use the energy of the material or system they monitor to increase (or reduce) the kinetic energy of the device. The ordinary household thermometer is an example. Mercury or other liquids in the bulb of a thermometer expand with the increase of kinetic energy. By observing the rising distance of the liquid in the tube, you can know the temperature of the measured object.

Because temperature is one of the most important parameters of materials, many instruments have been developed to measure temperature. One detector used is the resistance temperature detector (RTD). RTD is used in many nuclear facilities to measure the temperature of the monitored process or material [3].

RTD contains pure metals or certain alloys, and its resistance increases with the increase of temperature. On the contrary, the resistance decreases with the decrease of temperature. RTD is a bit like an electronic sensor, which converts temperature changes into voltage signals by measuring resistance. The metal most suitable for RTD sensor is pure metal with uniform quality, stable within a given temperature range, and can give reproducible resistance temperature readings [4].

Fig. 6.1 Electrical resistance–temperature curves

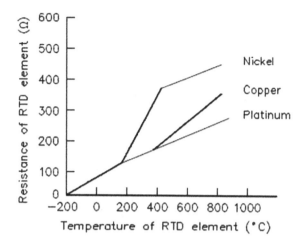

Only a few metals have the characteristics required for RTD components. RTD elements are usually composed of platinum, copper or nickel. These metals are best suited for RTD applications because of their linear resistance temperature characteristics (as shown in Fig. 6.1), high resistance coefficient and the ability to withstand repeated temperature cycles.

The resistance coefficient is the change in resistance per degree of temperature change, usually expressed as a percentage of each degree of temperature. The material used must be able to be drawn into filaments so that the element can be easily constructed.

RTD elements are usually long spring-shaped wires surrounded by insulators and encapsulated in a metal sheath. Figure 6.2 shows the internal structure of RTD [5].

This special design has a platinum element surrounded by a porcelain insulator. Insulators prevent short circuits between conductors and metal sheaths.

Inconel is a nickel iron chromium alloy, which is usually used to manufacture RTD sheath due to its inherent corrosion resistance. When placed in liquid or gas medium, the inconel sheath quickly reaches the medium temperature. A change in temperature will cause the platinum wire to heat or cool, resulting in a proportional change in resistance. The resistance change is then measured by an accurate resistance measuring device calibrated to provide the correct temperature reading. The device is usually a bridge circuit, which will be described in detail later in this chapter.

Figure 6.3 shows the RTD protection well and terminal head. The well protects the RTD from being damaged by the measured gas or liquid. At temperatures up to 1100 °C, protected wells are usually made of stainless steel, carbon steel, inconel or cast iron [6].

Fig. 6.2 Internal
construction of a typical
RTD

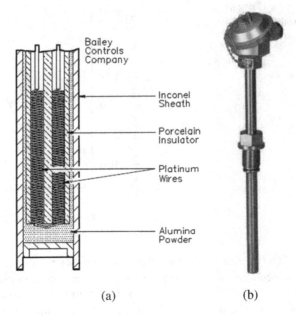

Bailey
Controls
Company

— Inconel
Sheath

— Porcelain
Insulator

— Platinum
Wires

— Alumina
Powder

(a)   (b)

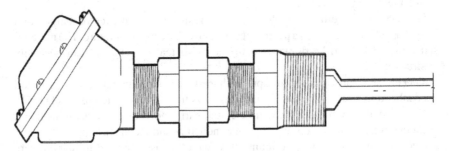

Fig. 6.3 RTD protective well and terminal head

6.1.2 *Thermocouple*

The thermocouple consists of two different wires connected at one end. When one end
of each wire is connected to the measuring instrument, the thermocouple becomes
a sensitive and high-precision measuring device. Thermocouples can be made of
several different combinations of materials. The properties of thermocouple materials
are usually determined by using this material with platinum. The most important
factor to consider when selecting a pair of materials is the "thermoelectric difference"
between the two materials. The significant difference between the two materials will
lead to better thermocouple performance. Figure 6.4 illustrates the characteristics of
more commonly used materials when used with platinum.

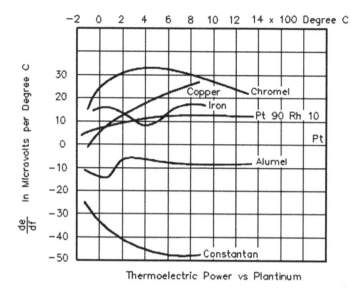

Fig. 6.4 Thermocouple material characteristics when used with platinum

In addition to the materials shown in Fig. 6.4, other materials can be used. For example, chromel constantan is suitable for temperatures up to 1200 °C; Nickel/Nickel-Molybdenum sometimes replaces Chromel–Alumel alloy; The service temperature of Tungsten-Rhenium is as high as 2800 °C. Some combinations for special applications are Chromel-White Gold, Molybdenum-Tungsten, Tungsten-Iridium, and Iridium/Iridium-Rhodium.

Figure 6.5 shows the internal structure of a typical thermocouple. The thermocouple leads are wrapped in a rigid metal sheath. The measuring joint is usually formed at the bottom of the thermocouple housing. Magnesium oxide surrounds the thermocouple wire to prevent vibration that may damage the thin wire and strengthen the heat transfer between the measuring joint and the medium around the thermocouple.

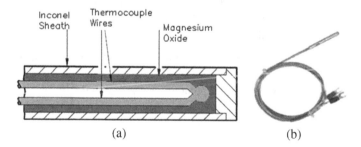

(a)   (b)

Fig. 6.5 Internal construction of a typical thermocouple

Fig. 6.6 Simple
thermocouple circuit

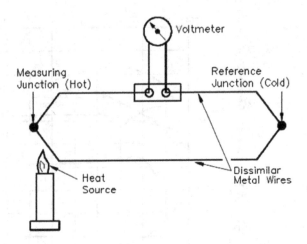

When the temperature changes, the thermocouple will cause an electric current
to flow in the connected circuit. The amount of current to be generated depends on
the temperature difference between the measurement and the reference junction, the
characteristics of the two metals used, and the characteristics of the attached circuit.
Figure 6.6 shows a simple thermocouple circuit.

The voltage generated by the measuring junction of the heating thermocouple is
greater than the voltage on the reference junction. The difference between the two
voltages is proportional to the temperature difference and can be measured on the
voltmeter (in millivolts). For the convenience of the operator, some voltmeters are
set to read directly at temperature using electronic circuits.

Other applications only provide millivolt readings. In order to convert the millivolt
reading to the corresponding temperature, you must refer to tables like the one shown
in Table 6.1. These tables are available from the thermocouple manufacturer and list
the specific temperatures corresponding to a range of millivolt readings.

Example 6.1 If the material shown in Table 6.1 is used for thermocouple, the
measured thermoelectric EMFs 7.2 mV, reference point temperature is 20 °C, what
is the measured temperature?

Solution: According to Table 6.1, reference point temperature is 20 °C, measured
temperature of 700 °C, thermoelectric EMFs is 6.486 mV. Measured temperature of
800 °C, thermoelectric EMFs is 7.563 mV.

So 7.2 mV means that the temperature of the measured point is between 700 and
800.

It can be calculated by linear interpolation, according to the linear relationship,
there are

$$\frac{t - t_1}{V - V_1} = \frac{t_2 - t_1}{V_2 - V_1}$$

Among them, the subscript 1 on behalf 700 °C, 2 on behalf of 800 °C.

Table 6.1 Temperature-vs-voltage reference (mV)

Reference temperature/°C

Temperature/°C	0	10	20	30	40	50	60	70	80	90	100
−0	0.000	−0.053	−0.103	−0.150	−0.194	−0.236					
+0	0.000	0.055	0.113	0.173	0.235	0.299	0.365	0.432	0.502	0.573	0.645
100	0.645	0.719	0.795	0.872	0.950	1.029	1.109	1.190	1.273	1.356	1.440
200	1.440	1.525	1.611	1.698	1.785	1.873	1.962	2.051	2.141	2.232	2.323
300	2.323	2.414	2.506	2.599	2.692	2.786	2.880	2.974	3.069	3.164	3.260
400	3.260	3.356	3.452	3.549	3.645	3.743	3.840	3.938	4.036	4.135	4.234
500	4.234	4.333	4.432	4.532	4.632	4.732	4.332	4.933	5.034	5.136	5.237
600	5.237	5.339	5.442	5.544	5.648	5.751	5.855	5.960	6.064	6.169	6.274
700	6.274	6.380	6.486	6.592	6.699	6.805	6.913	7.020	7.128	7.236	7.345
800	7.345	7.454	7.563	7.672	7.782	7.892	8.003	8.114	8.225	8.336	8.448
900	8.448	8.560	8.673	8.786	8.899	9.012	9.126	9.240	9.355	9.470	9.585
1000	9.585	9.700	9.816	9.932	10.048	10.165	10.282	10.400	10.517	10.635	10.754
1100	10.754	10.872	10.991	11.110	11.229	11.348	11.467	11.587	11.707	11.827	11.947
1200	11.947	12.067	12.188	12.308	12.429	12.550	12.671	12.792	12.913	13.034	13.155
1300	13.155	13.276	13.397	13.519	13.640	13.761	13.883	14.004	14.125	14.247	14.368
1400	14.368	14.489	14.610	14.731	14.852	14.973	15.094	15.215	15.336	15.456	15.576
1500	15.576	15.697	15.817	15.937	16.057	16.176	16.296	16.415	16.534	16.653	16.771
1600	16.771	16.890	17.008	17.125	17.243	17.360	17.477	17.594	17.711	17.826	17.942
1700	17.942	18.058	18.170	18.282	18.394	18.504	18.612				

$$t = t_1 + \frac{t_2 - t_1}{V_2 - V_1}(V - V_1) = 700 + \frac{800 - 700}{7.563 - 6.486} \times (7.2 - 6.486) = 766.3\,^{\circ}\mathrm{C}$$

6.1.3 Temperature Detection Circuitry

Bridge circuits are used whenever very accurate resistance measurements (such as RTD measurements) are required. Figure 6.7 shows a basic bridge circuit consisting of a voltage source, a sensitive ammeter, three known resistors R_1, R_2 and R_3 (variable) and an unknown variable resistor R_X (RTD).

Resistors R_1 and R_2 are the ratio arms of the bridge. They ratio the two variable resistors for current flow through the ammeter. R_3 is a variable resistor called a standard arm that is adjusted to match an unknown resistor. The sensing ammeter visually displays the current flowing through the bridge circuit. The analysis of the circuit shows that when R_3 is adjusted to make the ammeter reading zero current, the resistance of the two arms of the bridge circuit is the same. Equation 6.1 shows the resistance relationship between the two arms of the bridge.

$$\frac{R_1}{R_3} = \frac{R_2}{R_x} \tag{6.1}$$

Since the values of R_1, R_2 and R_3 are known, the only unknown is R_x. The R_X value of the bridge can be calculated under the condition of zero current of the ammeter. Knowing this resistance value can provide a reference point for the calibration of instruments connected to bridge circuits. The unknown resistance R_x is given by Eq. 6.2.

$$R_x = \frac{R_2 R_3}{R_1} \tag{6.2}$$

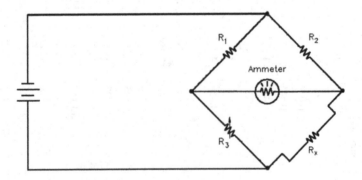

Fig. 6.7 Bridge type circuit for accurately measuring resistance

As shown in Fig. 6.7, R_x is placed in the circuit to operate the bridge, and then R_3 is adjusted to make all current flow through the arm of the bridge circuit. When this condition exists, there is no current in the ammeter, which is called bridge balance. When the bridge is balanced, the current through each arm is exactly proportional. In most cases, $R_1 = R_2$, then R_X and R_3 are equal.

When the balance exists, R_3 will be equal to the unknown resistance, even if the voltage source is unstable or inaccurate. A typical Wheatstone bridge has several dials for changing resistance. After the bridge is balanced, the dial can be read to find the value of R_3. The bridge circuit can be used to measure resistance with an accuracy of one tenth or even hundredths of a percent accuracy. When used to measure temperature, some Wheatstone bridges with precision resistors can be accurate to about +0.1 °F.

The temperature detection circuit of resistance thermometer adopts two bridge circuits (unbalanced and balanced). The unbalanced bridge circuit (Fig. 6.8) uses a millivoltmeter to calibrate according to the temperature unit corresponding to the RTD resistance.

The battery is connected to two opposite points of the bridge circuit. Connect the millivoltmeter to the other two points. The rheostat regulates the bridge current. The regulating current is distributed between the branch with fixed resistor and range resistor R_1 and the branch with RTD and range resistor R_2. With the change of RTD resistance, the voltage at point X and point Y also changes. The millivoltmeter detects voltage changes caused by uneven current distribution in the two branches. Since the only variable resistance value is the RTD resistance value, the instrument can be calibrated in temperature units.

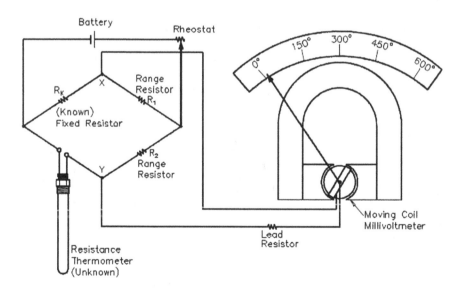

Fig. 6.8 Unbalanced bridge circuit

The balanced bridge circuit (Fig. 6.9) uses a galvanometer to compare the RTD resistance with the resistance of the fixed resistor. When the resistance of the arms is not equal, the pointer used by the galvanometer deflects on either side of zero. Adjust the resistance of the sliding contact wire until the ammeter indicates zero. The sliding resistance value is then used to determine the temperature of the system being monitored.

Sliding wire resistors are used to balance the bridge arms. As long as the resistance value of the sliding contact line does not allow current to flow through the galvanometer, the circuit will remain balanced. For each temperature change, there is a new value; therefore, the slider must be moved to a new position to balance the circuit.

Figure 6.10 is a block diagram of a typical temperature detection circuit. This indicates that the balanced bridge temperature detection circuit has been modified to eliminate the pneumometer.

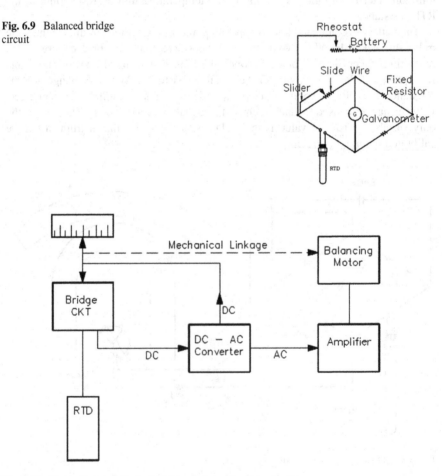

Fig. 6.9 Balanced bridge circuit

Fig. 6.10 Block diagram of a typical temperature detection circuit

Fig. 6.11 Resistance
thermometer circuit with
precision resistor in place of
resistance bulb

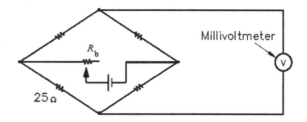

The block consists of a temperature detector (RTD), which measures temperature. The detector is resistant to the bridge network. The bridge converts this resistance to a DC voltage signal.

An electronic instrument has been developed where the DC voltage of the power meter or the bridge is converted to AC voltage. The AC voltage is then amplified to a higher (available) voltage used to drive a bi-directional motor. The bi-directional motor places the slider on the sliding line to balance the circuit resistance.

If the RTD is opened in an unbalanced or balanced bridge, the resistance will be infinite and the meter will indicate a very high temperature. If it becomes shorted, the resistance will be zero and the meter will indicate a very low temperature.

When calibrating the circuit, replace the resistance bulb with a precise resistor for the known value, as shown in Fig. 6.11. The battery voltage is then adjusted by varying R_b until the meter indicates that the known resistance is correct.

6.2 Pressure Detector

Many processes are controlled by measuring pressure. This chapter describes the detectors associated with measuring pressure.

Pressure gauges are used to measure the pressure of gas or liquid industrial automation instruments, which work according to the type of liquid column, elasticity type, load type and electrical type.

6.2.1 Bellows-Type Detectors

Pressure sensing elements that are extremely sensitive to low pressure are required and power is provided for the activation of recording and indication mechanisms, thus developing metal bellows pressure sensing elements. Metal bellows are most accurate when measuring pressures from 0.5 to 75 psig. However, when used with heavy range springs, some bellows can be used to measure pressures over 1000 psig. Figure 6.12 shows a basic metal bellows pressure sensing element.

Fig. 6.12 Basic metallic bellows

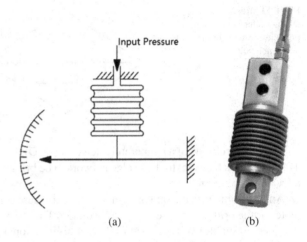

Input Pressure

(a)   (b)

The bellows are a single-piece, collapsible, seamless metal unit with deep folds formed by very thin wall tubes. The bellows range in diameter from 0.5 to 12 in and can be up to 24 folds. System pressure is applied to the internal volume of the bellows. The bellows will expand or contract due to changes in the inlet pressure of the instrument. The mobile end of the bellows is connected to the mechanical connection assembly. When the bellows and connecting components move, either an electrical signal is generated or a direct pressure indication is provided. The flexibility of metal bellows is similar in character to that of a helical, coiled compression spring. The relationship between load increment and deflection is linear under the elastic limit of the bellows. However, this relationship exists only when the bellows are compressed. It is necessary to build bellows so that all travel occurs on the compressed side of the balance point. Therefore, in practice, bellows must always be opposed by springs, and the deflection characteristic will be the resulting force of springs and bellows.

Compared to spring tubes and diaphragms, bellows are better suited for low pressure (3000 ~ 0.5 MPa) and can be used with rotating pointer or recording pen forces is relatively large. When a spring is added, the pressure range of the multi-layer bellows can reach MPa (measuring the secondary side steam pressure of the pressurized water reactor nuclear power plant).

6.2.2 Bourdon Tube-Type Detectors

The bourdon tube pressure instrument is one of the oldest pressure sensing instruments in use. The bourdon tube [7] (Fig. 6.13) consists of a thin-walled tube that is amputated flattened diametrically on opposite sides, forming a cross-sectional area elliptical in shape with two long flat edges and two short round faces. The tube is bent into an arc of 270–300°. The pressure applied to the inside of the tube causes distention of the flat sections and tends to restore its original circular cross-section.

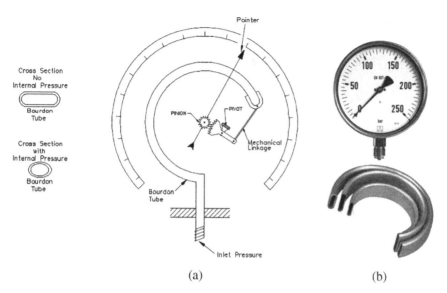

Fig. 6.13 Bourdon tube

This change in the cross-section causes the tube to straighten slightly. Because the tube is permanently fixed at one end, the tip of the tube traces a curve that is the result of changes in angular position with respect to the center. Within limits, the movement of the tip of the tube can be used to position a pointer or develop an equivalent electrical signal (discussed later in the text) to indicate the value of the internal pressure applied.

6.2.3 Resistance-Type Transducers

Any pressure detector discussed earlier can be connected to electrical equipment to form a pressure transducer. Transducers can produce changes in resistance, inductance, or capacitance.

Transducers in this category include strain gauges and moving contacts (slide wire variable resistors). Figure 6.14 illustrates a simple strain gauge. The strain gauge measures the external force (pressure) applied to the fine wire. The fine wire is usually arranged in the form of grids. Pressure changes can cause resistance changes due to deformation of the wires. The value of pressure can be found by measuring changes in wire grid resistance. Equation 6.3 shows the pressure to the resistance relationship [8].

Fig. 6.14 Strain gauge

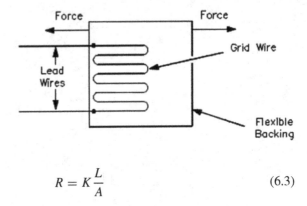

$$R = K\frac{L}{A} \tag{6.3}$$

where

  $R =$ resistance of the wire grid in ohms
  $K =$ resistivity constant for the particular type of wire grid
  $L =$ length of wire grid
  $A =$ cross sectional area of wire grid.

As the wire grid is deformed by elastic deformation, its length increases and the cross-sectional area decreases. These changes result in an increase in the resistance of the wire of the strain gauge. This change in resistance is used as a variable resistance in a bridge circuit that provides an electrical signal indicating pressure. Figure 6.15 illustrates a strain gauge pressure transducer.

Increased pressure at the inlet of the bellows causes the bellows to expand. The expansion of the bellows moves a flexible beam, and the strain gauge is connected to the beam. The motion of the beam causes the resistance of the strain gauge to change. The temperature compensation gauge compensates for the heat generated by the current flowing through the strain gauge wire. Other resistive transducers combine bellows or bourdon tubes with variable resistors, as shown in Fig. 6.16. As

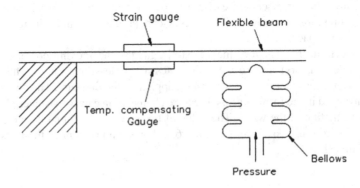

Fig. 6.15 Strain gauge pressure transducer

Fig. 6.16 Bellows
resistance transducer

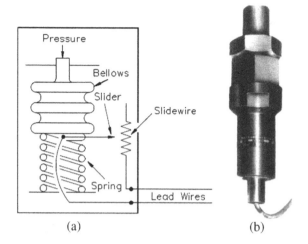

(a)   (b)

the pressure changes, the bellows expand or contract. This expansion and contraction cause the attachment slider to move along the sliding wire, increasing or decreasing resistance, indicating an increase or decrease in pressure.

Inductive transducers consist of three parts: coils, movable magnetic cores, and pressure sensing elements. The element is attached to the core, and, depending on the pressure, the element causes the core to move within the coil. The AC voltage is applied to the coil and the inductance of the coil changes as the core moves. The current passing through the coil will increase as the inductance decreases. To increase sensitivity, coils can be divided into two coils by using a center tap, as shown in Fig. 6.17. As the core coil moves in, the inductance of one coil increases and the inductance of the other coil decreases.

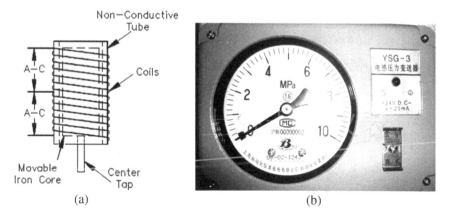

(a)   (b)

Fig. 6.17 Inductance-type pressure transducer coil

Fig. 6.18 Differential
transformer

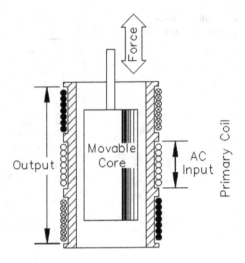

Another inductive inductance shown in Fig. 6.18 uses two coils wound on a single tube, commonly referred to as differential transformers.

The primary coil is wound in the center of the tube. The secondary coil is divided with one-half wound around each end of the tube. Each end is wound in the opposite direction, causing voltages induced to be opposite to each other. The core positioned by the pressure element is movable in the tube. When the core is lower, the lower half of the secondary coil provides output. When the core is in the upper position, the upper part of the secondary coil provides output. The size and direction of the output depends on how much the core moves out of its center position. When the core is in the middle, there is no secondary output.

The capacitive transducers shown in Fig. 6.19 consists of two flexible conductive plates and a dielectric. In this case, the dielectric is fluid.

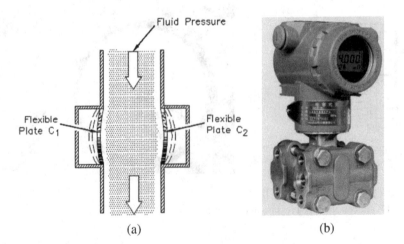

(a)   (b)

Fig. 6.19 Capacitive pressure transducer

As the pressure increases, the flexible conductive plate moves further, changing the capacitance of the transducer. This change in capacitance is measurable and proportional to changes in pressure.

6.3 Level Detector

Liquid level measurement devices are classified into two groups: (a) direct method and (b) inferred method. An example of a direct method is a dipstick in a car that measures the height of the oil in a oil pan. An example of an inferred method is the pressure gauge at the bottom of the tank, which measures the static head pressure at the height of the liquid.

6.3.1 Gauge Glass

A very simple way to measure the level of liquid in a container is the gauge glass method (Figs. 6.20). In the gauge glass method, connect the transparent tube to the bottom and top of the monitored tank (no top connection is required in a tank open to atmosphere). The height of the liquid in the pipe will be equal to the height of the water in the water tank.

Figure 6.20a shows a gauge glass for containers where the liquid is in ambient temperature and pressure conditions. Figure 6.20b shows a gauge glass, which is used for containers where the liquid is at high pressure or partially vacuumed. Note that the gauge glass in Fig. 6.20 effectively forms a "U" tube manometer, and the liquid seeks its own level due to the pressure of the liquid in the container.

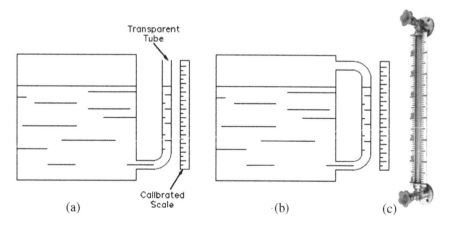

Fig. 6.20 Transparent tube

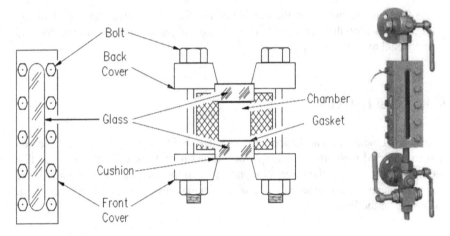

Fig. 6.21 Gauge glass

Gauge glass [9] made of tubular glass or plastic is used for service up to 2.5 MPa and 200 °C. If the level of a container with higher temperatures and pressures needs to be measured, use different types of gauge glass. The gauge glass type used in this example is made of metal with heavy glass or quartz parts for visual observation of liquid levels. The glass part is usually flat to provide strength and safety. Figure 6.21 shows a typical transparent gauge glass.

Another type of gauge glass is reflector glass (Figs. 6.22). In this type, one side of the glass part is prism-shaped. The glass is molded so that one side has a 90° angle and can run for long periods of time. Light shines on the outer surface of the glass at a 90° angle. Light passes through the glass and hits the inside of the glass at a 45° angle. The presence or absence of liquid in the chamber determines whether light is refracted into the chamber or reflected back onto the outer surface of the glass.

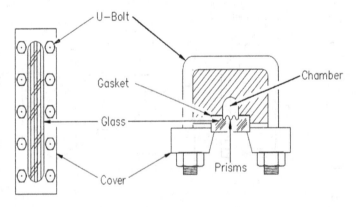

Fig. 6.22 Reflex gauge glass

When the liquid is at the middle level of the instrument glass, light encounters an air glass interface in one part of the chamber and a water glass interface in another part of the chamber. When an air-glass interface is present, light reflects back to the outer surface of the glass because the critical angle at which light is transmitted from air to glass is 42°. This causes the instrument glass to be silvery white. In the chamber section with a water glass interface, light is refracted into the chamber through a prism. Light reflected back to the outer surface of the instrument glass does not occur because the critical angle of light from the glass to the water is 62°. This causes the glass to be black because the walls of the room can be seen through water, which are painted black.

The third type of gauge glass is the refractive type (Figs. 6.23). This type is particularly useful in areas with reduced illumination: lights are usually attached to the gauge glass. The operation is based on the principle that light bends or refracts differently as it passes through various media. Light bends or refracts more in water than steam. For chamber parts containing steam, the light is relatively direct and the red lens is illuminated. For chamber parts containing water, the light bends, causing the green lens to be illuminated. The gauge portion containing water is green and the gauge portion up this level is red.

Fig. 6.23 Refraction gauge glass

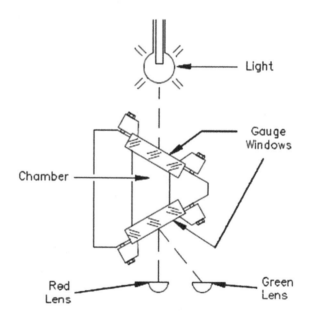

6.3.2 Ball Float

Spherical floating is a mechanism for reading liquid levels directly. The most practical design for floats is a hollow metal ball or sphere. However, there are no restrictions on the size, shape, or material used. The design consists of a ball float attached to the rod, which in turn is connected to the rotary shaft, indicating the level on the calibration scale (Fig. 6.24). The operation of the float is simple. The ball floats on the liquid in the tank. If the liquid level changes, the float follows and changes the position of the pointer connected to the axis of rotation.

The travel of the float is limited by its design, so that it is within ±30° of the horizontal plane, resulting in optimal response and performance. The actual horizontal range is determined by the length of the connecting arm.

The stuffing box is incorporated to form a waterproof seal around the shaft to prevent the container from leaking.

The diameter of the chain float gauge can be up to 12 inches and is used to exceed the small-level limit imposed by the float. The horizontal range measured will be limited only by the size of the container. The operation of a chain float is similar to a spherical float, except in the method of positioning the pointer and its connection to the position indication. The float is connected to the rotating element by a chain of weights connected to the other end to maintain chain tightness during horizontal changes (Figs. 6.25).

The magnetic bond method was developed to overcome the problem of cages and stuffing boxes. The magnetic bond mechanism consists of magnetic floats, which rise and fall with the change of level. The float travels outside a non-magnetic tube with an inner magnet attached to the level indicator. When the float rises and falls, the outer magnet attracts the inner magnet, causing the inner magnet to follow the level inside the container (Figs. 6.26).

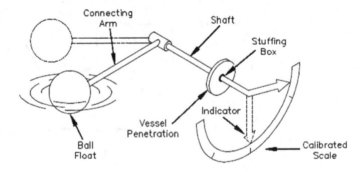

Fig. 6.24 Ball float level mechanism

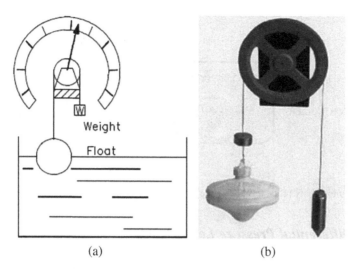

Fig. 6.25 Chain float gauge

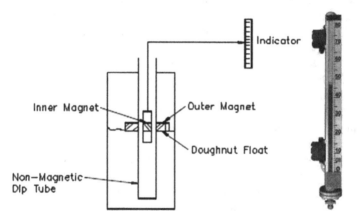

Fig. 6.26 Magnetic bond detector

6.3.3 Conductivity Probe

Figure 6.27 illustrates the conductive probe level detection system. It consists of one or more levels of detectors, an operating relay, and a controller. When a liquid comes into contact with any electrode, the current flows between the electrode and the ground. The current energies the relay, causing the relay contact to turn it on or off depending on the state of the process involved. Relays in turn will activate alarms, pumps, control valves, or all three. A typical system has three probes: a low-level probe, an high-level probe, and a high-level alarm probe.

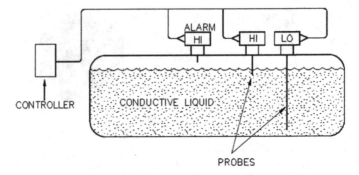

Fig. 6.27 Conductivity probe level detection system

6.3.4 Differential Pressure Level Detectors

The differential pressure (ΔP) detector method for liquid level measurement uses a ΔP detector connected to the bottom of the monitored tank. The higher pressure caused by the liquid in the tank is compared to the lower reference pressure (usually the atmosphere). This comparison occurs in the ΔP detector. Figure 6.28 illustrates a typical differential pressure detector connected to an open tank.

Tanks are open to the atmosphere: therefore, it is necessary to use only high-pressure (HP) connections on the ΔP transmitter. Low pressure (LP) side is vented to the atmospheric ventilation: therefore, the pressure difference is the water stationary head or weight of the liquid in the tank. The highest level measured by the transmitter is determined by the maximum height of the liquid above the transmitter. The minimum level that can be measured is determined by the point at which the transmitter is connected to the tank.

Not all tanks or vessels are open to the atmosphere. Many are completely enclosed to prevent steam from escaping, or to allow pressurizing the contents of the tanks. When measuring the level in a pressurized tank, the high pressure side and low pressure side of the ΔP transmitter must be connected (Figs. 6.29).

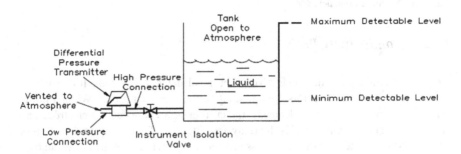

Fig. 6.28 Open tank differential pressure detector

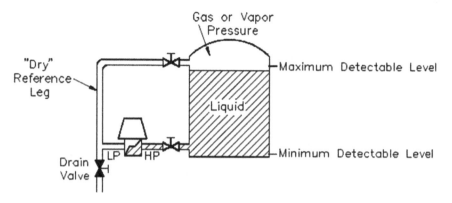

Fig. 6.29 Closed tank, dry reference leg

High-pressure connections are connected at or below the range value to be measured. The low-pressure side is connected to the "reference leg", which is connected to or above the upper range value to be measured. The reference leg is pressurized by gas or steam pressure, but liquid is not allowed in the reference leg. The reference leg must be kept dry so that there is no fluid head pressure on the low-pressure side of the transmitter.

The high-pressure side is exposed to the hydrostatic head of the liquid and the pressure of gas or steam applied to the surface of the liquid. Gas or steam pressure is also applied to both low-pressure side and high-pressure side. Therefore, the output of the ΔP transmitter is proportional to the hydrostatic head pressure (i.e. the level in the tank).

If the tank contains condensable fluid (e.g. steam), a slightly different arrangement is taken. In applications with condensable fluids, condensation in the reference legs is greatly increased. To compensate for this effect, the reference leg is filled with the same liquid as the tank. The liquid in the reference leg applies the hydrostatic head to the high-pressure side of the transmitter, and the value of this level is constant as long as the reference leg remains full. If this pressure remains constant, any change in ΔP is related to a change in the low-pressure side of the transmitter (Fig. 6.30).

The filled reference leg applies static water pressure to the high-pressure side of the transmitter, which is equal to the highest level to be measured. When the fluid level reaches its maximum, the ΔP transmitter is exposed to the same pressure on the high- and low-pressure sides; Therefore, the differential pressure is zero. As the tank level drops, the pressure applied to the low-pressure side decreases and the differential pressure increases. Therefore, the differential pressure and transmitter output are inversely proportional to the tank level.

When measuring levels at high temperatures in pressurized tanks, some additional consequences must be considered. As the temperature of the liquid in the tank increases, the density of the liquid decreases. As fluid density decreases, the liquid expands and takes up more volume. The liquid mass in the tank is the same, even if the density is small. The problem is that when the liquid in the tank is heated

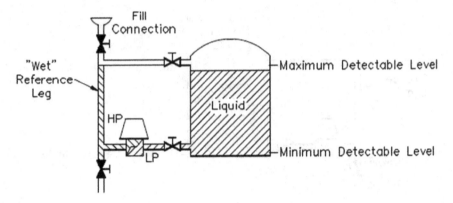

Fig. 6.30 Closed tank, wet reference leg

and cooled, the density of the liquid changes, but the reference leg density remains relatively constant, resulting in the indicated level remaining the same. The density of the reference leg fluid depends on the ambient temperature of the room in which the tank is located: therefore, it is relatively constant and independent of the tank temperature. If the liquid in the tank changes temperature and thus density, some density compensation must be used in order to accurately indicate the tank level. This is a problem when measuring the water level of the pressurized water or steam generator in a pressurized water reactor, and when measuring the water level of the reactor vessel in a boiling water reactor.

Figure 6.31 shows a typical pressurizer level system. During normal operation, the pressurizer temperature remains fairly constant. The ΔP level detector is calibrated with the pressurizer hot, and the effects of density changes do not occur. The pressurizer is not always be hot. It may be cooled due to non-operational maintenance conditions, in which case the second ΔP detector (calibrated for level measurements at low temperatures) will replace the normal ΔP detector. Density has not been truly compensated; it has actually been aligned out of the instrument by calibration.

Density compensation can also be done via electronic circuits. Some systems automatically compensate for density changes through the design of level detection circuits. Other applications manually adjust the density of inputs to the circuit because the pressurizer cools and decompresses, or during heating and pressurizing. The calibration chart can also be used to correct indications of changes in reference leg temperature.

Figure 6.32 illustrates a typical steam generator level detection arrangement [4]. The ΔP detector measures the actual differential pressure. A separate pressure detector measures the pressure of saturated steam. Because saturation pressure is proportional to saturation temperature, pressure signals can be used to correct the difference in density. Electronic circuits use pressure signals to compensate for the density difference between reference leg water and the steam generator body fluids.

As the saturation temperature and pressure increase, the density of steam generator water decreases. The ΔP detector should now indicate a higher level, even though

Fig. 6.31 Pressurizer level system

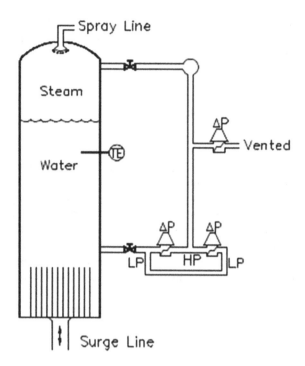

Fig. 6.32 Steam generator level system

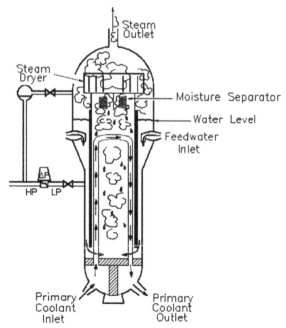

the actual ΔP has not changed. The increase in pressure is used to increase the output of the ΔP level detector to proportional to the saturation pressure to reflect changes in the actual level.

6.4   Flow Measurement

Flow measurement is an important process measurement to be considered in operating a facility's fluid systems. For efficient and economic operation of these fluid systems, flow measurement is necessary.

The principle of measuring the flow rate is the principle of the pressure difference generated by the flow of fluid through a particular component.

As shown in Fig. 6.33 to measurement of pipeline fluid mass flow (or volume). If in the middle of the pipe set a flow restrictor in the circulation area narrow, upstream and downstream flow restrictor will produce static pressure difference. If a U-shaped tube type differential pressure gauge connected, the mass flow rate can be measured by differential pressure.

According to the principle of local pressure drop introduced in the fourth chapter, there are

$$\Delta p_k = K \frac{\rho V_m^2}{2} \qquad (6.4.16)$$

In addition, mass flow rate q_m is

$$q_m = \rho V_m A \qquad (6.4)$$

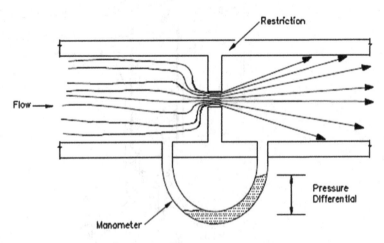

Fig. 6.33 Flow measurement by pressure difference

where, A is the pipeline flow area, V_m is the average velocity.

$$q_m = \sqrt{\frac{2\rho A^2}{K}\Delta p_k} = A\sqrt{\frac{2\rho}{K}\Delta p_k} \qquad (6.5)$$

where, K is the local loss coefficient.

6.4.1 Venturi Flow Meter

A Venturi flow meter belongs to differential pressure flow meter. Due to its high accuracy, it is applied widely. Its basic principle is the measurement method of flow rate based on the energy conservation law, Bernoulli equation and flow continuity equation (Fig. 6.34). Its basic principle is introduced in the fourth chapter, such as Fig. 6.34. As shown, the entrance of the fluid is a contraction of the conical tube, the middle is a straight, and export is an expansion of the conical tube. The conical tube section shrinkage fluid flow, the fluid is accelerated; pressure drop can be measured through the entrance pressure and the contraction pressure difference, carefully calibrated by the flow downstream of the expansion section. In order to restore the pressure, the pressure loss of the total pressure hole is measured at only about 10 ~ 25%. An advantage of this flow meter is that pressure difference can be carefully calibrated with high precision. Disadvantages of this type flow meter are: installation is complex, the installation cost and the difficulty are relatively high, and maintenance is not convenient.

In order to overcome the shortcomings of the Venturi flow meter and improved by Dall flow tube (Dall flow tube type flow meter), as shown in Fig. 6.35. Dole flow

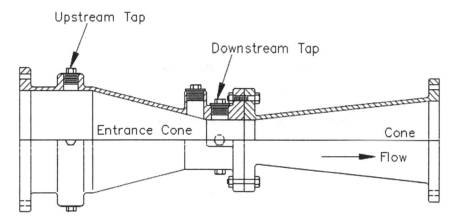

Fig. 6.34 Venturi flow meter

Fig. 6.35 Dall flow tube

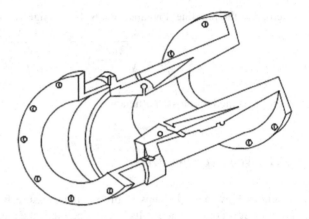

tube pressure head loss is only about 5%, usually for a large flow rate of large pipe flow measurement.

6.4.2 Pitot Tube

Pitot tube as illustrated in Fig. 6.36, is another type of differential pressure measurement for flow detection.

   Pitot tube opening is facing in the direction of flow (in air dynamics field, also known as the airspeed tube), due to the fluid kinetic energy into potential energy, the pressure is increased, Pitot tube connected end is connected to the high voltage end of the differential pressure transmitter. A Pitot tube is actually measuring velocity, but due to the velocity and flow between with specific relationship. Therefore, it can be used for flow measurement.

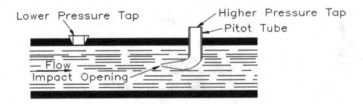

Fig. 6.36 Pitot tube

Fig. 6.37 Rotameter
schematic

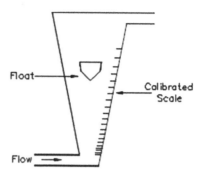

6.4.3 Rotameter

Moreover, some other forms of flow meter, float flow meter is a kind of widely used flow meter, its basic principle as shown in Fig. 6.37. Due to flow through the pipe flow and circulation area, therefore, a scale placed in the conical pipe float on the movement to change the flow area, the flow rate is scaled according to the height of the float position.

In order to improve the measurement precision and stability, float usually design a rotary float. Rotameter is commonly used for small flow measurement.

6.4.4 Steam Flow Measurement

Due to the small steam density, high velocity, measurement of steam flow rate and the water flow rate slightly different. Commonly used nozzle type throttle method of measuring flow rate, nozzle type throttle schematic diagram as shown in Fig. 6.38.

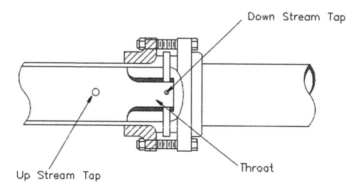

Fig. 6.38 Nozzle type throttle

Nozzle type throttle due to a streamlined design, the flow measurement orifice is 60% larger than a normal orifice plate. Due to the density of water vapor changes with the temperature and pressure of the more obvious, so through the nozzle type throttle volume flow and mass flow respectively are:

$$q_V = K\sqrt{\frac{\Delta p}{\rho}} \qquad (6.6)$$

$$q_m = \rho q_V \qquad (6.7)$$

where, K is a constant related to the specific structural parameters, for which the fluid can be approximated as an ideal gas.

$$\rho = \frac{pM}{RT} \qquad (6.8)$$

where, M is the molar mass, and R is the gas constant.

Therefore, the measurement of temperature and pressure can be used to determine the density, to determine the mass flow. For example, Fig. 6.39 shows a program uses temperature and pressure measurement values to calculate the flow rate of the measurement.

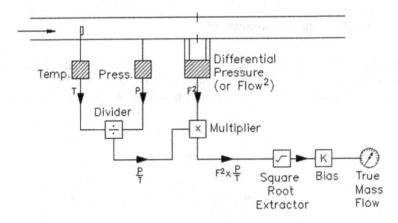

Fig. 6.39 Steam flow measurement using temperature and pressure

6.5 Position Measurement

In nuclear power plants, the position measurement is mainly used to measure the position of the control rod or the position of the valve, and the signal is transmitted to the main control room.

6.5.1 Synchro Equipment

Remote indicating positions can be adopted self-synchronous device (synchro equipment). Basic synchronization device includes an emission unit and a receiving unit. Divided according to function, there are five basic synchronization device: transmitter, differential transmitter, receiver, differential receiver and control transformer. Figure 6.40 shows the schematic diagram of the different synchronization equipment, including external connection head and internal coil position.

If a device needs more power than the device it is synchronized with, it needs a power amplifier to provide additional power.

Transmitter, consisting of a single coil of the rotor and 120° layout of three coils of the stator. When the moving mechanical parts, connected with the rotor will rotation, thus in the stator coil produced electromotive force (EMF), EMF depends on rotor position. Because of the size of the electromotive force reflects the displacement of the mobile machinery. Stator coil to generate a sense of electromotive force by

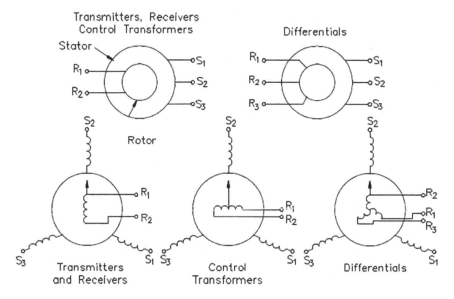

Fig. 6.40 Principle diagram of synchronization equipment

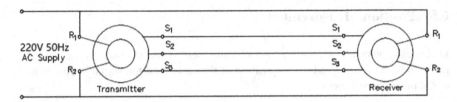

Fig. 6.41 Position synchronization device wiring diagram

wire transfer to the receiver, the receiver will display the position synchronization, as shown in Fig. 6.41.

The structure of receiver and transmitter is very similar. The transmitter output EMF functions on the stator coil of the receiver, the rotor of receiver will be turned to the corresponding location. Receivers and transmitters have a different point is that within the receiver to prevent excessive damping device to rotate. If there is no damping device, then the receiver tracking transmitter locations prone to overshoot, adjust and then drop back, after a few times to adjust to correspond to the location.

A different signal is input to differential synchro device between the transmitter and the receiver. The transmitter action signals operates differential signal with addition or subtraction firstly, and then sent to the receiver. This design enables the following characteristics of the receiver can be better designed. After the introduction of control principle, one will know that it is a feed forward control, which can be used to improve the following characteristics of system.

When a voltage signal is only required, the control transformer is connected with the transmitter without the receiver's display position, and the output of the control transformer is a voltage signal corresponding to the position.

6.5.2   Limit Switch

Limit switch, also known as travel switch, can be installed in the relatively static objects (such as fixed frame, control rod sleeve, etc.) or moving objects. When the moving parts is near the limit switch, the switch connecting rod drive contact caused by closing the switch or disconnect. Use the state of switch to change the action of the control system.

Figure 6.42 is a design to determine the position of the valve stem. According to the status of the two limit switches, you can determine the position of the stem, and can be used to display or control.

The fault of the limit switch is usually mechanical failure. If the predetermined control function is not implemented, the limit switch should be checked for mechanical failure.

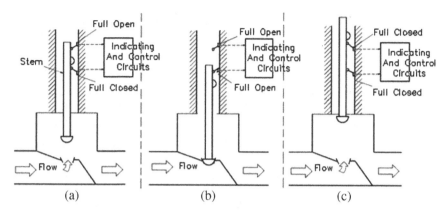

Fig. 6.42 Limit switches

6.5.3 Reed Switch

In order to overcome the shortcomings of high mechanical failure of limit switch, magnetic reed switch is designed. The magnetic reed switch has the advantages of simple structure; its reliability is higher than the limit switch. The working principle of the magnetic reed switch is composed of two pieces of magnetic reed (usually made of iron and nickel) which is equivalent to the role of reed a magnetic conductor. Before the operation, two reeds did not contact. By permanent magnet or electromagnetic coil generates a magnetic field, this magnetic field makes two reeds endpoint produces different polarities. When the magnetic reed exceeds the elastic force, the two reeds will contact to close a circuit. After the magnetic field is weakened or disappeared, the contact surface will be separated due to the elastic spring, thereby opening the circuit. Figure 6.43 shows the working principle. By using a large number of magnetic reed switches, incremental position can be measured. Nuclear power plant usually uses the magnetic spring switch to display the rod position of control rod.

6.5.4 Potentiometer

Potentiometer is also known as linear sensor. Commonly used displacement sensor comprises a potentiometer displacement sensors, inductive displacement sensor, capacitive displacement sensor, electric eddy current displacement sensor, Hall displacement sensor etc.

Potentiometer displacement sensors provide an accurate indication of position through the mechanical displacement of a valve or control rod with the resistance or voltage output. General linear and circular potentiometer can be respectively used as a linear displacement and angular displacement sensor. However, designed to achieve the purpose of measuring displacement of potentiometer, requires a determination of

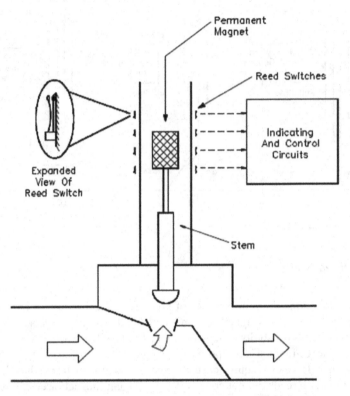

Fig. 6.43 Reed switch

the relation between the displacement change and resistance change. Potentiometric displacement sensor, a movable brush and the object to be measured is connected, as shown in Fig. 6.44.

The displacement of object causes resistance change. Resistance change reflects the displacement value and the resistance increase or decrease indicates that the displacement direction. Usually, by supplying a voltage, the resistance change is converted into voltage output signal. A major drawback of a potentiometer type sensor is easy to wear and tear. It has the advantages of simple structure, large output, easy to use, inexpensive. Displacement sensor failure is usually electrical failure; electrical short circuit or open circuit will make the sensor failure.

6.5.5 Linear Variable Differential Transformer

Linear variable differential transformer (LVDT) is a precise indication of the control rod or the position of the valve position measurement device, and its schematic diagram as shown in Fig. 6.45.

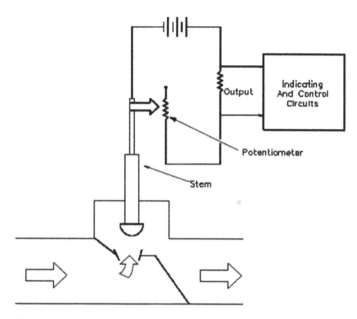

Fig. 6.44 Schematic diagram of displacement sensor

Moving iron core connected with a valve rod, iron core in between the primary coil and the secondary coil moving induced electromotive force is generated in the secondary coil, so the position of the secondary coil output and stem related. If only needs to show two positions, two secondary coils is enough. If you want to display multiple locations, just increase in the number of the secondary coil. This sensor has very high reliability; failure mode is mainly coil of the circuit breaker.

6.6 Radioactivity Measurement

Prior to the introduction of radioactive measurements, the first to be familiar with a few basic terms: ionization, ion pair, specific ionization and linear energy loss.

Ionization

Ionization is a physical process of atoms or molecules to form an ion by energy incited. It is according to the phenomena of ionization, we call those radiations as ionizing radiation. Radiation based on the ability to cause ionization can be divided into two categories: ionizing and non-ionizing radiation. Ionizing radiation include alpha, beta, gamma ray, neutron etc.; non ionizing radiation, including electromagnetic radiation, thermal radiation, microwave irradiation.

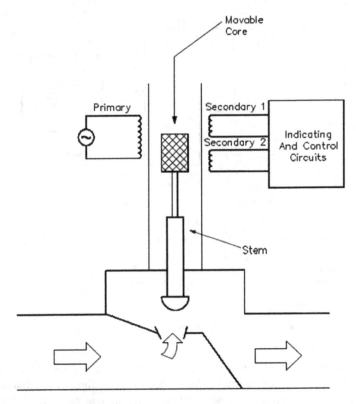

Fig. 6.45 Linear variable differential transformer

Electron–Ion Pair

Atoms or molecules lose an electron by foreign particles will take a unit of positive charge, the formation of positive ions. Ionization produced by positive ions and electrons known for electron ion, referred to as the ion pair.

Specific Ionization

High energy particle enters matter will cause the material to ionize, ion particles at unit distance is called as specific ionization, the value of $N/\Delta x$. Specific ionization is dependent with the mass, charge and energy of particle, and internal matters of the electron number density (number of electrons per unit volume). The mass of incident particle bigger, greater specific ionization is; of incident particle with charge more, greater specific ionization is; lower energy incident particles, greater specific ionization is.

The first few are more intuitive, the last one on the energy of incident particles we explain a little. Imagine such a scene, a table has a lot of very small staples, you get a magnet on your hand from the desktop position horizontal movement of a certain height, the staple will be sucked up by the magnet. The experiment is repeated to maintain the height of magnet, and the magnet movement speed is not the same, you will observe that the magnet moves more slowly, more staples will be sucked up. The motion of a charged particle in matter is some similar with this example, charged particle energy is low, the movement speed is small, in the unit distance will affect more atoms or molecules. This is because our definition is based on the unit distance, not on the number of ions per unit time.

Linear energy loss

Linear energy loss is a physical property quantity representing the blocking ability of a material. Linear energy loss is the energy loss of a particle within unit distance, its value for the $\Delta E/\Delta x$. Linear energy loss and specific ionization has linear relationship. The proportion coefficient is the average required energy for generating an ion. Namely, the specific ionization multiplied by the average energy required to ionize a pair of ions is the linear energy loss.

The working principle of radiation detector is based on the interaction of particles and material. When a particle through a substance, the substance will absorb some or all of its energy and produce ionization or excitation. If the particle is charged, the electromagnetic field and the atomic orbital electron material in direct interaction. If gamma ray or X ray then, after some intermediate process, produce photoelectric effect, Compton effect and electron pair effect, the orbit of electron energy in whole or in part to atoms, then produce ionization or excitation for uncharged neutral particles, such as neutrons, is produced by nuclear reactions of charged particles, then cause ionization or excitation radiation. Detectors are appropriate as detecting medium and particle effects of substances will be charged particle ionization particles produced in the detection medium or excited, into various forms with the direct or indirect information that can be felt by the human senses.

Before the introduction of various types of radiation detectors, the first to introduce ray types and a variety of radiation with matter. Here only introduce the necessary; more detailed content in the later nuclear physics chapters will be introduced.

6.6.1 Radiation Type

Because the ionization process is related to the energy, mass and charge quantity of the incident particles, we need to know what kind of radiation, which is the type of radiation.

Alpha ray

Alpha ray is radioactive substances emitted alpha particles. Alpha particle is composed of two protons and two neutrons, no electrons outside the nucleus, with two positive charges. Therefore, the alpha particle is ^{4}He without electrons outside the nuclei. It consists of a variety of radioactive substances (such as radium) emits to. Alpha radionuclide usually has many nucleons (protons and neutrons are collectively

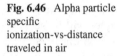

Fig. 6.46 Alpha particle specific ionization-vs-distance traveled in air

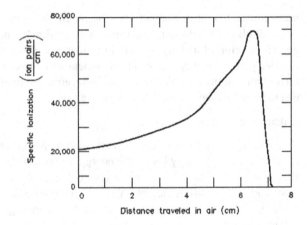

referred to as nucleons) and thus instability, through throw away two neutrons and two protons (alpha particles) into a lower energy state.

Because the mass of alpha particles is much larger than the electron, when it goes through a material, it is very easy to ionize atoms, so its ability to penetrate material is very weak; easy to be blocked by a thin layer of paper, but it has very strong specific ionization, in the air, as shown in Fig. 6.46.

Beta ray

Beta ray is refers to the beta decay of radioactive material released by the high energy electrons, its speed can reach to 99% of the speed of light. Beta ray is with a positive or negative charge on the electron.

In the process of beta decay, radioactive nuclei by electrons and neutrinos change for another nuclei; electron produced is called beta particle emission. In positive beta decay, a proton shifts to a neutron, and release a positron; in negative beta decay, a neutron changes to a proton, and releases an electron that is beta particle.

Due to the small mass, small charge and high speed, the specific ionization of beta ray is smaller than alpha ray.

Gamma ray

Gamma ray is the ray that is released during the transition of atomic nuclear energy level, and is a kind of electromagnetic wave with a very short wavelength, or as a photon.

There are three ways of interaction between gamma rays and matter (see Fig. 6.47): photoelectric effect (a), Compton effect (b) and electron pair effect (c).

The photoelectric effect is in the incident photon energy higher than a certain value; some of the material inside the electron will be photon stir up, the light electricity. Photoelectric phenomena was found by the German physicist Heinrich Hertz in 1887, and correctly interpreted by Einstein. Photoelectric effect schematic diagram shown in Fig. 6.47 a.

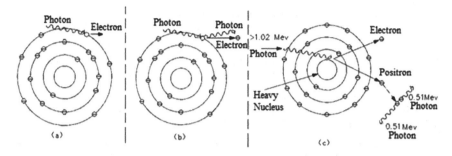

Fig. 6.47 Interaction between γ ray and matter

In 1923, American physicist Compton in X-ray studies through the material scattering experiment, found a new phenomenon, namely light scattering in addition to the original wavelength of X-ray, also produced wavelengths greater than X light with the wavelength of the incident light, the wavelength of incremental changes with different scattering angle. This phenomenon is known as Compton effect, as shown in Fig. 6.47b.

Electron pair effect happens when the incident photon energy is high enough when passing the side of the nucleus, because of the nuclear Coulomb field, the incident photon may changes into an electron and a positron, the process is called electronic pair effects.

Which interact will happen is dependent with the incident photon energy. If the incident photon energy is low, the photoelectric effect happens with large probability. If incident photon energy is above 1 meV, rarely photoelectric effect happens. Incident photons with energies between 1 and 2 meV, Compton effect happens. The incident photon energy higher, the electronic effect will be accounted for the main position. When the incident photon energy is lower than the 1.02 meV, electron pair effect will not occur.

Gamma rays penetrate very strong, and even penetrate tens of centimeters thick concrete walls or a few meters thick air.

Neutrons are not charged, and the mass is almost same as protons. The mass of the neutron is about 1800 times that of the electron, and about quarter of alpha particle. The main source of neutrons is nuclear fission reactions, also may be obtained by the decay of radioactive nuclides. Since neutrons are not charged, and the mass is high, so the ability to penetrate is strong. Neutron in the material in the line energy loss is relatively large.

The interaction of neutrons with matter is mainly in the following ways: inelastic scattering, elastic scattering, trapping or fission.

In the inelastic scattering process, the neutron transfers a portion of the kinetic energy to the target nucleus, into the nuclear kinetic energy or target. In this process, the neutron retarded, while the target nucleus are excited into high energy. The excited target nucleus will release gamma ray. The process of inelastic scattering can be understood, as is a compound nucleus formation process of neutron into the target nucleus after formation of a compound nucleus, then the compound nucleus due to the instability, releasing neutrons and gamma rays in a lower energy. The inelastic scattering exists threshold energy. For a hydrogen atom, the threshold energy can be considered infinite, namely neutron inelastic scattering is impossible for hydrogen atom. The oxygen atom threshold is about 6 meV, the threshold energy of uranium atoms is less than 1 meV.

Elastic scattering is the main process of high-energy neutrons with low mass nuclear. In this process, the neutron and the target nuclear elastic collisions, the nuclear target will not be excited. Capture reaction has compound nucleus formation same as inelastic scattering, but the composite nuclear mass generated is not high enough to launch a neutron, can only emit gamma rays. The fission reaction also generates a compound nucleus, but due to the compound nucleus too unstable, is split into two or more smaller mass number of nuclei, a bit like the nuclear target was broken by neutron. Because smaller nucleus need less neutron, it will release the excess neutrons. Fission reaction releases tremendous amounts of energy; an average of about 200 meV fission energy can be released per action.

Neutron is difficult to be detected directly because it is non-charged. It is usually required to measure the neutron by the interaction between the neutron and matter, for example, the following reactions can be used in the B-10 and the neutron:

$$ {}^{10}_{5}B + {}^{1}_{0}n \rightarrow {}^{7}_{3}Li^{3+} + {}^{4}_{2}He^{2+} + 5e^- \qquad (6.9) $$

6.6.2   Gas Ionization Detector

Gas ionization detector [10] uses gas as interaction material with radiation ray. The reason for using gas, because gas flows easily after ionized, thus forms an electric current or pulse. Commonly used gas is argon and helium; and measuring the neutron with boron trifluoride gas. Pulse type gas ionization detector schematic as shown in Fig. 6.48.

The gas detector is a closed container inside a cylindrical container filled with gas; there are two mutually insulated electrode, metal cylinder as cathode. The wire cylinder is in the middle of the anode, with DC voltage between the poles. When there is no ray incident to the gas chamber, no gas is ionized, no current in the circuit. After the switch is closed, a high voltage is formed on the anode by the adjustable high voltage; an electric field is formed inside the cylinder. When a ray incident to the ionization chamber, the gas is ionized, under the action of electric field force, positive ions to the cathode movement, negative ions or electrons to the anode movement. The voltage across the capacitor C is decreased because of these charges collected; resulting in an electrical pulse signal can be measured in the external circuit. When the adjustable voltage increases, the electric field becomes strong enough, strong electric field can make the movement of Electrons or ions get enough energy to

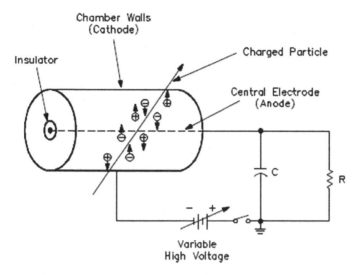

Fig. 6.48 Pulse type gas ionization detector schematic

further ionization (called secondary ionization), thereby increasing the number of ion pairs.

For capacitors, there are

$$Q = CV \tag{6.10}$$

That charge is equal to the product of capacitance and voltage.

The change of charge (ΔQ) is in direct proportion to the change of voltage (ΔV). ΔV is the voltage of the electric pulse height.

$$\Delta V = \frac{\Delta Q}{C} \tag{6.11}$$

Anode collecting to the number of electrons decided to charge changes ΔQ, collected the number of electrons are not necessarily all is primary ionization electrons, and a series of secondary ionization (if the electric field strength is large enough). Usually amplification coefficient A is used to describe the ratio of total number of electrons and primary ionization electrons.

$$\Delta V = \frac{Ane}{C} \tag{6.12}$$

where, A is the amplification factor; n is the number of the primary ionized ions; E is the charge of the electron, 1.602×10^{-19} C; C is the capacity, F.

So, by measure the height of ΔV and amplification factor A, the initial ionization number can be calculated.

Fig. 6.49 Relationship
between the number of Ions
and input voltage

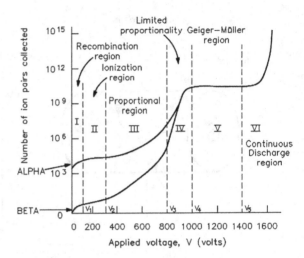

Due to the amplification factor and electric field strength, also is the input voltage related, so the relationship between input voltage and amplification factor need to be understood. Figure 6.49 is a gas ionization chamber detector to collect ion on the relationship between quantity and voltage. A is the alpha particle and another is beta particles, they all have similar characteristics can be divided into several areas.

When the voltage is low, is recombination region (I). At this region, because there is no enough electric field force to separate charged particles produced by ionization, in their own field force, recombination of positive and negative charge occurs. Thus collected charge quantity is less than the number of ion pairs. Ionization chamber usually will not work in this region.

When the voltage reaches $V_1 \approx 100$ V, into the ionization region (II), sometimes also called for the saturated zone. At this region, the amount of charge collected is very close to the number of ion pairs. The amount of charge collected is insensitive to input voltage, in a kind of saturated zone, so as a horizontal line.

When the voltage exceeds the $V_2 \approx 300$ V, entered the proportional region (III). At this region, the amount of charge collected increases with input voltage. Charged particles in electric field get enough energy, so secondary ionization happens. The slope is the amplification. Gas amplification factor can be $1000 \sim 10,000$. Reason that is proportional region is that the amplification coefficient is proportional to input voltage. Proportional counter is working in this region.

When the voltage exceeds the $V_3 \approx 800$ V, entered the limited proportional region (IV). Gas amplification continues to increase; making positive ions near the anode gathered a large number of secondary ionization, resulting in a large number of positive ions retention around the anodic formation of space charge. They produce electric field is partially offset by the applied electric field, and hence the relationship between amplification factor and input voltage is no longer a linear relationship. Detector generally does not work in this area.

The voltage reaches the $V_4 \approx 1000$ V, into the Geiger Muller region (V). In this region, amplification factor and the incident particle type are independent, and have nothing to do with the number of primary ionization. The collected charge number does not change with the change of input voltage, also known as Geiger platform region. G-M counter works in this region. The pulse height of this region can be up to several volts.

If the voltage is increased to $V_5 \approx 1400$ V, gas breakdown, continuous gas discharge. The radiation detector does not work in this area.

Radiation detectors are usually designed to be detecting a particular type of radiation particles. Due to the detector response is mainly the quantity of primary ionization, so it is related to the incident particle energy and quantity. Each kind of detector will define a certain range according to the type of ray detected. In the field of nuclear engineering using various types of detectors, and some can distinguish species, and some not. Some detectors can detect the number of incident particles, and some can detect number and energy of the incident particle.

6.6.3 Proportional Counter

A proportional counter works in the proportional region (Fig. 6.49 in region III). In order to be able to measure a single incident particle, the primary ionization ion pairs must be amplified. Therefore, it is necessary work in the proportional region. In this region, amplification factor of up to 10,000, through reasonable design, adjustment and calibration, the proportional counter can be used to measure alpha, beta, gamma or neutron.

The type of gas inside the chamber determines what types of particles, argon and helium are commonly used to measure alpha, beta, gamma particles, boron fluoride gases is used to detect neutron.

When a single gamma particle injected into the gas ionization chamber, it will produce a high-energy electron, the high-energy electron can produce about 10,000 primary ionization of the electron in the ionization chamber. If the gas amplification is 4, the anode collected 40,000 electrons. If the magnification is 10^4, collected 10^8 electrons.

Proportional counter is very sensitive; electrons generated within 0.1 ms will all have been collected, so every pulse corresponds to an incident particle. The collected charge number is proportional to the primary ionization; the coefficient of proportionality is the amplification factor. The number of primary electrons also related to incident particles energy.

Whenever an electron collected, it will leave a positive charged ion in the gas space. The mass of these positive ions are much larger than the electron, so the velocity is much smaller. Finally, these positively charged ions will move to the cathode. Electric neutralization happens when the electrons are obtained from the cathode. In the process of electro neutralization, because there is a surplus of energy, so it also ionized the surrounding air, will generate electrons. These electrons moving to anode, and the same process and other electronics. These electrons due to the electron arrival time are different, it will produce additional electric pulse, and the pulse is not a direct relationship with the incident particles, usually want to get rid of.

A method of eliminating this additional electric pulse is to add a small amount of gas (~10%) in organic gases, such as methane gas. The binding force of organic gas and electron is relatively small, so positive charged ions could quickly gain electrons from organic gas molecules. Then the positively charged organic gas ions move to the cathode, and neutralization at the cathode. The excess energy of the electric neutralization process will not ionize gas but the organic gas molecules (dissociation). The additional pulses are eliminated. However, organic gas will be consumed in the process, so as to shorten the service life of the proportional counter. If the organic gases can be replenished, the life can be prolonged. Some of the counters can be continuously added organic gas, called the gas flow counter.

Proportional counter produces an electrical pulse signal; the electric pulse is caused by incident particles caused by ionized gas in the process of the charge induced. The signal schematic diagram as shown in Fig. 6.50. This signal need to be dealt by a circuit to become scale readings. The measuring circuit of the proportional counter generally include preamplifier, amplifier, single channel pulse amplitude analyzer (referred to as single channel analyzer), counters, timers, etc.

Electric pulse amplitude analyzer classifies electric pulse signals according to the magnitude of the size, and records the number of each type of signal. Electric pulse amplitude analyzer is commonly used in analysis of output signal of X-ray detector to get radiation energy spectrum. Pulse amplitude analyzer includes single channel pulse amplitude analyzer and multi-channel pulse amplitude analyzer.

Single channel pulse amplitude analyzer is within a certain range of the input pulse counting at a time record. However, the multichannel pulse amplitude analyzer is the whole analysis. Its range is divided into several intervals (the size of the interval that

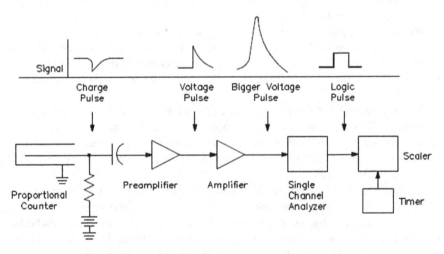

Fig. 6.50 Measuring circuit for proportional counter

Fig. 6.51 Schematic
diagram of single channel
analyzer

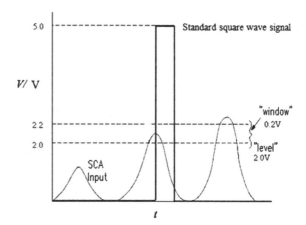

is channel width and interval number known as number of channel); a measurement
can get input pulse amplitude spectrum distribution.

Shown in Fig. 6.51 is single channel pulse amplitude analyzer works. Single
channel pulse amplitude analyzer has two parameters setting, level (threshold) and
channel width. As shown in the figure set level as 2.0 V, the channel width is 0.2 V,
is only between 2.0 and 2.2 V input pulse will produce an output pulse. The output
pulse is usually a standard with a certain width and height of the square wave signal.

Due to the single channel analyzer can be set through the level and the width of
channel, only the selected electric pulse will be outputted. Therefore, you can count
a particular ray in hybrid ray, while filtering out the other electrical pulse. Such as
Fig. 6.51, only one electric pulse will be outputted, and small or too large will not
produce the output.

The output signal of the single channel analyzer is a standard square wave signal
that can be counted by the counter. The counter can be controlled by a timer, so you
can count within a specific period of time. The count in unit time is called count rate.

Proportional counters can also be used to measure neutron. At this case, the gas
need with boron trifluoride. Measuring neutron counting in a reactor, because they
often have accompanied by gamma rays, so one hopes that the counter only sensitive
to neutrons and insensitive to gamma. Due to the release of alpha particles by the
effect of neutron and boron trifluoride, the specific ionization of alpha particles is
much higher than gamma rays, so you can set a level to filter out the gamma ray.

You can also filter out the gamma ray using a screening device. A screening device
is very similar to the single channel analyzer, but only need to set one parameter,
level (threshold), without the need to set the channel width, as shown in Fig. 6.52.

In a nuclear power plant, BF_3 proportional counter is mainly used for neutron
measurements under low power. In nuclear power plant, this is often referred to
as a "start channel" or "source range measurement". This counter is not used for
measurement of power range, because this kind of detector is a pulse detector. Usually
each pulse width of about 10 ~ 20 ms. In the power range, due to the very large

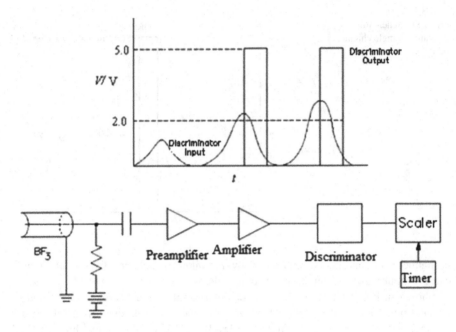

Fig. 6.52 Schematic diagram of filter gamma ray

number of neutron, so multiple pulse overlap, unable to distinguish, then need to use an ionization chamber detector.

6.6.4 Ionization Chamber

Although various radiation detectors are designed using ionization principle, but not all of them are called ionization chamber detector. The ionization chamber detector refers specifically to work in the ionization region of Fig. 6.49 (region II).

In the ionization region, the amount of charge collected almost has almost no relationship with input voltage, but only with the incident particle types. An ionization chamber working in this region, compared with the previous proportional counter, has two obvious disadvantages: low sensitivity and slow response time [11].

According to the different way of counting, there are two types of ionization chamber: pulse counting ionization chamber and integrating ionization chamber. Pulse counting ionization chamber counts the ionization number in the chamber, and integrating ionization chamber accumulate the number of pulse within the specified time interval. By adjusting the timer, any one of the ionization chamber can complete the same function. Therefore, integral type ionization chamber is generally used.

Ionization chamber can be made of flat or coaxial cylinder wall. Due to the uniform of the active area of the plate type ionization chamber, ion is not easy to aggregate

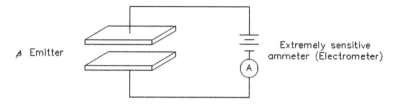

Fig. 6.53 Ionization chamber detector

in insulation region (will distort the local electric field), so engineering used more. Figure 6.53 is the schematic diagram of the ionization chamber detector.

If there is a beta particle emitting body close to the ionization chamber, beta particles enters the gas space between the flat. If the beta particle energy is high enough, gas is ionized, electrons and positively charged ions generated. A beta particle in air through 1 cm can ionize 40–50 electronic. Beta particles generated by the incident electrons will continue to ionize gas molecules. The total electronic quantity of an incident beta particle depends on the energy of incident beta particle and gas species.

In normal cases, the specific ionization of 1 meV beta particles is about 50 ions/cm, and the specific ionization of 0.05 meV beta particles is about 300 ions/cm. The lower energy beta particles, the greater specific ionization, because in unit distance gas molecules can have more collision. Each beta particle generated by ionization, when goes through the air, will further ionization thousands of electronic. 0.1 μA current needs 10^{12} electron /sec.

If 1 V voltage is added between the two plates, free electrons produced by ionization will move to anode under the action of electric field force. This can be measured in a sensitive galvanometer current. But it's not all electrons are able to reach the anode plate, because there are a large number of positive ions in ionization chamber, once an electronic meets with a positive ions, recombination happens. So sensitive galvanometer records only part of the electrons.

If the voltage is increased, the electric field intensity increased, the attractiveness of the anode plate to free electron is becomes strong. They will be at a faster speed with increase of voltage, and positive ion recombination opportunities will become smaller.

Figure 6.54 shows the change of the number of electrons with the increase of the voltage.

You can see in the beginning, as the voltage increases, the number of electrons detected increased. After the voltage reaches 40 V, continue to increase the voltage, the number of electrons stops to increased, in the ionization zone. According to this characteristic, some researchers called this region as saturation region. Hear "saturation" means a maximum value. In this sense, "saturation" is appropriate. However, here, as the voltage increases, the total number of electrons produced by incident particles does not change; what is changed is the number of electronic recombination. After the voltage reaches 40 V, all electrons were "separated" to be

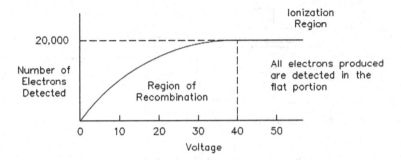

Fig. 6.54 Recombination region and ionization region

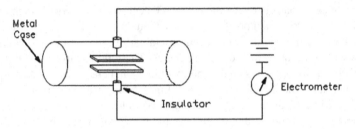

Fig. 6.55 Ionization chamber for measuring gamma rays

collected at anode. Therefore it makes the essential meaning of the ionization zone as "separation of ionization zone", namely all electrons are separated.

Figure 6.53 can also be used to measure the ionization chamber of gamma rays, due to the strong penetration of gamma rays, so you can use a metal container to enclose the ionization chamber, as shown in Fig. 6.55.

When gamma ray enters the ionization chamber, Compton scattering, photoelectric effect and electronic effect occurs. A high-energy electron will be released during these processes. This electron under the action of electric field force will continue to ionize the surrounding gaseous medium.

Metal container has great benefits, on the one hand can shielding electric field interference, on the other hand can be filled with special gas. If you want to measure alpha and beta particles, the container must be specially designed due to small penetration. Generally, a hole will be opened, with very thin material, and can be used as "window". However, a thin "window" will also block some alpha particles.

Neutron can be detected by ionization chamber. However, due to the neutron itself is not charged, cannot ionize gas molecules. Therefore, it is necessary to add special materials to induce neutron reaction. The ionization chamber wall is usually coated with thin layer of boron. Between the neutron and boron with such as reaction:

$$\ce{^{10}_{5}B} + \ce{^{1}_{0}n} \rightarrow \ce{^{7}_{3}Li^{3+}} + \ce{^{4}_{2}He^{2+}} + 5e^{-} \tag{6.9}$$

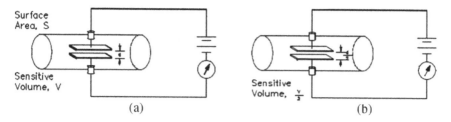

Fig. 6.56 Appropriately reducing the volume of an ionization chamber

This reaction captures a neutron, and then releases an alpha particle, which produces free electrons.

In addition to the wall surface coating method, some fill BF_3 gas in to chamber, the principle is the same.

When using ionization chamber to measure neutron, the interference of beta particles is easily to filter out because of metal wall shielding. However, the interference of gamma ray is not very easy to filter out. The measured results contain effect of gamma ray, which is not hope. To reduce the effect of gamma rays, there are several ways to deal with this problem.

We know that 1 meV alpha particles in the air through the distance about only a few centimeters, and 1 meV electron generated by gamma rays in the air can travel long distance. This gives us a good method. If one can appropriately reduce the volume of ionization chamber without changing the area of coated boron, it is possible to reduce the influence of gamma ray. Figure 6.56 shows this idea.

In Fig. 6.56b, plate spacing becomes $d/2$ makes sensitive area of the gas volume is reduced by half. The boron-coated area does not change, so the number of alpha particle is the same. Due to the gamma ray, induced free electrons between the plates are collected fewer, thereby reducing the effect of gamma rays.

By reducing the sensitive volume, the collection of gamma ray induced electron is reduced. So, it can also by reducing the pressure to reduce the density of the gas to reduce the effects of gamma rays. In addition to the above methods, by increasing the number of alpha particles to reduce the relative impact of low gamma ray is also possible. This can be achieved by increasing the boron-coated area.

In a neutron ionization chamber of nuclear power plant, a variety of above approaches are used to reduce the influence of gamma ray. Gamma ray effect can only be reduced, but not completely eliminated. When reactor are running with full power, neutron flux level is very high; almost all of the current are derived from neutron. This kind of ionization chamber used in high power levels is also called non-compensated ionization chamber. They are not suitable for low power level. In order to meet the low power of neutron measuring, gamma compensated ionization chamber is designed.

The basic principle of the ionization chamber with gamma compensation is shown in Fig. 6.57.

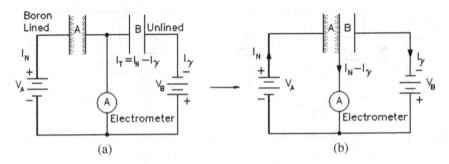

Fig. 6.57 γ compensated ionization chamber

Gamma compensated ionization chamber is made by two ionization chambers (A and B). The structure and internal filling gas are exactly the same and the difference is the chamber wall of A is coated with boron and the chamber wall of B without boron coated. Boron coated chamber A is connected to the positive voltage, not coated boron chamber B, connected with negative voltage, and the middle is a common pole. Boron coated chamber both to neutron sensitive is also sensitive to gamma rays, and not coated boron ionization chamber only for gamma ray sensitive. Such gamma rays produced current can just be subtracted, called gamma compensated ionization chamber.

It is possible to use pure gamma rays to test whether the current of the ionization chamber is zero or not, if it is not zero, it may be caused by the inaccuracy of the supply voltage, or the geometry of the two ionization chambers.

A, B two ionization chambers is difficult to be exactly the same. Actually, A, B two ionization chambers often deliberately designed to be different, as shown in Fig. 6.58, is a manufacturing into concentric ring shape of the compensated ionization chamber. The outer ring is coated by boron ionization chamber. The inner is not coated boron ionization chamber.

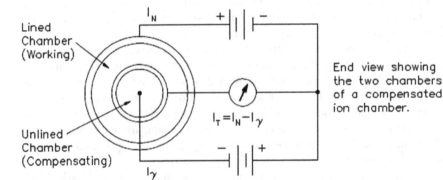

Fig. 6.58 Concentric ring γ compensation ionization chamber

The concentric ring is the advantage of two ionization chamber exposed to radiation field is almost the same. Even if the two chamber is not exactly the same, but can eliminate the gamma ray induced current by adjusting two chamber voltage. The working voltage of boron-coated chamber is provided by the manufacturer, the ionization chamber works in saturation region. The power supply voltage of boron-coated chamber is fixed, and then adjust another power supply voltage. While the pure gamma ray gets zero current, the influence of gamma ray is eliminated. The power supply voltage is calibrated during shutdown of nuclear power plant, because almost no neutrons during shutdown. A well calibrated ionization chamber responses to neutrons only.

The volume of not boron-coated ionization chamber is usually a little bigger than the boron coated chamber. The results of this design is that the compensation current will be slightly larger, known for over compensation. When the compensating current is less than that caused by the gamma ray, known for lack of compensate.

To describe over-compensation or lack of compensation, you can use the degree of compensation, defined as:

$$p = \left(1 - \frac{I_c}{I_{nc}}\right) \times 100\% \qquad (6.13)$$

where, I_c is the current measured by pure gamma ray, and I_{nc} is the current measured when the voltage of the compensation chamber is zero.

If the I_c is zero, the compensation degree is 100%; if the I_c is greater than zero, then the degree of compensation is less than 100% (lack of compensation); if the I_c is less than zero, the degree of compensation is greater than 100% (over compensation).

Figure 6.59 is a compensation curve of an ionization chamber. The working voltage of the ionization chamber is 420 V, in 1 Curie ^{60}Co source (pure gamma ray

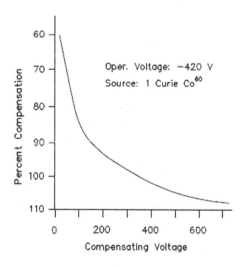

Fig. 6.59 Typical compensation curve

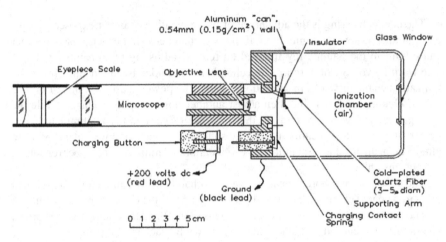

Fig. 6.60   Quartz fiber electroscope ionization chamber

source) irradiation in the compensation voltage under different degree of compensation with the increase of compensation voltage increases. For each chamber, this kind of compensation curve need to be made, in order to get reasonable compensation. It is very important for nuclear power plants, because the compensated ionization chamber in a nuclear power plant is mainly used for nuclear power detection, reactor power and neutron flux are related. If the voltage is too high, over compensation will happen. Part of I_n (current caused by neutron) will be compensated with I_γ (current caused by gamma ray) together. This will cause the power measured less than the actual power. Conversely, if inadequate compensation, power measured will larger than the actual power. When the reactor power level is relatively high, due to $I_n \gg I_\gamma$, therefore, insufficient compensation or over compensation not to be too serious. However, in low power level, it becomes very important.

There is another type of ionization chamber. It is electroscope ionization chamber, such as Fig. 6.60.

The ionization chamber early use of gold leaf electroscope was the first to be used for detecting radiation ionization chamber. The latest design and electroscope general use quartz fiber instead of gold leaf, makes a lot of performance is improved. For example: smaller size, sensitivity higher. Quartz fiber capacitance is about 0.2 pF, voltage sensitivity is about each discrimination scale 1 V. The detector sensitive element is quartz fiber coated with gold (diameter $3 \sim 5$ μμ), installed in a parallel metal support.

A quartz fiber without coated gold is installed with the coated one crossly. The relative position of the two fibers can be observed by microscope. When the charge button is pushed, quartz fiber will be charged by DC voltage. Full scope needs 200 V of DC. When the gas is ionized, quartz fiber collected charge to move from point 0. The moving distance is related with the amount of incident particles (integral effect). The scale can be reset to zero by recharge. Through the quartz fiber position scale, the integral dose can be directly read from the microscope.

The electroscope ionization chamber, because it is portable and easy to read, high precision, high sensitivity and other advantages, has been widely used.

6.6.5 Geiger-Miller Counter

Geiger-Miller counter [12] (also known as Geiger counter) is a gas ionization detector invented by H. Geiger and P. Miller in 1928. Similar to proportional counter, but voltage is higher, in the Geiger Miller region of Fig. 6.49 (region V). In this region, the magnification factor is not related with the type of incident particles, and the number of ion pairs of primary ionization. Independent of the number of charge quantity collected not only has nothing to do with the particle type, and does not change with the voltage. The advantages of the Geiger counter is of high sensitivity, high pulse amplitude, so the counting circuit is simple. The disadvantage is not fast counting, i.e. between two particles, the interval time is not too short. In addition, due to the different particle generating electric pulse height is the same, so Geiger counter cannot resolve particle energy, also cannot distinguish between particle types.

Geiger counter is one of the most commonly used counter using gas ionization. The structure of Geiger counter is usually filled with rare gas, in an enclosed tube with insulated metal in both ends (usually rare gases, adding a halogen such as helium, neon and argon.). An electrode is arranged along the axis of the tube. High voltage slightly lower than the voltage of gas breakdown is supplied between the metal wall and the metal wire electrode. When a certain high-speed particle enters into the tube, gas molecular ionization happens, free electrons move to metal wire under the action of the voltage. These electrons will ionize gas molecules along the way, and release more electrons. More and more electrons then will successive ionization more and more, finally to the gas discharge phenomenon. A pulse current is output to the input of the amplifier and is received by the counter at the output of the amplifier. The number of particles can be detected by this way.

Geiger counter is often used for alpha particle detection, low energy gamma rays and beta particles. For high-energy gamma rays, due to the gas density in Geiger counter is usually small, often shots out Geiger counter before it could be detected. Therefore, the detection sensitivity of high-energy gamma ray is low. Owing to the Geiger counter neither resolution particle energy nor particle species, it is generally not used for reactor neutron measurements.

6.6.6 Scintillation Counter

Scintillation counter refers to detection device using induced scintillation luminescence by ray or particle and using optoelectronic devices to record the intensity and energy of ray. E. Rutherford observed with microscope that a single alpha particle in zinc sulfide induced luminescence in 1911. He observed alpha particle bombardment of nitrogen generate oxygen and proton with fluorescent screen detector in 1919. This is the embryonic form of the scintillation counter. Scintillation counter consists of three parts: scintillator, light collection system and optoelectronic devices. By photoelectric device outputs electric pulse after pre electronics system (amplification, shaping, screening, etc.) into the particle data acquisition system and data processing and analysis.

In order to understand the scintillator, it is necessary for us to try to understand the physics of energy band theory. Energy band theory is used to discuss the state and motion of electron in crystal. It puts the motion of each electron in crystal as an independent movement in an equivalent potential field, namely single electron approximation theory.

Single electron approximation theory, it is assumed that atoms in solids is fixed, and according to certain rules periodically arrayed, and further, that each electron is in the potential field of periodic atoms and other electrons. The whole problem is simplified into a single electron problems and energy band theory is a kind of single electron approximation theory.

Crystal has a large number of atoms and the atoms are very close to each other's. Take silicon as an example, there are 5×10^{22} atoms per cubic centimeter volume. The shortest distance between two atoms is only 0.235 nm in silicon. Which makes overlapping of electronic shell far from the nuclear. The outer electrons are no longer confined to an atom, may be transferred to similar shell of adjacent atoms and even farther atoms. This phenomenon is called electronic sharing. Electronic sharing generates small energy differences between same energy states. The corresponding energy level is expanded to energy band. With a range of energy, with a certain width.

Allowed band is a band allows electron to take. Forbidden band is a band does not allow electrons to occupy, as shown in Fig. 6.61. The inner layer of atomic shell is always full electronic preemption, and then occupy higher energy outside a layer. Has been electronics full filled called full band. Atomic electrons in the outermost shell, said as the valence electrons, which can band to valence band. Above the valence band is known as the conduction band.

Incident particle to crystal, if the accepting energy of valence band electron reaches or exceeds the binding energy of electrons and nuclei, the valence band electrons will be excited to the conduction band, in the conduction band produced a free electron, and leaves a hole in the valence band. If energy is smaller than the binding energy, unable to transitions of electrons to the conduction band, but into excited band, the same in the valence band leave a hole.

By adding specific substances (activator) in crystal, this method could be used to make a capture with the active area of the material in the forbidden band. Activation can capture an electron in the forbidden band, the energy from the ground state G jumped to state E by excitation. With de-excited, they will release a photon. Scintillator crystals within the activation region known for fluorescence, photon emission in the range of visible light.

Fig. 6.61 Electron energy band and activation region of scintillation crystal

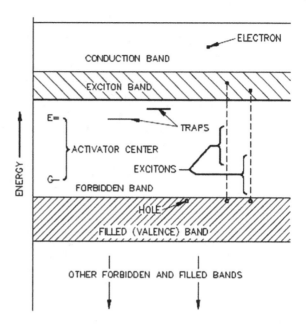

This kind of scintillator materials, when subjected to irradiation is capable of emitting light. Scintillation materials can be divided into inorganic scintillators and organic scintillators, a kind of any type can exist in solid, liquid or gaseous state, and to design brings great flexibility.

Inorganic scintillators used zinc sulfide, sodium iodide, lithium iodide, $C_S I$, Germanium Bismuthate, calcium fluoride and tungstic acid lead. Silver and thallium can be used as activator. Ray scintillation crystals in the valence band electronic excitation and de-excitation emit light. If the band directly back to the valence band, a photon attenuation time is short (1 ~ 10 ns) and high photon energy (ultraviolet). While emission of photons from the activation region below the conduction band, luminescence decay for a long time (about μs), and lower photon energy (wavelength from ultraviolet to yellow light). Figure 6.62 shows a thallium iodide detector.

The density of NaI is high, the efficiency of gamma ray detection is good, and it is suitable for the measurement of high-energy gamma rays and other charged particles, and has a certain energy resolving power.

ZnS scintillator is translucent material, can only be made thin, stop ability of heavy charged particles is very big, is not very sensitive of gamma ray. Applied to detect alpha particles in the beta and gamma background field, to detect slow neutron in proton and boron mixed situation.

Organic scintillation has naphthalene and other; the decay time of the photons emitted is about 10 ns, commonly used to detect beta ray.

The scintillator material can be mixed into plastic. Plastic scintillation luminescence decay time is the shortest; close to 1 ~ 2 ns is one of the fastest particle detector.

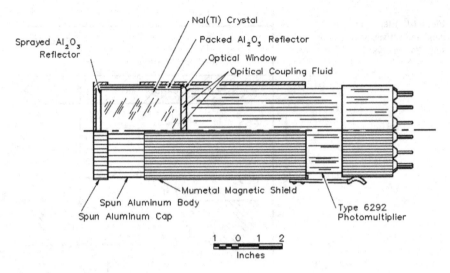

Fig. 6.62 NaI scintillation detector

Because of a large number of hydrogen atoms in plastic scintillation, it can be used to detect fast neutron using the recoil effect.

The light collection system is the connection between the scintillator and the light detector. It needs both sides respectively with the scintillator light output part and the shape of the light input part light detector conforms to the shape, to collect as much as possible so that the light and uniform light distribution. Photomultiplier tube and optical refractive index between two and the similar optical glass grease or optical glue sealing in order to achieve the most effective optical transmission. For most of the inorganic scintillator, because of its larger refractive index, not easily matched with the light detector, so often use Magnesium Oxide or alumina powder packaging scintillator and light guide, in order to improve the light collection efficiency of the diffuse reflection. Recently also developed some other light collecting system, such as optical collector and large wavelength shift light collector BBQ.

Figure 6.63 shows a schematic cross-section of a photomultiplier tube. A photomultiplier is a vacuum tube with a glass envelope containing a photocathode and a series of electrodes called dynodes. The light emitted by scintillation phosphor frees electrons from the photocathode through the photoelectric effect. The amount or energy of these electrons is not sufficient to be reliably detected by conventional electronics. However, in photomultiplier tubes, they are attracted by a voltage drop of about 50 V from the nearest dynode.

The photoelectrons hit the first dynode with enough energy to release several new electrons for each photoelectron. In turn, second-generation electrons are attracted to the second dynode, and third-generation electronics were emitted more. This amplification lasts through 10–12 stages. At the last dynode, there are enough electrons to form a current pulse suitable for further amplification of the transistor circuit. Voltage

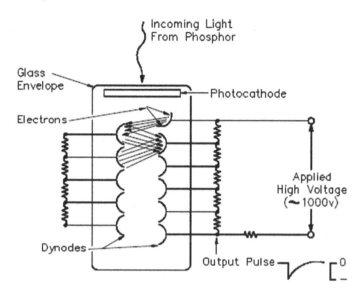

Fig. 6.63 Photomultiplier tube schematic diagram

drops between dynodes are established by a single external deviation (approximately 1000 V DC) and an external network of resistors (to balance voltage drops).

The advantage of scintillation counter is that their efficiency and high accuracy and counting rate are possible. These latter attributes are the result of a very short duration of light flashes, from approximately 10^{-9} to 10^{-6} s. The intensity of the light flash and the amplitude of the output voltage pulse are proportional to the energy of the particle responsible for the flash. As a result, scintillation counters can be used to determine the energy and number of exciting particles (or gamma photons). The photomultiplier tube output is very useful in the radiation spectrometry (the determination of the energy level of incident radiation).

In short, the scintillation detector has the advantages of high efficiency, high precision and high count rate. It can not only identify the energy, but also distinguish between the species and the number of particles.

6.6.7 Gamma Spectroscopy

Gamma spectroscopy is a radiochemical measurement method that determines the energy and count rate of gamma rays emitted by radioactive material. The gamma-ray energy spectrum is analyzed in detail to determine the identity and number of gamma emitters in the material. Gamma spectroscopy is an extremely important measurement.

The equipment used in gamma spectroscopy includes detectors, pulse sorters (multi-channel analyzers), and related amplifiers and data reading devices. Detectors

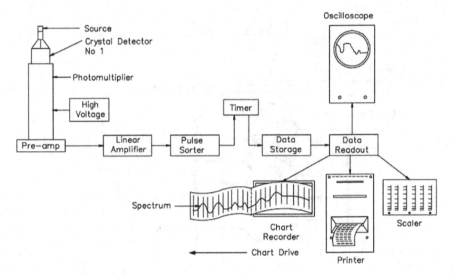

Fig. 6.64 Gamma spectrometer block diagram

are usually sodium iodide (NaI) scintillation counters. Figure 6.64 shows a block diagram of a gamma spectrometer.

A multi-channel pulse height analyzer is a device that separates pulses according to the height of the pulse. Each energy range at the pulse height is called a channel. The pulse height is proportional to the energy lost by gamma rays. The energy spectrum of the emitted gamma rays is shown according to the separation of the pulse based on pulse height.

Multi-channel analyzers typically have 100 or 200 channels with energy ranges from 0 to 2 meV. The output is a graph of pulse height and gamma activity, as shown in Fig. 6.65. By analyzing the emitted gamma ray spectrum, the user can identify the elements that cause the gamma pulse.

Figure 6.65 shows the output results of an experiment of multi-channel gamma spectrometer. Multichannel spectrometer generally has 100 ~ 200 channels, energy range for 0 ~ 2 meV. In Fig. 6.65, the horizontal axis is the pulse height. Pulse height is proportional to the energy of the gamma ray. According to several peak energy, 1.34, 1.75, and 2.75, 2.25 meV, is can determine the types of gamma ray. According to the relative activity and can analysis the relative content.

6.6.8 Miscellaneous Detectors

There are other types of radiation detectors: the self-powered neutron detector, wide range fission chamber, flux wire, and photographic film.

In very large reactor installations, neutron fluxes in the core need to be continuously monitored. This allows for quick detection of instability in any section of

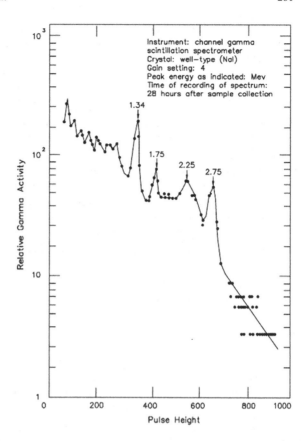

Fig. 6.65 Multichannel analyzer output

Instrument: channel gamma scintillation spectrometer
Crystal: well—type (NaI)
Gain setting: 4
Peak energy as indicated: Mev
Time of recording of spectrum: 28 hours after sample collection

the core. This requires the development of small, inexpensive, rugged, self-powered neutron detectors that can withstand the core environment. Self-powered neutron detectors do not require voltage supply. Figure 6.66 shows a simplified drawing of a self-powered neutron detector.

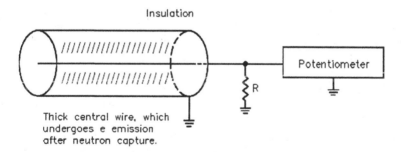

Fig. 6.66 Self-powered neutron detector

Self-powered neutron detector is in accordance with the nuclear battery principle. The neutron detector in response to fast speed, small size, can be tolerated reactor core inside strong radiation environment. Self-powered neutron detector without an external voltage supply, so there is no air space, so named for self-powered detector. Figure 6.68 is self-powered detectors schematics.

The central wire of the self-powered neutron detector is made of material that absorbs neutrons and undergoes radioactive decay by emitting electrons (β decay). Typical materials used for central wire are cobalt, cadmium, rhodium, and vanadium. Place a good insulation material between the central wire and the detector casing. Each time a neutron interacts with the central wire, it transforms an atom of the wire into a radioactive nucleus. The nucleus eventually decays as a result of the emission of electrons. As a result of the emission of these electrons, the charge of the wires is becoming more and positively charged. The positive potential of the wire causes the current to flow through the resistor, and the R. A millivolt meter measures the voltage drop of the resistor. Electronic current from β decay can also be measured directly by electricity meters.

Self-powered neutron detectors have two distinct advantages: (a) very few instruments are required—only one millimeter volt or an electrometer, and (b) the lifetime of the emission material is much larger than that of boron or U235 linings (used in wide range fission chambers).

One disadvantage of self-powered neutron detectors is the decay of the emission material, which has a typical half-life. In the case of rhodium and vanadium, the two most useful materials, the half- lives are 1 min and 3.8 min, respectively. This means that the detector cannot respond immediately to changes in neutron flux, but it can take up to 3.8 min to reach 63% of the steady-state value. This disadvantage is overcome with cobalt or cadmium emitters, which emit electrons within 10–14 s of neutron capture. Self-powered neutron detectors using cobalt or cadmium are called fast self-powered neutron detectors.

The relationship between the output voltage of the detector and the neutron flux is

$$V(t) = I(t)R = K\sigma q N\varphi\left(1 - e^{-0.693t/T}\right)R \qquad (6.14)$$

where K is a constant determined by the detector and the shape and the material; σ is thermal neutron absorption cross section of emitting material, cm^2; q is charge of emitter beta decay, C; N is the total atomic number of emitter; T is half-life of beta decay, φ is neutron flux, $n/(cm^2 s)$.

Such as vanadium, uses [51]V to absorb a neutron to become [52]V (absorption cross section 4.9b, half-life of [52] V beta decay is 226 s. this means that about 4 min is needed to reach 63% the steady-state values. Rhodium is a little better, [103]Rh absorption cross section is 150b, after absorbing a neutron [104]Rh has two kind of isotope, ground state [104]Rh (92.7%) has a half-life of 42s. The half-life of the metastable state of [104]Rh (7.3%) is 264 s. Cobalt and cadmium should be able to compensate for this problem also, because their child with a half-life of only about 10–14 s, therefore almost measuring the instantaneous.

Vanadium self-powered detectors due to absorption cross section is relatively small, it is generally made of relatively long detectors. So often used for detection of the radial power distribution, and for the distribution of the axial resolution is poor. Rhodium self-detectors, due to the large cross section can be used to detect the distribution of axial.

There is also a self-powered detector using gamma rays rather than beta decay, the composition of the detector structure is shown in Fig. 6.67.

The working principle of this kind of detector is that nuclei Pt is excited to the excited state of the compound nucleus after absorption of neutron, in the deexcitation process of the compound nucleus emit gamma rays, using gamma ray detector materials interaction of free electron. The electron current formed is proportional to neutron flux. The characteristics of this kind of detector is response speed, can be used for power regulation and protection, but the disadvantage is that are sensitive to both neutron and gamma rays, so the measuring precision of neutron flux is affected by local gamma ray.

The fission chamber uses neutron-induced fission to detect neutrons. The chamber is usually structurally similar to the ionization chamber, but the coating material is highly rich, ^{235}U. Neutrons interact with ^{235}U, causing fission. One of the two fission fragments enters the chamber, while the other fission fragment is embedded in the chamber wall.

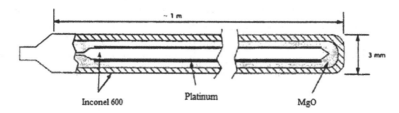

Fig. 6.67 Platinum self-energy neutron detector

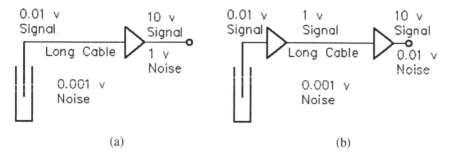

Fig. 6.68 Single and two-stage amplifier circuits

One advantage of using a ^{235}U coating instead of boron is that the energy level of the fission fragment is much higher than that of the alpha particles from the boron reaction. Neutron-induced fission fragments produce much more ionization in each interaction in the chamber than neutron-induced α particles. This allows the fission chamber to operate in a gamma field higher than the uncompensated ion chamber with boron lining. Fission chambers are commonly used as both current indicators and pulse devices simultaneously. They are particularly useful as pulse chambers because of the very large difference in pulse size between neutrons and gamma rays.

Due to the dual use of fission chambers, it is often used in "wide range" channels of nuclear instrumentation systems. The fission chamber can also operate at the neutron level of the source and intermediate range.

Whenever a reactor neutron flux profile needs to be measured, a piece of wire or foil is inserted directly into the reactor core. The wire or foil remains in the core for the length of time required for activation to the desired level. The cross-section of the flux line or foil must be known in order to obtain an accurate flux profile. Once activated, the flux wire or foil is quickly removed from the reactor core and the activity counted.

Activated foil can also discriminate energy levels by placing a cover over the foil to filter (absorb) neutrons at certain energy levels. Cadmium covers are commonly used for this purpose. The cadmium covers effectively filters out all thermal neutrons.

Photographic film can be used for X-ray work and dose measurements. When exposed to radiation, the film tends to darken. This general darkening of the film is used to determine the overall radiation exposure. Neutron scattering produces individual proton recoil tracks. Counting the tracks yields the film's exposure to fast neutrons. Filters are used to determine the energy and type of radiation. Some of the typical filters used are aluminum, copper, cadmium, or lead. These filters provide different amounts of shielding for the attenuation of different energies. An approximate spectrum is determined by comparing exposures under different filters.

6.6.9 Circuitry and Circuit Elements

Throughout the reactor operation, there are three ranges for monitoring the power level of the reactor: source range, intermediate range and power range. The source range typically uses a proportional counter, while the intermediate and power ranges use an ionization chamber. The intermediate range uses a compensated ion chamber. The power range uses an uncompensated ion chamber. All three different ranges use some or all of the following types of components.

The radiation detector output signal is usually weak and needs to be magnified to use. In radiation detection circuits, the nature of input pulses and discriminator determines the characteristics that preamplifiers and amplifiers must have. Most detection circuits use two stages of amplification to increase the signal-to-noise ratio.

The radiation detector is located at a distance from the reading. A shielded coaxial cable transmits the detector output to the amplifier. The detector's output signal may be as low as 0.01 V. The total gain required to increase this signal to 10 V, which is an usable output pulse voltage, is 1000. There is always a pick-up of noise in the long cable run; this noise can reach 0.001 V.

If all amplification were done on the remote amplifier, the 0.01 V pulse signal is 10 V and the 0.001 noise signal is 1 V. This is a signal-to-noise ratio of 10, which can be significantly reduced by dividing the total gain into two stages of amplification. A preamplifier located near the detector and a remote amplifier can be used. The preamplifier device virtually eliminates cable noise due to its short cable length. If the total gain of 1000, the gain of the preamplifier is 100, and the output signal of the preamplifier is 1 V. Signals transmitted over long cables still receive 0.001 V of noise. The amplifier amps the 1.0 V pulse signal and the 0.001 V noise signal by 10 times. The result is a 10 V pulse signal and a 0.01 V noise signal. This gives the signal-to-noise ratio of 1000.

Signal-to-noise ratio is the ratio of the electrical output signal to the electrical noise generated in the cable run or in the instrumentation.

This signal refers to the electronic signal needs to be processed, the noise refers to additional signal without rules does not exist in the original signal (or information), and the signal does not change with the change of the original signal. The signal-to-noise ratio of the measurement unit is dB (dB), the calculation method is $10\lg(P_s/P_n)$, the effective power of which P_s and P_n respectively represent the signal and noise, if converted to voltage amplitude ratio is $20\lg(V_s/V_n)$. V_s and V_n respectively represent the effective voltage signal and noise voltage. In nuclear electronics circuit, we hope that the amplifier amplifies the signal, should not add any other extra things. Therefore, the signal-to-noise ratio should be better. So set the preamplifier in the transmission cable, is conducive to improve the signal-to-noise ratio (in this example, the signal-to-noise ratio is a single amplifier 10, namely 20 dB; while in the two-stage amplification, Signal to noise ratio increased to 1000, that is, 60 dB).

Radiation detection circuit currents or pulse rates vary over a wide range of values. The current output of the ionization chamber may vary by 8 orders of magnitude. For example, the range might range from 10^{-13} amps to 10^{-5} amps. The most accurate way to display this range is to use a linear current meter with multiple scales and the ability to switch between them. This is not practical. A single scale that covers the entire range of values, is used. This scale is referred to as logarithmic. The logarithmic output meter must provide a signal proportional to the logarithm of the input signal. This is easy to do with diodes when the input signal comes from the ionization chamber. The voltage passing through the diode is equal to the logarithm of the current through the diode. Using this principle, the simplified circuit shown in Figs. 6.69 is used to convert the ionization chamber current to a voltage proportional to the logarithm of this current.

Discriminatory circuits choose the minimum pulse height. When the input pulse exceeds the preset level of the discriminator, the discriminator generates the output pulse. Discriminatory input is usually amplified and shaped by the detector signal. This signal is analog because the amplitude is proportional to the energy of the incident particle.

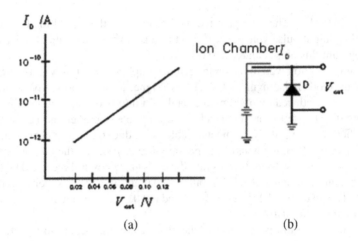

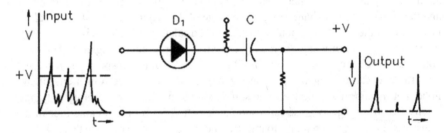

Fig. 6.69 Log count rate meter

Fig. 6.70 Biased diode discriminator

Bias diode circuits are the simplest form of discriminator. Figure 6.70 shows the biased diode discriminator circuit with its associated input and output signals.

Diode D_1 shows that its cathode is connected to a positive voltage source, +V. The diode cannot conduct unless the voltage of the anode is positive with respect to the cathode. As long as the voltage of the anode is less than that of the cathode, diode D_1 does not conduct and there is no output. At some point, the anode voltage exceeds the bias value of +V, and the diode conducts. The input signal is allowed to be passed to the output. Figure 6.70 illustrates the input and output signals and how the discriminator acts to eliminate all pulses below the preset level. The output pulse of this circuit has the same relative amplitude as the input pulse.

In many applications, it is important to know the rate of power changes. This rate usually increases or decreases exponentially with time. The time constant for this change is called the period. A five-second period means that the value changes by a factor of e (2.718) in five seconds. Figure 6.71 shows a basic period meter circuit.

Placing a signal through an RC circuit results in a voltage proportional to period reciprocity. If the current output of the ionization chamber is constant, the current does not flow through resistor R and the output voltage is zero. This is corresponding

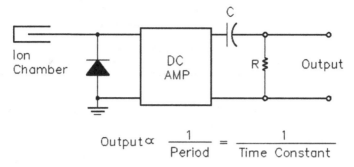

$$\text{Output} \propto \frac{1}{\text{Period}} = \frac{1}{\text{Time Constant}}$$

Fig. 6.71 Period meter circuit

to an infinite period. As the output current of the ion chamber changes, the voltage flows instantaneously on capacitor C and the current flows through resistor R. The faster the transient, the greater the voltage drop in resistance R and the shorter the period.

6.6.10 Detect of Neutron Flux in Reactor

In order to improve the reactor power density and fuel element burnup must accurately detect reactor core neutron flux. The main characteristics of reactor core neutron flux detector is compact structure, and can adapt to the harsh working environment (high level of irradiation, high temperature, high pressure). Through the measurement of reactor core neutron flux, the core design can be validated, the core supervision safety degree and deviate from the nucleate boiling ratio (DNBR) measured fuel burnup, in order to ensure the reactor safe and economic operation.

Core neutron flux measurement method mainly has two kinds: the reactor core detector to directly measure; the second is the activation method to measure indirectly. Pressurized water reactor nuclear power plant widely used in core neutron flux measurement system, is through the mechanical driving device, the selected channel in core fission chamber inserted into the reactor core of the measuring holes.

The direct measurement device includes a core fission chamber, a micro ionization chamber, a corresponding mechanical device, or a self-powered neutron detector fixed in the reactor.

(1) Reactor core fission chamber.

In pressurized water reactor and boiling water reactor, most mobile reactor core neutron flux measurement system are the core fission chamber as a neutron sensitive element. The fission chamber is uranium fuel consumption is quite small. On the pulse counting, the mean square voltage and average current (DC), fission chamber can satisfy all these three kinds of basic modes. Therefore, in the source range channel (using pulse counting), the middle range channel (using mean square technique) and

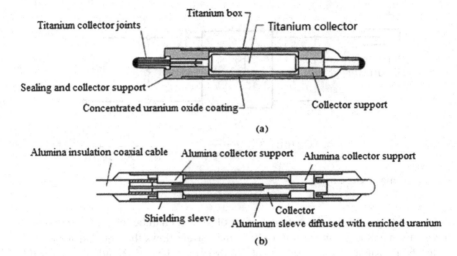

Fig. 6.72   Reactor core fission chamber

the power range channel (the average current technology), core fission chamber is appropriate. There are two basic types of core fission chamber, such as Fig. 6.72 shows.

One type is in the inner detector shell plating layer, enriched uranium, the formation of outer wall of the sensitive volume. Second type is enriched uranium-aluminum alloy sleeve on the outer surface of the sensitive volume. The weight and thickness are carefully controlled for uranium or uranium-aluminum coating sleeve. The more accurately control of weight and thickness the more sensitivity of neutron of the detector. The core fission chamber filled with a gas of a few tenths of 1 MPa, the most common is argon, helium and other nitrogen or argon and nitrogen mixed gas. The fission chamber neutron sensitivity depends on the emitter and collector gap between the current gaps, the greater gap the current is great. In order to improve the signal to noise ratio, the gap must be reduced when the neutron flux is high. The best way is to increase the uranium enrichment, and increasing uranium surface area by changing the length and diameter of the fission chamber. The outer diameter of the fission chamber inside core is about 6 mm, and the sensitivity length is about 12 ~ 25 mm.

(2)   Micro ionization chamber.

Boron coated ionization chamber can be used as a mobile core neutron flux measurements of the sensitive element. Generally speaking, the core fission chamber with full power of 9 months later, the sensitivity, down from its initial value to 50% or so; and micro ionization chamber in a month and a half, the sensitivity is reduced by 50% (due to thermal neutron cross section of ^{10}B is six times as ^{235}U, resulting in consumption is too large). As a mobile core measuring device, the time to through the core required less than 3 min, and through the core frequency was barely more than a month time. Therefore, the boron-coated ionization chamber can be satisfied to work for many years.

(3) Core neutron flux measurement system.

Including detector and its driving mechanism, measuring pipeline selector, measurement pipeline and other mechanical devices, and signal processing equipment and other parts. The operator control selector, select the corresponding measuring pipeline, by measuring pipe driving mechanism into the core from the bottom of the predetermined channel, and along the core for from bottom to top and from top to bottom, measure the current signal in the process of movement and the tail cable is transmitted to the detector signal processing equipment. A 900 MW pressurized water reactor nuclear power plant at the bottom of the reactor vessel is provided with 50 channels and 50 core neutron flux measuring pipe is connected with the driving mechanism 5 sets of detectors, each detector in order through 10 channels, repeatedly inserted. A neutron flux distribution measurement takes about 2 h.

(4) Self powered neutron detector.

Self-powered neutron detector is a basic detector by using the neutron activation of radioactive decay produces a signal current, does not require external voltage. No air ionization region, but to be used instead of solid structure of neutron sensitive materials of neutron. Sensitive materials and wires, and the ceramic insulator tightly with wire and neutron sensitive materials and detectors separated. Coat formed like a detector based on inorganic insulation coaxial cable, small volume and compact. Simple structure makes this kind of detector has many advantages, including low cost, readout device simple, low fuel consumption, long service life and sensitivity and good reproducibility.

A typical self-powered neutron detector is composed of 4 parts: emitter, insulator, conductor and coat (or collector). The emitter is very high activation cross section material for thermal neutron. Later, through the launch of high-energy beta rays with appropriate half-life decay, electron is released in this decay process. The insulator is solid, at the temperature in the reactor core and nuclear radiation environment; it must maintain a high resistance performance. According to the ideal situation, it should not emit beta (electronic wire and coat or collector must emit only few beta or electrons), so that the background noise signal to the minimum. Figure 6.73 shows the

Fig. 6.73 Self-powered neutron detector with Inconel wire

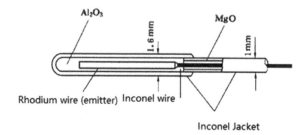

principle diagram of a self-powered neutron detector. The neutron sensitive electrode is fixed in Inconel wire, the emitter and the wire through the insulator. Magnesium Oxide Inconel coat sleeve in insulator. The entire assembly is extruded into a long cylindrical part with a diameter of 1.5 mm.

Self-powered detectors are mainly the following three types: ① emission pole (such as Rh) beta decay occurred after neutron capture, beta flow neutron detector; ② emission pole (such as Co, Sc or cadmium) release instantaneous gamma rays after neutron capture, and then eject electrons; ③ captured or scattered gamma rays in emission, Compton electrons and photoelectrons produced, so that the emitter positively charged. Rh and V is the most popular emission materials.

6.6.11 Nuclear Power Measurement

Nuclear power plant is usually used to observe directly and fission rate linked to the "radiation" to measure the reactor power level. Fission reactions associated with neutrons and gamma rays penetrate some distance then can still be detected, so nuclear power measurement technology is based on neutron detection, gamma rays or both at the same time. In order to reduce the gamma effect, set in reactor core around (for pressurized water reactor, set in pressure vessel lateral) nuclear power detector is usually used neutron detector, which is the main means of measurement of nuclear power. Neutron detector readings need to heat power calibration, said thermal power calibration.

The range of reactor power is great (from a few watts to several hundred megawatts), so despite the wide range of the detector, using a set of detectors and circuits is impossible to meet the requirements. The most common method is using three ranges: source range and intermediate range and power range. In order to make control and security functions by a set of detectors smoothly transferred to another group of detector and a portion of the range id required to be covered by another group of detector. The typical amount of overlap is one to two orders of magnitude.

Source range instruments typically consist of two redundant counting rate channels, each consisting of a high sensitivity proportional counter and a related signal measurement device.

These channels are typically used over a count range of $0.1–10^6$ counts per second, but vary depending on the reactor design. Its output is displayed on the gauge in terms of the logarithm of the count rate.

The source range instrument also measures the rate of change of the count rate. The change rate is displayed on the gauge at a startup rate of -1 to $+10$ decades per minute. Due to the inherent limitations of this range, the protection function is not usually associated with source range instruments. However, interlocks can be incorporated.

Many reactor plants have found it necessary to place source-range proportional counters in lead shielding to reduce gamma flux on detectors. This has two functions: (a) it increases the low-end sensitivity of the detector and (b) it increases the life of the detector. Another way to extend the detector's life is to disable the detector's high-voltage power supply and shorten the signal lead when neutron flux enters the intermediate range. Some reactor plants have made provisions for moving the source range detectors from their operating positions to a position of reduced neutron flux level, once the flux level increases above the source range.

The source range corresponding to a subcritical reactor. From the starting state for nuclear power to critical state measurement. This exposure to the neutron flux on the detector is usually very low, in fact, to pick up a single neutron situation. In this case, because the detector may be at a relatively high gamma ray background field so, pulsed neutron count detector is used. The lower limit of the source range is achieved by requirement of safety conditions. The lowest rate of count rate is generally 1 ~ 10 count/s. in order to ensure a reliable measure, must make the reactor neutron counting in the subcritical state rate more than this value. Therefore, an artificial neutron source usually should be set in the core (see Chap. 11). Usually the source neutron flux range covers 10^{-1} ~ two $\times 10^5 n/(cm^2 s)$, equivalent to 10^{-11} ~ 10^{-5} of rated full power.

The intermediate range depends on the source range, the typical maximum count rate is 10^6 counts/s, and allowing resolution loss is less than 10%. Corresponding to the reactor from the critical state to the rated power of about 10% of nuclear power. Range covers the neutron flux is 2×10^2 ~ 2×10^{10} n/(cm^2 s), the rated power of 10^{-8} ~100% full power.

The intermediate-range nuclear instrument consists of at least two redundant channels. Each channel consists of a boron-lined or boron gas-filled compensatory ion chamber and associated signal measurement equipment, where the output is a steady current generated by neutron flux.

The compensated ion chamber is used in the intermediate range because the current output is proportional to the relatively stable neutron flux and it compensates for the signal from the gamma flux. This range indication also provides a measure of the rate of change at the neutron level.

This rate of change is displayed on the meter at a startup rate of several decades per minute (-1 to $+10$ per minute). A high startup rate for either channel may initiate protective operations. This protective effect may take the form of a control rod withdrawal inhibit and alarm, or in the form of high start-up rate reactor trip.

Power range. The output of each power range channel is directly proportional to reactor power and typically covers a range from 0 to 125% of full power, but varies with each reactor. One can accurately, in linear proportion to read out the reactor power. The output of each channel is displayed on the meter at the power level of the full set power percentage. The gain per instrument is adjustable, which provides a means of calibrating the output. This adjustment is usually determined by using plant thermal balance. Protection actions can be initiated by high power levels across either channel: this is called a coincidence operation.

In the power range interference radiation, usually do not bring much difficulty. No compensation gamma neutron detectors can also meet the requirements, but for the sake of consistent, using a gamma compensation neutron ionization chamber. Measurement range is 5×10^2 ~ 5×10^{10} n/(cm^2 s), equal to 10^{-8} ~150% FP.

Power range nuclear instruments typically consist of four identical linear power level channels derived from eight uncompensated ion chambers. The output is a steady current generated by neutron flux. Since gamma compensation is unnecessary, uncompensated ion chambers are used in the power range: the flux ratio between neutrons and gamma is high. Neutrons have a higher flux than gammas, which means that the number of gammas is negligible compared to neutrons.

In addition, with the concentration of ^{16}N in gamma ray detectors for the measurement of reactor coolant loop to measure power reactor is also studying. ^{16}N is produced by neutron activation from oxygen, the concentration and reactor flux is directly proportional. That is to say and nuclear power is proportional to the concentration of ^{16}N.

6.7 Principles of Process Control

Control systems integrate elements whose function is to maintain a process variable at a desired value or within a desired range of values.

Process control in industry refers to the process parameters such as temperature, pressure, flow rate, liquid level and composition as controlled variable.

A control system is an integrated elemental system whose function is to maintain process variables based on expectations or within the desired value range. The control system monitors process variables or variables and then causes some operations to occur to maintain the required system parameters. Take the central heating device, which uses a thermostat to monitor the temperature of the house. When the house temperature drops to a preset, the stove opens to provide a heat source. The temperature of the house increases until the switch in the thermostat causes the stove to shut down.

The two terms that help define a control system are input and output. The control system input is the stimulation applied to the control system from an external source, and a specific response is generated from the control system. In the central heating unit, the control system enters the temperature of the house monitored by the thermostat.

The control system output is the actual response from the control system. In the example above, a drop in temperature to a preset value of the thermostat causes the furnace to turn on, providing heat to raise the temperature of the house.

In the case of nuclear facilities, inputs and outputs are determined by the purpose of the control system. A knowledge of the input and output of the control system enables the components of the system to be identified. The control system may have multiple inputs or outputs.

The control system is classified by control action, which is the quantity responsible for activating the control system to produce the output. The two general categories are open-loop and closed-loop control systems.

The open-loop control system is a system whose control operation is independent of the output. An example of an open-loop control system is a chemically added pump

Fig. 6.74 Open-loop control
system

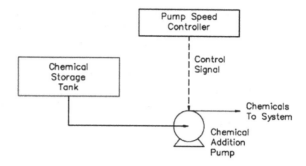

with variable speed control (Fig. 6.74). The feed rate of chemicals that maintain the
proper chemical composition of the system is determined by operators who are not
part of the control system. If the chemical composition of the system changes, the
pump cannot respond by adjusting the feed rate (speed) without operator operation.

A closed-loop control system is a system in which the control action depends
on the output. Figure 6.75 shows an example of a closed-loop control system. The
control system maintains the water level in the tank. The system performs this task
by continuously sensing the water level in the tank and adjusting the supply valve to
add more or less water to the tank. The required level is preset by the operator, who
is not part of the system.

Feedback is information about process variable conditions in a closed-loop control
system. This variable compares to the expected conditions that result in appropriate
control operations on the process. Information is constantly "feedback" to the control
circuit in response to control operations. In the previous example, the actual tank
water level perceived by the level transmitter is fed back to the water level controller.

Fig. 6.75 Closed-loop
control system

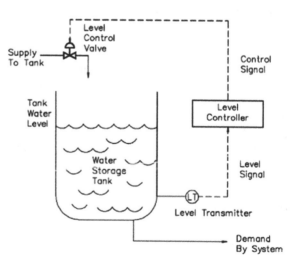

This feedback is compared to the required level to generate the required control action, positioning the level control as required to maintain the desired level.

The automatic control system is a preset closed-loop control system that requires no operator operation. This assumes that the process is still within the normal range of the control system. The automatic control system has two process variables associated with it: a controlled variable and a manipulated variable.

A controlled variable is a process variable that is maintained by a specified value or within a specified range. In the previous example, the tank level was a controlled variable.

A manipulated variable is a process variable that a control system performs to maintain a controlled variable within a specified value or range. In the previous example, the flow rate of water supplied to the water tank is a manipulative variable.

In any automatic control system, the four basic functions that occur are:

- measurement
- compare
- compute
- correction.

In the water tank water level control system in the example above, the level transmitter measures the water level in the water tank. The level transmitter sends a signal to the level control device representing the tank position, comparing it to the desired tank level. The level control device then calculates the extent to which the power valve is opened to correct any difference between the actual and desired tank levels.

The three functional elements required to perform the functions of the automatic control system are:

- Measurement elements
- Error detection element
- The final control element.

Figure 6.76 shows the relationship between these elements and the functions they perform in the automatic control system. Measurement elements perform measurement functions by sensing and evaluating controlled variables. The error detection element first compares the value of the controlled variable to the expected value, and then sends the error signal when there is a deviation between the actual value and the expected value. The final control element responds to the error signal by correcting the manipulation variables of the process.

6.7.1 Control Loop Diagrams

A block diagram is a pictorial representation of the cause and effect relationship between the input and output of a physical system. A block diagram provides a

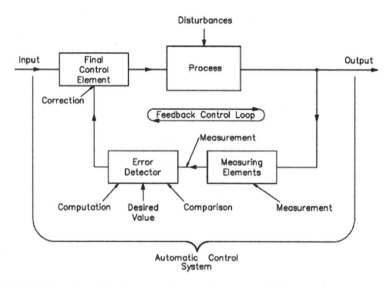

Fig. 6.76 Relationships of functions and elements in an automatic control system

means to easily identify the functional relationships among the various components of a control system.

A loop diagram is a "roadmap" that tracks process fluids through the system and specifies variables that could disrupt the balance of the system.

The simplest form of a block diagram is a block and arrow chart. It consists of a single block, an input, and an output. Blocks usually contain the name of the element (Fig. 6.77a) or the mathematical operation symbol (Fig. 6.77b) will be executed on the input to get the desired output. Arrows identify the direction of the information or signal flow [13].

The box is usually written in the name of the module (a), or the mathematical operation of the module (b). The arrow indicates the direction of the signal or the transmission of information.

Although blocks are used to identify multiple types of mathematical operations, addition and subtraction operations are represented by a circle called summing points. As Shown in Fig. 6.78, the summing point may have one or more inputs. Each input has its own appropriate positive and negative symbols. The summing point has only one output, equal to the sum of the algebraics of the input.

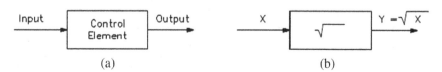

(a)   (b)

Fig. 6.77 Block and arrows

Fig. 6.78 Summing points

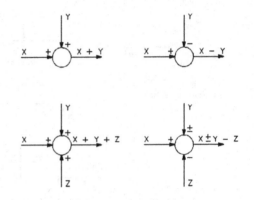

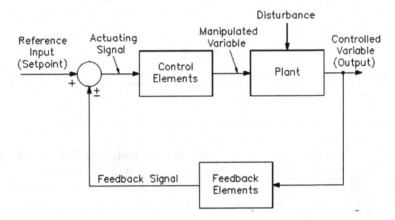

Fig. 6.79 Feedback control system block diagram

Figure 6.79 shows the basic elements of the feedback control system represented by the block diagram. The functional relationship between these elements is easy to see. An important factor to keep in mind is that block diagrams represent the flow path that controls the signal, but not the flow of energy through the system or process.

Here are a few terms related to closed-loop block diagrams.

A plant is a system or process that controls a specific number or condition. This is also known as a control system. Control elements are the components required to generate the appropriate control signal for the plant. These elements are also known as controllers. Feedback elements are the components required to determine the functional relationship between the feedback signal and the controlled output. The reference point is an external signal applied to the summing point of the control system, causing the plant to produce specific operations. This signal represents the expected value of the controlled variable, called a set point.

Controlled output is the number or condition of the plant, which is controlled. This signal represents a controlled variable. A feedback signal is a function of the output signal. It is sent to the summing point and algebra is added to the reference

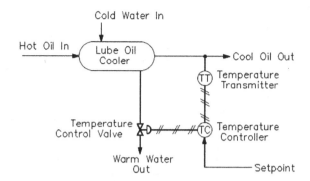

Fig. 6.80 Lube oil cooler temperature control system

input signal to obtain the drive signal. The drive signal represents the control function of the control ring, which is equal to the algebraic sum of the reference input and feedback signals. This is also known as an "error signal."

The manipulated variables are variables in the process of maintaining output (controlled variables) as expected. Perturbation is a undesirable input signal that upsets the value of the controlled output.

Figure 6.80 shows the typical application of a block diagram to identify the operation of the temperature control system for lubricants. Figure 6.80 shows a diagram of the lubricant cooler and its associated temperature control system.

Lubricants reduce friction between moving mechanical components and remove heat from components. As a result, the oil becomes hot. This heat is removed from the lubricant through the cooler to prevent oil failure and damage to mechanical components.

The lubricant cooler consists of a hollow shell through which several tubes pass. Cooling water flows into the shell of the cooler and outside the tube. The lubricant flows through the tube. Water and lubricants never make physical contact.

When water flows through the shell side of the cooler, it heats through the tube from the lubricant. This cools the lubricant and heats the cooling water as it leaves the cooler.

The lubricant must be kept in a specific operating band to ensure optimum equipment performance. This is achieved by controlling the flow of cooling water through the temperature control loop.

The temperature control loop consists of a temperature transmitter, a temperature controller, and a temperature control valve. The diagonally crossed lines indicates that the control signal is air (pneumatic).

The oil temperature is a controlled variable because it remains at the desired (set point). The cooling water flow rate is a variable of manipulation because it is regulated by the temperature control valve to maintain the lubricant temperature. The temperature transmitter senses the temperature of the lubricant as it leaves the cooler and sends an air signal proportional to the temperature controller. Next, the temperature controller compares the actual temperature of the lubricant to the set point (the desired value). If there is a difference between the actual temperature and the

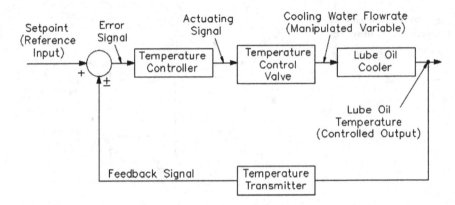

Fig. 6.81 Equivalent block diagram of temperature control system

expected temperature, the controller changes the control air signal to a temperature control valve. This causes it to move in the direction and as much as is needed to correct the difference. For example, if the actual temperature is greater than the set point value, the controller changes the control air signal, causing the valve to move in the open direction.

This results in more cooling water flowing through the cooler and lowers the temperature of the lube oil leaving the cooler.

Figure 6.81 represents the lubricant temperature control loop as a block diagram. The lubricant cooler is the plant in this example and the control output is the lubricant temperature. The temperature transmitter is the feedback element. It senses controlled output and lubricant temperatures and generates feedback signals.

Feedback signals are sent to summing points and added algebra to the reference input (set point). Note that the set point signal is positive and the feedback signal is negative. This means that the resulting driver signal is the difference between the set point and the feedback signal.

The actuating signal passes through two control elements: the temperature controller and the temperature control valve. The temperature control valve responds by adjusting the operating variable (cooling water flow rate). The temperature of the lubricant varies according to the water flow rate, and the control loop is complete.

In the last example, the control of the lubricant temperature may initially seem easy. Obviously, the operator simply measures the lubricant temperature, compares the actual temperature to the required (setpoint), calculates the amount of error, if any, and adjusts the temperature control valve to correct the error accordingly. However, the process is characterized by delays and delays in changes in the values of process variables. This feature greatly increases the difficulty of control.

Process time lag is a general term for the delay and retardations of these processes. Process time lag is caused by the three properties of the procedure. They are capacitance, resistance, and transportation time.

Capacitance are the ability to store energy in the process. For example, in Fig. 6.81, the pipe walls, cooling water, and lubricants in the lubricant cooler can store heat.

This energy storage property has the ability to slow down change. If the cooling water flow rate increases, it will take some time to remove more energy from the lubricant to reduce its temperature.

Resistance is part of the process against the transfer of energy between capabilities. In Fig. 6.81, the walls of the lubricant cooler oppose the transfer of heat from the oil in the tube to the cooling water outside the tube.

Transportation time is the time it takes for a process variable to go from one point to another. If the temperature of the lubricant (Fig. 6.81) is reduced by increasing the flow rate of cooling water, it will take some time before the lubricant is transferred from the lubricant cooler to the temperature transmitter. If the transmitter moves further away from the lubricant cooler, the transport time will increase. This time lag is more than just a slowdown or delay in change: it is an actual time delay during which nothing changes.

All control modes described earlier can return process variables to stable values after disturbances. This feature is called stability.

Stability is the ability of the control loop to return controlled variables to stable, non-cyclic value, following a disturbance.

The control loop can be stable or unstable. Instability is caused by the combination of process time lag (i.e. capacitance, resistance, and transport time) discussed earlier and the time lag inherent in the control system. This results in a slow response to changes in controlled variables.

As a result, controlled variables cycle continuously around the setpoint value. Oscillations describe this circular characteristic. Three types of oscillations can occur in the control loop. They are reducing amplitude, constant amplitude, and increasing amplitude. Each diagram is shown in Fig. 6.82.

Fig. 6.82 Types of oscillations

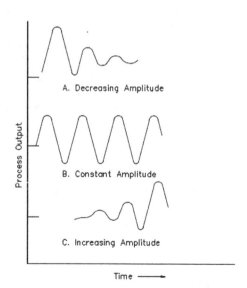

Decreasing amplitude (Fig. 6.82a). These oscillations decrease in amplitude and eventually stop with a control system that opposes the change in the controlled variable. This is the condition desired in an automatic control system.

Constant amplitude (Fig. 6.82b). Action of the controller sustains oscillations of the controlled variable. The controlled variable will never reach a stable condition; therefore, this condition is not desired.

Increasing amplitude (Fig. 6.82c). The control system not only sustains oscillations but also increases them. The control element has reached its full travel limits and causes the process to go out of control.

6.7.2 Two Position Control Systems

Two-position controllers are the simplest type of controller [14].

The controller is the device that generates the output signal based on the received input signal. The input signal is actually an error signal, which measures the difference between the variable and the desired value (or setpoint).

This input error signal indicates the degree of deviation between where the process system actually runs and where the process system needs to run. The controller provides an output signal to the final control element, which adjusts the process system to reduce this deviation. The characteristics of this output signal depend on the type or mode of the controller. This section describes the simplest type of controller, the two locations, or the ON OFF mode controller (Fig. 6.83).

A dual-position controller is a device that has two operating conditions: fully on or off.

Figure 6.84 shows the input to output, characteristic waveform of the dual-position controller switching from the OFF state to the ON state when the measuring variable is above the set point. Instead, when the measurement variable is below the set point, it switches from the "ON" state to the "OFF" state. This device provides output determined by whether the error signal is above or below the set point. The size of the error signal is above or below the set point. The magnitude of the error signal past that point is of no concern to the controller.

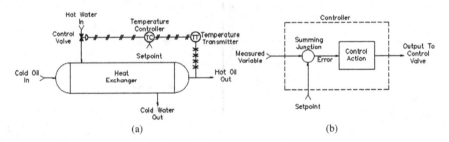

(a)   (b)

Fig. 6.83 Process control system operation

Fig. 6.84 Two position
controller input/output
relationship

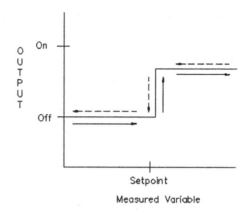

Fig. 6.85 Two position
control system

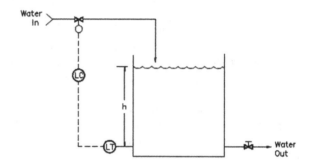

A system using a two-position controller is shown in Fig. 6.85.

The control process is the amount of water in the water tank. The controlled variable is the level in the tank. It is measured by a level detector that sends information to the controller. The output of the controller is sent to the final control element, which is a solenoid valve that controls the flow of water to the water tank.

When the water level initially drops, the measuring variable reaches a point below the set point. This creates a positive error signal. The controller fully opens the final control element. The water is then injected into the water tank and the water level rises. Once the water level exceeds the set point, a negative error signal is generated. A negative error signal causes the controller to shut down the final control element. This opening and closing of the final control element results in a cycling characteristic of the measured variable.

6.7.3 Proportional Control

Control mode is how the error between the control system's desired value (setpoint) and its actual value relative to the controlled variable is corrected. The control mode of a particular application depends on the characteristics of the process under control. For example, some processes can operate in the wide band, while others must remain very close to the set point. In addition, some processes change relatively slowly, while others change almost immediately.

Deviation is the difference between the set point of a process variable and its actual value. This is a key term used to discuss various control modes.

Four modes of control commonly used for most applications are:

Proportional
Proportional plus reset (PI)
Proportional plus rate (PD)
Proportional plus reset and rate (PID).

Each mode of control has characteristic advantages and limitations. The modes of control are discussed in this and the next several sections of this module.

In proportional (throttling) mode, there is a continuous linear relationship between the controlled variable value and the position of the final control element. In other words, the amount of valve movement is proportional to the amount of deviation.

Figure 6.86 shows the relationship between the valve position and the proportional mode control variable (temperature) characteristics. Please note that the valve position varies in exact proportions with deviation. In addition, the proportional mode responds only to the amount of deviation and is insensitive to the rate or duration of the deviation. Under the 2 and 4 min mark, when the temperature returns to its setpoint value, the valve returns to its original position. There is no valve correction without deviation.

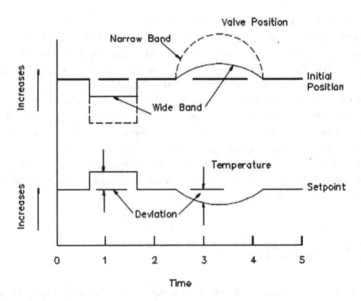

Fig. 6.86 Valve position and controlled variable at proportional mode

Three terms commonly used to describe the proportional mode of control are proportional band, gain, and offset.

Proportional band, (also called throttling range), is a change in the value of the controlled variable, resulting in the complete travel of the final control element. Figure 6.14 shows the relationship between the valve position and the temperature band of two different proportional bands.

The proportional band of a particular instrument is expressed as a percentage of the full range. For example, if the instrument has a full range of 200 °C and a temperature change of 50 °C to cause full valve travel, the percentage band of 200 °C is 50 °C, or 25%. Proportional bands can range from less than 1% to more than 200%. However, a proportional band of more than 100% does not cause full valve travel even if a full range change is made to the control variable.

Gain, also known as sensitivity, compares the ratio of the amount of change in the final control element to the amount of change in the controlled variable. Mathematically, gain and sensitivity are equivalent to proportional bands.

Offset, also known as droop, it is a deviation that persists after the process stabilizes. Offset is an inherent feature of the proportional control mode. In other words, the proportional control mode does not necessarily return controlled variables to their setpoints.

With proportional control, the final control element has a definite position for each value of the measurement variable. In other words, the output has a linear relationship to the input. The proportional band is the input change required for a full range of output changes due to proportional control. Or simply put, it is a percentage change in the percentage of the input signal required to change the output signal from 0 to 100%.

The proportional band determines the output value range of the controller that runs the final control element. The final control element acts on the manipulated variable to determine the value of the controlled variable. The controlled variable remains within the specified control point segment around the set point.

To demonstrate, let's look at Fig. 6.87. In the case of this proportional level control system, control the water supply flow to the inlet tank to keep the water tank level within the specified limits. The demand that disturbances placed on the process system are such that the actual flow rates cannot be predicted. Therefore, the system is designed to control tank levels in narrow bands to minimize the chance of significant demand disturbance caused by overflow or exhaustion. The fulcrum and lever assembly are used as proportional controllers. The float chamber is a level measuring element, and the four-inch stroke valve is the final control element. The level change in the fulcrum set to 4-in causes the valve to be fully 4-in stroke. Therefore, the 100% change in the controller output is equal to 4 in.

The controller has a proportional band of 100%, which means that the input must change 100% to cause a 100% change in the controller output.

If the fulcrum setting changes so that a change in the input level of 2 in or 50% results in a full 3-in stroke (or 100% of the output), the proportional band becomes 50%. The proportional band of the proportional controller is important because it determines the output range of a given input.

Fig. 6.87 Proportional
system controller

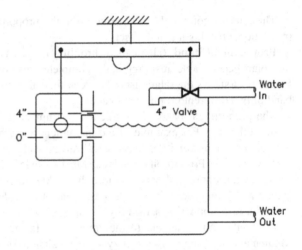

Figure 6.88 illustrates a process system using a proportional temperature controller for providing hot water.

Steam enters the heat exchanger to increase the temperature of the cold water supply. The temperature detector monitors the hot water outlet and produces a 20–100 kPa output signal, representing a controlled variable range from 100 to 300 °C. The controller compares the measured variable signal to the set point and sends the output of the 20–100 kPa to the final control element (i.e. 3-in control valve).

The controller is set to 50% proportional band. As a result, a 50% change in the span of 200 °C, or 100 °C, results in a 100% change in the controller output.

The proportional controller is reverse action, allowing the control valve to throttle to reduce steam flow as the hot water outlet temperature increases: as the water temperature decreases, the control valve opens further to increase the steam flow.

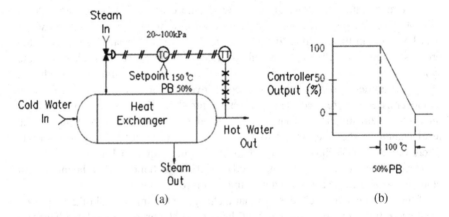

Fig. 6.88 Proportional temperature control system

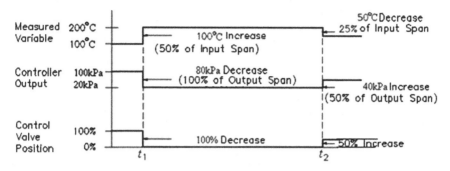

Fig. 6.89 Combined controller and final control element action

The combination of controller and control valve for different variations in the measuring variables is shown in Figs. 6.89.

Initially, the measured variable value is equal to 100 °C. The controller is set so that this value of the measurement variable corresponds to the 100% output (i.e. 100 kPa), which in turn corresponds to the "fully open" control valve position.

At time t_1, the measured variable increases by 100 °C, which is 50% of the measured variable span. Because the controller's proportional band is 50%, this 50% controller input change results in a 100% controller output change. The direction of the controller output changes because the controller is acting in reverse. A 100% reduction is equivalent to a reduction in output from 100 to 20 kPa, which causes the control valve to go from fully open to fully shut.

At time t_2, the measured variable is reduced by 50 °C, which is 25% of the measured variable span. A 25% reduction in controller input results in a 50% increase in controller output. As a result, the controller output is increased from 20 to 100 kPa and the control valve is increased from full shut-off to 50% open.

The purpose of the system is to provide hot water at the set point of 150 °C. The system must be able to handle demand disturbances that may cause the outlet temperature to rise or fall from the set point. Therefore, the controller's settings enable the system to look like it is shown in Fig. 6.90.

If the measured variable is below the set point, a positive error occurs and the control valve opens further. If the measured variable is above the set point, a negative error occurs and the control valve throttle is lowered (opening is reduced). A 50% proportional band causes the valve to make a full stroke between the error of +50° C and the error of −50° C. When the error is equal to zero, the controller provides a 50% or 90 kPa signal to the control valve.

When the error is above and below this point, the output generated by the controller is proportional to the size of the error, determined by the value of the proportional band. The control valve can then be positioned to compensate for any disturbances that may cause the process to deviate from the set point in either direction.

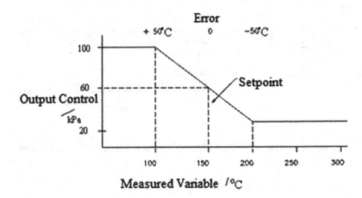

Fig. 6.90 Controller characteristic curve

6.7.4 Integral Control Systems

The output rate of the integral controller depends on the magnitude of the input.

Integral control describes a controller where the rate of output change depending on the magnitude of the input. Specifically, a smaller amplitude input can cause the output to change more slowly. This controller is called an integral controller because it approximates an integrated mathematical function. The integral control method is also known as reset control.

Devices that perform integration mathematical functions are called integrators. Integration mathematical results are called integral. The integrator provides linear output, the rate of change is directly related to the amplitude of the step change input, and a constant that specifies the function of integration.

For example, as in Fig. 6.91, the step changes at an amplitude of 10%, and the constant of the integrator causes a 0.2% change in the output for every 1% of the input. The role of the integrator is to convert step changes into gradually changing signals. As you can see, the input amplitude is repeated in the output every 5 s. As long as the input remains unchanged at 10%, the output continues to increase every 5 s until the integrator is saturated.

When the controlled variable of the deviation is large, the input signal will be, at this time changes in the controller output rate, the control unit changes quickly. In turn, if the deviation is small, the controlled variable close to the set value, response of the unit control is relatively stable.

Figure 6.92 shows an example of an integral flow control system.

Initially, the flow requirements for the system are expected to be 0.5 m³/s, equivalent to 50% of the control valve opening. The set point is equal to 0.5 m³/s and the actual flow measurement is 0.5 m³/s, sending a zero error signal to the input of the integral controller. The controller output is initially set to 50%, or 60 kPa, and the output position 6-in control valve to 3-in open position.

If the measured variable drops from an initial value of 0.5 m³/s to the new value of 0.45 m³/s shown in Fig. 6.93, a positive error of 5% is generated and applied to

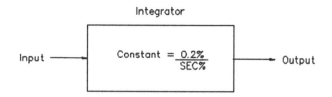

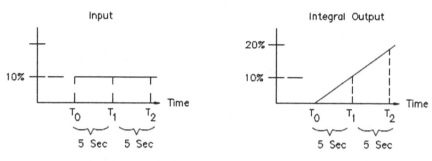

Fig. 6.91 Integral output for a fixed input

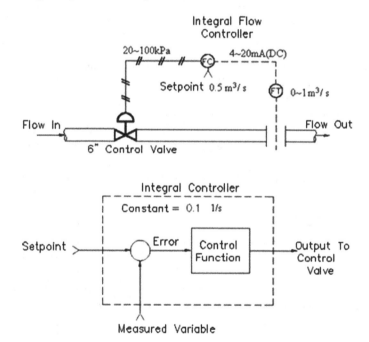

Fig. 6.92 Integral flow rate controller

Fig. 6.93 Integral controller response

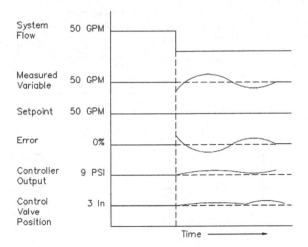

the input of the overall controller. The controller has a constant of $0.1 \ s^{-1}$, so the controller output changes at a rate of 0.5%/s.

A positive 0.5%/s indicates that the controller output has increased from 50% of the initial point to 0.5%/s. This causes the control valve to open further at 0.5%/s, increasing flow.

The controller returns the procedure to the set point. This is done through the repositioning of the control valve. When the controller causes the control valve to reposition, the measured variable approaches the set point and generates a new error signal. The cycle repeats itself until no error exists.

The integral controller responds to both the amplitude and time duration of the error signal. Some error signals are large or present for a long time, which may cause the final control element to reach its "full on" or "full off" position before the error is reduced to zero. If this happens, the final control element is still in extreme position and errors must be reduced in other ways in the actual operation of the process system.

The major advantage of integral controllers is that they have the unique ability to return the controlled variable back to the exact setpoint following a disturbance.

The disadvantage of the integral control mode is that it reacts relatively slowly to the error signal, and it initially allows for large deviations at the moment of error. This can cause the system to become unstable and cyclic operation. Therefore, the integral control mode is usually not used alone, but is combined with another control mode.

6.7.5 *Proportional Plus Integral Control Systems*

Proportional plus integral control is a combination of proportional control mode and integral control mode.

This type of control is actually a combination of the two control modes discussed earlier. Combining these two modes gives you an advantage and compensates for the disadvantages of both individual patterns.

The main advantage of proportional control mode is that whenever there is an error signal from the controller, the proportional output is generated immediately, as shown in Fig. 6.94. The proportional controller is considered a fast-acting device. This instant output change allows the proportional controller to reposition the final control element based on errors in a relatively short period of time.

The main disadvantage of the proportional control mode is that there is a residual offset error between the measurement variable and the set point, except for a set of system conditions. The main advantage of the integral control mode is that the controller output continues to reposition the final control element until the error is reduced to zero. This eliminates the remaining offset errors allowed by the proportional mode.

The main disadvantage of integral mode is that the controller output does not immediately direct the final control element to a new location in response to the error signal. The controller output changes at a defined change speed and takes time to reposition the final control element.

The combination of the two control modes is called the proportional plus integral (PI) control mode. It combines the direct output characteristics of the proportional control mode with the zero residual offset characteristics of the integral mode.

Let's refer again to our heat exchanger example (see Fig. 6.94a). This time, we will apply a proportional plus integral controller to the process system.

The response curve shown in Fig. 6.94b represents only the demand and measurement variables, which represent the hot water outlet temperature.

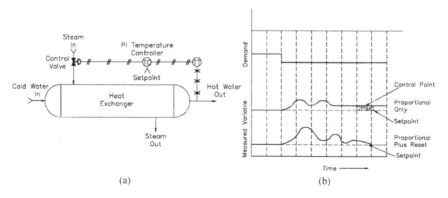

Fig. 6.94 Response of proportional plus integral control

Suppose this process experiences demand disturbance, which reduces the flow of hot water from the heat exchanger. The temperature and flow of steam into the heat exchanger remain the same. As a result, the temperature of hot water will begin to rise.

Proportional plus reset controller proportionality, if operated separately, will respond to disturbances and reposition the control valve, returning hot water to a new control point, as shown in the response curve. You'll notice that the residual error still exists.

By adding the reset action to the proportional action, the controller generates a larger output for a given error signal and causes a larger adjustment of the control valve. This causes the process to return to the set point more quickly. In addition, the reset eliminates offset errors over time.

The proportional plus reset controller eliminates offset errors found in proportional control by continuing to change output after the proportional action is complete and returning controlled variables to the set point.

The inherent disadvantage of proportional plus reset controllers is the possible adverse effects of large error signals. Large errors can be caused by a large number of demand deviations or when the system is initially started. This is a problem because a large persistent error signal eventually causes the controller to drive to its limits, resulting in a "reset windup".

Due to the end of the reset, this control mode is not suitable for processes that are frequently shut down and started.

6.7.6 Proportional Plus Derivative Control Systems

Proportional plus rate control is the control mode in which derivative parts are added to the proportional controller.

The proportional plus rate describes the control mode in which the derivative part is added to the scale controller. This derivative partially responds to the rate of change of the error signal, not the amplitude: this derivative operation responds to the rate of instantaneous change at which it begins. This causes the controller output to initially be directly related to changing the error signal rate. The higher the error signal rate, the faster the final control element is positioned to the expected value. The added derivative reduces the initial overshoot on the measurement variable, which helps stabilize the process faster.

This control mode is called proportional plus rate (PD) control because the derivative part responds to the rate of change of the error signal.

The device that generates the derivative signal is called the differentiator. Figure 6.95 shows the input–output relationship of the differentiator.

The output provided by the differentiator is directly related to the rate of change of the input, and the constant specifies the differentiated functionality. The derivative constant is represented in seconds and defines the differential controller output.

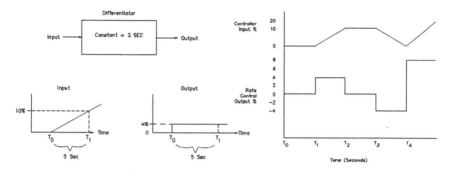

Fig. 6.95 Derivative output for a constant rate of change input

The role of the differentiator is to convert the changed signal to the constant magnitude signal shown in Fig. 6.95c. As long as the input rate of the change is constant, the amplitude of the output is constant. Changes in the new input rate will give a new output magnitude.

Derivatives cannot be used as control modes alone. This is because the steady state input produces zero output in the differentiator. If the differentiator were used as a controller, the input signal it would receive is the error signal. As mentioned earlier, the steady-state error signal corresponds to any number of necessary output signals used to locate the final control element.

Therefore, the derivative effect is combined with the proportional effect, so that the proportional part of the output acts as a derivative segment input.

The proportional plus rate controller utilizes the proportional and rate control mode. As seen in Fig. 6.96, the output provided by the proportional action is proportional to the error. If the error is not a one-step change, but a slow change, the proportional action is slow. When added, the rate action responds quickly to errors.

To illustrate proportional and rate control, we will use the same heat exchanger processes analyzed in the previous (see Figs. 6.97). However, for this example, the temperature controller used is either a proportional plus rate controller or derivative controller.

As shown in Fig. 6.97, the proportional control mode responds to the decrease in demand, but due to the inherent characteristics of the proportional control, the residual offset error still exists. Adding derivative actions affects the response by allowing only a small overshoot and quick stabilization to the new control point. Therefore, the derivative action provides greater stability for the system, but does not eliminate offset errors.

Proportional plus rate control is typically used in high-capacity or slow-response processes, such as temperature control. The leading role of controller output makes up for the lag characteristics of large capacity and slow process.

Rate action is not typically used for rapid response processes, such as flow control or noise processes, because derivative operations respond to any change rate of the error signal, including noise.

Fig. 6.96 Response of
proportional plus rate control

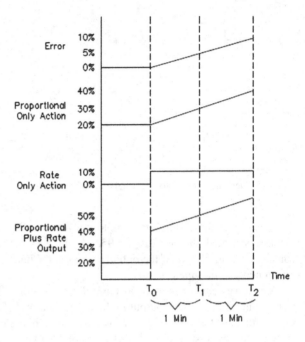

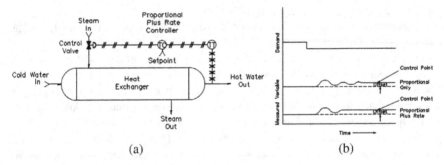

(a)   (b)

Fig. 6.97 Heat exchanger process

Proportional plus rate controllers are useful for processes because they are
typically started and shut down without being affected by reset windup.

6.7.7   Proportional-Integral-Derivative Control Systems

Proportional plus integral plus derivative controllers combine proportional control
actions with integral and derivative actions.

For processes that can be cycled continuously, a relatively inexpensive dual-
position controller is sufficient. Proportional controllers are commonly used for

processes that do not tolerate continuous cycling. For processes that neither continuous cycling nor offset errors can be tolerated, you can use proportional plus reset the controller. Proportional plus rate controllers are used for processes that require improved stability and tolerate offset errors.

However, some processes do not tolerate offset errors, but require good stability. The logical solution is to use a control mode that combines the benefits of proportional, reset, and rate action. This section describes patterns that are determined as proportional plus reset plus rate, commonly referred to as Proportional-Integral-Derivative (PID) [15].

When an error is introduced into the PID controller, the controller responds with a combination of proportional, integral, and derivative actions, as shown in Fig. 6.98.

Suppose the error is due to the slow increase in the measurement variable. As the error increases, the proportional action of the PID controller produces output proportional to the error signal. The controller reset action produces an output whose rate of change is determined by the magnitude of the error. In this case, as the error continues to increase at a steady rate, the reset output continues to increase its rate of change. The controller's rate action produces output determined by the rate of change. If combined, these actions produce the output shown in Fig. 6.98.

As you can see from the combined action curve, the resulting output immediately responds to an error, with a signal proportional to the magnitude of the error, and as long as the error continues to increase, the signal continues to increase.

Fig. 6.98 PID controller response

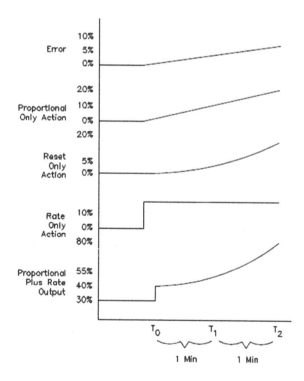

Fig. 6.99 PID control action
responses

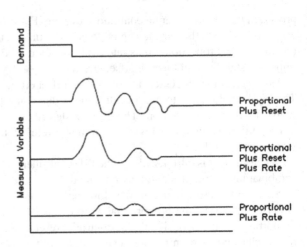

You must remember that if the control system does not take corrective action, these response curves are plotted. In fact, once the output of the controller begins to reposition the final control element, the error magnitude should begin to decrease. Eventually, the controller turns the error to zero and the control variable returns to the set point.

Figure 6.99 demonstrates the controller's combined response to demand disturbances. The controller's proportional action stabilizes the process. The reset action is combined with the proportional action to return the measurement variable to the set point. The combination of rate action and proportional action reduces initial overshoot and cycle period.

6.7.8 Controllers and Valve Actuators

Mechanical "watchdogs" called controllers are installed in a system to maintain process variables within a given parameter.

The controller is the control element of the control loop. Their function is to maintain a process variable (pressure, temperature, level, etc.) at a certain expected value. This value may or may not be constant.

This function is done by comparing the setpoint signal (expected value) with the actual value (controlled variable). If the two values are different, an error signal is generated. The error signal is amplified to produce the controller output signal. The output signal is sent to the final control element, which changes the manipulated variable and returns the controlled variable to the set point.

This sextion describes the two controllers that are common in the nuclear facility control room. Although the factory may have other types of controllers, the information provided here generally applies to those controllers as well.

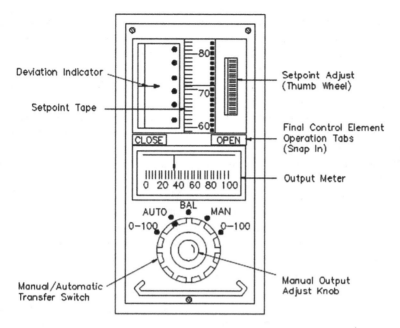

Fig. 6.100 Typical control station

Control stations performs the functions of the controller and provides additional controls and indicators that allow the operator to manually adjust the controller output to the final control element.

Figure 6.100 shows the front panel of a typical control station. It contains multiple metrics and controls. Each issue will be discussed.

The setpoint indicator is located in the center of the upper half of the controller and represents the set point (expected value) selected for the controller. Scales can be marked as 0–100%, or directly correspond to controlled variables.

The setpoint adjustment is located to the right of the set point indicator and is a thumbwheel adjustment dial that allows the operator to select the set point value. By rotating the thumb wheel, the scale moves below the setpoint index line.

The deviation indicator is located on the left side of the setpoint indicator and shows any error between the set point value and the actual controlled variable value (+10 to −10%). If there are no errors, and the deviation indicator remains mid-scale, consistent with the setpoint index mark. If the controlled variable is below the set point, the deviation indicates a downward deflection. If higher, the indicator deflects upward.

The output meter is a horizontal positioning meter below the deviation and setpoint indicator. It represents the percentage of the controller's output signal. The current range for this particular controller is from zero to 100%. However, this will correspond to the air signal of the pneumatic controller.

The Snap-in tab above both ends of the output meter indicates the direction in which the final control element moves to change the output signal. Tabs typically read "open–close" of the control valve and "slow–fast" of the variable speed motor or other appropriate designations.

The manual automatic (M-A) transmission switch selects the operating mode of the controller. The manual output adjustment knob is located in the center of the M-A transmission switch and changes the controller output signal in manual operating mode. The knob rotates clockwise to increase the signal and counterclockwise to decrease the signal.

The M-A transmission switch has five positions to change the operating mode. The indication is provided by the deviation meter.

AUTO. This is the normal position of the M-A transmission switch. It puts the controller in automatic operating mode. In addition, the deviation meter represents any deviation between the controlled variable and the setpoint.

0–100 (AUTO side). In this location, the controller is still in automatic mode. However, the deviation meter now represents an approximation of the controlled variable. The deviation meter is completely deflected downwards by a zero variable value and completely deflected to a 100% variable value.

MAN. This position places the controller in manual operation mode. The controller output will now be different by adjusting the manual output adjustment knob. This adjustment is indicated on the output gauge. The deviation meter represents any deviation between the controlled variable and the set point.

0–100 (MAN side). The controller is still in manual operating mode, and the deviation meter represents a controlled variable value (0–100%), as it is at the 0–100 (AUTO side) position.

BAL. In many cases, the controller output signal in automatic mode and manual mode may not be the same. If the controller is transferred directly from automatic to manual or manual to automatic, the controller output signal may suddenly change from one value to another. As a result, the final control element will experience a sudden change in position or "bump". This can cause the process variable values to move and may damage the final control element.

Bumpless transfer is the smooth transfer of the controller from one mode of operation to another. This smooth transfer is provided by the Balancing (BAL) position when the controller is transferred from automatic mode to manual mode. At the BAL position, the controller is still in automatic operating mode, but the deviation meter now represents the difference between the output of the manual and automatic control modes. Adjust the manual output until the deviation meter shows no deflection. The controller can now be transferred automatically and smoothly to manual.

To ensure manual-to-automatic bump transmission, the manual output signal indicated by the output table is adjusted to match the controlled variable values to set the point. The deviation meter does not deflect. After matching, the M-A transmission switch can be switched from manual (MAN) to automatic (AUTO) control.

Remote operation of the valve is easily managed by one of the four actuators.

The valve itself has no control over the process. The manual valve requires the operator to position it to control the process variables. Valves must be operated remotely and special equipment is automatically required to move them.

These devices are called actuators. The actuator may be a pneumatic, hydraulic or electric solenoid valve or motor.

A simplified diagram of pneumatic actuators is shown in Fig. 6.101. It operates in combination with the forces created by the Air Force and the Spring Force. The actuator locates the control valve by transmitting its motion through the stem. The rubber diaphragm separates the actuator housing into two air chambers. The upper room receives supply air through an opening at the top of the housing.

The bottom chamber contains a spring that forces the diaphragm to stop mechanically in the upper chamber. Finally, connect the local indicator to the stem to indicate the position of the valve.

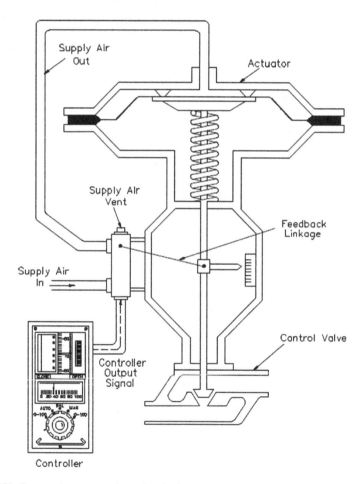

Fig. 6.101 Pneumatic actuator: air-to-close/spring-to-open

The position of the valve is controlled by different supply air pressures in the upper chamber. This causes different forces at the top of the diaphragm. Initially, when air is not supplied, the spring turns the diaphragm upward against the mechanical stops so that the valve is fully open. As the supply pressure increases from zero, the force at the top of the diaphragm begins to overcome the opposing forces of spring. This causes the diaphragm to move downwards and the control valve to close. As the supply pressure increases, the diaphragm continues to move downwards and compress the spring until the control valve is completely closed. Conversely, if the supply air pressure is reduced, the spring will start to force the diaphragm to open the control valve upward. In addition, if the supply pressure remains at a certain value between zero and maximum, the valve will be in the middle position. As a result, the valve can be located anywhere between full opening and full closure in response to changes in supply pressure.

Pneumatic actuators are typically used to control processes that require rapid and accurate response because they do not require a lot of power. However, hydraulic actuators are usually used when a lot of force is required to operate a valve, such as a main steam system valve. Although hydraulic actuators have many designs, piston types are the most common.

Typical piston hydraulic actuators are shown in Fig. 6.102. It consists of cylinders, pistons, springs, hydraulic supply and return line, and stem. The piston slides vertically into the cylinder, dividing the cylinder into two chambers. The upper chamber contains springs and the lower chamber contains hydraulic oil.

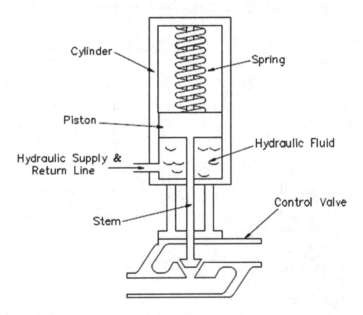

Fig. 6.102 Hydraulic actuator

The hydraulic supply and return line are connected to the lower chamber and allow hydraulic oil to flow in and out of the lower chamber of the actuator.

Stem transmits the movement of the piston to the valve.

Initially, the spring force keeps the valve in the closed position when there is no hydraulic oil pressure. When liquid enters the lower chamber, the pressure in the chamber increases. This pressure causes the force at the bottom of the piston to be opposite to the force caused by the spring. When the fluid pressure is greater than the spring force, the piston starts to move upward, the spring compresses, and the valve opens. As the hydraulic pressure increases, the valve continues to open. Conversely, when hydraulic oil is discharged from the cylinder, the fluid pressure becomes less than the spring force, the piston moves downwards, and the valve closes. By adjusting the amount of oil supplied or discharged from the actuator, the valve can be between full opening and full closure.

Hydraulic actuators operate in a similar way to pneumatic actuators. Each uses some power to overcome the spring force to move the valve. In addition, hydraulic actuators can be designed to fail on or off to provide fail safety feature.

Figure 6.103 shows a typical electric solenoid actuator. It consists of coils, armature, springs and stem.

The coil is connected to the external current supply. The spring rests on the armature to force it downward. The armature moves vertically within the coil and transmits its movement to the valve through the stem.

When current flows through the coil, a magnetic field forms around the coil. The magnetic field draws the center of the coil. When the armature moves upward, the spring collapses and the valve opens. The magnetic field collapses when the circuit opens and the current stops flowing to the coil. This allows the spring to expand and close the valve.

Fig. 6.103 Electric solenoid actuator

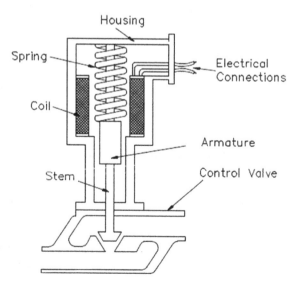

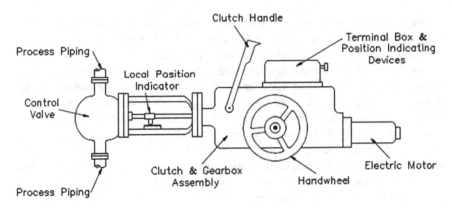

Fig. 6.104 Electric motor actuator

One of the main advantages of solenoid actuators is their fast operation. In addition, they are easier to install than pneumatic or hydraulic actuators. However, the solenoid actuator has two disadvantages. First, they have only two positions: fully open and completely closed. Second, they do not produce much force, so they usually operate only relatively small valves.

The design and application of motor actuators vary widely. Some motor actuators are designed to operate in only two positions (fully on or off). Other motors can be located between two positions. Typical electric motor actuators are shown in Fig. 6.104. Its main components include motors, clutches and gearbox assemblies, manual handwheels and stem to valves.

The motor moves the stem through the gear assembly. The motor rotates in reverse to open or close the valve. The clutch and clutch levers disconnect the motor from the gear assembly and allow the valve to operate the handwheel manually.

Most motor actuators are equipped with a limit switch, torque limiter or both. When the valve reaches a specific position, the limit switch de-energizes the motor. The torque limiter de-energizes the motor when the turning force reaches a specified value. When the valve reaches a position where it is fully open or closed, the turning force is usually the maximum. This function also prevents damage to the actuator or valve if the valve is combined in the middle position.

Exercises

1. Describe the principle of temperature measurement and the main material of RTD.
2. Describe the principle of thermal couple and compared with RTD.
3. When measuring the water level, why do you need to make the density compensation?
4. How to achieve the long distance water level display?
5. Describe the principle of Venturi flow meter, and attentions to use of it.
6. What are the characteristics of float type flow meter compared with other forms?

7. Brief description of the characteristics of PID controller.
8. According to Fig. 6.1, in the range of 200 ~ 400 °C, what metal as RTD material better? Why?
9. The thermocouple materials used in Table 6.1 are used to measure the electric potential of 3.48 mV, the reference point is 25 °C. How much is the temperature of the measuring point?
10. What is the compensation degree of γ compensated ionization chamber? How to set up the compensation current?
11. Describes the measurement range and characteristics of the three kinds of measurement range in nuclear power plant power measurement.
12. Describe the basic principles of nuclear power measurement using self-powered detectors in nuclear power plants briefly.
13. What is the signal to noise ratio? Why preamplifier could be used to improve the signal to noise ratio? What method can improve the signal to noise ratio?

References

1. Guofa L, Wenqi G. Nuclear power plant instrumentation and control. Beijing: Atomic Energy Press; 2014.
2. Blakemore SJ. Solid state physics. Cambridge UK: Cambridge University Press; 1985.
3. Strouse GF. Standard platinum resistance thermometer calibrations from the Ar TP to the Ag FP. Gaithersburg, M D: National Institute of Standards and Technology; 2008.
4. Academic Program for Nuclear Power Plant Personnel, vol. IV. General Physics Corporation, Library of Congress Card #A 397747, April 1982.
5. Rhodes TJ, Carroll GC. Industrial instruments for measurement and control. 2nd ed. McGraw-Hill Book Company.
6. U. S. Department of Energy. Instrumentation and control. Washington, D.C.: Lulu.com; 2016.
7. Kirk FW. Rimboi NR. Instrumentation. 3rd ed. American Technical Publishers, ISBN 0-8269-3422-6.
8. Fozard B. Instrumentation and control of nuclear reactors. London: ILIFFE Books Ltd.
9. Wightman EJ. Instrumentation in process control. Cleveland, Ohio: CRC Press.
10. Gollnick DA. Basic radiation protection technology. Temple City, California: Pacific Radiation Press.
11. Cork JM. Radioactivity and nuclear physics. 3rd ed. D. Van Nostrand Company, Inc.
12. Knoll GF. Radiation detection and measurement. Wiley; 1979. ISBN 0-471-49545-X.
13. Anderson NA. Instrumentation for process measurement and control. 2nd ed. Philadelphia, PA: Chilton Company; 1972.
14. Distefano JJ. III, Feedback and control systems, Schaum's Outline Series; 1967
15. Process measurement fundamentals, vol. I. General Physics Corporation; 1981. ISBN 0-87683-001-7.

Chapter 7
Chemistry and Chemical Engineering

Chemistry refers to the systematic investigation of the properties, structure and behavior of matter and the changes that occur in matter [1]. This general definition raises many questions. These questions have been answered in chemical research. This chapter will discuss terms and basic concepts that help to understanding chemistry.

The word "Chemistry" means science of change from the literal. Chemistry researches material composition, properties, structure and changes in molecular and atomic level in the laws of science. The world consists of matter; chemistry is one of the main methods and means of recognizing and rebuilding the physical world. In addition, chemical engineering issues related subject in engineering.

7.1 Chemical Basis

Chemistry studies the composition, structure, nature and change of matter at the atomic level, so it is necessary to first understand the structure of matter and the structure of atoms.

7.1.1 The Atom Structure

The physical form of matter can be divided into solid, liquid and gas.

Solid objects called solid, solid has a definite shape and volume. In a solid, interaction between molecules or atoms is very strong, so the solid does not need external support to maintain a certain shape. Interaction between atoms or molecules in the liquid is smaller than the solid; the liquid can determine the volume but no definite shape. The shape of the liquid depends on the shape of the container. Between the atoms or molecules of the gas force is smaller, neither shape nor the determination

© Tsinghua University Press 2022
J. Yu, *Fundamental Principles of Nuclear Engineering*,
https://doi.org/10.1007/978-981-16-0839-1_7

volume. The gas can be compressed or expanded; the volume depends on the size of the container.

Different forms of material have one thing in common; they are composed of the most basic particles—atoms. The atom is a basic particle chemical reaction cannot be divided, the atom is electrically neutral. Discovered atom has a long and interesting history, we will not go to the review. Although cannot be divided in chemical reactions, the atom has internal structure, to understand the internal structure of the atom for understanding the chemical process is very important. In the chap. 5, we introduced that the atom is composed of negative charged electrons and positive charged nucleus. The core of the atom composed of protons and neutrons (hydrogen atom is an exception, only one proton, and no neutrons). The neutron is electrically neutral. A proton is positive charge, so the charge of a nuclear core is equal to the number of protons.

The modern quark theory considers the neutrons and protons have internal structure, known as quark. Neutron although as a whole is not charged, but has charge distribution internally. For example, there are a theory that neutron outer surface is negatively charged, and positive charge internally, as a whole is not charged. Therefore, neutrons and protons within the nucleus of an atom can rely on electric field force to attract each other; this theory can help us to explain the structure of the atomic nucleus.

The mass of proton and neutron is very close, but different charge. Neutron is generally by a proton and a single electron electrical and combination in together. Therefore, the mass of a neutron and a proton plus electron is about the same. The mass of the electron is only about 1/1835 of the mass of the proton, an electron with a unit of negative charge.

Table 7.1 lists some useful parameters of atoms and elementary particles. Due to unit kg to measure the mass of the atomic, value is really too small, very inconvenient. For example, of a hydrogen atom it is 1.674×10^{-27} kg, the mass of one atom of oxygen is 2.657×10^{-26} kg. So we take into account a ^{12}C atom mass of 1.993×10^{-26} kg, with atomic mass unit (amu) to measure the atomic mass, it is defined as 1/12 of the mass of the ^{12}C atom.

Within the atom, electron in orbit around the nucleus of the orbit, like the earth, Mars and other planets move around the sun. The number of electrons is equal to the number of protons in the nucleus, so will the whole atom is electrically neutral. The electrons are constrained by the electric force between the nucleus and the electrons, like the earth by the sun's gravity attracted. If an electron is stripped by other force, the atom will show a positive charge. The atomic diameter is about the range of electrons can reach, only about 10^{-10} m. The diameter of nuclear is smaller, only

Table 7.1 Parameters of atoms and their basic particles

Name	Atomic mass unit/amu	Charge
Electron	0.0005486 or 1/1835	−1
Proton	1.007277	1
Neutron	1.008665	0

1/10000 of the diameter of the atom ($10^{-15} \sim 10^{-14}$ m). The nucleus concentrated most of the quality, but only a very small volume. Although the electron itself is very small, but the track occupied volume has accounted for the majority of atoms in the nucleus of an atom. Scene and the solar system is somewhat similar.

7.1.2 Chemical Elements and Molecules

A chemical element is a class of nuclear charged with the same number (i.e. nuclear proton number). The same atoms have similar chemical properties, so the same number of protons of the nuclide called chemical element or element shortly. Some elements examples are of hydrogen, nitrogen and carbon. Until 2015, a total of 118 elements were found, of which 94 species are naturally present in the earth. The atomic number greater than 83 (BI), the elements are unstable and will undergo radioactive decay. The instability does not mean nature does not exist, because even atomic number up to 95 of the elements can be found in nature, is due to the natural decay of uranium and thorium. It can also be caused by that universe in evolving; unstable elements are by short half-life gradually decay away.

At one time chemists used a variety of symbols to represent different elements. These symbols and names of elements of sketch is somewhat similar, so these symbols are very difficult to identify and use. Now the element's name initials to denote the element symbols is generally its English name with one or two letters. If two letters, capitalize the first letter, the second letter lowercase. For example, Fe stands for iron (ferrum) and Cu for copper (cuprum). The first letter of the chemical symbol is always capitalized. If the symbol has two letters, the second letter is always lowercase.

The number of protons in the nucleus plays such an important role in identifying atoms that it is given a special name, the number of atoms. The symbol Z is commonly used for the number of atoms (or protons). Hydrogen has 1 atomic number and lawrencium has 103. The atomic number is also equal to the number of electrons.

The sum of the total number of protons, Z, neutrons, N, is called the number of atomic mass. The symbol is A. Not all atoms of the same element have the same atomic mass number because N and A are different even though Z is the same. Atoms with the same element with different atomic mass numbers are called isotopes. The isotope and method of isotope separation will be introduced later.

In Table 7.1, the mass of atomic particles is given in atomic mass units (amu). These units represent a relative size in which the mass of the isotope carbon-12 is used as a standard, and all others are related to it. Specifically, 1 amu is defined as 1/12 of the atomic mass of carbon-12. Since the mass of protons or neutrons is about 1 amu, the mass of a particular atom will be roughly equal to its atomic mass number Z. The atomic weight of an element is usually more useful than the amount of mass of isotope. The atomic weight of an element is defined as a weighted average of all naturally occurring isotope mass. The atomic weight of the element is listed in Table 7.2. The element that have their atomic weight elements in parentheses are unstable.

Table 7.2 Table of elements

Atomic number	Name	Symbol	Atomic weight(amu)
1	Hydrogen	H	1.008
2	Helium	He	4.003
3	Lithium	Li	6.939
4	Beryllium	Be	9.012
5	Boron	B	10.810
6	Carbon	C	12.011
7	Nitrogen	N	14.007
8	Oxygen	O	15.999
9	Fluorine	F	18.998
10	Neon	Ne	20.183
11	Sodium	Na	22.990
12	Magnesium	Mg	24.312
13	Aluminum	Al	26.982
14	Silicon	Si	28.086
15	Phosphorus	P	30.974
16	Sulfur	S	32.064
17	Chlorine	Cl	35.453
18	Argon	Ar	39.946
19	Potassium	K	39.102
20	Calcium	Ca	40.080
21	Scandium	Sc	44.956
22	Titanium	Ti	47.900
23	Vanadium	V	50.942
24	Chromium	Cr	51.996
25	Manganese	Mn	54.938
26	Iron	Fe	55.847
27	Cobalt	Co	58.933
28	Nickel	Ni	58.710
29	Copper	Cu	63.546
30	Zinc	Zn	65.374
31	Gallium	Ga	69.723
32	Germanium	Ge	72.594
33	Arsenic	As	74.921
34	Selenium	Se	78.963
35	Bromine	Br	79.909
36	Krypton	Kr	83.798

(continued)

Table 7.2 (continued)

Atomic number	Name	Symbol	Atomic weight(amu)
37	Rubidium	Rb	85.468
38	Strontium	Sr	87.621
39	Yttrium	Y	88.906
40	Zirconium	Zr	91.224
41	Niobium	Nb	92.906
42	Molybdenum	Mo	95.942
43	Technetium	Tc	(99)
44	Ruthenium	Ru	101.072
45	Rhodium	Rh	102.905
46	Palladium	Pd	106.421
47	Silver	Ag	107.870
48	Cadmium	Cd	112.401
49	Indium	In	114.818
50	Tin	Sn	118.690
51	Antimony	Sb	121.750
52	Tellurium	Te	127.603
53	Iodine	I	126.904
54	Xenon	Xe	131.293
55	Cesium	Cs	132.905
56	Barium	Ba	137.337
57	Lanthanum	La	138.905
58	Cerium	Ce	140.116
59	Praseodymium	Pr	140.907
60	Neodymium	Nd	144.242
61	Promethium	Pm	(145)
62	Samarium	Sm	150.350
63	Europium	Eu	151.964
64	Gadolinium	Gd	157.253
65	Terbium	Tb	158.925
66	Dysprosium	Dy	162.500
67	Holmium	Ho	164.930
68	Erbium	Er	167.259
69	Thulium	Tm	168.934
70	Ytterbium	Yb	173.044
71	Lutetium	Lu	174.967
72	Hafnium	Hf	178.490

(continued)

Table 7.2 (continued)

Atomic number	Name	Symbol	Atomic weight(amu)
73	Tantalum	Ta	180.948
74	Tungsten	W	183.850
75	Rhenium	Re	186.207
76	Osmium	Os	190.230
77	Iridium	Ir	192.217
78	Platinum	Pt	195.085
79	Gold	Au	196.967
80	Mercury	Hg	200.570
81	Thallium	Tl	204.372
82	Lead	Pb	207.190
83	Bismuth	Bi	208.980
84	Polonium	Po	(210)
85	Astatine	At	(210)
86	Radon	Rn	(222)
87	Francium	Fr	(223)
88	Radium	Ra	226.030
89	Actinium	Ac	(227)
90	Thorium	Th	232.038
91	Protactinium	Pa	231.035
92	Uranium	U	238.024
93	Neptunium	Np	237.050
94	Plutonium	Pu	239.050
95	Americium	Am	(243)
96	Curium	Cm	(243)
97	Berkelium	Bk	(245)
98	Californium	Cf	(246)
99	Einsteinium	Es	(250)
100	Fermium	Fm	(253)
101	Mendelevium	Md	(256)
102	Nobelium	No	(259)
103	Lawrencium	Lr	(258)
104	Rutherfordium	Rf	(261)
105	Dubnium	Db	(268)
106	Seaborgium	Sg	(266)
107	Bohrium	Bh	(274)
108	Hassium	Hs	(265)

(continued)

Table 7.2 (continued)

Atomic number	Name	Symbol	Atomic weight(amu)
109	Meitnerium	Mt	(266)
110	Darmstadtium	Ds	(271)
111	Roentgenium	Rg	(272)
112	Copernicium	Cn	(285)
113	Ununtrium	Uut	(287)
114	Flerovium	Fl	(289)
115	Ununpentium	Uup	(288)
116	Livermorium	Lv	(293)
117	Hawkinium	Hw	(291)
118	Ununoctium	Uuo	(294)

For these elements, the atomic weight of the longest-lived isotope, rather than the average of all the isotope mass quantities is used.

Molecules are clusters or groups of atoms that held together through chemical binding. There are two types of molecules: molecules of an element and that of a compound.

In some cases, two individual atoms of an element can form molecules by attracting each other with a bond such as hydrogen, oxygen and bromine. These molecular formulas are H_2, O_2, and Br. Most gaseous elements exist as molecules of two atoms.

A compound is formed by two different element atoms that are held together by bonds. Molecules are the primary particles of compounds. Some examples of this molecule include hydrogen chloride (HCl), water (H_2O), methane (CH_4), and ammonia (NH_3).

The weight of a molecule is the total mass of the individual atoms. Therefore, if you know the formula of any molecule (i.e. the elements and the number of each that make up the molecule), it is fairly simple to calculate its mass. Note that the terms mass and weight are used interchangeably in chemistry.

Compound water has a formula of H_2O. This means that there is one oxygen atom and two hydrogen atoms. H_2O has a molecular weight of $15.999 + 1.008 \times 2 = 18.015$ amu.

7.1.3 Avogadro's Number

Consider an oxygen atom and a sulfur atom and compare their atomic weight. The atomic weight of oxygen is 15.999 amu. The atomic weight of sulfur (32.06 amu) is about twice that of oxygen atoms.

Since a sulfur atom weighs twice as much as an oxygen atom, one gram of oxygen sample contains twice as many atoms as one gram of sulfur. Therefore, two grams of sulfur samples contain the same number of atoms as one gram of oxygen samples. From the previous example, you can see a relationship between the weight of the sample and the number of atoms in the sample. In fact, scientists have determined a relationship between the number of atoms in the sample and the weight of the sample.

Experiments have shown that for any element, a sample containing atomic weight in grams contains 6.022×10^{23} atoms. As a result, 15.999 g of oxygen contains 6.022×10^{23} atoms, and 32.06 g of sulfur contains 6.022×10^{23} atoms. This number (6.022×10^{23}) is called as the Avogadro's number. The importance of Avogadro's number to chemistry should be clear. It represents the number of atoms in X grams of any element, where X is the atomic weight of the element. It allows chemists to predict and use the exact amounts of elements required to lead to an expected chemical reaction.

Rarely encounter a single atom or several atoms. Instead, larger macroscopic quantities are used to quantify or measure a collection of atoms or molecules, such as a glass of water, a gallon of alcohol, or two aspirins. Chemists have introduced a large unit of matter, moles, to process macro samples of the substance.

One mole represents a certain number of objects, substances, or particles. (For example, one mole of atoms, one mole of ions, one mole of moleculars, and even, in theory, one mole of elephant.) One mole is defined as the amount of pure matter containing 6.022×10^{23} units (atoms, ions, molecules, or elephants). In other words, one mole is the number of Avogadro's of anything.

For any element, the molar mass of the element's atom is the atomic mass in grams. For example, to calculate the mass of one mole of copper atom, you only need to find the atomic mass of copper in grams. Since copper has an atomic mass of 63.546 amu and one mole of copper has a mass of 63.546 g. The atomic mass of gold atom is 196.967 amu. As a result, the mass of one mole of gold is 196.967 g. The mass of one mole of atom is called as the gram atomic weight (GAW). The concept of moles allows the conversion of a substance's grams into moles and vice versa.

Figure 7.1 contains one ball of gold and one ball of copper. The two balls have different mass and size, but each ball contains the same number of atoms.

Example 7.1 A silver bar has a mass of 1870 g. How many moles of silver are in the bar?

Solution: Since the atomic mass of silver (Ag) is 107.87 amu, one mole of silver has a mass of 107.87 g. Therefore, there is one mole of Ag per 107.87 g of Ag. There are 1870 g of silver, $1870/107.870 = 17.336$ mol.

The mass of one mole of substance molecule is the molecule mass expressed in grams. For example, oxygen molecule (O_2) has a molecular mass equivalent to 32.0 g because each oxygen atom has a molecular mass of 16.0 g. (Recall that in order to obtain molecular mass, the atomic mass of all atoms that appear in the formula is added.) If you add the atomic mass of carbon in methane and the four hydrogen

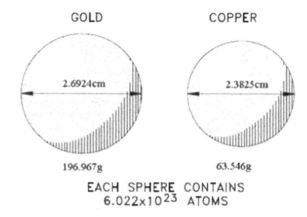

Fig. 7.1 A mole of gold compared to a mole of copper

atoms CH_4, the total is 16 amu. Therefore, the mass of a mole CH_4 is 16 g. The mass of molar molecules is called molar mass or gram molecular weight (GMW).

7.1.4 The Periodic Table

All known elements are in a pattern when placed in a periodic table, and the position in this pattern is determined by the atomic number of the element. This section will discuss the importance of this fact.

After years of chemical research, scientists have discovered the distinctive features of these elements. If the elements are arranged by their atomic number, the chemical properties of the elements are repeated periodically. To a lesser extent, physical properties are also repeated periodically. This periodic repetition can be seen in Table 7.3. Compare the properties of lithium, sodium and potassium as well as beryllium, magnesium and calcium. In the list of elements shown in Table 7.3, the property repeats every eight elements [2].

A table with chemically similar elements grouped together is called a periodic table. One of the most common versions is shown in Fig. 7.2. In this table, the elements are arranged in order to increase the atomic number in subsequent rows. Each horizontal line is called a period. Note that some periods are longer than others. Elements with similar chemical properties appear in vertical columns called groups. Each group is specified by Roman numerals and capital letters, with the exception of the group on the far right (inert gas). At the bottom of the periodic table are two long rows of elements, identified as the lanthanide series and the actinide series. They are separated from the table, mainly to prevent it from getting too wide. In addition, each element in both series exhibits similar chemical properties.

The number directly below each element is its atomic number, and the number above each element is its atomic weight. In several cases, the atomic weight is in parentheses. This suggests that these elements do not have stable isotopes: that is,

Table 7.3 Description of the properties of the first twenty elements

Atomic number	Symbol	Atomic weight	Description of properties
1	H	1.008	Colorless gas, reacts readily with oxygen to form H_2O; forms HCl with chlorine
2	He	4.003	Colorless gas, very non-reactive chemically
3	Li	6.939	Silvery white, soft metal, very reactive chemically, forms Li_2O and LiCl readily
4	Be	9.012	Grey metal, much harder than lithium, fairly reactive chemically, forms BeO and $BeCl_2$ easily
5	B	10.810	Yellow or brown non-metal, very hard element, not very reactive, but will form B_2O_3 and BCl_3
6	C	12.011	Black non-metal, brittle, non-reactive at room temperature. Forms CO_2 and CCl_4
7	N	14.007	Colorless gas, not very reactive, will form N_2O_5 and NH_3
8	O	15.999	Colorless gas, moderately reactive, will combine with most elements, forms CO_2, MgO etc
9	F	18.998	Green-yellow gas, extremely reactive, irritating to smell, forms NaF, MgF_2
10	Ne	20.183	Colorless gas, very non-reactive chemically
11	Na	22.990	Silvery soft metal, very reactive chemically, forms Na_2O and NaCl
12	Mg	24.312	Silvery white metal, much harder than sodium. Fairly reactive, forms MgO, $MgCl_2$
13	Al	26.982	Silvery white metal, like magnesium but not as reactive. Forms Al_2O_3, $AlCl_3$
14	Si	28.086	Gray, non-metallic, non-reactive at room temperature, forms SiO_2, $SiCl_4$
15	P	30.974	Black, red, violet, or yellow solid, low melting point, quite reactive, forms P_2O_5, PCl_3
16	S	32.064	Yellow solid with low melting point. Moderately reactive, combines with most elements, forms SO_2, MgS
17	Cl	35.453	Greenish-yellow gas, extremely reactive, irritating to smell, forms NaCl, $MgCl_2$
18	Ar	39.946	Colorless gas, very non-reactive chemically
19	K	39.102	Silver soft metal, very reactive chemically, forms K_2O, KCl
20	Ca	40.080	Silver-white metal, much harder than potassium, fairly reactive, forms CaO, $CaCl_2$

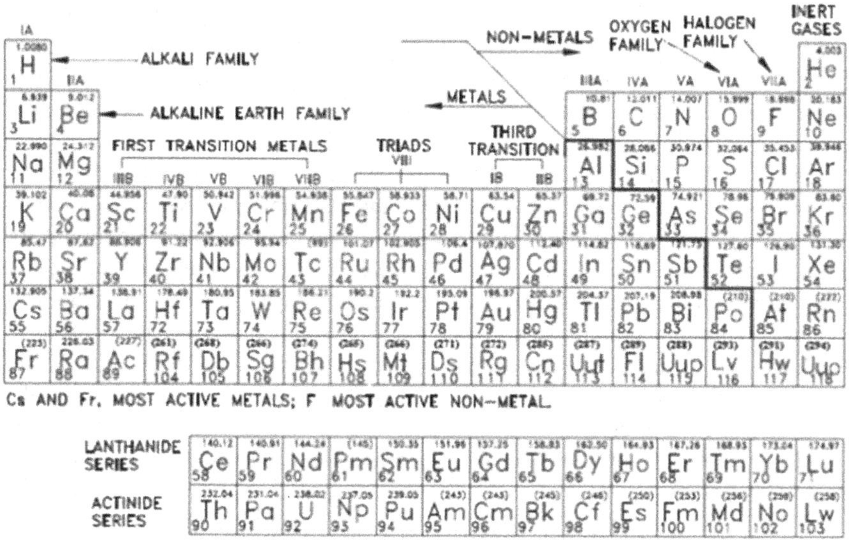

Periodic table labels: ALKALI FAMILY, ALKALINE EARTH FAMILY, FIRST TRANSITION METALS, TRIADS VIII, THIRD TRANSITION, NON–METALS, OXYGEN FAMILY, HALOGEN FAMILY, INERT GASES, METALS

IA	IIA	IIIB	IVB	VB	VIB	VIIB	VIII	IIIB	IIB	IB	IIB	IIIA	IVA	VA	VIA	VIIA	0
H 1 (1.0080)																	He 2 (4.003)
Li 3 (6.939)	Be 4 (9.012)											B 5 (10.81)	C 6 (12.011)	N 7 (14.007)	O 8 (15.999)	F 9 (18.998)	Ne 10 (20.183)
Na 11 (22.990)	Mg 12 (24.312)											Al 13 (26.982)	Si 14 (28.086)	P 15 (30.974)	S 16 (32.064)	Cl 17 (35.453)	Ar 18 (39.948)
K 19 (39.102)	Ca 20 (40.08)	Sc 21 (44.956)	Ti 22 (47.90)	V 23 (50.942)	Cr 24 (51.996)	Mn 25 (54.938)	Fe 26 (55.847)	Co 27 (58.933)	Ni 28 (58.71)	Cu 29 (63.54)	Zn 30 (65.37)	Ga 31 (69.72)	Ge 32 (72.59)	As 33 (74.921)	Se 34 (78.96)	Br 35 (79.909)	Kr 36 (83.80)
Rb 37 (85.47)	Sr 38 (87.62)	Y 39 (88.905)	Zr 40 (91.22)	Nb 41 (92.906)	Mo 42 (95.94)	Tc 43 (99)	Ru 44 (101.07)	Rh 45 (102.905)	Pd 46 (106.4)	Ag 47 (107.870)	Cd 48 (112.40)	In 49 (114.82)	Sn 50 (118.69)	Sb 51 (121.75)	Te 52 (127.60)	I 53 (126.90)	Xe 54 (131.30)
Cs 55 (132.905)	Ba 56 (137.34)	La 57 (138.91)	Hf 72 (178.49)	Ta 73 (180.95)	W 74 (183.85)	Re 75 (186.2)	Os 76 (190.2)	Ir 77 (192.2)	Pt 78 (195.09)	Au 79 (196.97)	Hg 80 (200.59)	Tl 81 (204.37)	Pb 82 (207.19)	Bi 83 (208.98)	Po 84 (210)	At 85 (210)	Rn 86 (222)
Fr 87 (225)	Ra 88 (226.03)	Ac 89 (227)	Rf 104 (261)	Db 105 (268)	Sg 106 (266)	Bh 107 (274)	Hs 108 (265)	Mt 109 (266)	Ds 110 (271)	Rg 111 (272)	Cn 112 (285)	Uut 113 (287)	Fl 114 (289)	Uup 115 (288)	Lv 116 (293)	Hw 117 (291)	Uuo 118 (294)

Cs AND Fr, MOST ACTIVE METALS; F MOST ACTIVE NON–METAL.

LANTHANIDE SERIES: Ce 58 (140.12), Pr 59 (140.91), Nd 60 (144.24), Pm 61 (145), Sm 62 (150.35), Eu 63 (151.96), Gd 64 (157.25), Tb 65 (158.83), Dy 66 (162.50), Ho 67 (164.93), Er 68 (167.26), Tm 69 (168.93), Yb 70 (173.04), Lu 71 (174.97)

ACTINIDE SERIES: Th 90 (232.04), Pa 91 (231.04), U 92 (238.02), Np 93 (237.05), Pu 94 (239.05), Am 95 (243), Cm 96 (245), Bk 97 (245), Cf 98 (248), Es 99 (250), Fm 100 (253), Md 101 (256), No 102 (250), Lw 103 (258)

Fig. 7.2 The periodic table

they are radioactive. The values attached in parentheses and used for atomic weight are the atomic mass of the most stable isotopes known, as indicated by the longest half-life.

There are three broad classes of elements. These are metals, non-metallic and semi-metals. The three classes are grouped on the periodic table, as shown in Fig. 7.3.

Fig. 7.3 Regional schematic of periodic table

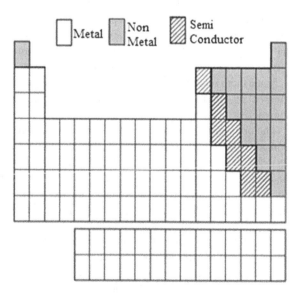

Metal ☐   Non Metal ■   Semi Conductor ▨

Metals make up the largest category of elements, located on the left and center of the periodic table, as shown in Fig. 7.3. In Fig. 7.2, a heavy line from boron (B) to astatine (At) usually separate metals from other elements (elements in the actinide and lanthanide series are metals). Metals tend to lose electrons to form positive ions, rather than obtaining electrons to become negative ions.

Most people are familiar with the physical properties of metals. They are usually hard and robust, mechanically molded (ductile and malleable), and have good thermal and electrical conductivity with shiny surfaces when cleaned. More important for chemical classification is the chemical properties of metals, as physical properties are not common to all metals. For example, mercury (Hg) is a metal, although it is a liquid at room temperature, sodium is a metal, although it is not hard or strong at all. Metals can participate in a wide range of chemical reactions. Their reactions to water ranged from sodium and potassium to gold and platinum. Metals fall into the following two categories.

1. Light metal is soft, low density, high chemical reaction, unsatisfactory as structural materials.

2. Transition metals are hard, dense, unreactive and are useful structural materials.

Metals in Category 1 are located on the far left of the table (groups IA and IIA). Metals in Category 2 are located in the middle of the table (Group B).

Non-metals occupy part of the periodic table, on the right side of the heavy, step line. (See Figs. 7.2 and 7.3).

In general, the physical properties of non-metals are contrary to the physical properties of metals. Non-metals are usually gases at room temperature. Solid non-metallic is not lustrous, does not have plasticity or ductility, is a poor conductor of heat and electricity. Some non-metals are very reactive, but the nature of the reaction is different from that of metals. Nonmetallics tend to obtain electrons to form negative ions rather than lose electrons to form positive ions.

The six elements in Group 0 represent special subclasses of non-metallics. They are very unreactive gases, so they are called inert gases. For years, it has been thought that inert gases do not and cannot participate in chemical reactions. In 1962, the true compounds of the inert gases XeF_4 and $XePtF_6$ were first identified.

Since then, several other compounds have been prepared. The preparation of these compounds requires special conditions: in general, inert gases may be considered nonreactive.

The obvious trend in the periodic table is that, from left to right, elements change from obvious metals (group IA) to distinct non-metals (group VIIA) across any period. This change in character is not a sharp definition, but a gradual one. In general, the element to the left of the heavy diagonal line is metal, while the element on the right is non-metallic. However, some elements near the line exhibit metal properties under certain conditions and non-metallic properties under other conditions. These elements are called semimetallic and include boron (B), silicon (Si), germanium (Ge), arsenic (As) and tellurium (Te). They are commonly classified as semiconductors of electricity and are widely used in electrical components.

Each set of elements displayed in the vertical column of the periodic table is called a group and represents a series of elements with similar physical and chemical properties. Group IA is the alkali family; IIA group is the alkaline earth family; Group VIA is the oxygen family; group VIIA is the halogen family. On the left side of the table are group IA elements (except hydrogen), which are soft metals that undergo similar chemical reactions. Elements in the IIA group form similar compounds, which are much harder than the neighbors in the IA group.

As shown in the previous section, there are some exceptions to the generalization of chemical properties and periodic tables. The most accurate observation is that all elements in a particular group have similar physical and chemical properties.

This observation is most accurate on both sides of the table. All elements in Group 0 are unreactive gases, and all elements in the VIIA group have similar chemical properties, although the physical properties are gradually changing. For example, fluorine (F) is a gas, while iodine (I) is a solid at room temperature.

Groups with a B designation (IB via VIIB) and group VIII are called transition groups. In this area of the table, exceptions are beginning to occur. In any group in this region, all elements are metals, but their chemical properties may vary. In some cases, elements may be more similar to neighbors within its period than to elements in groups. For example, iron (Fe) is more similar to cobalt (Co) and nickel (nickel) than to ruthenium (Lu) and osmium (Os). Most of these elements have multiple charges, and their ions in the solution are colored (all other elements are colorless).

Line that separate metal from non-metallic cuts across several groups. In this area of the table, the group similarity rule has lost much of its usefulness. For example, in the IVA group, carbon (C) is non-metallic; silicon (Si) and germanium (Ge) are semimetallic; tin (Sn) and lead (Pb) are metals.

Chemical activities can also be determined from the position in the periodic table. The most active metals are members of the alkali family, such as cesium (Cs) and francium (Fr). The most active non-metals are members of the halogen family, such as fluorine (F) and chlorine (Cl). The noble gases in Group 0 are inert. When you go right in the periodic table, the activity of the metal decreases and when you go to the left, the activity of the non-metallic decreases.

Why is the element in the element periodic table to show periodic change? With intra-atomic electronic structure of the secret has been very good interpretation. In steady state (the ground state) of the atomic, nuclear electronics will as far as possible according to the principle of minimum energy arrangement. In addition, due to the electronic impossible are crowded together, they must also comply with the Pauli incompatible principle and the Hund's rule. In general, under the guidance of these three rules can derive atoms extra nuclear electron configuration.

Arrangements of the electrons in outer nuclear atoms, as far as possible to make the electronic energy minimum. How to make the electronic energy minimum? For example, we stood on the ground, do not feel what is dangerous; if we stand on the 20 floor of the balcony, and then out of the window to see the ground you will feel fear. This is because the object in high has more potential energy; the object with potential energy will fall potentially, like free fall. The electron can be viewed as a kind of material, but also has the same properties; it is in general to in a relatively safe (or stable) state (ground state), which is the lowest energy state. When external force exist, electronic also can absorb energy to a higher energy state (excited state),

Fig. 7.4 Electronic
hierarchical model of atomic
nuclei

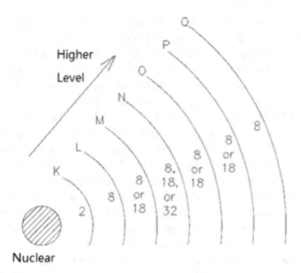

but it will always return to the ground state potentially. In general, from the electron nuclear closer with lower energy, with the increase of the number of electrons, the electron energy is more and higher (see Fig. 7.4).

In the atomic nucleus, located in the central atoms, electrons in the nucleus around the core for high speed movement. Because the electrons in the region for different nuclear motion, we can be seen as the electron is outside the nucleus. A hierarchical arrangement of 3 principles according to the arrangement of extra nuclear electrons will extra nuclear electrons. All the atoms around the atomic nucleus discovered extra nuclear electrons. Abide by the following rules: electronic outer nuclear as much as possible in the electronic distribution layer of lower energy on (nearer the nucleus); according to the principle of quantum mechanics, if the electronic layer is n, the number of electrons of this layer is at most $2n^2$; regardless of the number of layers. As for the outermost electron layer, the electronic number cannot be more than 8 layers, as for the last second layer, the number of electronic cannot be more than 18. The outermost layer of the same electron configuration are classified, they are in the same column of the periodic table of elements.

Therefore, decided to the chemical properties of the atoms is mainly for the arrangement of the electrons in the outermost shell. When the outermost layer contains eight electrons, usually a very stable. In addition, the outermost layer of only 1–2 electron atom will be very easy to lose the outermost electrons.

Internal atomic electrons can be excited into the higher level by external heat, light or particles. If atoms from the outside to obtain enough energy, can make the valence electrons excited into free electron. It is the free electron forms current in metal. If electrons are excited into free electrons, it will leave a positively charged atoms, said for the atom ionized. With charged called ionization in the positive, if nuclear won the excess electrons and become negatively charged, is called the negative ionization.

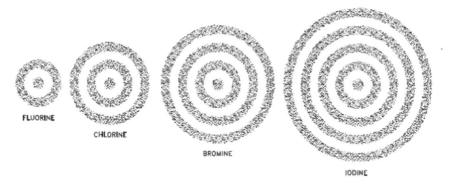

Fig. 7.5 Electron shells of atoms

According to experimental data, chemical reactions are known to involve only electrons in atoms. In fact, only a few electrons are involved. Because chemical properties are periodic, electrons must also have periodic characteristics. This feature is the way electrons are arranged in atoms. Electrons move around the nucleus. They have both kinetic and potential energy, and their total energy is the sum of the two. Total energy is quantified: that is, the total energy that atomic electrons can possess has a clear discrete value. These energy states can be visualized as spherical shells around the nucleus of cells, separated by prohibited areas where electrons do not exist steadily. Figure 7.5 illustrates this arrangement.

Usually said to be the electron shell around the nucleus, the shell is pointed by numbers. The first, or 1, the shell is the closest one to the core; the second, or 2, the shell is the next; then the third, or 3; and so on in numerical order. In general, electrons close to the nucleus have a lower energy state. Atomic electrons always seek the lowest energy state.

The electron shell represents the primary energy state of the electron. Each shell contains one or more subshells called orbits, each with slightly different energy. In order of increasing energy, the orbits are designated by lower case letters s, p, d, f, g, h.

No two shells consist of the same number of orbits. The first shell contains only one orbit, an s orbit. The second shell contains s and p orbits. In general, each higher shell contains a new type of orbit.

The first shell contains an s orbit,

The second shell contains s and p orbits,

The third shell contains s, p, and d orbits,

The fourth shell contains s, p, d, and f orbits,

And so on. Each orbit can hold a definite maximum number of electrons. There is also a limit to the number of electrons in each shell and the limit increases as one goes to higher shells. The numbers of electrons that can occupy the different orbits and shells are shown in Table 7.4.

More specific statements can now be made as to which electrons are involved in chemical reactions. Chemical reactions mainly involve electrons in the outermost

Table 7.4 Electrons, orbital, and shell relationships in atomic structure

Shell number	Types of orbitals	Maximum number of electrons in each orbital	Maximum total electrons in shell
1	s	2	2
2	s	2	8
	p	6	
3	s	2	18
	p	6	
	d	10	
4	s	2	32
	p	6	
	d	10	
	f	14	
5	s	2	50
	p	6	
	d	10	
	f	14	
	g	18	

shell of atoms. The outermost shell is the outermost shell from the nucleus, which allocates some or all of the number of electrons. Some atoms have more than one partially filled shell. All partially filled shells have some effect on chemical behavior, but the outermost shells have the greatest effect on chemical behavior. The outermost shell is called the valence shell, and the electrons in the shell are called valent electrons. The term valence (of an atom) is defined as the number of electrons gained or lost by an element, or the number of electrons it shares when interacting with other elements.

Schedule periodic charts to easily determine the valence of atoms. For elements in Group A of the periodic graph, the number of valente electrons is the same as the group number; carbon (C) is in the IVA group, and there are four valent electrons. Noble gas (Group 0) has eight in its valence shell, except for helium, which has two.

The arrangement of the outermost shell is either fully filled (e.g. He and Ne) or contains eight electrons (e.g. Ne, Ar, Kr, Xe, Rn) is called an inert gas configuration. The inert gas configuration is exceptionally stable because these inert gases are the least reactive of all the elements.

The first element in the periodic table, hydrogen, does not satisfactorily place it in any group of properties. Hydrogen has two unique characteristics: (a) the highest energy shell of a hydrogen atom can hold only two electrons, while all others (except helium) can hold eight or more: and (b) when hydrogen loses its electrons, ions form, $H+$, is a bare nucleus. Hydrogen ions are very small compared to positive ions of any other element, because there must be some electrons around the nucleus of any other element. Hydrogen can gain or lose electrons. It has properties similar to group IA elements, and some properties are similar to VIIA group elements.

The number of electrons in the outer layer or shell determines the relative activity of the element. Elements are arranged in the periodic table so that the same set of elements have the same number of electrons in the shell (except for transition groups). The arrangement of electrons in the shell explains why some elements are chemically active, some are not very active, and some elements are inert. In general, the less electrons an element must lose, gain, or share in order to achieve a stable shell structure, the greater the chemical activity of the element. The likelihood of elements forming compounds is strongly influenced by the stability of this shell and the resulting molecules. The more stable the molecules, the more likely they are to form.

7.2 Chemical Bonding

The development of matter, regardless of form, is the result of assumptions, theories and laws applied by chemists from the study of the nature of matter, energy and change. This section will discuss some of these theories and laws. The discussion will discuss how chemical bonds and atoms combine to form molecules. In addition, organic chemistry has been introduced.

As described in the previous, the number of electrons in the outer or outer layer determines the relative activity of the element. The arrangement of electrons in the shell explains why some elements are chemically active, some are not very active, and some elements are inert. In general, in order to achieve a stable shell structure, the less electrons an element must lose, gain, or share, the greater the chemical activity of the element. The likelihood of elemental formation of compounds is strongly influenced by the completion of the valence shell and the resulting molecular stability. The higher the stability of molecular production, the greater the likelihood of formation. For example, an atom "needs" two electrons to fully fill the valence shell, preferring to react with another atom, which must give up two electrons to satisfy its valence.

This is most likely in the case of H^{++} Br^- because the exchange will meet the needs of both atoms. Although it is not just the number of valence electrons that are considered, this is a good rule of thumb.

If an atom needs two electrons and picks up only one electron, it will still actively look for an extra electron. The H^+ + Te^{-2} reaction is unlikely to occur because the resulting molecules still have an incomplete valence shell. Of course, when two atoms want to release or acquire electrons, a combination of two atoms (e.g. H_2 or O_2) may occur, but this is unlikely to happen when other atoms are available.

Atoms are connected or bonded together through this interaction of electrons. There are several types of chemical bonds that bring atoms together: three that will be discussed, ionic, covalent, and metallic.

7.2.1 Ionic Bond

When one or more electrons are transferred completely from one element to another, ion bonds are formed and elements are attracted together due to the opposite charge. Examples of ion bonding are shown in sodium chloride (salt) Fig. 7.6a.

Sodium atoms transfer electrons from the shell to chlorine atoms, which fill the shell with electrons. When this happens, the sodium atom leaves a charge of +1 and a chlorine atom of −1 charge. Ion bonds are formed because of the attraction of two opposite charged particles. No single negatively charged ion is more likely to bind to a particular positive charge ion than any other ion.

As a result, positive and negative ions are arranged in three dimensions, as shown in Fig. 7.6b, to balance the charge between several ions. For example, in sodium chloride, each chloride ion contains as many sodium ions as possible, and it is easy to gather around it, that is, six. Similarly, each sodium ion is surrounded by six chlorine ions. As a result, each chlorinated ion binds to six of the nearest sodium ions and a smaller degree of distant sodium ions. Therefore, an ion bond is a force that brings together many atoms or ions, not a bond between two individual atoms or ions.

Some ionic bond strength, some other weak. The strength of the ionic affects the melting point of compound, boiling point and solubility properties. Ionic bond stronger, its melting point is high. Ionic radius smaller or more charges, anion and cation effect is stronger. For example, sodium ion radius is smaller than the radius of the potassium ion, NaCl ionic bonds stronger than in KCl ionic, so the melting point of sodium chloride is higher than potassium chloride.

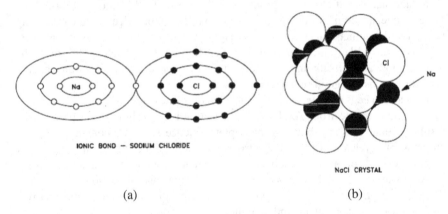

IONIC BOND — SODIUM CHLORIDE

NaCl CRYSTAL

(a)   (b)

Fig. 7.6 Ionic bond, sodium chloride NaCl

7.2.2 Covalent Bonds

When one or more electrons in an atom pair with one or more electrons of another atom to form an overlapping electron shell, a convalescent bond is formed in which two atoms share the paired electron. Unlike ion bonds, covalent bonds bring specific atoms together. The covalent bond can be unit covalent, double covalent, or triple covalent, depending on the shared electronic pair. Figure 7.7 shows the bonding that occurs in a methane molecule consisting of four single covalent bonds between one carbon atom and four hydrogen atoms.

When carbon dioxide is formed by one carbon atom and two oxygen atoms, two double covalent bonds are formed. Carbon atoms share four pairs of electrons, and as shown in Fig. 7.8, two oxygen atoms share two each. The combination of two electrons forms a lower energy combination than the energy at separation. This energy difference represents the force that binds a particular atom together.

When two shared electrons in a covalent bond come from the same atom, the bond is called the coordinate covalent bond. Although both shared electrons come from the same atom, coordinate covalent bonds are individual bonds of a similar nature to covalent bonds. Figure 7.9 illustrates the bonds of negatively charged chlorate ions. The ions consist of chlorine atoms and three oxygen atoms, with a net charge of -1, consisting of two coordinate covalent bonds and one covalent bond. Chlorine atoms efficiently acquire electrons through covalent bonds, resulting in a negative overall charge.

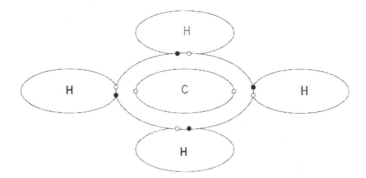

Fig. 7.7 Covalent bond, methane CH_4

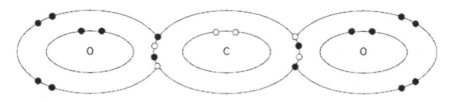

Fig. 7.8 Formation of the carbon dioxide molecule CO_2

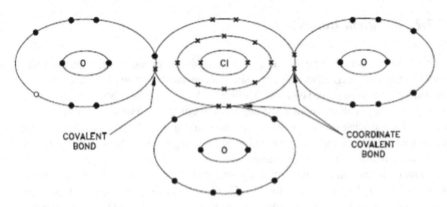

Fig. 7.9 Coordinate covalent bond, chlorate ion ClO_3

Covalent bonds can be polar or non-polar. When a shared pair of electrons is not equal, one end of the bond is positive and the other end is negative. This produces a bond with the pole, called the polar covalent bond.

Molecules with polar covalent bonds are called polar or polar molecules. Water is an example of polar molecules. When two atoms of the same element share one or more pairs of electrons (such as H or N), each pair has the same attraction to the shared electron pair or pairs. When electron pairs are evenly distributed or shared like atoms, bonds are called nonpolar covalent bonds. If all the bonds in a molecule are of this kind, the molecule is called a non-polar covalent molecule.

7.2.3 Metallic Bonds

Another chemical bonding mechanism is metal bonding. In metal bonds, atoms achieve a more stable configuration by sharing electrons in the valence shell with many other atoms. Metal bonds prevail, mainly in elements where valent electrons are not closely bound to the nucleus (i.e. metals), and are therefore called metal bonding. In this type of bond, each atom in a metal crystal contributes all electrons in its valence shell to all other atoms in the crystal.

Another way to study this mechanism is to imagine that high-energy electrons are not closely related to individual atoms, but move between atoms in crystals. As a result, individual atoms can "slip" each other, but electrostatic electricity applied by electrons is firmly held together. This is why most metals can be hammered into thin sheets (plasticity) or drawn into thin lines (extensions). When potential differences are applied, electrons flow freely between atoms and a current flows.

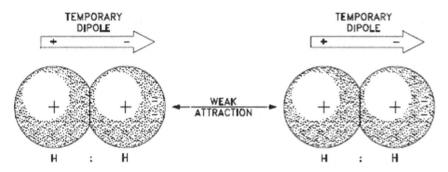

Fig. 7.10 Van der Waals forces

7.2.4 Van Der Waals Forces

In addition to the chemical binding between atoms, there is another attractive force between atoms, ions, or molecules, called van der Waals forces.

These forces occur between molecules of non-polar covalent substances such as H_2, Cl_2, and He. These forces are often thought to be caused by temporary dipole or unequal charge distributions because electrons move constantly in atoms, ions, or molecules. At a given moment, one region may have more electrons than the other, as shown in Fig. 7.10 [3].

Temporary dipoles induce similar temporary dipoles on nearby atoms, ions, or molecules. Hundreds of millions of these temporary dipoles form, break apart and reform, because weak electrostatic force of attraction known as van der Waals forces. It is worth noting that van der Waals forces exists among various molecules. Some molecules may have these forces, as well as dipole or other intermolecular forces. However, van der Waals forces is the only intermolecular bond between nonpolar covalent molecules such as H_2, Cl_2, and CH_4. The number of electrons in a substance increases as the gram molecular mass (mass in grams of one mole of compound) increases. As a result, the force between van der Waals forces materials increases with the increase in gram molecular mass.

Van der Waals forces are small compared to the forces of chemical bonding and are significant only when the molecules are very close together.

7.2.5 Hydrogen Bond

A hydrogen bond is the electrostatic attraction between two polar groups that occurs when a hydrogen (H) atom covalently bound to a highly electronegative atom such as nitrogen (N) or oxygen (O) experiences the electrostatic field of another highly electronegative atom nearby.

Hydrogen bonds can occur between molecules (intermolecular) or in different parts of individual molecules (intramolecular). Depending on the geometry and environment, the free energy content of the hydrogen bond is between 1 and 5 kcal/mole. This makes it more interactive than van der Waals interaction, but weaker than covalent or ion bonds. This type of bond may occur in inorganic molecules such as water and organic molecules such as DNA and proteins.

Hydrogen bonding is different from the van der Waals force, it has saturated and direction. Because of the hydrogen atom is very small compared with the atoms X and Y, so in X-H, hydrogen atoms can bond with only one Y atom with hydrogen bond. At the same time due to the repulsion between the negative ions, another large electronegativity of atom Y is hard to get close to hydrogen and the hydrogen bond is saturation.

Hydrogen bond has the orientation of the interaction between the X-H and the atomic Y due to the electric dipole moment, only when X-H... Y in a straight line is the strongest, and Y atoms generally not contain shared electrons, in the direction of the hydrogen bonds within the range and not sharing a pair of electrons of the symmetry axis aligned, the Y atoms negative charge distribution of the largest part of the closest to the hydrogen atom, so that the formation of hydrogen bonding is the most stable.

Hydrogen bonds in the binding energy is about 2 ~ 8 kcal. Hydrogen bond is stronger than the van der Waals force slightly; and is much weaker than covalent bond and ionic bond; and the stability is weaker than covalent bond and ionic bond.

7.3 Organic Chemistry

Organic chemistry is defined as the chemical composition of carbon compounds. Carbon compounds are biological forms, synthetic fabrics and plastics. Organic chemistry is a broad theme that is often subdivided into smaller areas. How carbon binds, and what it binds to determine the breakdown of specific compounds. These parts are called families or classes.

Carbon atoms can be combined into straight chains, rings, or branch chains. The bonds between carbon atoms can be single, double, and three, or a combination of these bonds. Other atoms (H, O, N, S, P) and halogens can be attached to carbon atoms to produce derivatives.

Large family of organic compounds contains only carbon and hydrogen, called hydrocarbons. These can be further divided into two main categories, aliphatic (fatty) and aromatic (fragrant).

Aliphatic hydrocarbons are divided into saturated and unsaturated, divided into alkanes, alkenes and alkynes. These subdivisions reflect the type of bond between carbon atoms.

Alkanes are saturated compound that have single bonds between carbon atoms and contains as many hydrogen atoms as possible. Each carbon has four covalent bonds on both sides, each hydrogen atom sharing a pair of electrons with the carbon atom, as shown in Fig. 7.11.

The general formula for alkanes is C_nH_{2n+2}. The alkanes are colorless, almost tasteless, insoluble in water, and easily dissolved in non-polar solvents such as benzene or ether. Alkanes are less reactivity. The reactions they occur are called halogenation, thermal decomposition (cracking) and combustion. These summaries are below.

Halogenation occurs when hydrogen atoms are replaced by halogen atoms. This is referred to as a substitution reaction. There is no limit to how many hydrogen atoms a molecule can replace.

$$CH_4 + Br_2 \rightarrow CH_3Br + HBr \tag{7.1}$$

Thermal decomposition or cracking is the process of breaking down large molecules into small molecules. Using heat as a catalyst, propane can be broken down into methane and ethylene:

$$C_3H_8(\text{heat}) \rightarrow CH_4 + C_2H_4 \tag{7.2}$$

When combustion occurs, the alkane combustion, the product is carbon dioxide gas, water and heat. These reactions are highly exothermic, so hydrocarbons are often used in fuels.

$$CH_4 + 2O_2 \rightarrow CO_2 + 2H_2O + 890kJ \tag{7.3}$$

Alkenes are hydrocarbon containing two fewer hydrogen atoms than the corresponding alkane. The general formula for Alkens is C_nH_{2n}. These molecules will have a double bond, as shown in Fig. 7.12a.

Alkenes are unsaturated hydrocarbons because hydrogen atoms are less than maximum. The main source of Alkenes is the cracking of alkanes.

$$C_2H_6(\text{heat}) \rightarrow H_2 + C_2H_4 \tag{7.4}$$

The third of the aliphatic hydrocarbons are alkynes. These compounds are unsaturated, like alkenes. They contain two less hydrogens than the corresponding alkane,

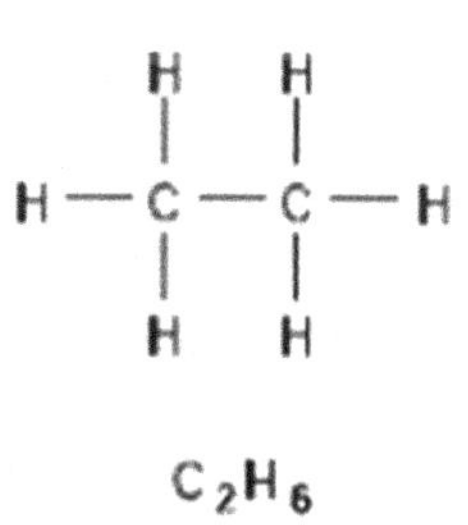

Fig. 7.11 Alkane

$$C_2H_6$$

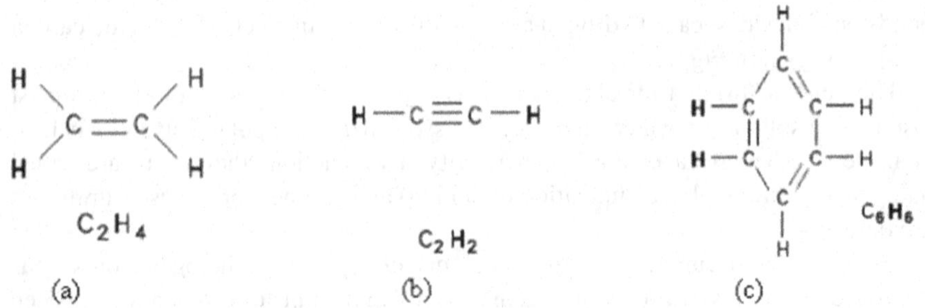

Fig. 7.12 Alkenes, alkynes and aromatic

C_nH_{2n-2}. As shown in Fig. 7.12b, alkyne hydrocarbons contain triple bonds between at least one set of carbon atoms.

Another major hydrocarbon is aromatics. These are not arranged in straight chains, but circulating, such as benzene. The derivatives of cyclic hydrocarbons have a pleasant (and sometimes toxic) smell. Benzene in rubber cement is a familiar odor. Circulating compounds have alternating single-double bonds, as shown in Fig. 7.12c.

Aromatics are chemically stable, much like alkanes. They will undergo substitution reactions rather than additions.

Alcohol is an aliphatic hydrocarbon with a hydroxyl, and the hydrocarbon (OH) group replaces one or more of the hydrogens shown in Fig. 7.13.

The -OH functional group does not work in an ionic manner in the case of alcohol. Alcohol is a molecular, not ionic, in nature. Alcohol is a versatile compound commonly used to make almost all other types of aliphatic compounds.

Aldehyde is one of the oxidizing products of alcohol. Each compound contains a carbon substrate (a double combination of carbon and oxygen atoms), as shown in Fig. 7.13b.

The term "aldehyde" is the contraction of the term "alcohol dehydrogenation", which means that when aldehyde is prepared from raw alcohol, two hydrogen atoms are removed from the end carbon. The functional group (-C = O) is always at the end of the carbon chain.

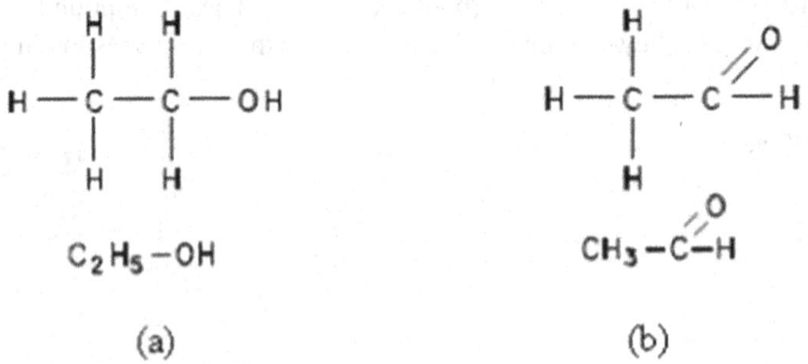

Fig. 7.13 Alcohol and aldehyde

7.4 Chemical Equations

As mentioned earlier, all matter is made up of atoms, which can join together to form compounds. From a chemical point of view, various forms of matter can be summarized as follows [4].

1.   Molecules are groups or clusters of atoms that are firmly bound together by chemical binding. There are two general types of molecules.
(a) The molecule of an element—two individual atoms of the same element, in some cases fixed by chemical bonds to form molecules with each other. Examples of this are hydrogen (H_2), oxygen (O_2) and bromine (Br_2). Most gaseous elements exist as molecules of two atoms.
(b) Compound molecules—Compounds contain at least two different types of atoms. Molecules are the ultimate particles of chemical compounds. Examples of compounds are hydrogen chloride (HCl), water (H_2O), methane (CH_4), and ammonia (NH_3).
2.   Elements are substances that cannot be decomposed by ordinary types of chemical changes, nor can they be decomposed by chemical combinations.
3.   A compound is a substance containing a variety of elements and is of a different nature than a single element. The composition of a particular compound is always clear.
4.   The mixture consists of a mixture of two or more substances with no constant percentage composition. Each component retains its original properties.

Chemistry and all other sciences are based on facts determined by experiments. The laws of science are condensed statements of fact and discovered experimentally. There are three basic laws that apply to chemical reactions. They are the Law of Conservation of Mass, the Law of Definite Proportions, and the Law of Multiple Proportions. These laws are described here to help readers understand the causes of the behavior of elements and compounds as they do so.

1.   The Law of Conservation of Mass

This law states that in a chemical reaction, the total mass of the product is equal to the total mass of the reactant. French chemist Antoine Lavoisier found that when tin reacts with air in a closed container, the weight and content of the container are the same as before. Scientists later discovered that whenever energy (heat, light, radiation) is liberated during a reaction, there is a small change in mass, but in ordinary chemical reactions, this change can be ignored.

2.   The Law of Definite Proportions

This law states that a given compound always contains the same elements in the same mass ratio, regardless of how it is prepared. British physicist John Dalton has discovered that when metals burn or oxidize in the air, they are always proportional to their weight. For example, the oxygen weight of a part is always combined with the weight of 1.52 parts of magnesium or 37.1 parts of tin. As a result of this approach, compounds are a combination of a certain number of atoms in one element and a certain number of atoms in another.

3.   The Law of Multiple Proportions

This law states that if two elements are combined into multiple compounds, the mass of one element in combination with another fixed mass is a simple proportion. Carbon and oxygen, for example, form two common compounds: carbon monoxide and carbon dioxide. Carbon monoxide (CO) is combined with 1.33 g of oxygen and 1 g of carbon. Together with carbon dioxide (CO_2), 2.67 g of oxygen is combined with 1 g of carbon. Therefore, the ratio of oxygen to fixed mass carbon is 2:1.

The Laws of Definite Proportions and Multiple Proportions as well as the relevant parts of atomic theory, form the basis of most quantitative calculations involving chemical reactions. Applying the basic laws of chemistry to chemical binding will help the reader understand the probability and proportion of chemical reactions. Regardless of the type of bond (ionic, covalent, coordinate covalent, or metallic), a specific amount of an element reacts with a specific amount of the element with which it is bound.

If two substances are placed in a container, regardless of the proportion, the result is a mixture. When a teaspoon of sugar is added to a cup of water, it slowly dissolves into the water and disappears from view. As a result, sugar molecules are evenly distributed in water and mixed with water molecules. Because the mixture of sugar and water is uniform, it is said to be homogenous. A homogenous mixture of two or more substances is called a solution. The solution is classified as a mixture rather than a compound because the composition is not proportionally fixed.

All solutions include solvents and one or more solutes. Solvents are materials that dissolve other substances. It is a dissolved medium. In aqueous solutions, water is a solvent. Substances dissolved in a solution are called solutes. Sugar is a solute in solutions. It is not always easy to determine which one is a solvent and which is a solute (e.g., a solution of semi-water and semi-alcohol).

A solution can exist in any of three states of substance, solid, liquid, or gas. The Earth's atmosphere is a gas solution of nitrogen, oxygen and lesser amounts of other gases. Wine (water and alcohol) and beer (water, alcohol and carbon dioxide) are examples of liquid solutions. Metal alloys are solid solutions (14-karat gold is gold combined with silver or copper).

One factor that determines the degree and/or speed of reaction is solubility. Solubility is defined as the maximum amount of substance dissolved at a given temperature at a given solvent. At this point, the solution is said to be saturated. When the solute and solvent are balanced at a specific temperature, the solution is saturated. Equilibrium is when the reaction conditions remain the same, the forward and reverse reaction rate of the chemical reaction is completely equal.

Kinetics is the study of factors that affect the rate of chemical reactions. There are five main factors to consider concentration, temperature, pressure, reactants characteristics and catalyst.

7.4.1 Le Chatelier's Principle

The effect of temperature on solubility can be explained by the Le Chatelier's Principle. Le Chatelier's principle states that if pressure (e.g. heat, pressure, concentration of reactants) is applied to the balance, the system will adjust as much as possible to minimize the effects of stress. This principle is valuable for predicting how well the system reacts to changes in external conditions. Consider the case where the solubility process is endothermic (heat added).

The increase in temperature puts stress on the equilibrium condition and causes it to move to the right. The stress is reduced because the dissolving process consumes some heat. Therefore, solubility (concentration) increases with temperature. If the process is exothermic (heat given off).

Temperature rise reduces the solubility by shifting the balance to the left.

When a solution is used for a specific purpose, the degree of solubility is important. To say a little, a lot, or a little bit wouldn't be very accurate if a particular concentration is needed. There are some common and accurate ways to represent concentration. These are density, molarity, normality, and parts per million.

In gas reversible reaction, when the pressure is increased, balance is in the direction of moves to decrease the size. Such as in the reversible reaction of $N_2 + 3H_2 = 2NH_3$, to achieve a balanced, the system pressure, such as pressure is increased two times as much as that of the original. At this moment the old balance to be broken, balance to direction of reducing the volume, namely in the reaction to positive direction. Establish a new balance, the increase of the pressure even weakened, is no longer twice the original balance. However, this increased pressure is impossible completely offset, nor with the original balance the same, but in between the two.

7.4.2 Concentrations of Solutions

Solution can dissolve how much solute, is very important. In pressurized water reactor nuclear power plant, the boric acid dissolved in the coolant water for reactivity control (the concept of reactivity will be in later chapter). Therefore, coolant temperature changes, for the concentration of boric acid has obvious effect. When the temperature drops even occur boron crystallization phenomenon.

A useful way to express exact concentrations of solutions is molarity (mol/L). Molarity is defined as moles of solute per liter of solution. Molarity is symbolized by the capital letter M. It can be expressed mathematically as follows.

Notice that the moles of solute are divided by the liters of solution not solvent. One liter of one molar solution will consist of one mole of solute plus enough solvent to make a final volume of one liter.

The sign of the molar concentration is [], for example, [H +] indicates the molar concentration of H + .

Example 7.2 Prepare one molar solution of NaCl.

Solution: One mole is equal to the gram molecular weight, so one mole $= 58.442$ g, 58.442 g of NaCl is weighed out and sufficient water is added to bring the solution to one liter.

Another term used to describe solution specific concentrations is parts per million or ppm. The term ppm is defined as the concentration of a solution in units of one part of solute to one million parts solvent. One ppm equals one milligram of solute per liter of solution. Another term, one parts per billion (ppb), is defined as one-part of the solute per billion solvents. One ppb is equal to one-microgram solute per liter of solution. These two terms are commonly used in very diluted solutions.

At present, in the most scientific journals. Don't use ppm, and the use of 10^{-6} or one thousandth, namely "‰". ppm conversion into per thousand: 1 ppm $= 0.001‰$.

7.4.3 Chemical Equations

Chemical equations are expressions of chemical reactions in a chemist's shorthand. In chemical equations, primitive substances are called reactants, and new substances that are forming are called products. In a chemical reaction, the reactant is located on the left side of the arrow and the product is on the right side of the arrow.

The number of atoms or molecules per substance is shown by the coefficients in the equation. Since atoms cannot be produced or destroyed in a chemical reaction, the chemical equation must be balanced so that the number of atoms on both sides of the equation is exactly the same.

Explain the following chemical equation.

$$2Fe + 3H_2O \rightarrow Fe_2O_3 + 3H_2 \uparrow \qquad (7.5)$$

This chemical equation indicates that iron reacts with water to form iron oxide and hydrogen (vertical arrows indicate gas). This chemical equation also shows that three water molecules in every two reaction iron atoms are used to form iron oxide molecules and three hydrogen molecules. This is a balanced chemical equation. There are two iron atoms on either side of the equation; six hydrogen atoms on each side; three oxygen atoms on each side.

The equation weight in grams of a compound or element is defined as the gram molecular weight times the number of molecules of the compound, as shown by the coefficients of the chemical equation for the reaction. The sum of the weights on each side of the chemical equation must be equal. Chemical calculations are based on the fact that each fraction or equation weight of the reacting substance gives the corresponding fraction or multiple equation weights in the reaction product. In other words, if a 30 g equation weighing 15 g reacts with another substance to form a product with an equation weight of 20 g, then a 40 g product will be formed.

Example 7.3 How many grams of ferric oxide will be formed if 27.9 g of iron reacts with water.

Solution: The equation weight of iron equals the gram atomic weight of iron times the number of atoms shown reacting in the equation, which is two.

According to the above reaction, 2 mol iron will generate 1 mol rust. 27.9/55.8 = 0.5 mol. 27.9 g can therefore generate 1 mol Fe_2O_3, it is 159.6 g.

7.5 Acids, Bases, Salts, and pH

Different substances react differently in the solution. The way matter behaves in water is of particular interest to the power industry. The interaction of water with acids, bases or salts is particularly interesting because water is used in many industries. This section will introduce students to the general behavior of these substances.

Substances that form ions when dissolved in water are called electrolytes. The three types of electrolytes are acids, bases and salts.

Acids are substances that are separated from water to produce hydrogen ion (H^+). An example of a common acid is sulphuric acid, H_2SO_4.

Acids are chemical compounds that react with hydrogen ions in aqueous solution, such as hydrochloric acid, sulfuric acid, and so on. The acid can react with many metals to release hydrogen, for example:

$$Zn + 2HCl \rightarrow ZnCl_2 + H_2 \uparrow \qquad (7.6)$$

Acid in the pH test paper will show red. Acid can conduct electricity, acid and alkali reaction will produce salt, for example:

$$HNO_3 + KOH \rightarrow KNO_3 + H_2O \qquad (7.7)$$

Acid and carbonate reactions release CO_2, for example:

$$2HCL + CaCO_3 \rightarrow CaCl_2 + CO_2 \uparrow + H_2O \qquad (7.8)$$

Bases are substances that produce hydroxide ions (OH^-) in water solutions. Alkali in a pH test paper appear blue. Typical alkali such as amine compounds (including ammonia, chemical formula: $NH_3 \cdot H_2O$) and caustic soda (sodium hydroxide, chemical formula: NaOH), slaked lime (calcium hydroxide, chemical formula: $Ca(OH)_2$).

When acids react with base, two products are formed: water and salt. Salt is an ion compound consisting of positive and negative ions. Ion bonds are molecular forms that maintain salt. Some compounds look like salt, but are actually covalent compounds (with covalent bonds).

Some soluble salts (mainly sodium, potassium, magnesium, and calcium) that have the properties of binding to acids to form neutral salts are called alkalis. The two common salts are sodium chloride and calcium chloride. Unlike acids and bases, salts are very different in all properties except ion properties. The taste of salt may be salty, sour, bitter, astringent, sweet, or tasteless. The solution of salt may be acidic, basic or neutral. Fused salt and aqueous solutions can conduct electrical currents. Salt reacts in a variety of ways.

There are other reactions that can produce salts, such as replacement reactions. The solution of soluble salts is conductive because there are free swimming ions in the solution, which can be used as an electrolyte.

Many compounds dissolve in water, changing the concentration of hydrogen ions. A compound that directly produces hydrogen ions when dissolved in water is called an acid, and a compound that directly produces hydroxyl ions when dissolved in water is called a base. In order to treat these aspects of chemistry more precisely, a quantitative system for expressing acidity or alkalinity is required. This need can be met by using an $[H^+]$ value, expressed in moles/litre, as a measure of acidity. However, in most cases, $[H^+]$ is in the range of 10^{-1} to 10^{-14} mol/litre. Since such numbers are not easy to work with, an alternative system is designed to represent the acidity of the diluted solution. This system is based on a quantity called a pH value. The pH is defined as a negative logarithm of the hydrogen concentration, expressed as $[H^+]$ in moles/litre.

$$pH = -\log[H^+] \qquad (7.9)$$

or

$$[H^+] = 10^{-pH} \qquad (7.10)$$

A negative logarithm is specified because the logarithm of any number less than 1 number is negative: multiplying -1 causes the pH to be positive within the range of our interest. (The term pH was originally defined by Danish chemists and comes from p for Danish "Power" and H for hydrogen.)

At 25 °C, an equilibrium exists between pure molecular water and its ions. The $[H^+]$ equals the $[OH^-]$ and both have values of 1×10^{-7} mol/liter. Using the pH definition, it follows that the pH of pure water at 25 °C is 7. pH values less than 7 indicate an acidic solution and values greater than 7 indicate a basic or alkaline solution.

The product of ionic concentrations, $K_w = [H^+][OH^-]$, is called the Ion Product Constant for water, or more frequently, the Ionization Constant or Dissociation Constant. At 25 °C, Kw equals 1×10^{-14}. K varies with temperature and, at 37 °C (body temperature), the value is about 3.4×10^{-14}.

Experimental values of K_w at various temperatures are listed in Table 7.5. Notice in Table 7.5 that the pH of pure water changes with temperature. For pure water at any temperature, however, $[H^+] = [OH^-]$. It should be noted that the equation pH + pOH = 14 is true only at or near 25 °C.

Table 7.5 Ion product constant and neutral pH for Water

Temperature/°C	Ion product constant	pH of pure water
18	0.64×10^{-14}	7.1
25	1.0×10^{-14}	7
60	8.9×10^{-14}	6.54
100	6.1×10^{-13}	6.10
150	2.2×10^{-12}	5.83
200	5.0×10^{-12}	5.65
250	6.6×10^{-12}	5.59
300	6.4×10^{-12}	5.60
350	4.7×10^{-12}	5.66

7.6 Corrosion

Uncontrolled corrosion in nuclear facilities may cause serious problems. Corrosion occurs constantly, and every metal in every facility is subjected to some type of corrosion. Even though corrosion cannot be eliminated, it can be controlled.

In nuclear facilities, particularly nuclear power plants, many precautions have been taken to control the corrosion of metals used in primary and secondary systems. Uncontrolled corrosion of the nuclear reactor system is detrimental for the following reasons.

Rapid local corrosion can lead to metal penetration of pipe or vessel containing the coolant. Radioactive coolants can leak from the system, endangering safe operation.

Corrosion of the nuclear fuel cladding can lead to brittle cladding and reduced ductile. Swelling caused by fission gas in the fuel can cause the cladding to rupture or blister, and then the highly radioactive fission product may be released into the coolant.

Some metal oxide corrosion products released into the coolant may be deposited on the surface of the reactor core. The high neutron flux in the core produces a nuclear reaction in corrosive products and is highly radioactive. These active corrosion products can then be released from the core and re-deposited on surfaces outside the core region. The radiation fields produced by this repositioning material may then significantly increase radiation levels and complicate maintenance and access capabilities. Corrosion product oxides can also cause fouling of the heat transfer surface and in the accelerated wear of moving parts by corrosion products trapped in or between them.

7.6.1 Corrosion Theory

The current is the electron flowing through the medium. Current can flow through metal conductors, and the metal does not show any noticeable chemical changes. This electrical conduction is called metal conduction.

Ionization [5] is the process of adding electrons to or removing electrons from atoms or molecules to produce ions. High temperatures, discharges and radiation can cause ionization. Many metals tend to lose electrons to atoms or ions, and atoms or ions tend to acquire electrons. The current can be carried out by the motion of these ions. A compound that conducts electrical current through ion motion is called an electrolyte, and this ion motion is called electrolytic conduction. Conductivity is an indicator of the ability of matter to allow electrons to flow. In terms of corrosion, conductivity represents the amount of ions in the solution, which is directly related to the potential of corrosion.

Corrosion [6] is the deterioration process of a material due to interaction with its environment. Corrosion can take many forms, both wet and dry. Electrolysis is the decomposition of current. This section focuses on metal corrosion (electrolytes) in water-based environments. An electrolyte is defined as a conductor: it has positive and negative ions that can move and form currents. Pure water is a relatively poor conductor, with a limited number of separated H^+ and OH^- ions. Addition of acids, alkalis, or salts that dissociate into ions will enhance the current carrying capacity of water (electrolyte).

Structures are made of metallic materials and are referred to as corrosion in the natural environment or under working conditions due to deterioration and destruction caused by the chemical or electrochemical action of their environmental mediums. Include the above factors and mechanical or biological factors. Some physical effects, such as the physical phenomenon of metal material dissolved in certain liquid metals, can also be classified as metal corrosion. It refers to the rusty steel and iron alloy which is concerned, corrosion rust products are formed mainly composed of hydrous ferric oxide in the form of oxygen and water. Non-ferrous metals and alloy corrosion may occur but do not rust, but form corrosion is similar to rust. Surfaces such as yew copper and copper alloys are occasionally referred to as rust.

There are many types of corrosion. Under different environmental conditions, metal corrosion factors are different and the influencing factors are very complex. In order to prevent and reduce corrosion damage, by changing certain conditions and factors, barrier and anti-corrosion control process, so as to develop methods, technology and corresponding engineering measures, become anti-corrosion engineering technology.

Here, we mainly focus on metal corrosion in water environments. Although both H^+ and OH^- ions are in pure water, the concentration is very low, so it is not easy to conduct electricity. If water is added to a little bit of acid, alkali, or salt, this will greatly improve the conductivity (electrolyte) of the water. According to the cell theory of corrosion, the nature of the corrosion process is electrochemical because of

Fig. 7.14 Formation of ferrous (Fe^{++}) ions in the corrosion of iron

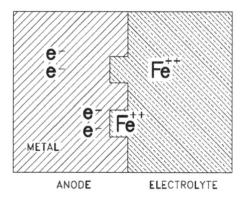

ANODE   ELECTROLYTE

the chemical reactions associated with charge transfer. Figure 7.14 shows the transfer process of charge of ferric contacts with acid water.

Iron enters the solution as Fe^{++} ions. When these ions enter the solution, the metal becomes negatively charged (by the electrons left behind) with respect to the electrolyte. Potential differences will be produced between electrolytes and metals. The process of discarding electrons and forming positive metal ions is called oxidation. The area where oxidation occurs on the metal surface consists of two different substances of electrochemical cells composed of microelectrode components: metals and electrolytes.

These microelectrodes produce many microcells connected by the bulk of the metal. If a different metal is used, it will go into solution to a greater (or lesser) extent producing a larger (or smaller) potential difference between the metal and electrolyte than was the case for iron. For example, magnesium and zinc enter the solution to a greater extent than iron, and these metals have a greater negative effect on electrolytes than iron. Nickel, lead and copper are unlikely to enter the solution, resulting in small potential differences. Table 7.6 lists the potential differences between the various metals in the water. The order of the series can be changed for different electrolytes (e.g., different pH values, ions in the solution).

Electrochemical cells and oxidation potentials are important in understanding most corrosion processes. Examples of electrochemical cells include galvanic cells (cells made up of electrodes of two different substances) and concentration cells (cells containing electrodes of the same substance at different concentrations).

The surface of any metal is a complex of large numbers of microelectrodes, as shown in Fig. 7.15. In order to corrode, microcells must also be connected through some conductive pathways outside the metal. Usually the external connection is provided by water or aqueous solution, and the battery generates an electric current that allows the chemical reaction responsible for corrosion to continue.

Consider the iron in the water. If the surface of the iron and aqueous solution is uniform, the iron enters the solution with Fe^{++} ions until the potential difference between the positively charged solution and the negatively charged metal prevents the iron ions from leaving the surface. However, in practice, impurities and imperfections

Table 7.6
Electromotive—force series
(25 °C)

Element	Electrode reaction	Standard electrode potential/V
Sodium	Na → Na$^+$ + e	−2.712
Magnesium	Mg → Mg^{2+} + 2e	−2.34
Beryllium	Be → Be^{2+} + 2e	−1.70
Aluminum	Al → Al^{3+} + 3e	−1.67
Manganese	Mn → Mn^{2+} + 2e	−1.05
Zinc	Zn → Zn^{2+} + 2e	−0.762
Chromium	Cr → Cr^{3+} + 3e	−0.71
Iron	Fe → Fe^{3+} + 3e	−0.44
Cadmium	Cd → Cd^{2+} + 2e	−0.402
Cobalt	Co → Co^{2+} + 2e	−0.277
Nickel	Ni → Ni^{2+} + 2e	−2.250
Tin	Sn → Sn^{2+} + 2e	−0.136
Lead	Pb → Pb^{2+} + 2e	−0.126
Copper	Cu → Cu^{2+} + 2e	+ 0.345
Copper	Cu → Cu$^+$ + e	+ 0.522
Silver	Ag → Ag$^+$ + e	+ 0.800
Platinum	Pt → Pt^{2+} + 2e	+ 1.2
Gold	Au → Au^{3+} + 3e	+ 1.42

Fig. 7.15 Metal surface
showing arrangement of
micro-cells

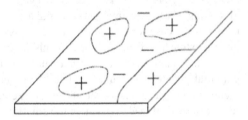

(e.g. oxide coatings) first cause metals to be removed from certain parts of the surface and may vary, for example in two metal systems. Corrosive cells vary with the surface and solution, leading to overall corrosion. If the cells do not move, the result is a pit.

    Metal corrosion (i.e. chemical conversion that is recognized as destructive to the metal) is an oxidation step throughout the oxidation process. Oxidation is the process of losing electrons; reduction is the process of obtaining electrons. Metal atoms release electrons (oxidation) and become positive ions. The location where this occurs is called a node. Typical oxidation half-reactions include the following.

$$Zn \rightarrow Zn^{2+} + 2e^- \qquad (7.11)$$

$$Al \rightarrow Al^{3+} + 3e^- \tag{7.12}$$

$$Fe \rightarrow Fe^{2+} + 2e^- \tag{7.13}$$

The cations (positive ions) can enter the solution, or they may bind to any available ion (negative ion) or water to form an ion compound. The exact fate of the cations is important for the subsequent process, but the main effect is that the atoms leave the metal state and the metal deteriorates.

An oxidation process cannot happen without a simultaneous reduction (gain of electrons) process. The nature of the corrosion reduction step sometimes varies with the metal and its exposed environment. For most metals in a water environment, an important reduction in half-reaction is the reduction of hydrogen ions (hydrogen ions are simply hydrogen ions attached to water molecules).

$$H_3O^+ + e^- \rightarrow H + H_2O \tag{7.14}$$

Small changes in concentration in solutions in contact with metals may also affect the speed and properties of corrosion reactions. Therefore, it is often impossible to predict the exact nature of corrosion reactions. However, it is generally believed that for most metals exposed to water, the half-reaction involved in corrosion is the reduction reaction of Eq. (7.14) and the oxidation half -reaction of the types shown in Eqs. (7.11) to (7.13).

General corrosion is a slow, relatively uniform process of metal surfaces: the removal of materials. This occurs on the surface of a single metal, rather than dissimilar metals. In general corrosion, an almost unlimited number of microcells are built on the metal surface. Oxidation occurs in the anodic area and reduction in the cathodic region. Microcells are evenly distributed on the metal surface, and as the reaction progresses, the cells may migrate, disappear, and re-form. That is, any particular micro-region may be alternately anodic and cathodic. The result is a uniform attack on the surface of metal.

In some cases, relatively large areas become anodic or cathodic. Such regions have less tendency to migrate and may operate over a long term. In this case, a serious attack on the metal will occur in the anodic (oxidation) area. The result may be a visible pit on the metal surface.

Iron and steel are resistant to rapid corrosion in water despite the tendency of iron to oxidize as indicated by its standard electrode potential listed in Table 7.6. This resistance is due to the passivation effect of the oxide film and cathodic polarization due to atomic hydrogen that absorbs the surface of the oxide.

Metals that are often victims of corrosion sometimes exhibit a passivity to corrosion. Passivity is a characteristic of metals when they are inactive in corrosion reactions. Passivity is caused by the accumulation of stable and tenacious layers of metal oxides on the metal surface. This oxide layer is formed by corrosion of the clean metal surface, which is insoluble in the specific environment in which the metal is exposed. Once a layer or film is formed, it acts as a barrier separating the metal surface from the environment. For further corrosion, the reactant must be diffused through the oxide film. This diffusion is very slow or non-existent, so corrosion is either significantly reduced or stopped.

Metals, such as zirconium, chromium, aluminum, and stainless steel, form thin, tough oxide films when exposed to pure water in the atmosphere or at room temperature. In some cases, the film is very thin and may be invisible to unaided eyes, but it is still very effective in giving these metals a distinct passivity.

If the reactants in the system are netly converted to the product, the system will be chemically unstable and the reaction will continue to stabilize. This stable state is called equilibrium.

Active electrochemical cells (redox reactions) are unstable chemical systems. For example, as current flows and oxidation decreases, the potential associated with electrocells decreases. Eventually, the potential drops to zero, the battery no longer provides electrical energy, and no further net reaction occurs. The system is now in equilibrium. In electrochemical cells, the reduction in cell potential caused by cell operation (current flow) is called polarization.

This change in cell potential is determinable. Consider the zinc-copper galvanic cell shown in Fig. 7.16. When the reaction occurs, the Zn^{++} n (produced by zinc oxidation) enters the solution. When the copper plate is released, the Cu^{++} in the solution are reduced. As a result, the concentration of Zn^{++} in the solution increases and the concentration of Cu^{++} decreases according to the following overall reaction.

As Zn^{++} increases and Cu^{++} decreases, the potential decreases. This decrease in cell potential, which is the result of a change in concentration, is a polarized form called concentration polarization.

Now consider an galvanic cell with zinc and platinum electrodes, as shown in Fig. 7.17. The half-reactions in the cell are as follows.

Fig. 7.16 A galvanic cell

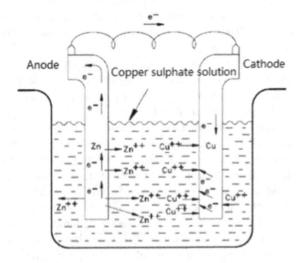

Fig. 7.17 A galvanic cell showing absorbed hydrogen atoms on a cathode

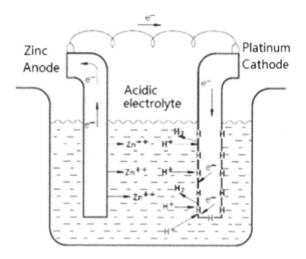

$$Zn \rightarrow Zn^{2+} + 2e^- \tag{7.15a}$$

$$H_3O^+ + e^- \rightarrow H + H_2O \tag{7.15b}$$

Similarly, when the cell is operating, the cell potential decreases. Part of the decrease is due to an increase in the concentration of Zn^{++} and a decrease in the concentration of H_3O^+, but another type of polarization also occurs in this cell. The second type is associated with a reduction in half-reactions.

The hydrogen atoms formed by the Eq. (7.15b) and this reaction absorbs the metal surface and remain in existence until they are removed by one of two processes: a combination of two hydrogen atoms forms molecular hydrogen, which is then released into gas or in reaction to dissolved oxygen to form water. In the absence of oxygen (deaerated solutions), the first procedure is applied. The net reduction half-reaction is:

$$Zn \rightarrow Zn^{2+} + 2e^- \tag{7.16a}$$

$$H_3O^+ + e^- \rightarrow H + H_2O \tag{7.16b}$$

$$2H \rightarrow H_2 \tag{7.16c}$$

They effectively block where the Eq. (7.14) reaction occurs before removing absorbed hydrogen atoms from the metal surface. At low temperatures, the reaction of the Eqs. (7.16a, 7.16b, 7.16c) is slow relative to the Eq. (7.14), because although this reaction is strongly favored, the combination of two hydrogen atoms requires a great deal of activated energy. Equations (7.16a, 7.16b, 7.16c) show the rate control steps for a net reduction of the half-reaction. Since oxidation half-reaction does not occur faster than reduction half-reaction, the rate of overall oxidation reduction reaction is controlled by reaction Eqs. (7.16a, 7.16b, 7.16c).

Both concentration and activation polarization reduce the reaction rate of net oxidation. Activation polarization usually has a greater effect during corrosion.

The layer of absorbed atomic hydrogen is said to polarize the cell. This type of polarization is called activation polarization and is sometimes referred to as hydrogen polarization or cathodic polarization because the polarization reaction occurs at the cathode.

Both concentration and activation polarization decrease the net oxidation–reduction reaction rate. In corrosion processes, activation polarization usually has the greater effect.

7.6.2 General Corrosion

Here we describe the general corrosion processes of iron and carbon steel (not stainless steels) in aqueous environments. Of particular interest is the formation of the oxide film and the effects of system variables on the corrosion process.

General corrosion is the process whereby the surface of a metal undergoes a slow, relatively uniform, removal of material. The two conditions typically required for a metal to undergo general corrosion are: (1) metal and water in the same environment, and (2) a chemical reaction between the metal and water that forms an oxide.

Unless noted otherwise, the following discussion applies to deaerated water at room temperature and approximately neutral pH. The effects of temperature, oxygen, and pH are discussed later. The oxidation and reduction half-reactions in the corrosion of iron are as follows.

$$Fe \rightarrow Fe^{2+} + 2e^- \tag{7.17a}$$

$$H_3O^+ + e^- \rightarrow H + H_2O \tag{7.17b}$$

The overall reaction is the sum of these half-reactions.

$$Fe + 2H_3O^+ \rightarrow Fe^{2+} + 2H + 2H_2O \tag{7.17c}$$

The Fe^{2+} ions readily combine with OH^- ions at the metal surface, first forming $Fe(OH)_2$, which decomposes to FeO.

$$Fe + 2\,OH^- \rightarrow Fe(OH)_2 \rightarrow FeO + H_2O \tag{7.17d}$$

Ferrous oxide (FeO) then forms a layer on the surface of the metal. Below about 500 °C, however, FeO is unstable and undergoes further oxidation.

$$2FeO + H_2O \rightarrow Fe_2O_3 + 2H \tag{7.18}$$

Atomic hydrogen then reacts to form molecular hydrogen, as described previously, and a layer of ferric oxide (Fe_2O_3) builds up on the FeO layer. Between these two layers is another layer that has the apparent composition Fe_3O_4. It is believed that Fe_3O_4 is a distinct crystalline state composed of O^{-2}, Fe^{+2}, and Fe^{+3} in proportions so that the apparent composition is Fe_3O_4. These three layers are illustrated in Fig. 7.18.

Once the oxide film begins to form, the metal surface is no longer in direct contact with the water environment. For further corrosion, the reactant must be diffused into the oxide barrier. It is believed that the oxidation step, Eq. (7.13), occurs in the metal oxide interface. The Fe^{+2} ions and electrons then spread through the oxide layer to the oxidizing water interface. Eventually, the Fe^{+2} ions encounter OH^- ions and form FeO. Electrons and hydrogen ions are involved in the reduction reaction. The latter reaction is thought to occur mainly in the oxide interface, but some reactions may occur in the oxide layer by diffusing H^+, OH^-, and H_2O into the layer.

Whatever the exact diffusion mechanism, the oxide layer represents a barrier to continuous corrosion and tends to reduce the rate of corrosion. The exact effect of this layer on corrosion rate depends on the uniformity and toughness of the film. If the film is loosely connected, defective, or removed, the metal surface is exposed to the environment again and corrosion is more likely to occur.

As with most other chemical reactions, corrosion rates increase as the temperature increases. The temperature and pressure of the medium control the solubility of corrosive species in liquids such as oxygen (O_2), carbon dioxide (CO_2), chlorides and hydroxides. As a rule of thumb, the reaction rate doubles as the temperature increases from 20 to 50 °C. This linear increase in temperature does not last indefinitely, in part because of changes in the oxide film.

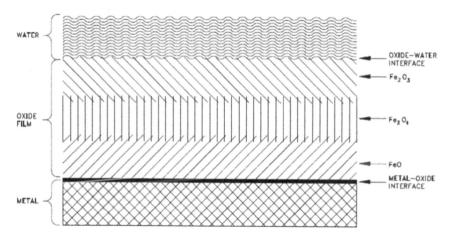

Fig. 7.18 Simplified schematic diagram of oxide corrosion film on the surface of a metal

When the water speed is very high, the effects of water tend to remove some of the metals (erosion) that protect the oxide layer and below it, resulting in more metal corrosion. Water speeds of 10–15 m/s are generally thought to cause erosion.

Oxygen exposed to iron in water increases corrosion rates. The increase is due to the rapid reaction between the oxygen and oxide layers absorbed by the atomic hydrogen polarization layer. The following reactions quickly eliminate the polarization layer.

Therefore, oxygen has two effects: it removes the polarization layer of atomic hydrogen, which reacts directly with metal or metal oxides: therefore, the corrosion rate increases. In this case, removing substances such as O_2 from absorbed atomic hydrogen is called a depolarizer. The depolarization effect of O is illustrated in Fig. 7.19.

The effect of pH of steel exposure to water is affected by temperature. The potential of hydrogen or symbol (pH) is defined as a negative logarithm of hydrogen concentrations, expressed in [H^+]moles/litres as.

First, consider exposing iron to aerated water at room temperature (aerated water will contain dissolved oxygen). The corrosion rate of iron as a pH function is shown in Fig. 7.20. In the range of pH 4 to pH 10, the corrosion rate of iron is relatively independent of the pH of the solution. Within this pH range, the corrosion rate is mainly controlled by the speed at which oxygen reacts with the absorption of atomic hydrogen, thus depolarizing the surface and allowing the reduction reaction to continue. Ferrous oxide (FeO) is soluble for pH below 4.0. As a result, oxides dissolve when formed, rather than depositing on metal surfaces to form films. In the absence of a protective oxide film, the metal surface comes into direct contact with the acid solution, and the corrosion reaction occurs faster than the high pH. Hydrogen has also been observed to be produced in acid solutions with pHs below 4, suggesting that corrosion rates no longer depend entirely on the depolarization of oxygen, but on a combination of these two factors (the evolution and depolarization of hydrogen). For pH above pH 10, the corrosion rate decreases as the pH increases. This is thought to be due to an increased in the rate of the reaction of oxygen with

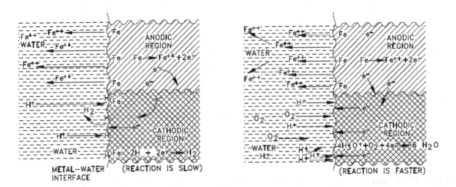

Fig. 7.19 Representation of cathodic depolarization by oxygen

Fig. 7.20 Effect of pH on the corrosion rate of iron in water

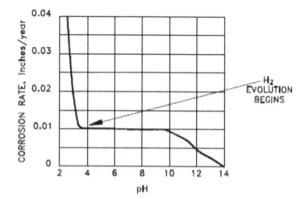

Fe $(OH)_2$ (hydrated FeO) forming a more protective Fe_2O_3 in the oxide layer (note that this effect is not observed in deaerated water at high temperatures).

But when temperature increases, this phenomenon will change. For example at a temperature of 300 °C, iron in the corrosion rate of the deoxidizing water at a pH of greater than 10 but will become larger. Therefore under high temperature and water environment, pH control in 8 ~ 10 (weak alkaline) is favorable.

Hydrogen, usually dissolved in reactor coolants, has no detectable direct effect on the corrosion rate of steel exposed to reactor coolants. However, it does have important indirect effects to prevent the accumulation of dissolved oxygen in reactor coolants, which will accelerate corrosion. Dissolved oxygen reacts with the protective hydrogen layer of the cathode to form water.

Conditions and composition of metal surfaces can affect corrosion rates. Deposits, scale or irregular surfaces form areas on the metal, and local corrosion can start and take place faster than normal. Some metal alloys are more corrosive than others.

When steel is exposed to hot water, the corrosion rate of the metal decreases with the early stages of exposure. After thousands of hours, the corrosion rate becomes relatively stable at low values. In the early stages of exposure, the oxide film on the metal surface thickens as corrosion rates decrease. However, the film grows at a slower rate over time. The thickness of the oxide film quickly reaches a relatively constant value, after which the thickness of the film does not change significantly with further exposure. Unsurprisingly, a relatively constant corrosion rate and oxide film thickness were achieved around the same time. Since tightly adhered corrosion film inhibit further corrosion, great care is required during the initial filling of the reactor plant to facilitate the formation of the optimum corrosion film. This process, called pretreatment, or pickling, involves carefully controlling the chemical composition and temperature of the reactor coolant water during pretreatment.

Plant chemistry is used to control corrosion. The type of corrosion determines the method used to prevent or minimize corrosion rates.

Passivation is a condition in which naturally active metals corrode at a very low rate, possibly due to oxide coatings or absorbed oxygen layers. Some chemicals, called passivators or inhibitors, can be added to water to provide this type of passivation by reducing it on the metal surface. A common inhibitor is potassium chromate.

Using cathodic protection to provide an external current to the iron so that it acts as a cathode with no anodic area, is another preventive chemical control method. This can be achieved by using an external voltage source or by using a acrificial anode, such as zinc, which corrodes and provides current.

Chemistry control in the form of removing corrosive agents from the system is a widely used method. One approach is to use a deaerators to remove dissolved oxygen and, to a lesser extent, carbon dioxide. Water is treated by softening and demineralization to remove dissolved solids and reduce electrical conductivity.

Adding chemical components to systems that alter chemical reactions or connect specific corrosion agents is a common control method. Filming amines (organic compounds used as ammonia derivatives) are protected by forming a adhering organic film on the metal surface to prevent contact between corrosive species in condensate and metal surfaces. Phosphate and sodium hydroxide are used to adjust the pH of the system and remove hardness.

The flow velocity of the fluid on the corrosion has great influence. Fluid flow velocity is larger when (i.e. more than 10 m/s), the formation of protective oxide film easily washed away, which will happen erosion phenomenon, greatly accelerate the corrosion rate.

In some of the nuclear power plant, the use of aluminum or aluminum alloy as fuel cladding material. Aluminum corrosion characteristics and iron basically almost. Just for aluminum, the formation of oxide film of pre processing is particularly important. In addition, with aluminum as structure material system, and usually remain weak acid water.

7.6.3 Crud and Galvanic Corrosion

A major potential problem is crud [7]. Crud can lead to increased background radiation levels. When two different metals approach, the chances of ion transfer are high. The result is a perfect environment for galvanic corrosion. Because of the wide variety of materials used in nuclear facilities, galvanic corrosion is a major problem.

In addition to corrosion film, corrosion products in the form of finely divided, insoluble oxide particles called crud suspended in reactor coolants or loosely adhered to metal surfaces. Crud has several unwelcome features. It can be transported throughout the reactor coolant system. As a result, it can accumulate and foul heat transfer surface or block flow channels. However, the most undesirable feature of crud is that it is activated when exposed to radiation. Because crud can be transported throughout the reactor coolant system, it can be clustered outside the reactor core, causing radiation hotspots to increase environmental radiation levels. Hot spots caused by the crud collection may occur at the entrance to the purifier and in other low-flow areas. Crud that is loosely adhered to the metal surface can suddenly become suspended in the reactor coolant.

Crud release may be due to increased oxygen concentrations, reduced pH (or significant changes), temperature changes (heat up or cooldown), or physical shock on the system. Physical shocks include starting, stopping, or altering pump speeds or other evolutions, such as reactor scram or pressure-relief valve lifts. The result is a sudden increase in reactor coolant activity. Releasing crud in this way is called as a crud burst. Crud bursts often result in the removal of protective corrosion film, making newly exposed metals more susceptible to additional corrosion. In addition to corrosion films and crud, some corrosion products are soluble and can be easily transported to the entire system.

The high crud concentration in the system also complicates the treatment of primary coolants. Many corrosion products have a relatively long half-life and represent significant biological hazards. Therefore, if the main coolant is discharged or leaked from the plant shortly after the rupture, additional procedures may be required to minimize the impact of the condition.

Therefore, if the conditions mentioned previously (O_2, pH) changed, the solubility of these corrosive products will change and can then be transported to any part of the reactor coolant system and deposited.

Another corrosive by-product is scale, which consists of the deposits of insoluble compounds formed by normally soluble salts on the surface. The most common is calcium or magnesium carbonate ($CaCO_3$ or $MgCO_3$).

Galvanic corrosion is corrosion when two different metals with different potentials are placed in electrical contact in an electrolyte.

Of all the different types of corrosion, galvanic corrosion best matches the electrochemical cells described earlier in this chapter because electrochemical corrosion occurs when two electrochemically dissimilar metals (in electrical contact) are connected in a conductive medium (electrolyte). It may also occur on a metal with heterogeneity (dissimilarities) (e.g., impurity inclusions, particles of different sizes, differences in grain composition, differences in mechanical stress); abnormal levels of pH; and high temperatures. Potential differences between metals are the driving force for current flowing through corrodant or electrolytes. These currents cause corrosion of one of the metals. The greater the potential difference, the greater the likelihood of galvanic corrosion. High corrosion can only cause one of these metals to deteriorate. Metals with poor resistance and strong activity become the site of anodic corrosion. Stronger, nobler metals are catholic and protected. Without electrical contact, the two metals would be attacked evenly by the corrosive medium, as if the other metal were absent. Two locations that are susceptible to galvanic corrosion are the transition of pipes from one metal to another and sacrificial anode such as zinc.

Figure 7.21 illustrates that galvanic corrosion occurs when two different metals are in contact and exposed to an electrolyte.

Figure 7.10 shows the junction of iron and copper pipes containing a solution of a copper salt. The oxidation potential of iron is sufficiently greater than that of copper so that iron is capable of reducing Cu^{++} ions to copper metal. In this case, iron corrodes near the junction, and additional copper builds up on the copper pipe near the junction.

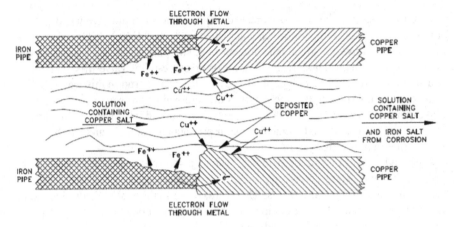

Fig. 7.21 Galvanic corrosion at iron-copper pipe junction

The solution to which the metal junction is exposed need not contain a salt of one of the metals for galvanic corrosion to occur. If the iron-copper junction were exposed to water without Cu^{++} ions. Again, iron would corrode near the junction, but in this case, hydrogen would be formed on the surface of the copper.

A method called catholic protection, previously discussed in this section, is commonly used to retard or eliminate galvanic corrosion. One of several ways to do this is to connect a third metal to the metal to be protected. The oxidation potential of this metal must be greater than that of the metal in order to be protected. The most active metals then tend to corrode to replace protected metals. Metals that corrode to protect another metal are called sacrificial anode. This method is applied to the original design of structural materials. Zinc is a common sacrificial anode and is commonly used in cooling water systems containing seawater.

Galvanic corrosion can also be limited by: (1) using only metals that are close on the activity series, (2) electrical insulation of dissimilar metals, and (3) using poorly conducting electrolytes (very pure water).

The relative surface area of the two metals is also important. A much larger surface area of the non-active metal, compared to the active metal, will accelerate the attack. The relative surface area has been determined to be the determining factor in corrosion rate. If the different metals are:

1. Separated by non-conducting junctions,
2. Separated from the conductive environment, and.
3. Is located in the conductive poor conductivity of the conductive medium (pure water).

7.6.4 Specialized Corrosion

Because of the unique environment that may exist in the nuclear industry, some specialized types of corrosion must be considered.

Pitting occurs when the anodic are fixed in a small area and holes (deep attacks) are formed in other unaffected areas. Gap corrosion is a kind of pit corrosion that occurs especially in the low flow area of the gap.

Pitting corrosion can develop in depth of pitting on the surface of metal parts, the rest of corrosion or slight corrosion. The corrosion morphology called pitting, also known as pitting corrosion. With steel, for example, the surface of the stainless steel tiny rust hole rapid increasing, is reasons to cause massive corrosion of stainless steel. Small changes in corrosion substance concentration or temperature can significantly accelerate the corrosion rate.

Crevice corrosion is a kind of occurred in local corrosion phenomenon. Was found in space, where the fluid is not flowing (can be understood as "stagnant") exists in the gap. Corrosion was uniform corrosion in the gap. Then the development of cathode and anode corrosion because of fluid filled gap, and gap for oxygen rich region (cathode reaction to losing electrons), internal clearance for anoxic zone (anodic reactions that gains electrons). Because of the gap in the liquid, internal clearance has been anoxic zone anodic reactions to occur, because have cathode, anode corrosion cell structure, corrosion rate will greatly improve. Crevice corrosion of schematic diagram as shown in Fig. 7.22. In stagnant zone, an anoxic zone formation, the formation of an accelerated corrosion of anode. In addition to oxygen, chlorine containing fluid will also have this phenomenon.

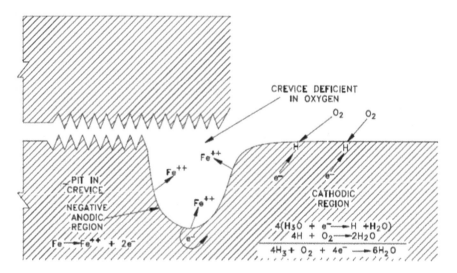

Fig. 7.22 Crevice corrosion schematic

The presence of oxygen can also promote pitting in areas where the metal surface is initially opposed to adjacent areas. For example, suppose that adjacent areas of the metal surface exhibit slightly different oxidation potentials. Oxidation or loss of metal is carried out in areas with high potential. Corrosion in areas with high potential leads to the formation of porous oxide films (at least initially). The thickness of the film formed in the adjacent cathode region will be much smaller. Oxygen in most solutions is easier to reach the cathode surface (with a thin film) than the nearby area of the anodic film surface (with a thicker oxide film). The cathode region (thin film) of the oxygen solution tends to maintain the cathode in this area, while the lack of oxygen under the thick porous corrosion film helps to maintain the anodic condition in the area. The overall result is metal that corrodes or wastes the area of the nasal membrane under the thinner film. As a result, a pit in the metal surface is formed under the mound of surface oxide, as shown in Figs. 7.23. This type of pitting is common in both low-temperature and high-temperature iron systems if precautions are not taken to remove oxygen from the water in the system.

Certain ions, especially chlorinated ions, have also been found to cause pitting in steel. The exact mechanism by which this happens is not known, but to some extent chlorinated ions can cause defects in the oxide layer of passivation on the metal surface. Defects are highly localized and surrounded by large passive areas, which are often cathodes. As a result, a small anodic (oxidation) area is surrounded by a large cathode (reduction) area. Then, at the anodic site, the current density will be

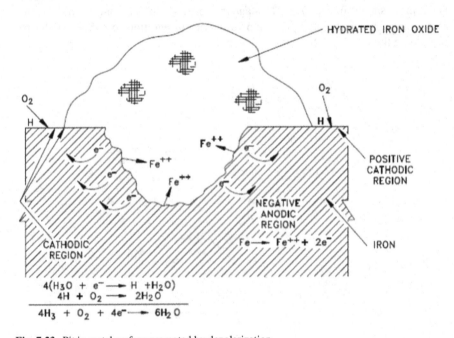

$$4(H_3O + e^- \longrightarrow H + H_2O)$$
$$4H + O_2 \longrightarrow 2H_2O$$
$$4H_3 + O_2 + 4e^- \longrightarrow 6H_2O$$

Fig. 7.23 Pit in metal surface promoted by depolarization

very large and the attack on the metal will be very rapid. In some test cases, deep pits were observed within a few hours.

Due to the rapid penetration of metals, although the overall loss of mass is little, pitting and crevice corrosion is a major risk to nuclear facilities. Nuclear facilities minimize pitting and crevices corrosion through the following actions.

1. Avoid stagnant or low-flow conditions.
2. Use metals and alloys that are less susceptible to the corrosion.
3. Avoid preparations in the medium that cause pitting (e.g. chloride and oxygen).
4. Design systems and components so that there are no crevices.

Stress Corrosion Cracking (SCC) is an interparticular attack corrosion that occurs at particle boundaries under tensile stress. If the alloy is stressed, SCC occurs in the susceptible alloy when the alloy is exposed to a specific environment. Stress corrosion cracks appear to be relatively independent of the general uniform corrosion process. As a result, the degree of general corrosion is essentially zero, and stress cracking can still occur. Most pure metals are immune to such attacks.

According to the most well-known theory, stress corrosion cracking is caused by a process called chemisorption. Unlike relatively weak physical absorption, such as hydrogen on platinum, chemical adsorption can be thought of as the formation of compounds between surface metal atoms, as a single molecular layer of chemical material, such as Cl^-, OH^-, Br^-, and other ions. The formation of this chemical layer greatly reduces the attractiveness between adjacent metal atoms. The initial defects grow under stress as metal atoms separate, and more chemical adsorption occurs, a process that continues. In very serious cases, this cracking takes only a few minutes.

Many stainless steels are susceptible to stress corrosion cracking. Stainless steels containing 18% chromium and 8% nickel are prone to cracking in environments containing chlorine ions and in concentrated corrosive environments (i.e. in environments with high hydroxyl ion concentrations). On the other hand, these types of stainless steel do not exhibit any tendency to crack when exposed to water containing nitrates (NO^{3-}), sulfates (SO^{3-}) and ammonium (NH_3^+) ions.

SCC is of great concern because it can easily crack metal of appreciable thickness. If the environment is severe enough, cracking can occur in a very short period of time. Cracks can then cause severe component or system failure, as well as all consequent results (e.g. contamination, coolant loss, and pressure loss).

The most effective way to prevent SCC is to properly design, reduce stress, eliminate critical environmental contributors such as hydroxides, chlorides and oxygen, and avoid stagnating areas and crevices in heat exchangers where chlorides and hydroxides may be concentrated. Low alloy steel is less susceptible than high alloy steel, but is affected by SCC in water containing chlorinated ions. Nickel-based alloys are not affected by chlorides or hydroxide ions.

Two types of SCC are major concerns for nuclear facilities.

Chloride Stress Corrosion Cracking (Stainless Steels)

The three conditions that must be present for chloride stress corrosion to occur are as follows.

1. Chlorinated ions are present in the environment.
2. Dissolved oxygen is present in the environment.
3. The metal is under tensile stress.

Stainless steel is a non-magnetic alloy consisting of iron, chromium and nickel with low carbon content. This alloy is highly corrosion resistant and has ideal mechanical properties. One type of corrosion that can attack stainless steel is chloride stress corrosion. Chloride stress corrosion is a kind of intergranular corrosion.

Chloride stress corrosion involves selective attack of metals along grain boundaries. In the formation of steel, chromium-rich carbide precipitates at grain boundaries, making these areas less protected and therefore vulnerable to attack. The study found that this is closely related to some heat treatments resulting from welding. This situation can be greatly reduced through an appropriate annealing process.

This form of corrosion is controlled by maintaining low chlorine ions and oxygen levels in the environment and by using low carbon steel. In auxiliary water systems, environments containing dissolved oxygen and chlorinated ions can be easily created. Chlorinated ions can enter these systems through leaks in condensers or through auxiliary systems associated with nuclear facilities that are cooled by unpurified cooling water. Dissolved oxygen can easily enter these systems with feed and make-up water. Therefore, chloride stress corrosion cracking is worthy of attention, control must be used to prevent its occurrence.

Caustic Stress Corrosion Cracking   Caustic stress corrosion, or caustic embrittlement, is another form of intergranular corrosion cracking. This mechanism is similar to chloride stress corrosion. Mild steel (steel with low carbon and low alloy content) and stainless steel will crack when exposed to concentrated corrosive (high pH) environments under the tensile stress of the metal. In stress cracking caused by corrosive environment, the presence of dissolved oxygen is not necessary for cracking.

For the first time in the operation of riveted steam boilers, corrosion cracks are encountered. These boilers are sometimes found to have failed along riveted seams. The failure is due to corrosive cracking near rivets and in areas of high stress in the lower layer. Boiler water can easily flow into the gap under the rivets.

Radioactive heating causes water in the crevices to boil. When steam forms, it escapes from the gap. More boiler water then flows into the crevice, boils, and then passes through the gap as steam. The end result of this continuous process is the caustic concentration under the rivet. The combination of high stress and high caustic concentration eventually leads to destructive cracking of boiler containers.

Where the steam generation (boiling) rate is high, it is difficult to eliminate the problem of solute concentration in the boiler area. Corrosive stress corrosion may be concentrated in areas such as rapid evaporation of water, but it is considered unlikely that the corrosion concentration induced by this mechanism will be high enough to induce stress cracking.

Available data indicates that caustic concentrations greater than 10,000 ppm, and probably up to 50,000 ppm, are required to induce caustic stress cracking (40,000 ppm NaOH is equivalent to 40 g per liter or 1 mol per liter). The pH of this solution is 14. Alkaline environments are generated and controlled by using solutions that have certain characteristics of buffers (i.e. tend to retardant or slow down reactions or force them in one direction or another).

7.7 Water Chemistry of Reactor

Radiation synthesis is a process that occurs in a reactor coolant system. This phenomenon is limited to reactor coolant systems, as high flux (radiation) levels in the core area further complicate chemical control of reactor plants.

When reactor coolant flows through the core region of an operating reactor, it is exposed to intense radiation. The main components of the radiation field are neutrons, protons, gamma rays, and high-energy electrons (β particles). These types of radiation interact mainly with coolant water through ionization processes, which have a significant effect on the chemical reaction between the water itself and the substances dissolved in the water. These effects are discussed in this section, in particular those involving gases dissolved in reactor coolants.

The interaction between radiation and matter produces ion pairs. Typically, the negative members of an ion pair are free electrons, and the positive members are polyatomic cation, depending on the specific substance being irradiated. For example, the interaction between radiation and water is illustrated by the following reactions.

$$H_2O \xrightarrow{\text{辐射}} e^- + H_2O^+ \tag{7.19}$$

Both species are chemically reactive, and each has several ways of reacting. Some of these mechanisms are complex and often have no real value to the reactor operator, who is more concerned with the overall, observable effect.

Reactions (7.19) show that irradiation of pure water produces electrons and H_2O^+ ions. Both species are highly reactive. H_2O^+ ions react rapidly with water molecules as follows.

$$H_2O^+ + H_2O \rightarrow OH + H_3O^+ \tag{7.20}$$

The OH species is an uncharged hydroxyl group. In neutral groups like this, not all chemical bonding capacity is satisfied and are intermediate species common in chemical reactions, known as radicals or sometimes free radicals.

The electrons produced by the reaction (7.19) first form a species called hydrate electrons, represented by e_{aq}^-. Hydrate electrons may be thought to be the result of positive interactions between negative electrons and polar water molecules. This is similar to the interaction of a positive proton (H^+) with the negative end of a water molecule to form a hydrogen ion. Since water molecules associated with hydrate electrons do not participate in subsequent chemical reactions, they are not displayed in chemical equations and are replaced by hydrate electrons (e_{aq}^-).

Hydrated electrons may interact with H_3O^+ ions in solution or with water molecules. Both reactions produce another reactive species, atomic hydrogen.

$$H_3O^+ + e_{aq}^- \rightarrow H + H_2O \tag{7.21a}$$

or

$$H_2O + e_{aq}^- \rightarrow H + OH^- \tag{7.21b}$$

Reaction (7.21a) usually predominates. Because Reactions (7.21a) and (7.21b) are slow compared to that in Reaction (7.20), there are three reactive species present at any one time: hydroxyl radicals (OH), hydrated electrons (e_{aq}^-), and hydrogen atoms (H). These species may undergo any of several possible reactions such as the following.

$$OH + OH \rightarrow H_2O_2 \tag{7.22}$$

$$OH + H \rightarrow H_2O \tag{7.23}$$

$$H + H \rightarrow H_2 \tag{7.24}$$

$$H + e_{aq}^- + H_2O \rightarrow H_2 + OH^- \tag{7.25}$$

$$H_2 + OH \rightarrow H_2O + H \tag{7.26}$$

Hydrogen peroxide, formed by Reaction (7.22), may also react with the original reactive species, but at high temperatures H_2O_2 is unstable, and the predominant reaction is decomposition.

$$2H_2O_2 \rightarrow O_2 + 2H_2O \tag{7.27}$$

In the case of water, the overall effect of irradiation is shown in the following reaction.

$$2H_2O \xrightarrow{\text{Radiation}} 2H_2 + O_2 \tag{7.28}$$

The end result of these reactions is simply the decomposition of water. If H_2 and O_2 are allowed to escape the solution as gases, the reaction will continue as written. However, if water is contained under pressure (e.g. reactor coolant systems), H_2 and O_2 are limited and balanced because radiation can also cause reactions (7.28) to reverse. Mainly neutron and gamma radiation induces the decomposition of water and the recombination of H_2 and O_2 to form water. Therefore, it is appropriate to write the reaction (7.28) as a balanced reaction caused by radiation.

$$2H_2O \xleftrightarrow{\text{Radiation}} 2H_2 + O_2 \tag{7.29}$$

In order to achieve the overall effect of radiation on water, the above process involves assuming that each reactive species has only one reaction pathway. This is done primarily for convenience, as adding every possible response to the summary process becomes quite cumbersome. Even when all reactions are taken into account, the net results are the same as those of the reactions (7.29), which is reasonable because inspections of the reactions (7.21) to (7.27) indicate that the only stable products are H_2 and O_2, and H_2O (H_3O^+ and OH^- combination, forming water and H_2O_2 decomposition at high temperatures). Perhaps not as obvious, more water is consumed than is produced in these reactions, and the end result is the initial decomposition of the water until the equilibrium concentrations of H_2 and O_2 are established.

Before discussing the effects of radiation on other processes, mentiones should be made to the chemical balance of ionizing radiation.

Radiation has an effect on the balance of water. In the absence of radiation, water does not decompose spontaneously at 300 °C. However, when irradiated, water does decompose, as shown in the figure above. In addition, H_2 and O_2 typically do not react at 300 °C because the reaction requires a lot of activation energy. In fact, radiation provides this activation energy, and reactions are easy to occur. As a result, radiation increases the rate of forward and reverse reactions, although not by the same factors.

In general, the effects of radiation on a given reaction balance cannot be quantified. The observed effect on balance may vary with the intensity of the radiation, complicating the situation. In nuclear facilities, the impact may vary depending on the power level of the facility. In most cases, this complication is not a serious problem because the effect is in the same direction; only the degree or magnitude of the effect varies with the intensity of the radiation.

As mentioned earlier, reactor coolants are kept at a basic pH (in facilities other than aluminum components or in facilities controlled by chemical shim reactivity control) to reduce corrosion processes. For the same reason, it is important to remove dissolved oxygen from reactor coolants. However, as shown in the previous section, the natural consequence of exposing pure water to ionizing radiation is the production of hydrogen and oxygen. Adding a base to control pH has little effect on this feature.

Hydrogen is added to prevent the formation of oxygen in the reactor coolant. Hydrogen inhibits the formation of oxygen mainly depends on its effect on the OH radicals produced by the reaction (7.20). In the presence of excess hydrogen, hydroxyl radicals reacts primarily by reaction (7.26) rather than reaction (7.22) to (7.24).

Hydrogen atoms from Eq. (7.26) subsequently react to form H_2 and H_2O. None of these reactions leads to O_2, or H_2O_2, which decomposes to form O_2 and H_2O at high temperatures. Thus, the addition of H_2 to reactor coolant largely eliminates production of free oxygen.

Reactor coolant makeup water usually contains a small amount of air, mainly composed of nitrogen and oxygen, the volume ratio of 4:1 (80% nitrogen, 20% oxygen). These gases experience radiation-induced reactions. In addition to the small amount of air normally dissolved in make-up water, air is also likely to be accidentally injected directly into the reactor coolant system. Whenever air enters the reactor coolant system and the reactor operates, the most immediate reaction is oxygen and hydrogen in the air, which is usually present in the coolant. The concentration of hydrogen normally kept in the reactor coolant is that a small amount of oxygen is consumed quickly before any excess oxygen causes serious corrosion problems.

After almost all oxygen is consumed by hydrogen, the nitrogen in the air is preserved. For small air supplements, some hydrogen is also retained: as a result, reactor coolants will contain both dissolved hydrogen and dissolved nitrogen. These two gases do not react at low temperatures and pressure in unirradiated solutions. However, when exposed to radiation, the gas does react according to the following reactions.

$$3H_2 + N_2 \overset{\text{Radiation}}{\longleftrightarrow} 2NH_3 \tag{7.30a}$$

Again, this is an equilibrium reaction, and radiation induces the reaction in both directions. Ammonia (NH_3) produced by this reaction combines with water to form ammonium hydroxide (NH_4OH).

$$NH_3 + H_2O \leftrightarrow NH_4^+ + OH^- \tag{7.30b}$$

Under the operating conditions of reactor coolant, Reaction (7.30a) is far from complete. In most cases, less than about 10 percent of the nitrogen will be converted to ammonia. If no additional base were added to reactor coolant, Reaction (7.30a) would be sufficient to cause the coolant to be mildly basic, pH 9. In the presence of added base, however, the reaction has only a very slight and negligible effect on pH.

If the base NH_3 is used to control the reactor coolant pH, the reverse reaction (7.30a) will be even more important. The reverse steps of this reaction require that some ammonia added to the coolant be broken down into N_2 and H_2. Since operating conditions favor this step of balancing, rather than the formation of NH_3, most of the ammonia added is expected to break down. However, ammonia decomposition is slow and the pH of the reactor coolant can be maintained within the specified range. It should also be noted that the decomposition of NH_3 will produce a high concentration of hydrogen in the reactor coolant (enough to meet normal H_2 requirements).

If a large amount of air is injected into the reactor coolant system, the stocks of dissolved hydrogen will be quickly depleted as a result of the reaction (7.29). If the amount of air injected is large enough, oxygen may remain in the coolant after the hydrogen is exhausted. In this case, there is another reaction between oxygen and nitrogen.

$$2N_2 + 5O_2 + 2H_2O \xrightarrow{\text{Radiation}} 4HNO_3 \qquad (7.31)$$

Nitric acid (HNO_3) produced by this reaction will neutralize any base contained in the coolant, and if sufficient acid is produced, the coolant will acquire an acidic pH.

Typically, the amount of hydrogen maintained in the reactor coolant, together with other precautions used, greatly reduces the likelihood that the amount of oxygen entering the coolant will be sufficient to cause a reaction (7.31). If a large amount of air is accidentally added to the reactor coolant, one solution is to add more hydrogen. The added hydrogen reacts with the remaining oxygen, disrupting the reaction balance (7.31), causing the reverse steps of the reaction to occur. When all oxygen is removed, H_2 and N_2 can react by reacting (7.30a) and helping to reconstruct a basic pH. Figure 7.24 illustrates the relationship between these reactions and the pH after the initial oxygen is added, and the subsequent hydrogen addition.

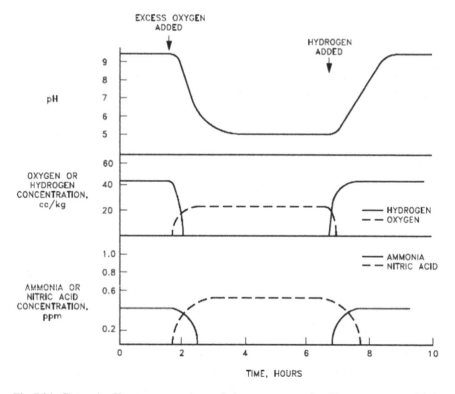

Fig. 7.24 Change in pH, gas concentration, and nitrogen compounds with excess oxygen added

The flux of neutrons and protons in a nuclear reactor core region leads to several important nuclear reactions with the constituent atoms of water. Most of these reactions involve oxygen isotopes and fast neutrons or protons.

In many cases, the absorption of a fast neutron by a nucleus is immediately followed by ejection of a proton. These reactions are called neutron-proton or n-p reactions and are commonly written (using the ^{16}O reaction to illustrate) in the following manner.

$$(n,p)(t_{1/2} = 7.13s) \tag{7.32}$$

In this notation, the original isotope that undergoes the reaction is written first, the product isotope is last, and the two are separated by, in order, the particle absorbed and the particle emitted. The isotope ^{16}N decays to ^{16}O with a 7.13-s half-life by emitting a beta particle and a high-energy gamma ray (6 meV predominantly).

$$^{16}_{7}N \rightarrow {}^{16}_{8}O + \beta^- + \gamma \tag{7.33}$$

^{17}O undergoes a similar reaction.

$$^{17}_{8}O(n,p)^{17}_{7}N(t_{1/2} = 4.1s) \tag{7.34}$$

$$^{17}_{7}N \rightarrow {}^{16}_{8}O + \beta^- + {}^{1}_{0}n + \gamma \tag{7.35}$$

Reactions (7.32) and (7.34) have no significant chemical effect on reactor coolants because the number of atoms experiencing these reactions is relatively small. However, they are of considerable importance, as radioactive species ^{16}N and ^{17}N are brought out of the core region by the flow of reactor coolant. Neutron and high-energy gamma rays emitted by these isotopes can easily penetrate pipes and components containing coolant and are important considerations in the shielding design of nuclear facilities. Because these isotopes have a short half-life, they decay to low levels quickly after shutdown, so there is little concern during this period.

Two other nuclear reactions with oxygen isotopes are shown below.

$$^{18}_{8}O(p,n)^{18}_{9}F(t_{1/2} = 112min) \tag{7.36}$$

$$^{16}_{8}O(p, \alpha)^{13}_{9}N(t_{1/2} = 10min) \tag{7.37}$$

The protons that cause these reactions are caused by the inelastic collision of fast neutrons with hydrogen atoms in water molecules. The radioactivity levels of these isotopes are much lower than the ^{16}N and ^{17}N levels during the operation of the reactor facility. However, ^{13}N and ^{18}F are the main sources of radiation in reactor coolant in most reactor facilities, from a few minutes to about five hours after the reactor is shutdown or after a coolant sample is taken from the system.

The only significant nuclear reaction that occurs with hydrogen involves deuterium (^{2}H), which comprises about 0.015% of natural hydrogen.

$$^2_1H(n, \gamma)^3_1H(t_{1/2} = 12.3\text{year}) \tag{7.38}$$

Tritium (^{3}H) decays by emission of a very weak particle (0.02 meV) and no gamma rays. Thus, tritium is not a radiological hazard unless it enters the body in significant amounts. Tritium can enter the body through inhalation or ingestion. It is also possible to absorb forms of tritium through the skin.

7.7.1 Chemistry Parameters of Reactor

The reasons for controlling the selected chemical parameters, as well as some of the more common ways to control them, are discussed. Since the reactor facilities involved is different, no attempt will be made to determine the specific values of any of the parameters discussed, but will include an overview of the base and the use of common methods. For operating values and specifications, users should refer to local facility publications. In addition, some information about tritium is provided.

Specific chemical parameters vary from facility to facility, but usually include the following: pH, dissolved oxygen, hydrogen, total gas content, conductivity, chloride, fluorine, boron and radioactivity. For the parameters indicated, control is usually implemented by one or more of the three basic procedures.

(1) Ion exchange in the primary system demineralizer(s) or by supplemental chemical additions.
(2) Oxygen scavenging by hydrogen or hydrazine addition.
(3) Degasification.

Table 7.7 lists the more common chemistry parameters measured and/or controlled, the reasons each is measured and/or controlled, and control methods utilized. Table 7.8 lists typical water quality indexes of pressurized water reactor nuclear power plant during the period of power operation.

The reason for controlling pH in reactor coolant systems is to minimize and control corrosion. The presence of excess H^+ in the solution can lead to an acidic state. Acid conditions are harmful to building materials in many ways in reactor facilities (except those containing aluminum). The processes in which acidic conditions in primary coolants lead to potential damage to the system are as follows. First, low pH promotes rapid corrosion by deteriorating or "stripping off" protective corrosion film; second, corrosion products such as iron oxide (Fe_3O_4), which dominates corrosion film, are highly soluble in acidic solutions. Figure 7.25 shows how corrosion increases when pH decreases. As a result, neutral or highly basic pH is less corrosive for facilities that do not use aluminum components.

In nuclear facilities that do not use chemical shim to control reactivity, pH is usually maintained at relatively high values, such as pH of about 10. In these facilities,

Table 7.7 Summary of reactor coolant chemistry control

Parameter	Reason	Method to control
pH	To inhibit corrosion To protect corrosion film To preclude caustic stress corrosion	Ion exchange Ammonium hydroxide addition Nitric acid addition
Dissolved Oxygen	To inhibit corrosion	Hydrogen addition Hydrazine addition
Hydrogen	To scavenge oxygen To suppress radio lytic decomposition of water To scavenge nitrogen To preclude hydrogen embrittlement	Hydrazine addition Degasification
Total Gas Content	To protect pumps To indicate air in leakage	Degasification Deaeration of makeup water
Conductivity	To minimize scale formation To indicate increased corrosion	Ion exchange Feed and Bleed
Chlorides	To preclude chloride stress corrosion	Ion exchange Feed and bleed
Fluorine	To preclude corrosion of Zr cladding	Ion exchange Feed and Bleed
Boron	To control reactivity	Boric acid addition
Radioactivity	To indicate increased corrosion To indicate a crud burst To indicate a core fuel defect To monitor effectiveness of demineralizer	Ion exchange Feed and bleed

Table 7.8 Typical water quality indexes of PWR

Items	Unit	Primary side	Secondary side
pH	25 °C	–	9.3 ~ 9.8
Dissolved	mg/kg	≤ 0.1	≤ 0.05
Oxygen	mg/kg	≤ 0.15	–
Chlorides	mg/kg	≤ 0.15	–
Fluorine	mL/kgH$_2$O	25–50	–
Hydrogen	mg/kg	≤ 1.0	< 1.0
Suspended	mg^{7}Li/kg	0.4–2.2	–
Solids	mgB/kg	0–2300	–
pH, ^{7}LiOH	mg/kg	≤ 1.0	< 1.0
Boron	mg/kg	≤ 0.05	–
SiO$_2$	mg/kg	≤ 0.05	–
Al	mg/kg	≤ 0.05	–
Ca	μs/cm(25 °C)	–	< 1.0
Mg	mg/kg	0.2	< 0.02
Conductivity	mg/kg	–	According to pH
Na	m^3/h	–	As much as possible
NH$_3$			
Discharge rate			

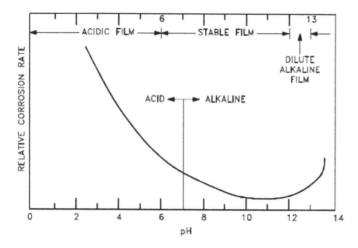

Fig. 7.25 Corrosion rate versus pH for iron

the upper limit of the pH is set based on corrosive stress corrosion considerations, as corrosive stress corrosion is more likely as the pH approaches higher.

In facilities controlled by chemical shim to control reactivity (involves adding boron in the form of boric acid), the pH is kept much lower. Low pH is necessary because a large amount of boric acid is added to the reactor coolant. Therefore, the pH in these facilities is kept as high as possible consistent with the reactivity requirements of nuclear facilities, and pH ranging from 5 to 7 is common.

In facilities using aluminum components, pH is maintained on the acidic side of the scale because of the corrosion characteristics of aluminum. In these facilities, pH may be controlled by the addition of a dilute nitric acid (HNO_3) solution to the reactor coolant system in conjunction with an ion exchange system of some type.

Most facilities use an ion exchange process to help control pH, regardless of the maintained pH range. For high pH facilities, the most common means of control is the use of lithium or ammonium forms of ions and hydroxyl to form ions. When using lithium, it must be ^7Li, as other lithium isotopes produce tritium, which poses a significant biological hazard to personnel. In facilities that use high pH chemical control and do not use chemical shim reactivity control, it is sometimes necessary to add a strong base solution, such as ammonium or lithium hydroxide. When chemical additives are used for pH control, facility design and operating procedures are used to eliminate excessive concentration at any point in the system, which can lead to corrosive stress conditions. Many reactions in reactor coolant systems affect pH, therefore, chemical controls must be carefully considered to eliminate the pH balance interruption provided by the ion exchanger.

Controlling dissolved oxygen levels in reactor facility systems is critical because it contributes to increased corrosion. The basic reactions to concerns about high concentrations of dissolved oxygen are as follows.

$$3Fe + 2O_2 \rightarrow Fe_3O_4 \qquad (7.39a)$$

$$4Fe + 3O_2 \rightarrow 2Fe_2O_3 \qquad (7.39b)$$

They depend on the concentration and temperature of oxygen. Reaction (7.39a) at high temperatures (>300 °C), oxygen concentrations are dominant at low levels. This corrosive film, iron oxide, also known as magnetite, is a black, usually tightly-adherent film that provides protective function on the surface of the facility. Reactions (3.39b) occur at temperatures below about 300 °C with higher oxygen concentrations. Iron oxide (Fe_2O_3) is commonly referred to as rust and is generally a red-dish color. This corrosion product is loosely adhered to the surface and is therefore easily removed and transported throughout the system for subsequent deposition and possible irradiation. In either reaction, an increase in dissolved O_2 concentrations accelerates corrosion rates, and other substances that may be present in the system may increase further.

In addition to its direct contribution to corrosion, oxygen reacts with nitrogen, reducing the pH of reactor water, which also leads to increased corrosion rates. Oxygen and nitrogen form nitric acid through the following reactions.

$$2N_2 + 5O_2 + 2H_2O \overset{radiation}{\longleftrightarrow} 4HNO_3 \qquad (7.40)$$

In all reactions presented, oxygen concentrations can be seen to promote corrosion. Therefore, if corrosion is to be minimized, oxygen concentrations must be kept as low as possible. In reactor coolant systems at most nuclear facilities, the limit on dissolved oxygen concentration is expressed in ppb (parts per billion). Concentrations can be continuously monitored by using an online analysis system or by regularly taking samples and analyzing samples. Oxygen levels are monitored not only to ensure that no oxygen is available for corrosion, but also to indicate that air enters the system.

Since the presence of dissolved oxygen contributes to most corrosion mechanisms, the concentration of oxygen can be controlled and reduced by adding scavenging agents in most facilities. Hydrogen (H_2) and hydrazine (N_2H_4) are scavenging agents commonly used to eliminate dissolved oxygen in reactor coolant systems. These substances remove oxygen through the following reactions.

$$2H_2 + O_2 \overset{radiation}{\longleftrightarrow} 2H_2O \qquad (7.41)$$

$$N_2H_4 + O_2 \rightarrow 2H_2O + N_2 \qquad (7.42)$$

Because hydrazine decomposes rapidly at temperatures above about 95 °C (forming NH_3, H_2, and N_2), hydrogen gas is used as the scavenging agent during hot operation and hydrazine is used when the reactor coolant system is cooled below 95 °C.

$$2N_2H_4 \xrightarrow{\text{High Temperature}} 2NH_3 + N_2 + H_2 \tag{7.43}$$

The decomposition reactions of hydrazine pose additional problems in chemistry control. Even if sufficient hydrazine were added to overcome the loss due to decomposition, instability of coolant pH would probably occur by the following reactions.

$$2N_2 + 5O_2 + 2H_2O \rightarrow 4HNO_3(\text{Acid}) \tag{7.44a}$$

$$3H_2 + N_2 + 2H_2O \rightarrow 2NH_4OH \text{ (Alkline)} \tag{7.44b}$$

The use of hydrogen gas at temperatures above 95 °C precludes the generation of the compounds formed by Reactions (7.44a) and (7.44b). In addition, hydrogen is compatible with the high flux levels present in the reactor core. Accordingly, advantage may be taken of the reversibility of the radiolytic decomposition of water.

Boiling water reactor (BWR) facilities are susceptible to corrosion from dissolved oxygen and react in the same way as pressurized water reactors (PTRs). However, due to the design of these facilities, the use of chemical additives is prohibited because of the continuous concentration of reactor vessels due to boiling. Boiling can lead to an electroplating process, and irradiation of these concentrated additives or impurities creates extreme environments with radiation levels and adverse corrosion positions.

From the operational nature of BWR facilities, the accumulation of dissolved oxygen at high concentrations can be prevented. Since boiling occurs in reactor vessels, the resulting steam is used in a variety of processes and then concentrated, so the removal of dissolved gases is an ongoing process. As mentioned earlier, boiling is an effective method of removing gas from the solution. If we compare the oxygen content of steam and water in BWR, we will find that the typical concentration in water is 100–300 ppb and the concentration in steam is 10,000–30,000 ppb. This concentration process is continuous during operation, with dissolved oxygen remaining in a gaseous state and subsequently removed in the condensation unit along with other noncondensible gases. Like PWR facilities, BWR facilities pretreat makeup water in some way to minimize the introduction of dissolved oxygen. The oxygen concentration measured in the steam system is mainly derived from the transmission analysis of water according to the reaction (7.29), and the equilibrium concentration of 100–300 ppb is established as the operation continues. This oxygen concentration is consistent with the goal of minimizing corrosion.

7.7.2 Water Treatment

Water usually contains many impurities, including trace minerals and chemicals. This section discusses the need to remove these impurities.

Water treatment is necessary to eliminate impurities contained in water, as found in nature. Controlling or eliminating these impurities is necessary to combat corrosion, scale formation, and heat transfer surface fouling throughout the reactor facility and support system [8].

Here are three reasons for using very pure water in reactor facility systems.

1. Minimize corrosion enhanced by impurities.
2. Minimize radiation levels at reactor facilities. Some natural impurities and most corrosion products become highly radioactive after exposure to neutron flux in the core region. If not removed, these soluble and insoluble substances can be carried to all parts of the system.
3. Minimize fouling of the heat transfer surface. Corrosive products and other impurities may be deposited on core surfaces and other heat transfer areas, reducing heat transfer capacity by fouling surfaces or blocking critical flow paths. High concentration areas of these impurities and corrosion products can also lead to extreme conditions in various corrosion processes, resulting in component or system failure.

Several processes in the reactor facility are used to purify water in the system and water used as makeup. Deaeration is used to remove dissolved gases, filtration effectively removes insoluble solid impurities, and ion exchange removes undesirable ions and replaces them with acceptable ions. Typical electroelectric impurities found in water are Ca^{2+}, Mg^{2+}, Na^+, K^+, Al^{3+}, Fe^{2+}, Cu^{2+}; NO_3^-, OH^-, SO_4^-, Cl^-, HCO_3^-, $HSiO_3^-$, $HCrO_3^-$.

One of the more common water treatment methods is the use of demineralizers and ion exchange.

Ion exchange is a widely used process in nuclear facilities to control the purity and pH of water by removing undesirable ions and replacing them with acceptable ions. Specifically, it is an ion exchange between a solid substance (called resin) and aqueous solution (reactor coolant or makeup water). Depending on the identification of the ions released into the water by the resin, this process may result in the purification of the water or control the concentration of specific ions in the solution. Ion exchange is a reversible exchange of ions between liquids and solids. This process is typically used to remove undesirable ions from liquids and replace acceptable ions from solids (resins).

Devices that occur for ion exchange are often referred to as demineralizers. This name derives from the term demineralize, the process of forming pure water by exchanging impure ions with H^+ and OH^- ions to remove impurities from the inlet fluid (water). H^+ and OH^- are found in the areas of resin beads contained in demineralizer tanks or columns.

There are two general types of ion exchange resins: resins that exchange positive ions (called cation resins) and resins that exchange negative ions (called anion resins). An cation is an ion with a positive charge. Common cations include Ca^{++}, Mg^{++}, Fe^{++}, and H^+. A cation Resin is a resin that exchanges positive ions. An anion is negatively charged ions. Common anions include Cl^-, SO_4^{2-}, and OH^-. An anion resin exchange negative ion. Chemically, both types are similar and belong to a group of compounds called polymers, which are very large molecules, made up of many molecules that produce one or two compounds in a long-chained, repeated structure.

A mixed bed demineralizer is a container that usually has a volume of several cubic feet containing resin. Physically, ion exchange resins are formed in very small beads, called resin beads, with an average diameter of about 0.005 mm. Wet resins have the appearance of moist, transparent, amber sand and are insoluble in water, acids and bases. The opening of the retaining element at the top and bottom or other suitable equipment is smaller than the diameter of the resin bead. The resin itself is a uniform mixture of acids and ion resins in a specific volume ratio, depending on its specific gravity. The ratio is usually 2 parts of the cation resin to 3 parts of the anion resin.

In some cases, chemical bonds may form between individual chain molecules at various points in the chain. The polymers are said to be cross-linked. This polymer forms the basic structure of the ion exchange resin. In particular, cross-linked polystyrene is a common polymer in ion exchange resins. However, the chemical treatment of polystyrene needs to give it the ability to exchange ions, which varies depending on whether the final product is an cation resin or an anion resin.

The chemical processes involved in the production of anion and cation resins are outlined in Figs. 7.26 and 7.27, starting with the formation of cross-linked polystyrene. The polymer itself is a covalent compound. According to the chemical reaction shown in Fig. 7.27, hydrogen atoms that are combined with the original polymer in some locations are replaced by functional groups such as SO_3H (sulfonic acid) and $CH_2N(CH_3)_3Cl$ (quaternary ammonium). Each such group binds to a polymer, but each group also contains an atom that binds to the radical group through an predominantly ionic bond. In the two examples above, the H^+ in SO_3H and Cl^- in $CH_2N(CH_3)_3Cl$ are ionizing bonding atoms. Sometimes, these are written as $SO3^-H^+$ and $CH_2N(CH_3)_3{}^+Cl^-$ to emphasize their ionic characters. These ions (H^+ and Cl^-) can be replaced by other ions. That is, the H^+ will be exchanged with other cations in the solution, and the Cl^- will be exchanged with other anions (Fig. 7.28).

In its final form, the ion exchange resin contains a large but limited number of sites occupied by exchangeable ions. All resins are inert during the exchange process except exchangeable ions. Therefore, it is customary to use symbols such as R-Cl or H-R for ion exchange resins. R represents part of the inert polymer base structure and the alternative base point that is not involved in the exchange reaction. The term R is inaccurate because it is used to represent the inert parts of the cation and anion resins, which are slightly different. In addition, the structure represented by R contains many sites of exchange, but only one site is displayed by symbols such as R-Cl. Despite these disadvantages, the word R is used for simplicity.

Specific resins can be prepared in different forms depending on the identity of the attached interchangeable ions. It is usually named after ions on an active site. For example, the resin represented by R-Cl is said to be in the chloride form of the anion resin. Other common forms are ammonium form (NH-R), hydroxyl 4 form (R-OH), lithium form (Li-R), and hydrogen form (H-R).

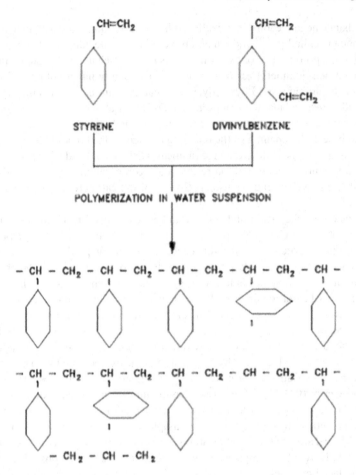

Fig. 7.26 Cross-linked polystyrene

The mechanism of the ion exchange process is somewhat complex, but the basic characteristics can be understood according to the concept of balance and recognize that the strength of the ion bond between the resin and the ion varies with the specific ion. That is, for a particular resin, different ions experience different attraction resins. The term affinity is often used to describe the attraction between a resin and a given ion. This affinity can be quantified by experimentally determining parameters called relative affinity coefficients. For qualitative discussions, attention needs to be paid to the relative affinity between resins and different ions.

In order of decreasing strength, the relative affinities between cation resin and various cations are as follows.

$$Ba^{2+} > Sr^{2+} > Ca^{2+} > Co^{2+} > Ni^{2+} > Cu^{2+} > Mg^{2+} > Be^{2+}$$

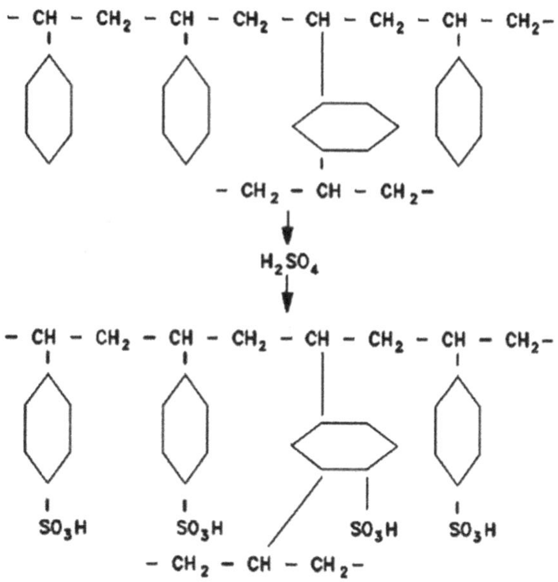

Fig. 7.27 Sulfonation to produce cation resin

$$- CH - CH_2 - CH - CH_2 - CH - CH_2 - CH - CH_2 -$$

$$- CH_2 - CH - CH_2 -$$

$$\downarrow H_2SO_4$$

$$- CH - CH_2 - CH - CH_2 - CH - CH_2 - CH - CH_2 -$$

$$SO_3H \qquad SO_3H \qquad SO_3H \qquad SO_3H$$

$$- CH_2 - CH - CH_2 -$$

$$Ag^+ > Cs^+ > Rb^+ > K^+ > NH^+ > Na^+ > H^+ > Li^+$$

Similarly, the relative affinities between an anion resin and various anions are as follows.

$$SO_4^{2-} > I^- > NO_3^- > Br^- > HSO_3^- > Cl^- > OH^- > HCO_3^- > F^-$$

Suppose a solution containing Na^+ ions is passed through hydrogen resin. From the relative affinities given, Na^+ ions are attracted to the resin more strongly than H^+ ions. Thus, Na^+ ions will displace H^+ ions from the resin or, in other words, Na^+ ions and H^+ ions exchange place between resin and solution. The process can be described by the following equilibrium reaction.

$$H - R + Na^+ \leftrightarrow Na - R + H^+ \tag{7.45}$$

If a solution containing Cl^- ions is passed through hydroxyl resin, the Cl^- ions will be removed according to the following reaction.

$$R - OH + Cl^- \leftrightarrow R - Cl + OH^- \tag{7.46}$$

Figure 7.29 shows the physical arrangement of an ion exchange container for purifying water. Ion exchange resins are contained in containers with a volume of several cubic feet. Retention elements at the top and bottom include a screen, slotted cylinders, or other suitable equipment with openings smaller than the resin beads to

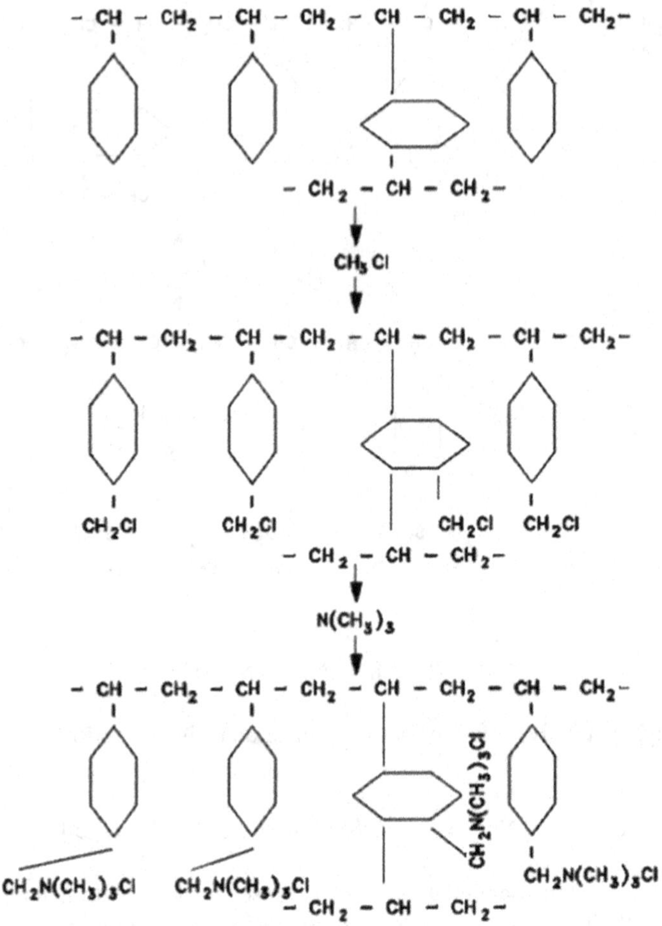

Fig. 7.28 Reactions to produce anion resin

prevent the resin from escaping from the container. The resin bed is a uniform mixture of cation and anion resin, with a volume ratio of 2 parts of cation resin and 3 parts of anion resin. This arrangement is called mixed bed resin, rather than arranging cation and anion resins in discrete layers or separate containers. The different volumes of the two resins used are due to differences in exchange capacity between cation and anion resins. Exchange capacity is the mass that can be removed by a given amount of resin, which has unit of moles/ml, equivalents/ml, or moles/gm. The anion resin is less dense than the cation resin; therefore, it has a smaller exchange capacity, and anion resins require a larger volume than cation resins to achieve an equal total exchange capacity.

The flow of the solution (impure water) flows from top to bottom due to the different densities of anion and cation resins. If the flow is reversed, the lighter anion

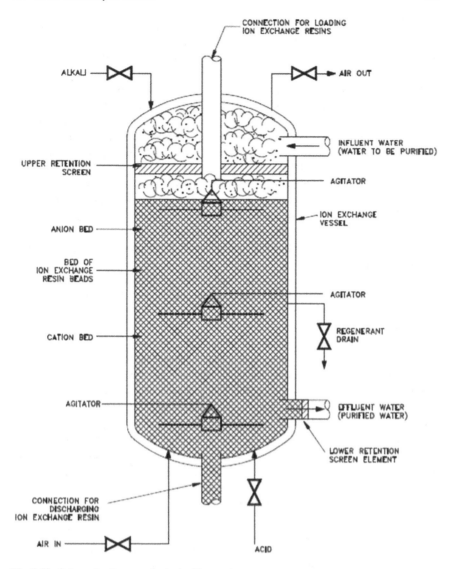

Fig. 7.29 Schematic diagram of a typical ion exchanger

resin will gradually rise to the top through a process called classification, forming a layer of anion resin on top of the cation resin, as shown in Fig. 7.29. In the illustration shown, the layering results come from regeneration and/or backwashing. In systems that do not use backwash, anion and cation resin beads are uniformly mixed. Many systems use backwash procedures if the resin is regenerated, removing the filtered collected solids and separating the resin from regeneration. They are remixed after regeneration.

For fixed amounts of anion and cation resins, removing impurities from the mixed bed resin is more efficient than layering. The main reason is that for layered resins, there may be a large pH gradient in the resin column. For example, if the hydroxyl form resin is on top, when the solution passes through it, the ion impurities are removed and replaced by OH^- ions: therefore, the pH increases. An increase in pH may reduce the efficiency of removing impurities in the lower part of the resin bed. It can also lead to some impurity precipitation, as the solute varies with the pH. Resin columns filter some unsolved material, but the filtration efficiency is usually significantly lower than that of ion exchange removal. Therefore, the overall efficiency is lower than that of the mixed bed resin.

The ion exchange process is reversible. If too much solution is passed through the ion exchanger (that is, the capacity of the resin has been exceeded) the exchange may reverse, and undesirable ions or other substances that were previously removed, will be returned to the solution at the effluent. Therefore, it is necessary to periodically monitor the performance of the ion exchanger and either replace or regenerate the resin when indicated.

The resin beds of ion exchangers are susceptible to malfunction from a number of causes. These causes include channeling, breakthrough, exhaustion, and overheating.

Channeling
Channeling is a condition in which the resin allows water to flow directly through the ion exchanger. A flow channel is established from the inlet to the exit of the ion exchanger, allowing water to flow through the resin through these paths with essentially unlimited restrictions. If channeling occurs, the water flowing through the resin bed is not in contact with the resin beads enough, resulting in a decrease in the effectiveness of the ion exchanger.

The channeling is most often caused by improper filling of the ion exchanger with resin. If there is not enough water to mix with the resin when adding the resin, the resin column may contain pockets or voids. These voids may then set the flow path for channeling. Improper design or failure of the water inlet connection (flow diffuser) may also result in a channeling.

Breakthrough and Exhaustion
The second kind of failure mode is depleted, as in Fig. 7.29 water from up to downstream, the above resin ball first ion exchange, and ion exchange capacity is gradually lost (known to consume). Container resin ball exchange capacity is not evenly consumed, but with the process, like a burning candle as gradually from top to down. If all resin ball in a container have lost the ability to exchange, called depleted. At this point, the outflow of water is basically the same as inflow water, without ion exchange.

Resin Overheating
Under most facility operating conditions, there is the possibility of temperature increase; We will examine in detail the process that occurs when the resin in the ion exchanger overheats. Although the basic structure of the inert polystyrene resin

stabilizes to a fairly high temperature (approximately 150 °C), the active exchange sites are not. The anion resin begins to decompose slowly at about 60 °C and quickly decomposes above 80 °C. The cation resin is stable at around 120 °C. Since these temperatures are much lower than normal reactor coolant temperatures, the coolant temperature must be lowered before it passes through the ion exchange resin.

7.7.3 Dissolved Gases and Suspended Solids

The presence of dissolved gases, suspended solids, and incorrect pH may damage water systems associated with reactor facilities. Therefore, these conditions must be minimized or eliminated in order to reduce corrosion in the facility system. This section discusses the ways in which these conditions are controlled and the difficulty of controlling them.

Dissolved gases [9] come from different sources, depending on the system we are inspecting. In the discussion below, we will discuss makeup water, reactor coolant systems, secondary facility water systems, sources of dissolved gases, and ways to reduce their concentrations to acceptable levels.

Many facilities use raw water as a source of makeup water systems. From distillation to the series of different processes shown in Fig. 7.30, the pretreatment of this water is done in a variety of ways. In a pretreatment system similar to Fig. 7.30, a resin column containing a cation resin (hydrogen form) is used to remove cations. The water entering the cations exchanger contains many ions, including sodium (Na^+),

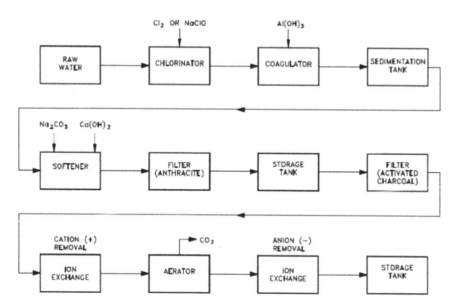

Fig. 7.30 A typical pretreatment system

bicarbonate (HCO_3^-), and others (HCO_3^- is one of the main impurities in many primary water systems). The Na^+ come from water softeners located upstream of the pretreatment system. In addition to HCO_3^- ions, raw water contains a large amount of magnesium (Mg^{++}) and calcium (Ca^{++}), as well as a small amount of other ionic impurities.

Reactions that occur in water softeners include the removal of both Mg^{++} and Ca^{++} ions. The water softener contains resins where the insoluble transposition point is an SO_3 molecule and the soluble ion connected to the exchange site is a Na^+ ion. When water containing Mg^{++}, Ca^{++}, and HCO_3^- ions passes through the resin in the softener, the ions are exchanged through the following reactions

$$2(R\text{-}SO_3Na) + Ca^{2+} \rightarrow (2R\text{-}SO_3)Ca + 2Na^+ \tag{7.47}$$

The cation exchanger contains a resin in the form of hydrogen. In this processing step, all ions entering the ion exchanger will basically be held at the exchange site, and H^+ will be released in the following typical reaction.

$$R\text{-}SO_3H + Na^+ \rightarrow R\text{-}SO_3Na + H^+ \tag{7.48}$$

The water leaving the resin is a bit acidic (depending on the incoming ion concentration) because it contains H^+ ions and any ions associated with incoming ions. After passing the cation resin, the HCO_3 ions are combined with the H^+ ions to form carbonic acid (H_2CO_3). Carbonic acid is a weak acid that is broken down into water and carbon dioxide by the following reactions.

$$H^+ + HCO_3^- \rightarrow H_2O + CO_2 \uparrow \tag{7.49}$$

Because carbonic acid is easy to separate, the aerator is used to remove carbon dioxide from the make-up water at this point in the system. If we aerate water in some way, such as by spraying water through a tower or blowing air through it, carbon dioxide is "stripped" from the water and released into the atmosphere. The removal of carbon dioxide forces the reaction (7.49) to move to the right, converting more HCO_3^- into carbon dioxide. If there is enough aeration, all bicarbonate (HCO_3^-) and carbon dioxide can be removed.

A similar reaction occurs in an anion exchanger. For example, an anion resin with hydroxide ions at a exchange sites reacts in accordance with the following typical reactions.

$$R\text{-}N(CH_3)_3OH + Cl^- + H^+ \rightarrow R\text{-}N(CH_3)_3Cl + H_2O \tag{7.50}$$

In this pretreatment system, the anion resin is located downstream of the cation resin, and the only cations are hydrogen ions. When hydroxyl ions are released from the anion exchange point, they bind to hydrogen ions to form water. As a result, pure water appears in effluent.

Another method sometimes used to remove dissolved gases from water is deaeration. During this process, water is stored in a vented tank with an electric heater or steam coil. Water is heated to a temperature that is high enough for slow boiling to occur. The boiling strip dissolves the gas in the stored water, which is then released into the atmosphere. Typically, the ventilation gas is guided by a small condenser to limit the loss of water vapor, which escapes with steam and gas. This method is particularly effective in removing dissolved oxygen and other constrained gases (CO_2, N_2 and Ar).

The removal of dissolved gas from the reactor coolant system is usually done by ventilating the steam space or the high point in the system. In pressurized water reactors (PWRs), this is usually done in pressurized reactors. Steam space is the high point of the system, and boiling and condensation cause the dissolved gas to continue to stripping of. Steam space is intermittently or continuously spewed out and gas is taken away in the process.

In addition to the above mechanical means, the use of waste scavengers in PWR prevents the presence of dissolved oxygen. Two methods are usually used in this regard. When the facility temperature is above approx. 93 °C, gaseous hydrogen is added and maintained in the main coolant to remove oxygen. Another scavenger is hydrazine (N_2H_4). Hydrazine is thermally unstable at temperatures above 93 °C and decomposes to form ammonia (NH_3), nitrogen (N_2), and hydrogen (H_2). Therefore, the use of hydrazine as an oxygen scavenger is limited to temperatures below 93 °C.

The presence of dissolved gases in PWR steam facilities is as harmful as the presence of these gases in reactor cooler systems. Because steam facility systems contain metals other than stainless steel, they are more susceptible to certain types of corrosion in the presence of oxygen and carbon dioxide. Dissolved gases are removed from steam systems in two ways: by mechanical means, such as air ejectors or mechanical pumps; and chemicals are used to remove oxygen.

Since boiling occurs in a steam generator, any dissolved gas immersed in water is stripped during boiling. These gases are transported with steam through turbines and auxiliary systems and end up in condensers. The condenser is designed to collect noncondensible gases (e.g. O_2, CO_2) and transport them to an air removal system (consisting of an air ejectors or mechanical pump) and then discharge them into the atmosphere.

Scavenging involves the use of solid additives and volatile chemicals. One commonly-used solid chemical additive is sodium sulfite (Na_2SO_3). Scavenging of oxygen occurs by the following reaction.

$$2Na_2SO_3 + O_2 \rightarrow 2Na_2SO_4 \tag{7.51}$$

As can be seen by Reaction (7.51), oxygen is consumed in the reaction resulting in the formation of sodium sulfate, Na_2SO_4 (a soft sludge). Addition of this scavenging agent is limited to drum type steam generators. Once Through Steam Generators (OTSG) do not use this method, but instead use controls that keep all scale-forming chemicals out of the steam generators.

Sodium sulfate reacts rapidly with oxygen and is a very effective scavenger. However, as a source of solids and another solid (Na_2SO_4), sodium sulfate produced during the reaction has the potential of fouling the heat transfer surface. Due to sodium sulfate problems, many facilities use volatile chemical controls of secondary steam systems to control dissolved gases in conjunction with air removal systems. This control uses hydrazine (reaction 7.52) and morpholine (reaction 7.53) to eliminate oxygen and carbon dioxide, respectively.

$$N_2H_4 + O_2 \rightarrow H_2O + N_2 \uparrow \qquad\qquad (7.52)$$

$$2C_4H_9NO + CO_2 + 2H_2O \rightarrow C_4H_9NO^{\cdot}C_4H_9COOH + HNO_3 + H_2 \qquad (7.53)$$

As can be seen by Reaction (7.52), no solids are formed; thus, the tendency of fouling heat transfer surfaces is reduced. An additional benefit results from the decomposition of hydrazine by the following reactions.

$$2N_2H_4 \rightarrow 2NH_3 + N_2 + H_2 \qquad\qquad (7.54)$$

$$NH_3 + H_2O \rightarrow NH_4OH \qquad\qquad (7.55)$$

These reactions result in an alkaline pH condition that decreases corrosion in the steam facility. As can be seen in Reaction (7.53), the consumption of CO_2 takes place. Two benefits result from this reaction; (1) the inventory of dissolved gases in the steam facility is reduced, and (2) is the reaction contributes to maintaining a higher pH by eliminating carbonic acid (H_2CO_3), thus reducing corrosion.

Referring back to Fig. 7.30 and examining the effluent of the softener, we find that both sodium salts and precipitates are present. These substances result from reactions that typically occur based on the presence of Ca^{++} and $Mg + +$ salts. The chemicals most commonly used for softening are soda ash or sodium carbonate (Na_2CO_3) and hydrated lime ($Ca(OH)_2$). Hard water (water containing Ca^{++} and Mg^{++} salts) contains calcium and magnesium bicarbonates ($Ca(HCO_3)_2$) and ($Mg(HCO_3)_2$), as well as calcium sulfate ($CaSO_4$) and magnesium chloride ($MgCl_2$). These impurities produce the following reactions.

$$Ca(HCO_3)_2 + Ca(OH)_2 \rightarrow CaCO_3 + 2H_2O \qquad\qquad (7.56)$$

$$Mg(HCO_3)_2 + 2Ca(OH)_2 \rightarrow Mg(OH)_2 + 2CaCO_3 + 2H_2O \qquad (7.57)$$

$$MgSO_4 + Ca(OH)_2 \rightarrow Mg(OH)_2 + CaSO_4 \qquad\qquad (7.58)$$

$$CaSO_4 + Na_2CO_3 \rightarrow CaCO_3 + Na_2SO_4 \qquad\qquad (7.60)$$

$$MgCl_2 + Ca(OH)_2 \rightarrow Mg(OH)_2 + CaCl_2 \tag{7.61}$$

$$CaCl_2 + Na_2CO_3 \rightarrow CaCO_3 + 2NaCl \tag{7.62}$$

It is clear from the above reaction that, although Ca^{++} and Mg^{++} ions can be removed from the solution, soluble sodium salts have been formed. Therefore, the solid content of total dissolved is basically unchanged. $CaCO_3$ and $Mg(OH)_2$ are precipitated and must also be removed from the solution. One way to do this is to filter. Filtration is the process of removing insoluble solids from water through a filter medium consisting of a porous material. This process removes suspended solids and sediments, but has no effect on dissolved solids. Many materials are used as filter media, including sand, activated charcoal, anthracite, diatomaceous earth, and resins in ion exchangers. Sand is not typically used in nuclear applications due to silicate ion (SiO_3^{2-}) associated. Silicate ions are not desirable because they hydrolysis in water and form weak acids, which tend to increase corrosion. Activated charcoal is commonly used in chlorinators in water treatment systems because it removes excess residual chlorine and suspended material.

There are two types of mechanical filters in use, gravity and pressure. Pressure filters are most widely used because they can be installed in pressurized systems without the need for additional pumps (gravity filters require pumps to provide power). In addition, in pressurized filtration systems, flow and other related parameters can be better controlled.

Another method used to remove suspended corrosion products from facility fluid systems is electromagnetic filters. These are becoming increasingly popular in PWR feed and condensate systems, and they have been shown to be effective in reducing the crud loading of these systems, thereby reducing the inventory of corrosive products in steam generators.

The ion exchanger also acts as a filtration unit through the size of the resin beads and the torturous path that water must follow when passing through the resin. However, the filtration efficiency is significantly lower than that of ion exchange (90% or less, compared to about 100% for most ion exchange reactions). The filtration efficiency mainly depends on the size of the suspended material, and the larger particle efficiency is higher. The adverse effects of this filtration process are similar to adverse reactions that occur in other types of filters. In a radioactive system, the accumulation of filtered particles (debris) can increase radiation to prohibitive levels or lead to reduced flow, which may require removal or backwash of the resin.

7.7.4 Water Purity

The less contaminants there are in the water, the less corrosion occurs. Water treatment methods have been explored previously. This section discusses how to quantify the purity of water. Measuring purity helps keep treatment effective.

The level of purity of water used in nuclear facilities must be consistent with the overall objective of chemical control at nuclear facilities.

There are many ways to obtain pure water, including distillation systems and pretreatment systems, similar to those mentioned earlier in this chapter. Regardless of the method used, the required purity must be achieved.

Water purification is defined in many different ways, but a generally accepted definition indicates that high-purity water is distilled and/or de-ionized water, so its specific resistance is 500,000 ohms (2.0 micromhos conductivity) or higher. This definition is satisfactory as a basis for work, but for more critical requirements, it is recommended that the breakdown shown in Table 7.9 represent purity [10].

Conductivity is a measure of the ease with which electricity through matter. The presence of ions greatly facilitates the passage of current. Pure water is only slightly ionized by the separation of water. At 25 °C, the concentrations of hydrogen and hydroxyl ions are 10^{-7} mol/L.

The numerical values of the equivalent conductance of several ions are listed in Table 7.10.

Here we calculate the absolute purity of the theoretical conductivity of the water. In the absolute pure water at 25 °C, H^+ and OH^- molar concentrations are 10^{-7} mol/L. and therefore the conductivity is

$$(350 + 192)\frac{\text{mhos} \cdot \text{cm}^2}{\text{mol}} \times 10^{-7}\frac{\text{mol}}{\text{L}} \times 10^{-3}\frac{1}{\text{cm}^3} \times 10^6\frac{\mu\text{mho}}{\text{mho}} = 0.054\frac{\mu\text{mho}}{\text{cm}}$$

Example 7.4 Conductivity will very quickly indicate the presence of any ionic impurities, even if the impurity concentration is extremely small. As an example, suppose 1.0 mg of NaCl impurity were deposited in 1 L of demineralized water. The normality of this solution would be as follows

Table 7.9 Water purity

Degree of purity	Maximum conductivity/(μmhos/cm)	Approximate concentration of electrolyte/(mg/l)
Pure	10	2 ~ 5
Very pure	1	0.2 ~ 0.5
Ultrapure	0.1	0.01 ~ 0.02
Theoretically pure	0.054	0

Table 7.10 Equivalent conductance of several ions

Ions	Equivalent conductance (mhos-cm^2/mole)
H^+	350
OH^-	192
Na^+	51
Cl^-	75

$$1 \times 10^{-3} \frac{g}{L} \times \frac{mol}{58g} = 1.7 \times 10^{-5} \frac{mol}{L}$$

The conductivity of the solution is

$$(51 + 75) \frac{mhos \cdot cm^2}{mol} \times 1.7 \times 10^{-5} \frac{mol}{L} \times 10^{-3} \frac{1}{cm^3} \times 10^6 \frac{\mu mho}{mho} = 2.2 \frac{\mu mho}{cm}$$

Which is well above the limit of $1 \mu mho/cm$.

7.7.5 Radiation Chemistry of Water

In the core of nuclear reactor, fuel, structural materials, shell materials, water and impurities in the water are under strong radiation, will occur in a variety of nuclear reactions, resulting in radionuclides. The specific can be divided into four cases:

(1) The water itself induced radioactivity: see Table 7.11. The radiation of ^{16} N and ^{17} N is very strong, the amount is very large, is the main factors to be considered in the primary loop of reactor shield design. However, the short half-life, does not lead to accumulation of the radioactivity in the coolant.

(2) The radioactive impurity in the water: refers to the impurity in the active area with the coolant, such as gas impurity argon, ion impurities, sodium and potassium, and so on, Table 7.12 lists the activation of the neutron produced radionuclides.

(3) Fission product: fuel cladding failure, will cause the escape of fission products. This will increase the radiation level of the loop, the effect of operation and

Table 7.11 The radioactivity produced by water itself

Nuclear reaction	Product	Isotope abundance/%	Half-life /s	γ or neutron energy /MeV
$^{16}O(n, p)^{16}$ N	^{16} N	99.76	7.13	6.13, 7.12(γ)
$^{17}O(n, p)^{17}$ N	^{17} N	0.037	4.17	1.12(n)
$^{18}O(n, p)^{18}$ N	^{18} N	0.204	0.63	1.07, 1.65, 1.98(γ)

Table 7.12 Induced radioactivity in water

Nuclear reaction	Product	Isotope abundance/%	Half-life/h	γ energy/MeV
$^{40}Ar(n, \gamma)^{41}Ar$	^{41}Ar	99.6	1.83 h	1.3
$^{23}Na(n, \gamma)^{24}Na$	^{24}Na	100	14.66 h	2.75, 1.37
$^{27}Al(n, \alpha)^{24}Na$	^{42} K	6.88	12.36 h	1.5
^{41} K$(n, \gamma)^{42}$ K				

Table 7.13 Main fission product of ^{235}U

Nuclide	^{235}U fission product/%	Half-life	γ energy MeV
^{131}I	3.1	8.04d	0.364, 0.637
^{137}Cs	6.2	30y	0.662
^{144}Ce	6.0	285d	0.134, 0.036
^{90}Sr	5.75	28.5y	(β decay)
^{85}Kr	0.29	10.7y	0.514
^{133}Xe	6.6%	5.25d	0.081

maintenance. ^{235}U fission products have different elements in 30 kinds of 200 kinds of radioactive isotope. Table 7.13 lists several kinds of main fission product. During the operation can be through the total gamma detection system, delay neutron detection system, fission gas and coolant radiochemical analysis to monitor the fuel shell is broken or damaged, the specific use what kind of method for power plant varies.

(4)   Activated corrosion products: the main component of stainless steel and nickel-based alloy is iron, nickel and chromium. The corrosion products are in four states emerged in the loop: ions dissolved in water; water insoluble suspended particulate debris or sediment; loose texture; corrosion product film material covered on the surface. The corrosion products after neutron irradiation will produce radioactive nuclide, become the corrosion activation products. Their half-life is generally longer, will form a radioactive substance accumulation is the main factor to the staff by irradiation. In Stainless Steel Co as an impurity exists, because ^{59}Co will be generated ^{60}Co radioactive neutron activation and it has a long half-life, and high energy gamma rays, so need to limit cobalt content in stainless steel. Other sources of cobalt with high cobalt alloy components of the control rod drive mechanism, coolant pump and valve rot Corrosion and abrasion of the product. Therefore, it has to be used with caution in the components in high cobalt alloys. Nickel from stainless steel and steam generator tube of nickel base alloy. Table 7.14 lists several activated corrosion product nuclides.

Table 7.14 Activated corrosion product nuclides

Nuclear reaction	Cross section/b	Product	Isotope abundance/%	Half-life	Γ energy/MeV
$^{58}Fe(n, \gamma)^{59}Fe$	0.68	^{59}Fe	0.33	44.5d	1.10, 1.29
$^{59}Co(n, \gamma)^{60}Co$	27.2	^{60}Co	100	5.27y	1.33, 1.17
$^{58}Ni(n, p)^{58}Co$	0.096	^{58}Ni	67.8	70.9d	0.81
$^{54}Fe(n, p)^{54}Mn$	0.068	^{54}Mn	2.86	312d	0.834
$^{55}Mn(n, \gamma)^{56}Mn$	12.1	^{56}Mn	100	2.58 h	0.85, 1.81
$^{50}Cr(n, \gamma)^{51}Cr$		^{51}Cr	4.35	27.7d	0.32

7.8 Extraction and Refinement of Uranium

Uranium extraction and purification is the impurities in uranium leaching and leaching for separation of liquid and get a part of uranium enrichment process. In the purification of uranium concentrate, made of uranium oxide process called uranium refining.

7.8.1 Leaching of Uranium

The leaching of uranium is the process of collecting uranium from the ore, which has two kinds of dry and wet processes. The process is generally used in industry, and the selection of the extraction agent depends on the nature of the ore, and can be used as a solution of sulfuric acid or carbonic acid.

(1) Sulfuric acid leaching method: leaching agent, MnO_2 or $NaClO_3$ as oxidant and iron ion catalyst together with the ore leaching to join, oxidation of U^{4+} to UO_2^{2+}. In order to promote mixing and oxidation, air mixing can be used. Acid and oxidant are respectively 1 ~ 100 g/L and 1 ~ 3 kg/ L. leaching from a few hours to a day of low grade uranium ore, but also can be used in heap leaching. The site will be heaped without water leakage, from the top spraying dilute sulfuric acid direct leaching of uranium oxide. A further method is the ore-leaching agent by drilling into the natural burial, selectively dissolved uranium solution extracted from leaching. This method is called in situ leaching or leaching processing.

(2) Alkali leaching method: for in calcite or the content of carbonate minerals in percent more than a few of uranium ore. The leaching process: in the ore adding sodium carbonate solution and uranyl carbonate complex ion. Then add alkali will precipitate diuranate. Alkali leaching method is generally performed in Pachuca tank. 70–85 °C, atmospheric pressure; leaching time from a few hours to 40 h. The shortcomings is necessary to the ore is ground into a fine powder.

7.8.2 Extraction of Uranium

The extraction of uranium from leaching solution of low concentration, high concentration of uranium or uranium ore pulp respectively by ion exchange and solvent extraction method.

Ion exchange method is the use of chemical replacement reaction between solid resin and leaching solution, containing uranium ion exchange or adsorption to a solid resin. Then acid solution and washing (or desorption), purified and enriched uranium solution.

Of carbonate or sulfate leaching solution, using such as Cl^- and NO_3 anion exchange resin containing uranyl complex anions. Then 1 M hydrochloric acid or sulfuric acid leaching. The reaction is as follows:

$$4KCl + UO_2(SO_4)_3^{4-} \rightarrow K_4UO_2(SO_4)_3 + 4Cl^- \tag{7.63}$$

$$K_4UO_2(SO_4)_3 + 4Cl^- \rightarrow 4KCl + UO_2^{2+} + 3SO_4^{2-} \tag{7.64}$$

Adsorption process is carried out in the adsorption tower. During adsorption, some such as Si, Mo, S and Ti ions accumulates in the resin, the resin exchange function decreased. This phenomenon known as resin poisoning. Poisoning of the resin can be used with different concentrations, such as 10% of the NaOH-NaCl solution regeneration.

Solvent extraction using organic solvent (organic phase) and containing uranium solution (water) mixed contact, so that the uranium extraction of the aqueous phase to organic phase, then back extracted into the aqueous phase, thereby separating the impurity to achieve the purpose of extracting. Typical commercial process AMEX and DAPEX method. They were using amine and phosphate extraction agent.

AMEX with alkyl tertiary amine (30 ~ 70%) as extractant, 3% of the aliphatic alcohol as stabilizer, kerosene as diluent and in a stirred tank for precipitation of countercurrent extraction. Extraction and back extraction reaction is as follows:

$$4R_3NHCl + \left[UO_2(SO_4)_3\right]^{4-} \rightarrow (R_3NH)_4UO_2(SO_4)_3 + 4Cl^- \tag{7.65}$$

$$(R_3NH)_4UO_2(SO_4)_3 + 5Na_2CO_3 \rightarrow 4R_3N + Na_4UO_2(CO_3)_3 \tag{7.66}$$

Saturated organic phase should be washed with acid water to remove impurities in the organic phase.

Amine solvent with high selectivity, separation of most of the impurities (including Fe^{3+}). But it will reduce the ability of the solvent extraction, heating and adding 10% Na_2CO_3 solution treatment, the organic phase regeneration.

ELUEX method using a combination of adsorption and extraction saturated uranium resin leaching liquid direct extraction to obtain the high purity uranium solution. This method for more impurities in the leaching solution, but also can be conveniently used for wastewater recycling, thereby reducing the consumption of reagents.

Chemical concentrates precipitation. For neutralization of sulfuric acid leaching solution the solution of adsorption and extraction process, and then NH_3, NaOH, Ca (OH) $_2$ and MgO, diuranate precipitation down:

$$2UO_2^{2+} + 2NH_4OH + H_2O \rightarrow (NH_4)_2U_2O_7 + 4H^+ \tag{7.67}$$

The precipitate by filtration, drying at the temperature of 120 ~ 175 °C, the chemical concentrate is called yellow cake. The content of U_3O_8 is 40 ~ 80%.

7.8.3 Refining of Uranium

Uranium refining refers to further separate of uranium hydrometallurgy products (chemical concentrate). Neutron poison with large thermal neutron absorption cross-section and influence product of uranium processing will be extracted. Purification of solvent extraction and ion exchange method. With tributyl phosphate (TBP) for solvent extraction methods are widely used in industry. The method to use nitrate dissolved uranium concentrate, after the solid–liquid separation of the solution into the extraction system. Extractant for kerosene 22 ~ 40%TBP, in pulsed column according to the following reaction were extracted:

$$UO_2^{2+} + 2NO_3^- + 2TBP \rightarrow UO_2(NO_3)_2 \cdot 2TBP \qquad (7.68)$$

Then on uranium bearing organic phase with 0.1 M HNO_3 aqueous wash. Finally, with deionized water or dilute nitric acid solution is heated to 60 °C for back extraction, pure uranyl nitrate solution. After filtering and according to the requirements of the follow-up process, or adding ammonia precipitation of ammonium diuranate (ADU) crystal; or the solution evaporated into hexahydrate $UO_2(NO_3)_2 \cdot 6H_2O$ (UNH), and then heating the denitrification of UO_3; or by the precipitation of NH_3 and CO_2 was ammonium uranyl tricarbonate (AUC) crystal.

7.9 Chemical Conversion of Uranium

Chemical conversion of uranium is the uranium refinery natural U_3O_8 (yellow cake) or UO_2 and other intermediate products made of uranium oxides, fluorides, and uranium metal process [11].

7.9.1 Preparation of Uranium Dioxide

Natural uranium is the production of heavy water reactor fuel bundle of important raw material, and preparation tetra fluoride uranium, uranium hexafluoride and uranium metal an important intermediate product. U_3O_8 (powder) in the horizontal rotary furnace heated to 800 ~ 870 °C by hydrogen direct reduction can be made of UO_2 (powder), the reaction:

$$U_3O_8 + 2H_2 \rightarrow 3UO_2 + 2H_2O \qquad (7.69)$$

Also available ammonium diuranate (ADU) or ammonium uranyl tricarbonate (AUC) as raw materials to get UO_2. UO_2 was dissolved in nitric acid and in pulsed sieve plate column with solvent extraction method high pure uranyl nitrate solution, and finally transformed nuclear pure ceramic grade UO_2 powder. The reaction:

$$2UO_2(NO_3)_2 \cdot 6H_2O \rightarrow 2UO_3 + 4NO_2 + O_2 + 12H_2O \qquad (7.70)$$

$$UO_3 + H_2 \rightarrow UO_2 + H_2O \qquad (7.71)$$

The above reactions were carried out at $300 \sim 450\ °C$ and $600 \sim 800\ °C$ respectively.

7.9.2 Preparation of UF₄

UF_4 is a key intermediate for the production of metal uranium and UF_6. The preparation method has two kinds of wet and dry methods. The raw materials are UO_2.

Wet process: with UO_2 material lining glue dissolving tank injecting hydrochloric acid and hydrofluoric acid production complex solution, then instilled with 40% hydrofluoric acid produce UF_4, reaction temperature is about 60–70 °C. The two-step reaction is

$$UO_2 + 4HCl + HF \rightarrow H(UCl_4F) + 2H_2O \qquad (7.72)$$

$$H(UCl_4F) + 3HF \rightarrow UF_4 + 4HCl \qquad (7.73)$$

The vacuum filter precipitation slurry, followed by pulping, washing, drying and calcining to obtain the low water content UF_4.

Dry process: the use of anhydrous hydrogen fluoride in the $350 \sim 600\ °C$ to the solid state UO_2 directly into UF_4, the reaction is

$$UO_2 + 4HF \rightarrow UF_4 + 2H_2O \uparrow \qquad (7.74)$$

Commonly used horizontal stirred bed, fluidized bed or moving bed in the reaction furnace, which is widely used in industry because of its short process and good economy.

7.9.3 Preparation of UF₆

Natural UF_6 is used for enrichment of uranium. Low enriched UO_2 fuel is an important fuel material of nuclear power plant. Industrial use oven, horizontal, vertical countercurrent reactor or flame furnace in the 300–400 °C or 900–1000 °C into pure direct fluoride, solid UF_4 system by gaseous UF_6 reaction equation for

$$UF_4 + F_2 \rightarrow UF_6 \tag{7.75}$$

Want to improve the enrichment, the enrichment of natural gas UF_6 to be enriched by gas diffusion or centrifugal enrichment, see Chap. 11 of the relevant isotope separation.

7.9.4 Preparation of Metallic Uranium

Metal uranium was prepared by using UF_4 as raw material and metal calcium or magnesium metal.

$$UF_4 + 2Ca \rightarrow U + 2CaF_2 \tag{7.76}$$

$$UF_4 + 2Mg \rightarrow U + 2MgF_2 \tag{7.77}$$

Process is in the lining of high purity CaF_2 or MgF_2 reduction device. With calcium metal as a reducing agent, powder mixing occurs instantaneous reaction released heat sufficient to enable the separation of uranium metal and slag. And using metal magnesium as a reducing agent, reducing device is arranged in the electric furnace heating to 600–700 °C, lead magnesium thermal reduction reaction. Then the reaction of heat release the temperature will quickly reach 1500–1600 °C the uranium and MgF_2 and melting. After cooling crude metal uranium deposit at the bottom of the reaction device, and MgF_2 slag floating on the surface. In addition, the removal of the slag, leaching by acid, washing, sandblasting surface cleaning, and metal uranium raw ingot is prepared.

Exercises

1. 1000 g gold, how many Mole of atoms?
2. If the temperature is 400 °C, the ion product constant of water is 4.1×10^{-12}, calculate the pH value of water.
3. Indicates the current direction of the battery in the following figure.

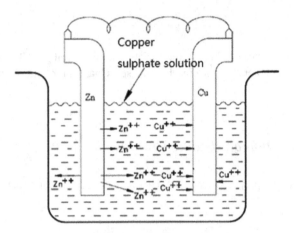

4.   Describe the principle of balanced movement, and its application in the analysis of chemical reaction.
5.   What is the relationship between the periodic table of elements and the distribution of nuclear electrons?
6.   Point out the error in the following chemical reactions: $Fe + H_2O \rightarrow Fe_2O_3 + H_2$.
7.   Compare the ratio of the nucleus to atomic and the sun to the entire solar system. If the mass of the earth is equivalent to the mass of an electron, then the mass of the sun is equivalent to the mass of a nucleus with how many mass number?
8.   Uranium mining from the beginning, until the isotope separation, which are the main process of the process?
9.   Calculation of the molecular weight of pig insulin.
10. If Mole to measure the total number of the world's population, about how many moles?
11. Lists four common semiconductor elements.
12. Why the boiling water reactor coolant caused by ^{17}O and ^{16}N is not too affected by the steam turbine overhaul?
13. Nuclear power plants are mainly concerned about the water chemical parameters which have?
14. Please arrange the relative affinity of Ni^{2+}, Mg^{2+}, Ba^{2+}, Ca^{2+}, Co^{2+}, Cu^{2+}.
15. If there is NaCl of 5 mg dissolved in $1m^3$ pure water, calculate the electrical conductivity.
16. In the main activation corrosion products, the half-life of the longest is what? How to control?
17. Why is it used UF_6 instead of UF_4 or metal?
18. Why should add hydrogen in the primary loop of nuclear power plant? How to add?

References

1. Donald H. Andrews and Richard J. Kokes, Fundamental Chemistry: Wiley; 1963.
2. Underwood. Chemistry for Colleges and Schools. 2nd ed. Edward Arnold, Ltd.; 1967.
3. General Physics Corporation. Fundamentals of Chemistry. General Physics Corporation; 1982.
4. Sienko MJ, Plane RA. Chemical principles and properties. 2nd ed. McGraw and Hill; 1974.
5. Dickerson G. Darensbourg and Darensbourg. Chemical Principles. 4th ed. The Benjamin Cummings Publishing Company; 1984.
6. Yu M, Yu Z, Chen L. Metal corrosion theory and corrosion control. Beijing: Chemical Industry Press; 2009.
7. Glasstone S, Sesonske A. Nuclear reactor engineering. 3rd ed. Van Nostrand Reinhold Company; 1981.
8. Day RA Jr., Johnson RC. General chemistry. Prentice Hall, Inc.; 1974.
9. Compressed Gas Association, Inc., Handbook of Compressed Gases. 2nd ed. Reinhold Publishing Corporation; 1981.
10. Steere NV and Associates. CRC handbook of laboratory safety. 2nd ed. CRC Press, Inc.; 1971.
11. Li G, Wu S. Nuclear fuel (M). Beijing: Chemical Industry Press; 2007.

Chapter 8
Material Science

Materials science is the subject of the relationship between the material composition, structure, processing, properties, and performance, as well as to provide scientific basis for the design, materials, manufacturing, process optimization and rational use of. This chapter introduces in detail all the knowledge and progress in material science, but also introduces the basic principle of material application in the field of nuclear engineering.

8.1 Structure of Metal

Metal is the main structural material used in the nuclear engineering system. It is very necessary to understand and master the basic structure of the metal for the understanding of the various properties of the metal materials.

In the last chapter, we introduced five kinds of chemical bonds, ionic bonds, covalent bonds, and metal bond, van der Waals bonding and hydrogen bonding. Metal bond is a chemical bond, found mainly in metal. In the metal, all atoms are, and many of the surrounding atoms shared valence electron energy state is lower and more stable.

8.1.1 Types of Crystal

In metals, and in many other solids, the atoms are arranged in regular arrays called crystals. A crystal structure consists of atoms arranged in a pattern that repeats periodically in a three-dimensional geometric lattice. The forces of chemical bonding cause this repetition. It is this repeated pattern which control properties like strength, ductility, density, conductivity (property of conducting or transmitting heat, electricity, etc.), and shape.

© Tsinghua University Press 2022
J. Yu, *Fundamental Principles of Nuclear Engineering*,
https://doi.org/10.1007/978-981-16-0839-1_8

Solids can be divided into two major categories of crystal and non-crystal. The so-called crystal, the atoms, ions, molecules in space for rules of three-dimensional periodic arrangement is the most basic structural features, such as Fig. 8.1 is NaCl crystal structure [1].

Any crystal can always find a with three-dimensional periodicity corresponding to the basic structure. In the NaCl crystal, around the Na atoms each have six Cl atoms (around). Similarly, around each Cl atoms have six Na atoms. Atoms formed a very regular arrangement, and achieve a stable of the lowest energy state.

In general, the three most common basic crystal patterns associated with metals are: (a) the body-centered cubic, (b) the face-centered cubic, and (c) the hexagonal close-packed. Figure 8.2 shows these three patterns.

NaCl crystal belongs to the BCC structure, in which the FCC structure and HCP structure of the volume ratio of the same, and higher than the BCC. The so-called volume ratio is the number of atoms in the unit volume can be loaded.

In a body-centered cubic (BCC) arrangement of atoms, the unit cell consists of eight atoms at the corners of a cube and one atom at the body center of the cube.

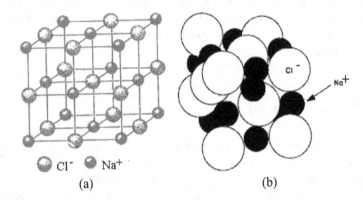

Cl^- Na^+

(a)   (b)

Fig. 8.1 NaCl crystal structure

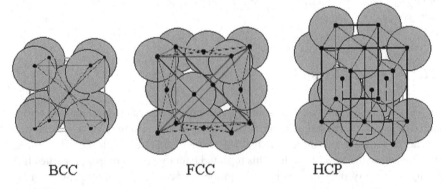

BCC   FCC   HCP

Fig. 8.2 Crystal structure

In a face-centered cubic (FCC) arrangement of atoms, the unit cell consists of eight atoms at the corners of a cube and one atom at the center of each of the faces of the cube.

In a hexagonal close-packed (HCP) arrangement of atoms, the unit cell consists of three layers of atoms. The top and bottom layers contain six atoms at the corners of a hexagon and one atom at the center of each hexagon. The middle layer contains three atoms nestled between the atoms of the top and bottom layers, hence, the name close-packed.

Most diagrams of the structural cells for the BCC and FCC forms of iron are drawn as though they are of the same size, as shown in Fig. 8.2, but they are not. In the BCC arrangement, the structural cell, which uses only nine atoms, is much smaller.

Metals such as α-iron (Fe) (ferrite), chromium (Cr), vanadium (V), molybdenum (Mo), and tungsten (W) possess BCC structures. These BCC metals have two properties in common, high strength and low ductility (which permits permanent deformation). FCC metals such as γ-iron (Fe) (austenite), aluminum (Al), copper (Cu), lead (Pb), silver (Ag), gold (Au), nickel (Ni), platinum (Pt), and thorium (Th) are, in general, of lower strength and higher ductility than BCC metals. HCP structures are found in beryllium (Be), magnesium (Mg), zinc (Zn), cadmium (Cd), cobalt (Co), thallium (Tl), and zirconium (Zr).

8.1.2   Grain Structure and Boundary

If you were to take a small section of a common metal and examine it under a microscope, you would see a structure similar to that shown in Fig. 8.3a. Each of the

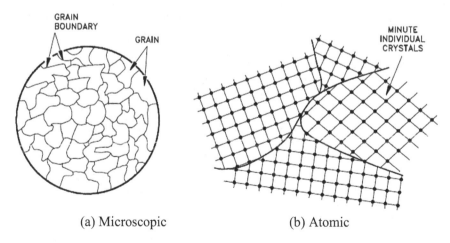

(a) Microscopic   (b) Atomic

Fig. 8.3 Grains and boundaries

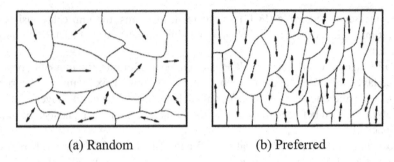

(a) Random   (b) Preferred

Fig. 8.4 Grain orientation

light areas is called a grain, or crystal, which is the region of space occupied by a continuous crystal lattice. The dark lines surrounding the grains are grain boundaries. The grain structure refers to the arrangement of the grains in a metal, with a grain having a particular crystal structure.

The grain boundary refers to the outside area of a grain that separates it from the other grains. The grain boundary is a region of misfit between the grains and is usually one to three atom diameters wide. The grain boundaries separate variously oriented crystal regions (polycrystalline) in which the crystal structures are identical. Figure 8.3b represents four grains of different orientation and the grain boundaries that arise at the interfaces between the grains.

A very important feature of a metal is the average size of the grain. The size of the grain determines the properties of the metal. For example, smaller grain size increases tensile strength and tends to increase ductility. A larger grain size is preferred for improved high-temperature creep properties. Creep is the permanent deformation that increases with time under constant load or stress. Creep becomes progressively easier with increasing temperature.

Another important property of the grains is their orientation. Figure 8.4a represents a random arrangement of the grains such that no one direction within the grains is aligned with the external boundaries of the metal sample. This random orientation can be obtained by cross rolling the material. If such a sample were rolled sufficiently in one direction, it might develop a grain-oriented structure in the rolling direction as shown in Fig. 8.4b. This is called preferred orientation. In many cases, preferred orientation is very desirable, but in other instances, it can be most harmful. For example, preferred orientation in uranium fuel elements can result in catastrophic changes in dimensions during use in a nuclear reactor.

8.1.3 Polymorphism

Polymorphism is the property or ability of a metal to exist in two or more crystalline forms depending upon temperature and composition. Most metals and metal

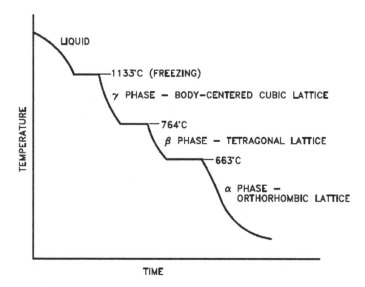

Fig. 8.5 Cooling curve for unalloyed uranium

alloys exhibit this property. Uranium is a good example of a metal that exhibits polymorphism. Uranium metal can exist in three different crystalline structures. Each structure exists at a specific phase, as illustrated in Fig. 8.5.

The alpha (α) phase is stable at room temperature and has a crystal system characterized by three unequal axes at right angles.

1. The alpha phase, from room temperature to 663 °C
2. The beta phase, from 663 to 764 °C
3. The gamma phase, from 764 °C to its melting point of 1133 °C

In the alpha phase, the properties of the lattice are different in the X, Y, and Z-axes. This is because of the regular recurring state of the atoms is different. Because of this condition, when heated the phase expands in the X and Z directions and shrinks in the Y direction. Figure 8.6 shows what happens to the dimensions (Å = angstrom, one hundred-millionth of a centimeter) of a unit cell of alpha uranium upon being heated.

As shown, heating and cooling of alpha phase uranium can lead to drastic dimensional changes and gross distortions of the metal. Thus, pure uranium is not used as a fuel, but only in alloys or compounds.

The beta (β) phase of uranium occurs at elevated temperatures. This phase has a tetragonal (having four angles and four sides) lattice structure and is quite complex.

The gamma (γ) phase of uranium is formed at temperatures above those required for beta phase stability. In the gamma phase, the lattice structure is BCC and expands equally in all directions when heated.

Two additional examples of polymorphism are listed below.

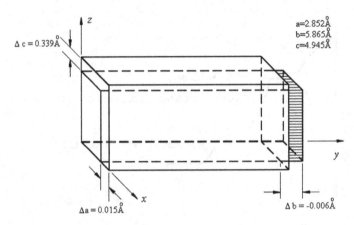

Fig. 8.6 Change in alpha uranium upon heating from 0 to 300 °C

1. Heating iron to 907 °C causes a change from BCC (alpha, ferrite) iron to the FCC (gamma, austenite) form.
2. Zirconium is HCP (alpha) up to 863 °C, where it transforms to the BCC (beta, zirconium) form.

The properties of one polymorphic form of the same metal will differ from those of another polymorphic form. For example, gamma iron can dissolve up to 1.7% carbon, whereas alpha iron can dissolve only 0.03%.

8.1.4   Alloy

An alloy is a mixture of two or more materials, at least one of which is a metal. Alloys can have a microstructure consisting of solid solutions, where secondary atoms are introduced as substitutions or interstitials in a crystal lattice. An alloy might also be a crystal with a metallic compound at each lattice point. In addition, alloys may be composed of secondary crystals imbedded in a primary polycrystalline matrix. This type of alloy is called a composite (although the term "composite" does not necessarily imply that the component materials are metals).

Alloys are usually stronger than pure metals, although they generally offer reduced electrical and thermal conductivity. Strength is the most important criterion by which many structural materials are judged. Therefore, alloys are used for engineering construction. Steel, probably the most common structural metal, is a good example of an alloy. It is an alloy of iron and carbon, with other elements to give it certain desirable properties.

As mentioned in the previous chapter, it is sometimes possible for a material to be composed of several solid phases. The strengths of these materials are enhanced by allowing a solid structure to become a form composed of two interspersed phases.

When the material in question is an alloy, it is possible to quench the metal from a molten state to form the interspersed phases. The type and rate of quenching determines the final solid structure and, therefore, its properties.

Type 304 stainless steel (containing 18%–20% chromium and 8%–10.5% nickel) is used in the tritium production reactor tanks, process water piping, and original process heat exchangers. This alloy resists most types of corrosion.

The wide variety of structures, systems, and components found in DOE nuclear facilities are made from many different types of materials. Many of the materials are alloys with a base metal of iron, nickel, or zirconium. The selection of a material for a specific application is based on many factors including the temperature and pressure that the material will be exposed to, the materials resistance to specific types of corrosion, the materials toughness and hardness, and other material properties.

One material that has wide application in the systems of DOE facilities is stainless steel. There are nearly 40 standard types of stainless steel and many other specialized types under various trade names. Through the modification of the kinds and quantities of alloying elements, the steel can be adapted to specific applications. Stainless steels are classified as austenitic or ferritic based on their lattice structure. Austenitic stainless steels, including 304 and 316, have a face centered cubic structure of iron atoms with the carbon in interstitial solid solution. Ferritic stainless steels, including type 405, have a body-centered cubic iron lattice and contain no nickel. Ferritic steels are easier to weld and fabricate and are less susceptible to stress corrosion cracking than austenitic stainless steels. They have only moderate resistance to other types of chemical attack.

Other metals that have specific applications in some DOE nuclear facilities are Inconel and zircaloy. The composition of these metals and various types of stainless steel are listed in Table 8.1.

Table 8.1 Typical composition of common engineering materials

Type	%Fe	%C Max	%Cr Max	%Ni Max	Others
304 stainless steel	Balanced	0.08	19	10	2% Mn Max, 1%Si Max
304L stainless steel	Balanced	0.03	18	8	2%Mn Max, 1%Si Max
316 stainless steel	Balanced	0.08	17	12	2.5%Mo, 2%Mn Max, 1%Si Max
316L stainless steel	Balanced	0.03	17	12	2.5%Mo, 2%Mn Max
405 stainless steel	Balanced	0.08	13	–	1%Mn Max, 1%Si Max
Inconel	8	0.15	15	Balanced	1%Mn Max, 0.5%Si Max
Zircaloy-4	0.21	–	0.1	–	Zr (balanced)

8.1.5   Imperfections in Metals

The discussion of order in microstructures in the previous section assumed ideal-
ized microstructures. In reality, materials are not composed of perfect crystals, nor
are they free of impurities that alter their properties. Even amorphous solids have
imperfections and impurities that change their structure.

Microscopic imperfections are generally classified as either point, line, or
interfacial imperfections.

1.   Point imperfections have atomic dimensions.
2.   Line imperfections or dislocations are generally many atoms in length.
3.   Interfacial imperfections are larger than line defects and occur over a two-
     dimensional area.

Point imperfections in crystals can be divided into three main defect categories.
They are illustrated in Fig. 8.7.

1.   Vacancy defects result from a missing atom in a lattice position. The vacancy
     type of defect can result from imperfect packing during the crystallization
     process, or it may be due to increased thermal vibrations of the atoms brought
     about by elevated temperature.
2.   Substitution defects result from an impurity present at a lattice position.

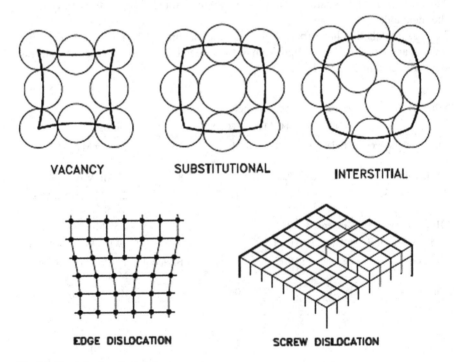

Fig. 8.7   Imperfections in metals

3. Interstitial defects result from an impurity located at an interstitial site or one of the lattice atoms being in an interstitial position instead of being at its lattice position. Interstitial refers to locations between atoms in a lattice structure. Interstitial impurities called network modifiers act as point defects in amorphous solids. The presence of point defects can enhance or lessen the value of a material for engineering construction depending upon the intended use.

Line imperfections are called dislocations and occur in crystalline materials only. Dislocations can be an edge type, screw type, or mixed type, depending on how they distort the lattice, as shown in Fig. 8.7. It is important to note that dislocations cannot end inside a crystal. They must end at a crystal edge or other dislocation, or they must close back on themselves.

Edge dislocations consist of an extra row or plane of atoms in the crystal structure. The imperfection may extend in a straight line all the way through the crystal or it may follow an irregular path. It may also be short, extending only a small distance into the crystal causing a slip of one atomic distance along the glide plane (direction the edge imperfection is moving).

The slip occurs when the crystal is subjected to a stress, and the dislocation moves through the crystal until it reaches the edge or is arrested by another dislocation. The slip of one active plane is ordinarily about 1000 atomic distances and, to produce yielding, slip on many planes is required.

Screw dislocations can be produced by a tearing of the crystal parallel to the slip direction. If a screw dislocation were followed all the way around a complete circuit, it would show a slip pattern similar to that of a screw thread. The pattern may be either left or right handed. This requires that some of the atomic bonds be re-formed continuously so that the crystal has almost the same form after yielding that it had before.

The orientation of dislocations may vary from pure edge to pure screw. At some intermediate point, they may possess both edge and screw characteristics. The importance of dislocations is based on the ease at which they can move through crystals.

Interfacial imperfections exist at an angle between any two faces of a crystal or crystal form. These imperfections are found at free surfaces, domain boundaries, grain boundaries, or interphase boundaries. Free surfaces are interfaces between gases and solids. Domain boundaries refer to interfaces where electronic structures are different on either side causing each side to act differently although the same atomic arrangement exists on both sides. Grain boundaries exist between crystals of similar lattice structure that possess different spatial orientations. Polycrystalline materials are made up of many grains, which are separated by distances typically of several atomic diameters. Finally, interphase boundaries exist between the regions where materials exist in different phases (i.e., BCC next to FCC structures).

Three-dimensional macroscopic defects are called bulk defects. They generally occur on a much larger scale than the microscopic defects. These macroscopic defects generally are introduced into a material during refinement from its raw state or during fabrication processes.

The most common bulk defect arises from foreign particles being included in the prime material. These second-phase particles, called inclusions, are seldom wanted because they significantly alter the structural properties. An example of an inclusion may be oxide particles in a pure metal or a bit of clay in a glass structure.

Other bulk defects include gas pockets or shrinking cavities found generally in castings. These spaces weaken the material and are therefore guarded against during fabrication. The working and forging of metals can cause cracks that act as stress concentrators and weaken the material. Any welding or joining defects may also be classified as bulk defects.

8.2   Properties of Metal

After introducing the microstructure of the metal crystal, we can observe the macroscopic properties of the metal, and there is a close relationship between the macroscopic properties and microstructure.

8.2.1   Stress and Strain

Any component, no matter how simple or complex, has to transmit or sustain a mechanical load of some sort. The load may be one of the following types: a load that is applied steadily ("dead" load); a load that fluctuates, with slow or fast changes in magnitude ("live" load); a load that is applied suddenly (shock load); or a load due to impact in some form. Stress is a form of load that may be applied to a component. Personnel need to be aware how stress may be applied and how it effects the component.

When a metal is subjected to a load (force), it is distorted or deformed, no matter how strong the metal or light the load. If the load is small, the distortion will probably disappear when the load is removed. The intensity, or degree, of distortion is known as strain. If the distortion disappears and the metal returns to its original dimensions upon removal of the load, the strain is called elastic strain. If the distortion disappears and the metal remains distorted, the strain type is called plastic strain. Strain will be discussed in more detail in the next chapter.

When a load is applied to metal, the atomic structure itself is strained, being compressed, warped or extended in the process. The atoms comprising a metal are arranged in a certain geometric pattern, specific for that particular metal or alloy, and are maintained in that pattern by interatomic forces. When so arranged, the atoms are in their state of minimum energy and tend to remain in that arrangement. Work must be done on the metal (that is, energy must be added) to distort the atomic pattern. (Work is equal to force times the distance the force moves.)

Stress is the internal resistance, or counterforce, of a material to the distorting effects of an external force or load. These counterforces tend to return the atoms to their normal positions. The total resistance developed is equal to the external load. This resistance is known as stress.

Although it is impossible to measure the intensity of this stress, the external load and the area to which it is applied can be measured. Stress (σ) can be equated to the load per unit area or the force (F) applied per cross-sectional area (A) perpendicular to the force as shown in Eq. (8.1).

$$\sigma = \frac{F}{A} \tag{8.1}$$

where: σ is stress, Pa; F is the applied force, N; A is cross-sectional area, m^2.

Such a definition of stress is an average value, rather than some textbooks used to define the limits of mathematics, so that the area tends to zero when the value of a certain point.

Stresses occur in any material that is subject to a load or any applied force. There are many types of stresses, but they can all be generally classified in one of six categories: residual stresses, structural stresses, pressure stresses, flow stresses, thermal stresses, and fatigue stresses.

Residual stresses are due to the manufacturing processes that leave stresses in a material. Welding leaves residual stresses in the metals welded. Stresses associated with welding are further discussed later in this module.

Structural stresses are stresses produced in structural members because of the weights they support. The weights provide the loadings. These stresses are found in building foundations and frameworks, as well as in machinery parts.

Pressure stresses are stresses induced in vessels containing pressurized materials. The loading is provided by the same force producing the pressure. In a reactor facility, the reactor vessel is a prime example of a pressure vessel.

Flow stresses occur when a mass of flowing fluid induces a dynamic pressure on a conduit wall. The force of the fluid striking the wall acts as the load. This type of stress may be applied in an unsteady fashion when flow rates fluctuate. Water hammer is an example of a transient flow stress.

Thermal stresses exist whenever temperature gradients are present in a material. Different temperatures produce different expansions and subject materials to internal stress. This type of stress is particularly noticeable in mechanisms operating at high temperatures that are cooled by a cold fluid.

Fatigue stresses are due to cyclic application of a stress. The stresses could be due to vibration or thermal cycling.

Fatigue stresses refers to the effect of cyclic stress; the material is subjected to alternating stress. Cyclic stress may be caused by rotation or vibration, may also is caused by the periodic change of temperature. Fatigue failure is one of the important reasons for the failure of mechanical parts. According to statistics, in the mechanical parts failure in about 80% belong to the fatigue failure, and fatigue failure and with no apparent deformation, so the fatigue failure often causes major accidents. So for shafts, gears, bearings, blades, a spring, etc. under alternating load parts to select materials with good fatigue strength to make.

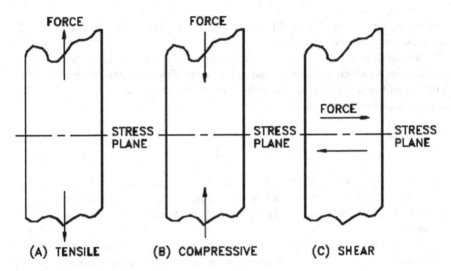

Fig. 8.8 Types of applied stress

The importance of all stresses is increased when the materials supporting them are flawed. Flaws tend to add additional stress to a material. In addition, when loadings are cyclic or unsteady, stresses can effect a material more severely. The additional stresses associated with flaws and cyclic loading may exceed the stress necessary for a material to fail.

Stress intensity within the body of a component is expressed as one of three basic types of internal load. They are known as tensile, compressive, and shear. Figure 8.8 illustrates the different types of stress. Mathematically, there are only two types of internal load because tensile and compressive stress may be regarded as the positive and negative versions of the same type of normal loading.

However, in mechanical design, the response of components to the two conditions can be so different that it is better, and safer, to regard them as separate types.

As illustrated in Fig. 8.8, the plane of a tensile or compressive stress lies perpendicular to the axis of operation of the force from which it originates. The plane of a shear stress lies in the plane of the force system from which it originates. It is essential to keep these differences quite clear both in mind and mode of expression.

Tensile stress is that type of stress in which the two sections of material on either side of a stress plane tend to pull apart or elongate as illustrated in Fig. 8.8a.

Compressive stress is the reverse of tensile stress. Adjacent parts of the material tend to press against each other through a typical stress plane as illustrated in Fig. 8.8b.

Shear stress exists when two parts of a material tend to slide across each other in any typical plane of shear upon application of force parallel to that plane as illustrated in Fig. 8.8c.

Assessment of mechanical properties is made by addressing the three basic stress types. Because tensile and compressive loads produce stresses that act across a plane, in a direction perpendicular (normal) to the plane, tensile and compressive stresses are called normal stresses. The shorthand designations are as follows.

For tensile stresses: "+S_N" (or "S_N") or "σ" (sigma)
For compressive stresses: "−S_N" or "−σ" (minus sigma).

The ability of a material to react to compressive stress or pressure is called compressibility. For example, metals and liquids are incompressible, but gases and vapors are compressible. The shear stress is equal to the force divided by the area of the face parallel to the direction in which the force acts, as shown in Fig. 8.8c.

Two types of stress can be present simultaneously in one plane, if one of the stresses is shear stress. Under certain conditions, different basic stress type combinations may be simultaneously present in the material. An example would be a reactor vessel during operation. The wall has tensile stress at various locations due to the temperature and pressure of the fluid acting on the wall. Compressive stress is applied from the outside at other locations on the wall due to outside pressure, temperature, and constriction of the supports associated with the vessel. In this situation, the tensile and compressive stresses are considered principal stresses. If present, shear stress will act at a 90° angle to the principal stress.

When stress is present, strain will be involved also. The two types of strain will be discussed in this chapter. Personnel need to be aware how strain may be applied and how it affects the component.

In the use of metal for mechanical engineering purposes, a given state of stress usually exists in a considerable volume of the material. Reaction of the atomic structure will manifest itself on a macroscopic scale. Therefore, whenever a stress (no matter how small) is applied to a metal, a proportional dimensional change or distortion must take place.

Such a proportional dimensional change (intensity or degree of the distortion) is called strain and is measured as the total elongation per unit length of material due to some applied stress. Equation 8.2 illustrates this proportion or distortion.

$$\varepsilon = \frac{\delta}{L} \tag{8.2}$$

where: ε is strain (m/m); δ is total elongation (m); L is original length (m).

Strain may take two forms, elastic strain and plastic deformation.

Elastic strain is a transitory dimensional change that exists only while the initiating stress is applied and disappears immediately upon removal of the stress. Elastic strain is also called elastic deformation. The applied stresses cause the atoms in a crystal to move from their equilibrium position. All the atoms are displaced the same amount and still maintain their relative geometry. When the stresses are removed, all the atoms return to their original positions and no permanent deformation occurs.

Plastic deformation (or plastic strain) is a dimensional change that does not disappear when the initiating stress is removed. It is usually accompanied by some elastic strain.

The phenomenon of elastic strain and plastic deformation in a material are called elasticity and plasticity, respectively.

At room temperature, most metals have some elasticity, which manifests itself as soon as the slightest stress is applied. Usually, they also possess some plasticity, but this may not become apparent until the stress has been raised appreciably. The magnitude of plastic strain, when it does appear, is likely to be much greater than that of the elastic strain for a given stress increment. Metals are likely to exhibit less elasticity and more plasticity at elevated temperatures. A few pure unalloyed metals (notably aluminum, copper and gold) show little, if any, elasticity when stressed in the annealed (heated and then cooled slowly to prevent brittleness) condition at room temperature, but do exhibit marked plasticity. Some unalloyed metals and many alloys have marked elasticity at room temperature, but no plasticity.

The state of stress just before plastic strain begins to appear is known as the proportional limit, or elastic limit, and is defined by the stress level and the corresponding value of elastic strain. The proportional limit is expressed in pounds per square inch. For load intensities beyond the proportional limit, the deformation consists of both elastic and plastic strains.

As mentioned previously, strain measures the proportional dimensional change with no load applied. Such values of strain are easily determined and only cease to be sufficiently accurate when plastic strain becomes dominant.

When metal experiences strain, its volume remains constant. Therefore, if volume remains constant as the dimension changes on one axis, then the dimensions of at least one other axis must change also. If one dimension increases, another must decrease. There are a few exceptions. For example, strain hardening involves the absorption of strain energy in the material structure, which results in an increase in one dimension without an offsetting decrease in other dimensions. This causes the density of the material to decrease and the volume to increase.

If a tensile load is applied to a material, the material will elongate on the axis of the load (perpendicular to the tensile stress plane). Conversely, if the load is compressive, the axial dimension will decrease. If volume is constant, a corresponding lateral contraction or expansion must occur. This lateral change will bear a fixed relationship to the axial strain. The relationship, or ratio, of lateral to axial strain is called Poisson's ratio after the name of its discoverer. It is usually symbolized by v.

8.2.2 Hooke's Law

If a metal is lightly stressed, a temporary deformation, presumably permitted by an elastic displacement of the atoms in the space lattice, takes place. Removal of the stress results in a gradual return of the metal to its original shape and dimensions. In 1678, an English scientist named Robert Hooke ran experiments that provided data that showed that in the elastic range of a material, strain is proportional to stress. The elongation of the bar is directly proportional to the tensile force and the length of the bar and inversely proportional to the cross-sectional area and the modulus of elasticity.

Hooke's experimental law may be given by Eq. (8.3).

$$\delta = \frac{F \cdot L}{A \cdot E} \tag{8.3}$$

where:

F force producing extension of bar (N)
L length of bar (m)
A cross-sectional area of bar (m^2)
δ total elongation of bar (m)
E elastic constant of the material, called the Modulus of Elasticity, or Young's Modulus (Pa).

This simple linear relationship between the force (stress) and the elongation (strain) was formulated using the following notation.

The quantity E, the ratio of the unit stress to the unit strain, is the modulus of elasticity of the material in tension or compression and is often called Young's Modulus.

Previously, we learned that tensile stress, or simply stress, was equated to the load per unit area or force applied per cross-sectional area perpendicular to the force measured in Pa.

With Eqs. (8.1), (8.2) and (8.3), then we can get another form of Hooke's law:

$$\varepsilon = \frac{\sigma}{E} \tag{8.4}$$

Namely, the strain is equal to the stress divided by young's modulus, or young's modulus is equal to the ratio of stress and strain:

$$E = \frac{\sigma}{\varepsilon} \tag{8.5}$$

Young's Modulus (sometimes referred to as Modulus of Elasticity, meaning "measure" of elasticity) is an extremely important characteristic of a material. It is the numerical evaluation of Hooke's Law, namely the ratio of stress to strain (the measure of resistance to elastic deformation). To calculate Young's Modulus, stress (at any point) below the proportional limit is divided by corresponding strain. It can also be calculated as the slope of the straight-line portion of the stress–strain curve. (The positioning on a stress–strain curve will be discussed later.)

We can now see that Young's Modulus may be easily calculated, provided that the stress and corresponding unit elongation or strain have been determined by a tensile test as described previously. Strain (ε) is a number representing a ratio of two lengths; therefore, we can conclude that the Young's Modulus is measured in the

Table 8.2 Properties of common structural materials	Material	E/MPa	Yield strength/MPa	Ultimate strength/MPa
	Aluminum	6.9×10^4	240–310	370–450
	Stainless steel	2.0×10^5	275–345	540–690
	Carbon steel	2.1×10^5	205–275	380–450

same units as stress (σ), that is, in pounds per square inch. Table 8.2 gives average values of the Modulus E for several metals used in nuclear facilities construction.

Example 8.1 What is the elongation of 5 m of aluminum wire with a 0.05 cm^2 area if it supports a force of 500 N?

Solution:

$$\delta = \frac{F \cdot L}{A \cdot E} = \frac{(500\,\text{N}) \times (5\,\text{m})}{\left(0.05 \times 10^{-4}\,\text{m}^2\right) \times \left(6.9 \times 10^4\,\text{MPa}\right)} = 0.0072\,\text{m}$$

The Bulk Modulus of Elasticity is the elastic response to hydrostatic pressure and equilateral tension or the volumetric response to hydrostatic pressure and equilateral tension. It is also the property of a material that determines the elastic response to the application of stress.

Bulk modulus is sometimes referred to as bulk change modulus, assuming that the volume of the material under P_0 is V_0, the volume becomes smaller if pressure becomes higher, and the coefficient is the bulk modulus.

$$K = \frac{\Delta P}{|\Delta V|/V_0} \tag{8.6}$$

The Shear Modulus of Elasticity is derived from the torsion of a cylindrical test piece. Its symbol is G.

Shear modulus is the ratio of shear stress and strain. Also known as, modulus of rigidity is one of the mechanical performance of the material. Shear modulus is the ratio of shear stress and strain when material in the shear stress, in the elastic deformation within the limits. Characterization of a material's ability to resist shear strain. Large shear modulus, said rigid material. Between the shear modulus and Young's modulus, there is a relationship:

$$G = \frac{E}{2(1 + \mu)} \tag{8.7}$$

where, μ is the Poisson's ratio of material.

8.2.3 Relationship Between Stress and Strain

Most polycrystalline materials have within their elastic range an almost constant relationship between stress and strain. Experiments by an English scientist named Robert Hooke led to the formation of Hooke's Law which states that in the elastic range of a material strain is proportional to stress. The ratio of stress to strain, or the gradient of the stress–strain graph, is called the Young's Modulus.

To determine the load-carrying ability and the amount of deformation before fracture, a sample of material is commonly tested by a Tensile Test. This test consists of applying a gradually increasing force of tension at one end of a sample length of the material. The other end is anchored in a rigid support so that the sample is slowly pulled apart. The testing machine is equipped with a device to indicate, and possibly record, the magnitude of the force throughout the test. Simultaneous measurements are made of the increasing length of a selected portion at the middle of the specimen, called the gage length. The measurements of both load and elongation are ordinarily discontinued shortly after plastic deformation begins; however, the maximum load reached is always recorded. Fracture point is the point where the material fractures due to plastic deformation. After the specimen has been pulled apart and removed from the machine, the fractured ends are fitted together and measurements are made of the now extended gage length and of the average diameter of the minimum cross section. The average diameter of the minimum cross section is measured only if the specimen used is cylindrical.

The tabulated results at the end of the test consist of the following.

(a) designation of the material under test
(b) original cross section dimensions of the specimen within the gage length
(c) original gage length
(d) a series of frequent readings identifying the load and the corresponding gage length dimension
(e) final average diameter of the minimum cross section
(f) final gage length
(g) description of the appearance of the fracture surfaces (for example, cup-cone, wolf's ear, diagonal, start).

A graph of the results is made from the tabulated data. Some testing machines are equipped with an autographic attachment that draws the graph during the test. (The operator need not record any load or elongation readings except the maximum for each.) The coordinate axes of the graph are strain for the x-axis or scale of abscissae, and stress for the y-axis or scale of ordinates. The ordinate for each point plotted on the graph is found by dividing each of the tabulated loads by the original cross-sectional area of the sample; the corresponding abscissa of each point is found by dividing the increase in gage length by the original gage length. These two calculations are made as follows.

Stress and strain, as computed here, are sometimes called "engineering stress and strain." They are not true stress and strain, which can be computed on the basis of the

Fig. 8.9 Typical ductile
material stress–strain curve

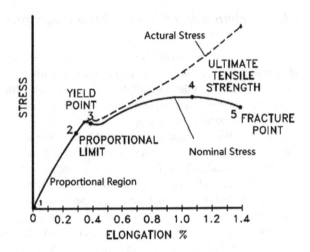

area and the gage length that exist for each increment of load and deformation. For example, true strain is the natural log of the elongation (ln (L/L$_0$)), and true stress is F/A, where A is area. The latter values are usually used for scientific investigations, but the engineering values are useful for determining the load carrying values of a material. Below the elastic limit, engineering stress and true stress are almost identical.

The graphic results, or stress–strain diagram, of a typical tension test for structural steel is shown in Fig. 8.9. The ratio of stress to strain, or the gradient of the stress–strain graph, is called the Modulus of Elasticity or Elastic Modulus. The slope of the portion of the curve where stress is proportional to strain (between Points 1 and 2) is referred to as Young's Modulus and Hooke's Law applies.

The following observations are illustrated in Fig. 8.9:

- Hooke's Law applies between Points 1 and 2.
- Hooke's Law becomes questionable between Points 2 and 3 and strain increases more rapidly.
- The area between Points 1 and 2 is called the elastic region. If stress is removed, the material will return to its original length.
- Point 2 is the proportional limit (PL) or elastic limit, and Point 3 is the yield strength (YS) or yield point.
- The area between Points 2 and 5 is known as the plastic region because the material will not return to its original length.
- Point 4 is the point of ultimate strength and Point 5 is the fracture point at which failure of the material occurs.

Figure 8.9 is a stress–strain curve typical of a ductile material where the strength is small, and the plastic region is great. The material will bear more strain (deformation) before fracture.

Fig. 8.10 Typical brittle
material stress–strain curve

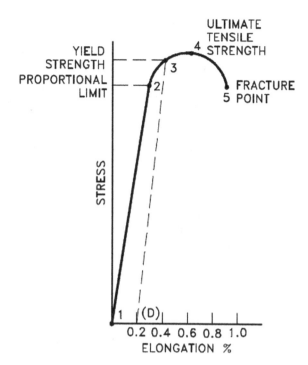

Figure 8.10 is a stress–strain curve typical of a brittle material where the plastic region is small and the strength of the material is high.

The tensile test supplies three descriptive facts about a material. These are the stress at which observable plastic deformation or "yielding" begins; the ultimate tensile strength or maximum intensity of load that can be carried in tension; and the percent elongation or strain (the amount the material will stretch) and the accompanying percent reduction of the cross-sectional area caused by stretching. The rupture or fracture point can also be determined.

8.2.4 Physical Properties of Material

Material is selected for various applications in a reactor facility based on its physical and chemical properties. This section discusses the physical properties of material.

Materials are selected according to their physical and chemical properties, including strength, ultimate tensile strength, yield strength, ductility, flexibility, stiffness and hardness.

Strength
Strength is the ability of a material to resist deformation. The strength of a component is usually considered based on the maximum load that can be borne before failure

is apparent. If under simple tension the permanent deformation (plastic strain) that takes place in a component before failure, the load-carrying capacity, at the instant of final rupture, will probably be less than the maximum load supported at a lower strain because the load is being applied over a significantly smaller cross-sectional area. Under simple compression, the load at fracture will be the maximum applicable over a significantly enlarged area compared with the cross-sectional area under no load.

This obscurity can be overcome by utilizing a nominal stress figure for tension and shear. This is found by dividing the relevant maximum load by the original area of cross section of the component. Thus, the strength of a material is the maximum nominal stress it can sustain. The nominal stress is referred to in quoting the "strength" of a material and is always qualified by the type of stress, such as tensile strength, compressive strength, or shear strength.

For most structural materials, the difficulty in finding compressive strength can be overcome by substituting the tensile strength value for compressive strength. This substitution is a safe assumption since the nominal compression strength is always greater than the nominal tensile strength because the effective cross section increases in compression and decreases in tension.

When a force is applied to a metal, layers of atoms within the crystal structure move in relation to adjacent layers of atoms. This process is referred to as slip. Grain boundaries tend to prevent slip. The smaller the grain size, the larger the grain boundary area. Decreasing the grain size through cold or hot working of the metal tends to retard slip and thus increases the strength of the metal.

Ultimate Tensile Strength

The ultimate tensile strength (UTS) is the maximum resistance to fracture. It is equivalent to the maximum load that can be carried by one square inch of cross-sectional area when the load is applied as simple tension.

If the complete engineering stress–strain curve is available, as shown in Fig. 8.9, the ultimate tensile strength appears as the stress coordinate value of the highest point on the curve. Materials that elongate greatly before breaking undergo such a large reduction of cross-sectional area that the material will carry less load in the final stages of the test (this was noted in Figs. 8.9 and 8.10 by the decrease in stress just prior to rupture). A marked decrease in cross-section is called "necking." Ultimate tensile strength is often shortened to "tensile strength" or even to "the ultimate." "Ultimate strength" is sometimes used but can be misleading and, therefore, is not used in some disciplines.

Yield Strength

A number of terms have been defined for the purpose of identifying the stress at which plastic deformation begins. The value most commonly used for this purpose is the yield strength. The yield strength is defined as the stress at which a predetermined amount of permanent deformation occurs. The graphical portion of the early stages of a tension test is used to evaluate yield strength. To find yield strength, the predetermined amount of permanent strain is set along the strain axis of the graph, to the right of the origin (zero). It is indicated in Fig. 8.10 as Point (D).

A straight line is drawn through Point (D) at the same slope as the initial portion of the stress–strain curve. The point of intersection of the new line and the stress strain curve is projected to the stress axis. The stress value, in Pa, is the yield strength. It is indicated in Fig. 8.10 as Point 3. This method of plotting is done for the purpose of subtracting the elastic strain from the total strain, leaving the predetermined "permanent offset" as a remainder. When yield strength is reported, the amount of offset used in the determination should be stated. For example, "Yield Strength (at 0.2% offset) = 348 MPa".

Alternate values are sometimes used instead of yield strength. Several of these are briefly described below.

The yield point, determined by the divider method, involves an observer with a pair of dividers watching for visible elongation between two gage marks on the specimen. When visible stretch occurs, the load at that instant is recorded, and the stress corresponding to that load is calculated.

Soft steel, when tested in tension, frequently displays a peculiar characteristic, known as a yield point. If the stress–strain curve is plotted, a drop in the load (or sometimes a constant load) is observed although the strain continues to increase. Eventually, the metal is strengthened by the deformation, and the load increases with further straining. The high point on the S-shaped portion of the curve, where yielding began, is known as the upper yield point, and the minimum point is the lower yield point. This phenomenon is very troublesome in certain deep drawing operations of sheet steel. The steel continues to elongate and to become thinner at local areas where the plastic strain initiates, leaving unsightly depressions called stretcher strains or "worms."

The proportional limit is defined as the stress at which the stress–strain curve first deviates from a straight line. Below this limiting value of stress, the ratio of stress to strain is constant, and the material is said to obey Hooke's Law (stress is proportional to strain). The proportional limit usually is not used in specifications because the deviation begins so gradually that controversies are sure to arise as to the exact stress at which the line begins to curve.

The elastic limit has previously been defined as the stress at which plastic deformation begins. This limit cannot be determined from the stress–strain curve. The method of determining the limit would have to include a succession of slightly increasing loads with intervening complete unloading for the detection of the first plastic deformation or "permanent set." Like the proportional limit, its determination would result in controversy. Elastic limit is used, however, as a descriptive, qualitative term.

In many situations, the yield strength is used to identify the allowable stress to which a material can be subjected. For components that have to withstand high pressures, such as those used in pressurized water reactors (PWRs), this criterion is not adequate. To cover these situations, the maximum shear stress theory of failure has been incorporated into the ASME (The American Society of Mechanical Engineers) Boiler and Pressure Vessel Code, Section III, Rules for Construction of Nuclear Pressure Vessels. The maximum shear stress theory of failure was originally proposed for use in the U.S. Naval Reactor Program for PWRs. It will not be discussed in this text.

Ductility

Ductility is the percent elongation reported in a tensile test is defined as the maximum elongation of the gage length divided by the original gage length.

The reduction of area is reported as additional information (to the percent elongation) on the deformational characteristics of the material. The two are used as indicators of ductility, the ability of a material to be elongated in tension. Because the elongation is not uniform over the entire gage length and is greatest at the center of the neck, the percent elongation is not an absolute measure of ductility. (Because of this, the gage length must always be stated when the percent elongation is reported.) The reduction of area, being measured at the minimum diameter of the neck, is a better indicator of ductility.

Ductility is more commonly defined as the ability of a material to deform easily upon the application of a tensile force, or as the ability of a material to withstand plastic deformation without rupture. Ductility may also be thought of in terms of bendability and crushability. Ductile materials show large deformation before fracture. The lack of ductility is often termed brittleness. Usually, if two materials have the same strength and hardness, the one that has the higher ductility is more desirable. The ductility of many metals can change if conditions are altered. An increase in temperature will increase ductility. A decrease in temperature will cause a decrease in ductility and a change from ductile to brittle behavior.

Cold-working also tends to make metals less ductile. Cold-working is performed in a temperature region and over a time interval to obtain plastic deformation, but not relieving the strain hardening. Minor additions of impurities to metals, either deliberate or unintentional, can have a marked effect on the change from ductile to brittle behavior. The heating of a cold-worked metal to or above the temperature at which metal atoms return to their equilibrium positions will increase the ductility of that metal. This process is called annealing.

Ductility is desirable in the high temperature and high pressure applications in reactor plants because of the added stresses on the metals. High ductility in these applications helps prevent brittle fracture.

Many metals have ductility under certain conditions, and when conditions change can be converted into brittle. For example, the majority of steel under high temperature, good ductility, but at low temperature will occur from brittle fracture to brittle ductile transition. The temperature is called the nil ductility transition temperature (NDT) of metallic materials. The nil ductility transition temperature with the original mechanical processing, heat treatment and contains impurities. It is of great significance for nil ductility transition temperature of reactor pressure vessel steel; this is because the value of the temperature increases with the irradiation of fast neutron. In order to prevent the brittle failure of the reactor pressure vessel, must requirements of pressure vessels in the pressure when the working temperature is greater than the nil ductility transition temperature.

Grain refinement can increase the material ductility. Therefore, the provisions of steel pressure vessel used for fine grain steel. Materials in the presence of defects, such as cracks, sharp cracks and other will promote the brittle development; we must pay attention to avoid. However, in fact, pressure vessel will have some defects that may allow, these defects in use period may be gradually expanded into the crack, prompting fracture. So in vessel design, also need to necessary fracture analysis, or in service inspection do necessary fracture evaluation.

Malleability

Where ductility is the ability of a material to deform easily upon the application of a tensile force, malleability is the ability of a metal to exhibit large deformation or plastic response when being subjected to compressive force. Uniform compressive force causes deformation in the manner shown in Fig. 8.11. The material contracts axially with the force and expands laterally. Restraint due to friction at the contact faces induces axial tension on the outside. Tensile forces operate around the circumference with the lateral expansion or increasing girth. Plastic flow at the center of the material also induces tension.

Therefore, the criterion of fracture (that is, the limit of plastic deformation) for a plastic material is likely to depend on tensile rather than compressive stress. Temperature change may modify both the plastic flow mode and the fracture mode.

Toughness

The quality known as toughness describes the way a material reacts under sudden impacts. It is defined as the work required to deform one cubic inch of metal until it fractures. Toughness is measured by the Charpy test or the Izod test.

Both of these tests use a notched sample. The location and shape of the notch are standard. The points of support of the sample, as well as the impact of the hammer, must bear a constant relationship to the location of the notch.

The tests are conducted by mounting the samples as shown in Fig. 8.12 and allowing a pendulum of a known weight to fall from a set height. The maximum

Fig. 8.11 Malleable deformation of a cylinder under uniform axial compression

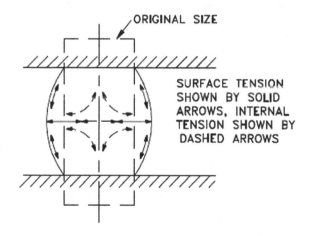

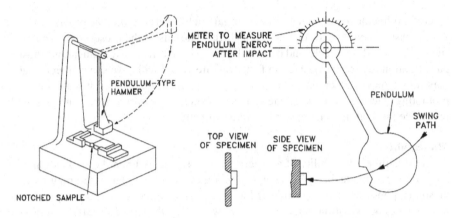

Fig. 8.12 Material toughness test

energy developed by the hammer is 16.59 kg m (120 ft lb) in the Izod test and 33.18 kg m (240 ft lb) in the Charpy test. By properly calibrating the machine, the energy absorbed by the specimen may be measured from the upward swing of the pendulum after it has fractured the material specimen as shown in Fig. 8.9. The greater the amount of energy absorbed by the specimen, the smaller the upward swing of the pendulum will be and the tougher the material is.

Hardness

Hardness is the property of a material that enables it to resist plastic deformation, penetration, indentation, and scratching. Therefore, hardness is important from an engineering standpoint because resistance to wear by either friction or erosion by steam, oil, and water generally increases with hardness.

Hardness tests serve an important need in industry even though they do not measure a unique quality that can be termed hardness. The tests are empirical, based on experiments and observation, rather than fundamental theory. Its chief value is as an inspection device, able to detect certain differences in material when they arise even though these differences may be undefinable. For example, two lots of material that have the same hardness may or may not be alike, but if their hardness is different, the materials certainly are not alike.

Several methods have been developed for hardness testing. Those most often used are Brinell, Rockwell, Vickers, Tukon, Scleroscope, and the files test. The first four are based on indentation tests and the fifth on the rebound height of a diamond-tipped metallic hammer. The file test establishes the characteristics of how well a file takes a bite on the material.

As a result of many tests, comparisons have been prepared using formulas, tables, and graphs that show the relationships between the results of various hardness tests of specific alloys. There is, however, no exact mathematical relation between any two of the methods. For this reason, the result of one type of hardness test converted to readings of another type should carry the notation "___converted from___" (for example "352 Brinell converted from Rockwell C-38").

Brinell hardness determination principle is with a fixed size of the force F and the diameter D of the quenching of steel or hard alloy ball pressed into the surface of the metal to be tested, to maintain a specified time after unloading force, reading microscope to measure average indentation diameter D, then according to the formula for Brinell hardness value, or according to D from the table ready Brinell hardness found hardness value.

Rockwell hardness is dimensionless, is a dimensionless mechanical performance index, the most common hardness scales have A, B, C, often credited as HRA, HRB, HRC, written as hardness and hardness of symbols, such as 50HRC.

The Rockwell hardness test is one of the common hardness test. The A, B and C in HRA, HRB and HRC are three kinds of different standards. The initial pressure of the three standard is 98.07 N, and finally the hardness value is calculated according to the indentation depth.

Standard A uses sphere conical diamond indenter and then pressurized to 588.4 N; Standard B uses diameter 1.588 mm ball as indenter, and then pressurized to 980.7 N; and Standard C uses the same sphere conical diamond indenter, but pressure is 1471 N. So standard B is suitable for relatively soft materials, and standard C for a hard material.

Several other hardness tests will not be introduced here.

8.3 Heat Treatment of Metal

Heat treatment and working of the metal are discussed as metallurgical processes used to change the properties of metals. Personnel need to understand the effects on metals to select the proper material for a reactor facility.

Heat treatment of large carbon steel components is done to take advantage of crystalline defects and their effects and thus obtain certain desirable properties or conditions.

During manufacture, by varying the rate of cooling (quenching) of the metal, grain size and grain patterns are controlled. Grain characteristics are controlled to produce different levels of hardness and tensile strength. Generally, the faster a metal is cooled, the smaller the grain sizes will be. This will make the metal harder. As hardness and tensile strength increase in heat-treated steel, toughness and ductility decrease.

The cooling rate used in quenching depends on the method of cooling and the size of the metal. Uniform cooling is important to prevent distortion. Typically, steel components are quenched in oil or water.

Because of the crystal pattern of type 304 stainless steel in the reactor tank (tritium production facility); heat treatment is unsuitable for increasing the hardness and strength.

Welding can induce internal stresses that will remain in the material after the welding is completed. In stainless steels, such as type 304, the crystal lattice is face-centered cubic (austenite). During high temperature welding, some surrounding metal may be elevated to between 260 and 538 °C. In this temperature region, the austenite is transformed into a body centered cubic lattice structure (bainite). When the metal has cooled, regions surrounding the weld contain some original austenite and some newly formed bainite. A problem arises because the "packing factor" (PF = volume of atoms/volume of unit cell) is not the same for FCC crystals as for BCC crystals.

The bainite that has been formed occupies more space than the original austenite lattice. This elongation of the material causes residual compressive and tensile stresses in the material. Welding stresses can be minimized by using heat sink welding, which results in lower metal temperatures, and by annealing.

Annealing is another common heat-treating process for carbon steel components. During annealing, the component is heated slowly to an elevated temperature and held there for a long period of time, then cooled. The annealing process is done to obtain the following effects.

(a)   to soften the steel and improve ductility
(b)   to relieve internal stresses caused by previous processes such as heat treatment, welding, or machining
(c)   to refine the grain structure.

Plastic deformation which is carried out in a temperature region and over a time interval such that the strain hardening is not relieved is called cold work. Considerable knowledge on the structure of the cold-worked state has been obtained. In the early stages of plastic deformation, slip is essentially on primary glide planes and the dislocations form coplanar arrays. As deformation proceeds, cross slip takes place. The cold-worked structure forms high dislocation density regions that soon develop into networks. The grain size decreases with strain at low deformation but soon reaches a fixed size. Cold working will decrease ductility.

Hot working refers to the process where metals are deformed above their recrystallization temperature and strain hardening does not occur. Hot working is usually performed at elevated temperatures. Lead, however, is hot-worked at room temperature because of its low melting temperature. At the other extreme, molybdenum is cold-worked when deformed even at red heat because of its high recrystallization temperature.

The resistance of metals to plastic deformation generally falls with temperature. For this reason, larger massive sections are always worked hot by forging, rolling, or extrusion. Metals display distinctly viscous characteristics at sufficiently high temperatures, and their resistance to flow increases at high forming rates. This occurs not only because it is a characteristic of viscous substances, but also because the rate of recrystallization may not be fast enough.

8.4 Hydrogen Embrittlement and Irradiation Effect

Personnel need to be aware of the conditions for hydrogen embrittlement and its formation process when selecting materials for a reactor plant. This chapter discusses the sources of hydrogen and the characteristics for the formation of hydrogen embrittlement.

Another form of stress-corrosion cracking is hydrogen embrittlement. Although embrittlement of materials takes many forms, hydrogen embrittlement in high strength steels has the most devastating effect because of the catastrophic nature of the fractures when they occur. Hydrogen embrittlement is the process by which steel loses its ductility and strength due to tiny cracks that result from the internal pressure of hydrogen (H_2) or methane gas (CH_4), which forms at the grain boundaries. In zirconium alloys, hydrogen embrittlement is caused by zirconium hydriding. At nuclear reactor facilities, the term "hydrogen embrittlement" generally refers to the embrittlement of zirconium alloys caused by zirconium hydriding.

Sources of hydrogen causing embrittlement have been encountered in the making of steel, in processing parts, in welding, in storage or containment of hydrogen gas, and related to hydrogen as a contaminant in the environment that is often a by-product of general corrosion. The latter concerns the nuclear industry. Hydrogen may be produced by corrosion reactions such as rusting, catholic protection, and electroplating. Hydrogen may also be added to reactor coolant to remove oxygen from reactor coolant systems.

Hydrogen embrittlement is a primary reason that the reactor coolant is maintained at a neutral or basic pH in plants without aluminum components.

Hydrogen embrittlement in carbon steel is refers to the soluble in steel in hydrogen or methane, polymerization for gas molecules, resulting in local stress concentration in the steel to form tiny crack. Hydrogen embrittlement usually appears to be under the action of force delayed fracture phenomenon. There have been car spring leaf pieces of zinc, within hours after the assembly gradually, the occurrence of fracture, fracture ratio up to 40%~50%. This had formulated strict to hydrogen technology. In addition, there are also some hydrogen embrittlement is not delayed fracture phenomena, such as electroplating rack because of the multi electroplating and pickling depleting, hydrogen permeation is more serious, in use often appear a fold occurs brittle fracture phenomenon.

In the field of nuclear engineering, pay attention to the absorption of hydrogen embrittlement is the fuel rods of zirconium alloy cladding after hydrogenation decreased ductility. Zirconium and water will occur corrosion reaction as follows:

$$Zr + 2H_2O \rightarrow ZrO_2 + 2H_2 \uparrow \qquad (8.8)$$

The corrosion reaction of hydrogen will form zirconium hydride fleck block ($ZrH_{1.5}$), zirconium hydride will make the alloy becomes brittle. Produced in the zirconium hydride and the formation of cracks, and will continue to spread, and thus the occurrence of brittle failure. Zr-2 alloy can absorb 50% of hydrogen produced by corrosion reaction, therefore easier to hydrogen embrittlement failure. According to a large number of observation and research, nickel content of zirconium alloys infects hydrogen embrittlement. Therefore, the development of the low nickel of Zr-4 alloy, and through appropriate additions of niobium can also further reduce the hydrogen absorption rate.

The effect of irradiation on the material after exposure to neutron and gamma (mainly neutron) will have a certain change. Here we will discuss the mechanism of irradiation effect qualitatively.

Materials by fast neutron irradiation energy greater than 1 meV, which is hit by a neutron atom, will produce off phenomenon. In the series after the collision, appear the peak from the group consisting of material defects will, at the same time vacancies and interstitials respectively through aggregation and collapse will form wrong ring, stacking fault and so on. Because of these, the stress field around the defect is relatively large, causing the material hardening, and associated embrittlement. At the same time, because of the material from the peak of interstitial atoms in the remaining very intensive. By the residual energy and collision energy, the local micro zone quickly rose to a high temperature. The thermal peak and off because of the heat. The associated peak volume is very small, and then the temperature dropped rapidly, as this will cause quenching, hardening and embrittlement. Metal materials by irradiation induced hardening and embrittlement effect, sometimes called radiation damage. It increases with the neutron flux. With the increase of temperature, but also increased the recovery.

The change of material properties can also be caused by thermal neutron irradiation. The thermal neutron can be absorbed by the atoms in the material to produce a nuclear reaction. For example, ^{10}B atoms by neutron bombardment occurred (n, alpha), generated by the reaction of Li and alpha particles; ^{58}Ni neutron absorption after the first into the ^{59}ni, and because of the (n, alpha), the reaction of ^{56}Fe and alpha particles, alpha particles trapped electron helium generating. Formation of defects and a set of helium bubbles will have influence on the material physical properties (density, conductivity, thermal conductivity, elastic modulus, etc.) and chemical properties (such as nickel into iron) and mechanical properties (strength, ductility, toughness, creep rupture strength, creep strength and fatigue strength).

The effect of irradiation on the properties of metal structure materials, different materials have different effects. The intensity of yield will increase in the zirconium alloy after irradiation, while the elongation decreased. The irradiation intensity and ductility changes depends on the temperature and irradiation dose. With the increase of irradiation significantly enhanced neutron flux increases, and as temperature increasing it is decreased. Irradiation strengthening plays a decisive role in the opposite effect. Generally, the intensity changes approximately in the fast neutron flux of 10^{21}–10^{22} n/cm^2 tends to be saturated; while the ductility value in fast neutron flux of 3×10^{19}–1×10^{20} n/cm^2 up to saturation; the total elongation at about fast neutron flux of 5×10^{20} n/cm^2 saturated. In addition, irradiation creep properties of zirconium alloy also had significant effect. The general trend is the creep rate, as the number of levels tend to increase.

For ferritic steel (such as carbon steel, low alloy steel), irradiation makes tensile strength increased slightly, the ductility decreases, the elongation rate decreased significantly, while the cross section shrinkage decreased less. The most important effect of irradiation on this kind of steel is the ductility, the nil ductility transition temperature increases, is extremely unfavorable for this kind of steel used for reactor vessel. In order to improve the use of this type of steel, besides the improvement of steel structure, the impurity and gas content in steel decreased and uniform microstructure, especially to control the three elements Cu, Ni, P. The general requirements of the vessel steel should be less than 0.1% Cu. Reduce the content of Ni is also favorable for nuclear vessel steel. In addition, to reduce the contents of P and S. For P, the recommended content is less than 0.012%, S content is less than 0.015%.

The effect of irradiation on graphite. Irradiation induced increase of graphite compressive and flexural strength, thermal conductivity and electrical conductivity decreased. Irradiation can accelerate creep. Graphite in irradiation, changes in anisotropy in size will also. If the irradiation temperature is very high, there will be enough to force the mobile atom displacement back balance position. Effect of irradiation of graphite is mainly produced in the low temperature condition: atomic displacement during irradiation will make the graphite accumulation of large amounts of energy. This energy in graphite heated to 500 °C, and the release of atomic displacement due to reset. When the irradiation flux reaches 10^{20} n/cm^2, this energy can reach 150 J/g. Suddenly released, will burn out the reactor internals. Keep working temperature above 500 °C in graphite (e.g. HTGRs) can avoid the radiation damage effect.

8.5 Thermal Stress

Thermal stresses arise in materials when they are heated or cooled. Thermal stresses effect the operation of facilities, both because of the large components subject to stress and because they are effected by the way in which the plant is operated. This section describes the concerns associated with thermal stress.

Thermal shock (stress) can lead to excessive thermal gradients on materials, which lead to excessive stresses. These stresses can be comprised of tensile stress, which is stress arising from forces acting in opposite directions tending to pull a material apart, and compressive stress, which is stress arising from forces acting in opposite directions tending to push a material together. These stresses, cyclic in nature, can lead to fatigue failure of the materials.

Thermal shock [2] is caused by no uniform heating or cooling of a uniform material, or uniform heating of no uniform materials. Suppose a body is heated and constrained so that it cannot expand. When the temperature of the material increases, the increased activity of the molecules causes them to press against the constraining boundaries, thus setting up thermal stresses.

If the material is not constrained, it expands, and one or more of its dimension increases. The thermal expansion coefficient (α) relates the fractional change in

Table 8.3 Coefficients of linear thermal expansion of common used materials

Material	$\alpha/(1/^\circ C)$
Carbon steel	10.4×10^{-6}
Stainless steel	17.3×10^{-6}
Aluminum	23.9×10^{-6}
Copper	16.7×10^{-6}
Lead	29.3×10^{-6}

length, called thermal strain, to the change in temperature per degree Δt.

$$\Delta L = \alpha \cdot L \cdot \Delta t \tag{8.9}$$

where, α is the linear thermal expansion coefficient, $1/^\circ C$. Δt is the change in temperature, $^\circ C$. Table 8.3 lists the coefficients of linear thermal expansion for several commonly encountered materials [3].

In the simple case where two ends of a material are strictly constrained, the thermal stress can be calculated using Hooke's Law.

$$\varepsilon = \frac{\Delta L}{L} = \alpha \cdot \Delta t \tag{8.10}$$

According to Hooke's law, in the elastic region, the stress and the strain should be proportional to the thermal stress.

$$\sigma = \varepsilon E = E \cdot \alpha \cdot \Delta t \tag{8.11}$$

where, E is the Young's modulus.

Example 8.2 Given a carbon steel bar constrained at both ends, what is the thermal stress when heated from 60–300 °C?

Solution: $\sigma = \varepsilon E = E\alpha\Delta t = (2.1 \times 10^5) \times (10.4 \times 10^{-6}) \times 240 = 524\,\mathrm{MPa}$

Which is higher than the yield point.

Thermal stresses are a major concern in reactor systems due to the magnitude of the stresses involved. With rapid heating (or cooling) of a thick-walled vessel such as the reactor pressure vessel, one part of the wall may try to expand (or contract) while the adjacent section, which has not yet been exposed to the temperature change, tries to restrain it. Thus, both sections are under stress. Figure 8.13 illustrates what takes place.

A vessel is considered to be thick-walled or thin-walled based on comparing the thickness of the vessel wall to the radius of the vessel. If the thickness of the vessel wall is less than about 1 percent of the vessel's radius, it is usually considered a thin-walled vessel. If the thickness of the vessel wall is more than 5%–10% of the vessel's radius, it is considered a thick-walled vessel. Whether a vessel with wall thickness

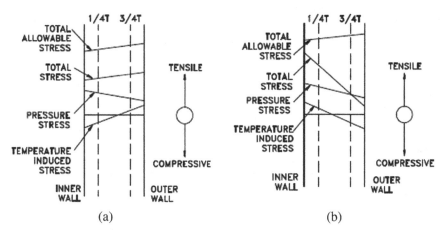

Fig. 8.13 Stress on reactor vessel wall

between 1 and 5% of radius is considered thin-walled or thick-walled depends on the exact design, construction, and application of the vessel.

When cold water enters the vessel, the cold water causes the metal on the inside wall (left side of Fig. 8.13) to cool before the metal on the outside. When the metal on the inside wall cools, it contracts, while the hot metal on the outside wall is still expanded. This sets up a thermal stress, placing the cold side in tensile stress and the hot side in compressive stress, which can cause cracks in the cold side of the wall. These stresses are illustrated in Fig. 8.13.

The heat up and cooldown of the reactor vessel and the addition of makeup water to the reactor coolant system can cause significant temperature changes and thereby induce sizable thermal stresses. Slow controlled heating and cooling of the reactor system and controlled makeup water addition rates are necessary to minimize cyclic thermal stress, thus decreasing the potential for fatigue failure of reactor system components.

Operating procedures are designed to reduce both the magnitude and the frequency of these stresses. Operational limitations include heat up and cooldown rate limits for components, temperature limits for placing systems in operation, and specific temperatures for specific pressures for system operations. These limitations permit material structures to change temperature at a more even rate, minimizing thermal stresses.

Finally, it should also be pointed out, thermal stress and mechanical stress are of great difference. The mechanical stress exceeds the yield limit will suddenly fracture. While the thermal stress exceeds the yield stress is limited. This is because the material once started even just a little bit of yield, yield process will make the thermal stress rapidly decreased. This is the so-called self-limiting thermal stress. So the general plastic containers or parts of material, harm of heat stress than under pressure or other mechanical load is small, the limit of it is relatively wide. However, the brittle materials, because there is no obvious yield process, thermal stress is difficult to ease, when it is easy to reach the limit of the strength of rupture.

8.6   Brittle Fracture

Brittle fracture occurs under specific conditions without warning and can cause major damage to plant materials.

8.6.1   Brittle Fracture Mechanism

Metals can fail by ductile or brittle fracture. Metals that can sustain substantial plastic strain or deformation before fracturing exhibit ductile fracture. Usually a large part of the plastic flow is concentrated near the fracture faces.

Metals that fracture with a relatively small or negligible amount of plastic strain exhibit brittle fracture. Cracks propagate rapidly. Brittle failure results from cleavage (splitting along definite planes). Ductile fracture is better than brittle fracture, because ductile fracture occurs over a period of time, whereas brittle fracture is fast, and can occur (with flaws) at lower stress levels than a ductile fracture. Figure 8.14 shows the basic types of fracture [4].

Brittle cleavage fracture is of the most concern in this module. Brittle cleavage fracture occurs in materials with a high strain-hardening rate and relatively low cleavage strength or great sensitivity to multi-axial stress.

Fig. 8.14 Basic fracture type

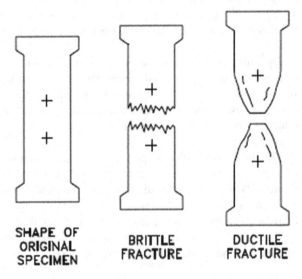

SHAPE OF
ORIGINAL
SPECIMEN

BRITTLE
FRACTURE

DUCTILE
FRACTURE

Many metals that are ductile under some conditions become brittle if the conditions are altered. The effect of temperature on the nature of the fracture is of considerable importance. Many steels exhibit ductile fracture at elevated temperatures and brittle fracture at low temperatures. The temperature above which a material is ductile and below which it is brittle is known as the Nil-Ductility Transition (NDT) temperature. This temperature is not precise, but varies according to prior mechanical and heat treatment and the nature and amounts of impurity elements. It is determined by some form of drop-weight test (for example, the Izod or Charpy tests).

Ductility is an essential requirement for steels used in the construction of reactor vessels; therefore, the NDT temperature is of significance in the operation of these vessels. Small grain size tends to increase ductility and results in a decrease in NDT temperature. Grain size is controlled by heat treatment in the specifications and manufacturing of reactor vessels. The NDT temperature can also be lowered by small additions of selected alloying elements such as nickel and manganese to low-carbon steels.

Of particular importance is the shifting of the NDT temperature to the right (Fig. 8.15), when the reactor vessel is exposed to fast neutrons. The reactor vessel is continuously exposed to fast neutrons that escape from the core. Consequently, during operation, the reactor vessel is subjected to a flux of fast neutrons, and as a result, the NDT temperature increases steadily. It is not likely that the NDT temperature will approach the normal operating temperature of the steel. However, there is a possibility that when the reactor is being shut down or during an abnormal cooldown, the temperature may fall below the NDT value while the internal pressure is still high. The reactor vessel is susceptible to brittle fracture at this point. Therefore, special attention must be given to the effect of neutron irradiation on the NDT temperature of the steels used in fabricating reactor pressure vessels. The Nuclear Regulatory Commission requires that a reactor vessel material surveillance program be conducted in water-cooled power reactors in accordance with ASTM Standards (designation E 185-73).

Pressure vessels are also subject to cyclic stress. Cyclic stress arises from pressure and/or temperature cycles on the metal. Cyclic stress can lead to fatigue failure.

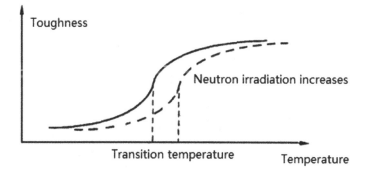

Fig. 8.15 Nil-ductility transition (NDT) temperature

Fatigue failure can be initiated by microscopic cracks and notches and even by grinding and machining marks on the surface. The same (or similar) defects also favor brittle fracture.

The nil ductility transition temperature of materials is not very clear; it depends on material processing, heat treatment and whether it contains impurities and other factors. It is usually obtained with orthogonal experiment method, commonly used such as the Izod and Charpy test.

The finer the grain is, the better the ductility of the material can be, so the reduction of grain size can be reduced by heat treatment, and the addition of nickel, magnesium and other trace elements in low carbon steel is also beneficial to the reduction of NDT [5].

8.6.2   Nil-Ductility Transition Temperature

Brittle fracture is the material has no obvious plastic deformation, the stress far from reaching the material tensile strength sometimes occur suddenly fracture. The nil ductility transition temperature, the material will lose its original possess excellent mechanical properties.

Irradiation on the nil ductility transition temperature effects is very important of; fast neutron irradiation will change of steel lattice structure and the mechanical properties of the steel changes. Usually, radiation will make the Fig. 8.15 curve shifts to the right, i.e., nil ductility transition temperature increased. Although the temperature not to rise to the operating temperature of the system, but if the system suddenly experiences transient process of cooling, once the temperature is lower than the NDT, and pressure of the system is still in operation pressure, it may happen brittle fracture failure. Thus, irradiation on the nil ductility transition temperature is the main factor to determine the service life of the pressure vessel.

Figure 8.16 shows the reactor pressure vessel of the nil ductility transition temperature with integral neutron fluence variation curve (by pressure vessel material irradiation test samples). The initial nil ductility transition temperature is $-27\,°C$, when

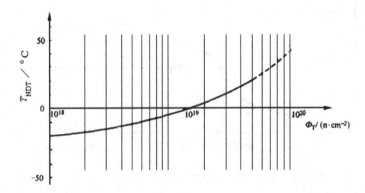

Fig. 8.16 NDT of reactor pressure vessel with integral neutron flux

the integral neutron fluence to 10^{20} n/cm^2, the nil ductility transition temperature can reach 50 °C to the right or to the left. After 25–30 years' operation, NDT may rise to 100–160 °C, and then consider the factors of defects in materials, the allowable temperature may reach 200 °C. That is to say, in this case, the system must rises to 200 °C before pressure lifting.

8.7 Materials in Nuclear Reactor

Reactor material refers to the material used to build reactors, including nuclear fuel, coolant material, moderator materials, construction materials, control material, shielding material; it has formed a system of materials. Reactor materials besides should have the performance of common engineering materials, nuclear physics performance should be good and very good with the reactor environment compatible characteristics.

The reactor materials research mainly in the nuclear physics properties and irradiation effect, chemical compatibility of research, and with a variety of application on the performance of. This study greatly broadens the materials science and technology development and application.

Nuclear fuel: reactor using fissile ^{235}U, ^{233}U (uranium) or ^{239}Pu (plutonium) of metal, alloy material or its oxide, carbide, nitride, and other ceramic materials, ceramic uranium dioxide is the most widely used fuel in light water reactor. In addition, dispersion fuel or liquid fuel is used at some cases.

Coolant material: materials to transport reactor core fission produced heat; there are two types of coolant, gaseous and liquid. Common liquid coolant material with water, heavy water and liquid metal (sodium, potassium sodium alloy, bismuth, lead bismuth alloy). Common gas cooling materials have carbon dioxide (CO_2), air and helium (He).

Moderator material: in the thermal neutron reactor, slow down the fast neutron to thermal neutron, also known as the moderator. Commonly used moderator materials are of two types of solid and liquid. Graphite, beryllium and beryllium oxide solid moderator materials. Commonly used liquid moderator material has light water and heavy water, in addition to organic moderator material. For moderator materials, in addition to excellent nuclear properties is required for the, also called the good engineering properties.

Structural materials: reactor structural materials including core structure materials, fuels (rod) clad material and reactor pressure vessel, driving mechanism of materials. Commercial reactor structural material selection should be considered when the strength, toughness and resistance to corrosion and anti-irradiation embrittlement performance. Nuclear grade high toughness and low synthetic steel, stainless steel, nickel base alloy and other widely used as core structural materials and materials for reactor pressure vessel reactor. Zirconium alloys are widely used in fuel rods clad materials and structural materials of fuel assembly package.

Control material: for manufacturing control rod of the reactivity control, such materials with strong neutron absorption properties such as hafnium, silver indium cadmium alloy, containing boron and rare earth materials of Sm, Er, europium, gadolinium and some oxides and carbides.

Shielding materials: the reactor structure used to reduce various rays, to avoid the staff and equipment suffered radiation damage to the facilities used in the material, mainly lead, iron, heavy concrete, water and other materials.

8.7.1 Nuclear Fuel

The elements that can be used as nuclear fuel are not much, and the nuclear fuel can be divided into three categories: fissile material, fissionable material and fertile material.

Fissile material can occur fission reactions under the action of a variety of energy neutron, such as ^{233}U, ^{235}U, ^{239}Pu these three kinds of nuclear material. The natural existence of fissile material is only one kind of ^{235}U.

Fissionable material is only when the energy is greater than a certain threshold of neutron bombardment of their nuclei, will cause the fission reaction of the radionuclides, such as ^{238}U.

Fertile material is a material in the energy below the threshold energy of fission neutrons are not fission reaction occurs, but after the neutron capture can convert to fissile material, for example, ^{232}Th and ^{238}U is good fertile material. However, such a division is not absolute, because for any atom, when the neutron bombardment with sufficient energy, in principle, will be the occurrence of fission [6].

^{235}U is present in natural uranium as fissile nuclear fuel. In natural uranium, the existence of a large number of is ^{238}U, accounted for approximately 99.28% and ^{235}U content accounts for only approximately 0.714% and the rest of about 0.006% ^{234}U. ^{233}U and ^{239}Pu are obtained in production reactor by artificial methods. They are formed by ^{232}Th and ^{238}U neutron capture.

Depending on the physical state of nuclear fuel, basic characteristics and design approach, it can be roughly divided into solid fuel, liquid fuel and dispersion fuel (see Table 8.4). A typical structure of solid fuel is cladding materials will fuel encapsulated into fuel elements. Cladding can prevent the corrosion of the fuel by the coolant, also can stop the release of fission products from fuel pellets.

Liquid fuel more in the form of a soluble from fuel, coolant and moderator together, and can be divided into aqueous solution, suspension liquid, liquid metal and molten salt. However, due to the irradiation resistant of liquid fuel materials is not stable, fuel processing is difficult, so there is no industrial application at present. Original design thought of dispersion fuel is in order to improve the heat transfer efficiency of the fuel element, and some also have fuel and graphite as a moderator together, is a promising form of fuel.

Table 8.4 Nuclear fuel classification

Fuel form	Form	Material	Suitable for reactor type
Solid	Metal	U	Graphite moderated reactor
	Alloy	U-Al	Faster reactor
		U-Mo	Faster reactor
		U-ZrH	Pulsed reactor
	Ceramics	U_3Si	Heavy water reactor
		$(U, Pu)O_2$	Faster reactor
		$(U, Pu)C$	Faster reactor
		$(U, Pu)N$	Faster reactor
		UO_2	Light water reactor, heavy water reactor
Dispersion	Metal–Metal	UAl_4-Al	Heavy water reactor
	Ceramics–Metal	UO_2-Al	Heavy water reactor
	Ceramics–Ceramics	$(U, Th)O_2$–(Pyrolysis graphite, SiC)–Graphite	High temperature gas cooled reactor
	Metal–Ceramics	$(U, Th)C_2$-(graphite, SiC)–Graphite, UO_2-W	High temperature gas cooled reactor
Liquid	Aqueous solution	$(UO_2)SO_4$-H_2O	Boiling water reactor
	Suspension	U_3O_8-H_2O	Uniform water reactor
	Liquid metal	U-Bi	
	Molten salt	UF_4-LiF-BeF_2-ZrF_4	Molten salt reactor

Nuclear fuel choice should consider above all is to neutron fission cross section, fission cross section, the bigger the better; followed by the need to consider is fuel density, usually want to fuel density to a large number; should also be considered, consisting of fuel element materials are easy to obtain, manufacturing, processing and post processing is difficult or not, and corrosion resistance, high temperature resistance, resistance to radiation performance etc.. Considering these factors, the current commercial nuclear power plants most of the compounds in the form of ceramic fuel, is by far the most widely used is UO_2. Here we discuss the physical properties of UO_2.

Density of UO_2

First, look at the density of theory. The so-called theory of density is 10.96 g/cm^3. However actually produced UO_2 pellets is produced by sintering of UO_2 powder. The theoretical density is obtained by the lattice constants of UO_2. Due to the manufacturing process caused by the inevitable presence of voids, computation in general take 95% of theoretical density value.

$$\rho = 95\% \, \rho_0 = 10.41 \text{ g/cm}^3 \tag{8.12}$$

Melting Point of UO₂

The melting point of UO_2 changes with O/U radio and impurities. As a result of UO_2 in high temperature will release oxygen, making O/U radio change in the heating process, so the melting point of UO_2 is difficult to determine. It is because of this reason, different researchers measured the melting point of each are not identical, but the majority around 2800 °C, some researchers measured UO_2 without irradiated $(2840 \pm 20\ °C)$, $(2860 \pm 30\ °C)$, $(2800 \pm 100\ °C)$, $(2760 \pm 30\ °C)$, $(2860 \pm 45\ °C)$, $(2865 \pm 15\ °C)$, $(2800 \pm 15\ °C)$. Usually engineering using non irradiated UO_2 2800 $\pm$ 15 °C as melting point.

After fuel pin were irradiated, along with the change of the ratio of solid fission product accumulation and O/U radio, the melting point of the fuel will decline. Usually the energy per unit mass of fuel from said for burnup, the unit is J/kg, engineering habits to mount the energy of per ton of uranium as fuel consumption depth units, i.e., MW·d/t(U). According to the continuous accumulation of reactor operating experience, burning consumption depth for each additional 10^4 MW·d/t(U) and the melting point decreased about 32°. For example, burnup of 50,000 MW·d/t(U) fuel, melting point for 2800–5 × 32 = 2640 °C.

Thermal Conductivity of UO₂

Thermal conductivity of UO_2 is very important in the calculation of heat transfer of fuel element, because the thermal conductivity will directly affect the temperature distribution and the maximum temperature in the fuel.

Figure 8.17 shows the relationship between the thermal conductivity and the temperature of UO_2 of 95% theoretical density. It can be seen that the thermal conductivity is minimum at about 1800 °C. The relationship can be expressed as

$$k_{95} = \frac{3824}{t + 402.4} + 6.1256 \times 10^{-11}(t + 273)^3 \tag{8.13}$$

where, the unit of k is W/(m °C), and the unit of t is °C.

The thermal conductivity under other density can use Maxwell-Euken formulas:

$$k_\varepsilon = \frac{1 - \varepsilon}{1 + \beta\varepsilon}k_{100} \tag{8.14}$$

where, ε is fuel porosity (volume fraction); β is a constant determined by the experiment, for greater than or equal to 90% of the theoretical density of UO_2, beta = 0.5, other density beta = 0.7.

Integral Thermal Conductivity

Thermal conductivity generally varies with temperature, and this function is often nonlinear, such as UO_2 fuel pin not only small, but also its value with the fuel temperature change is larger (see Fig. 8.17). In this situation, if the average temperature used to calculate thermal conductivity of fuel pellet, will produce certain error; however, if the thermal conductivity of temperature function is directly used to solve the problem,

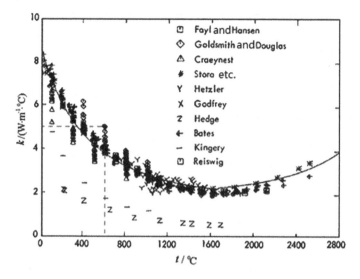

Fig. 8.17 UO_2 relationship between thermal conductivity and temperature

it will be very complicated. So the introduction of integral thermal conductivity is presented here. In dealing with the temperature t_2 to t_1 heat conduction problem, only integral thermal conductivity in the temperature range of the average value is used as the material of the average thermal conductivity.

$$\bar{k} = \frac{\int_{t_1}^{t_2} k(t)\mathrm{d}t}{t_2 - t_1} \qquad (8.15)$$

Specific Heat Capacity of UO_2

Specific heat capacity of UO_2 is a function of temperature, as shown in Fig. 8.18. It can be expressed as following equations.

When 25 °C < t < 1226 °C:

$$c_p = 304.38 + 0.0251t - 6 \times 10^6(t + 273.15)^{-2} \qquad (8.16a)$$

When 1226 °C ≤ t < 2800 °C:

$$c_p = -712.25 + 2.789t - 0.00271t^2 + 1.12 \times 10^{-6}t^3 - 1.59 \times 10^{-10}t^4 \quad (8.16b)$$

where, t is temperature in °C, c_p is specific heat capacity in J/(kg·°C).

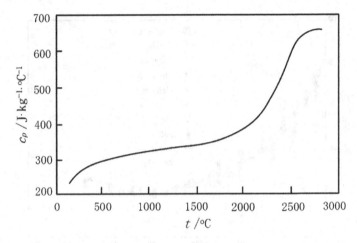

Fig. 8.18 Relationship of specific heat capacity of UO_2 with temperature

Expansion Coefficient of UO_2

In analyzing the behavior of nuclear fuel in reactor, expansion coefficient is an important character. Although the results of the test are not very consistent, but in less than 1000 °C the expansion coefficient is about 1×10^{-5} °C^{-1}. In more than 1000 °C, 13×10^{-5} °C^{-1}. Because of UO_2 over 2450 °C significantly evaporation, so high temperature expansion coefficient is just qualitative.

Mechanical Properties of UO_2

The mechanical properties [7] of UO_2 are found in Table 8.5.

Plutonium

Plutonium has 15 kinds of isotopes (^{232}Pu–^{246}Pu), the most important is ^{239}Pu and ^{238}Pu. ^{239}Pu half-life of the 2.4 million years, is fissile nuclide. Therefore, ^{239}Pu is a kind of nuclear material or nuclear weapons material. ^{238}Pu half-life of 86.4 years with high decay heat can be as power sources for space device and a pacemaker [8].

In raw uranium ore, plutonium contents only about 10^{-11}, without mining extraction value. Therefore, plutonium is by the artificial manufacture. Production reactor can produce plutonium containing less ^{240}Pu, called military plutonium, as material for nuclear weapon. Nuclear power plant also produces plutonium but with higher

Table 8.5 Mechanical properties of UO_2

The longitudinal elastic modulus/MPa	Shear modulus/MPa	Compression modulus/MPa	Flexural strength/MPa	Compressive strength/MPa	Poisson's ratio μ
1.75×10^5	0.75×10^5	1.645×10^5	98.0–112.0	420–980	0.303

content of ^{240}Pu and ^{241}Pu, which is called as industrial plutonium. Industrial plutonium can be used as fuel of nuclear power plant. The world's major industrial countries have been extracted a large number of industrial plutonium from production reactor and power reactor spent fuel. Plutonium will be a main source of fuel for fast neutron breeder reactor in the future.

There are important differences between industrial and military plutonium, mainly ^{240}Pu content is not the same. Once more than 7% of ^{240}Pu plutonium, it cannot be used in the manufacture of nuclear weapons. This is because ^{240}Pu will lead to the spontaneous fission, this will makes neutron background too high, and Plutonium is more likely to be split by neutron. So when containing ^{240}Pu to much, high compression is not sufficient in the initiation of explosion of atom bomb and will cause a large number of premature chain reaction, thus destroying the symmetry of compression, so that the overall incomplete combustion, the consequence is greatly reduced the explosion, or even failure, become a dirty bomb. Therefore, the plutonium for nuclear weapons should reduce the abundance of ^{240}Pu, less than 7%. If it is used in thermonuclear weapons, the requirement is even higher. Therefore, in civil nuclear facilities supervision of the International Atomic Energy Agency, supervision for residence time in nuclear reactor core for nuclear fuel is an extremely important. If a nuclear fuel assembly stays in a nuclear reactor core too short time, the content of ^{240}Pu will less than 7%. Moreover, after the content of ^{240}Pu once more than 7%, it is quite difficult to lower it and even impossible, because the isotope separation of ^{239}Pu and ^{240}Pu is much more difficult than the separation of ^{235}U and ^{238}U.

Plutonium is a silvery white metal with melting point 913 K, boiling point 3503 K. The critical mass of solid plutonium is small, only a few kilograms, which limits the amount of post processing of each operation process. Plutonium is highly toxic substances; it is a short range of alpha emitters. If plutonium is inhaled in to lung, it will be integrated into lung, skeletal and hematopoietic tissue, causing these organ and tissue damage. Plutonium is easy to form aerosol in the air of and inhalation is danger. High mass plutonium isotopes also have spontaneous fission, gamma ray and X ray radiation. Therefore, the operation of plutonium are required in vacuum or necessary protective gas charging in the glove box. When plutonium mixed with strong gamma emitters, remote operation is required with a shielding thick wall.

^{239}Pu and ^{241}Pu are fissile nuclides; ^{240}Pu is fissiable only in the fast neutron reactor. When ^{240}Pu absorbs neutrons can be converted into ^{241}Pu. ^{242}Pu are generally not involved in nuclear fission. The average number of neutron release per action of ^{239}Pu is dependent on neutron energy, for fast neutron, the value is 2.9, for thermal neutron is 2.07. Therefore, in the fast reactor, in addition to the neutron is required to sustain a chain reaction, remaining more neutrons will be converted from ^{238} to ^{239}Pu. Therefore, the value of plutonium in fast reactor is about 1.4 times of thermal reactor. At the same time, the use of plutonium can greatly improve the utilization of natural uranium resources rate. For example in fast breeder reactor nuclear power plant, the utilization of natural uranium from less than 1% to 60%–70%; in PWR also can be increased to 1%– 2%.

8.7.2 Structure Materials

The requirement of structure material changes with the types of reactors in some degree. In addition to the mechanical properties such as yield strength and stiffness must be able to meet the conditions for running, thermal conductivity should generally be relatively high, and heat expansion coefficient should be smaller or be compatible with other materials, must also be able to withstand heat stress. In addition, corrosion resistance is also very important. In addition to the physical and mechanical properties of, neutron properties of the inner core structure material must also be considered, apparently hoped to have a smaller neutron absorption cross section.

Because of neutron capture, many materials will become radioactive. As a result, neutron irradiation equipment in maintenance and repair become a difficult problem. If possible, in the choice of materials should choose those materials that without radioactive substance after neutron capture; or to choose radioactive is weak, not emit strong is γ ray, or with a short half-life substances.

Table 8.6 lists a variety of elements may exist in the structure of materials, including may exist as alloying elements and impurities elements [9]. The table lists the abundance in natural products and its neutron absorption cross section data, also lists the radionuclides produced, and their half-life and the most penetrating γ ray (the highest gamma ray energy). It should be pointed out that, in the data table only the longer half-life (greater than 1 h) of gamma ray (>0.3 meV) is selectively listed. Material with only beta particle radiation is not included, because they generally do not constitute serious maintenance problems, if to avoid the bremsstrahlung.

Table 8.6 Presence of metal induced radioactivity in structural material

Element	Mass number of isotope	Natural abundance/%	σ_a/bar	Radionuclide	Half-life	γ pay energy/MeV
Titanium	50	5.3	0.04	Ti-51	72d	1.0
Chromium	50	4.4	16	Cr-51	27h	0.32
Manganese	55	100	13.3	Mn-58	2.6h	2.1
Iron	58	0.33	0.8	Fe-59	46d	1.3
Cobalt	59	100	37	Co-60	5.3y	1.3
Nickel	64	1.9	3.0	Ni-65	2.5h	0.93
Copper	63	69	4.3	Cu-64	12.8h	1.35
Zinc	64	48.9	0.5	Zn-65	250d	1.12
	68	18.5	0.1	Zn-69	13.8h	0.4
Zirconium	94	17.4	0.1	Zr-95	65d	0.92
Molybdenum	98	23.8	0.13	Mo-99	67h	0.84
Tantalum	181	100	21.3	Ta-182	113d	1.2
Tungsten	186	28.4	34	W-187	24h	0.76

You can find a lasting sense radioactivity elements in main is chromium, manganese, cobalt, copper, zinc, tantalum and tungsten. These elements in a pure form will never be used as materials in reactor core. However, these metals may become other structural materials of impurity or alloying components, so we must pay much attention to them by neutron irradiation induced radioactivity.

For thermal neutron reactor materials, it is a prerequisite that material should have a fairly small thermal neutron absorption cross section. If the section were too large, this material even with relatively good mechanical properties or other benefits would not be chosen. For a middle neutron absorption cross section of the element, if there are other advantages can be compensated, may also be adopted. Because almost all the elements of the fast neutron absorption cross section is small, so the choice of fast neutron reactor material constraints is much less.

Metal elements can be roughly divided into three categories according to the value of the thermal neutron absorption cross section:

1. Elements of a small (<1 bar) thermal neutron absorption cross section;
2. Elements of a medium (1–10 bar) thermal neutron absorption cross section;
3. More than 10 bar elements.

In the first or second category, soft metal or melting point below 500 °C metal can be omitted, because they do not seem to have much use for structure material. The rest elements with easy to get are listed in Table 8.7.

You can see a much smaller thermal neutron cross section of metal aluminum often in a pure (>99.0%) form used as cladding material inside reactor, fuel element materials and other not to expose the material at high temperatures. Although magnesium has a section of smaller than aluminum, but it cannot compensate the high price and processing difficulty. For beryllium, in a certain extent also with magnesium as, although the metal may in some special types of reactors used as structural materials at the same time can also be used as moderator and reflector materials at the same time.

Zirconium alloy is one of the most promising reactor structure and cladding materials, especially in the high pressure and high temperature water thermal neutron

Table 8.7 Thermal neutron absorption cross-sections and melting point of elements

Low thermal neutron cross section			Medium thermal neutron cross section		
Metal	σ_A/bar	Melting point/°C	Metal	σ_a/bar	Melting point/°C
Beryllium	0.009	1280	Niobium	1.1	
Magnesium	0.069	650	Iron	2.4	2415
Zirconium	0.18	1845	Molybdenum	2.4	1539
Aluminum	0.22	660	Chromium	2.9	2625
			Copper	3.6	1083
			Nickel	4.5	1455
			Vanadium	5.1	1900
			Titanium	5.6	1670

reactors. Zirconium in addition to the advantages of thermal neutron capture cross section is very small, also has excellent mechanical properties and processing properties with good corrosion resistance. But not for its high melting point occurs to some misunderstanding, because it is about 400 °C mechanical strength will decrease, and become susceptible to corrosion in water.

Looking at medium thermal neutron cross section element, one can find it is still very limited in the choice of materials. Niobium has good mechanical properties, but it is very rare, also in more than 200 °C will be serious oxidation. Chromium and molybdenum in the processing difficulty and vanadium expensive, making these elements can only as alloy adding elements. In addition, iron, nickel, copper after adding other alloying elements will be greatly improved, rarely in the form of non-alloy alone.

Titanium caused great attention. The main advantage of titanium lies in the fact that it has a high strength to weight ratio at the range of 100–450 °C, which makes it the most suitable for the need to be light. In some circumstances, the corrosion resistance of titanium is superior, and it is used in high temperature water situation. Titanium's high cross section will limit it in the reactor core in use, but its corrosion resistance may make it be used in the reactor neutron absorption does not become a key problem. For example, used to transport water solution or chemical processing equipment.

In the most familiar alloy materials, all kinds of austenitic stainless steel has good corrosion resistance and good mechanical properties. The factors to be considered in reactor are similar to other chemical and high temperature applications. These include the following considerations: (1) after heat treatment, the possibility of intergranular corrosion occurs in a certain environment; (2) the SCC crack; (3) Brittle sigma phase formation, especially in the deposited layer of high iron content welding rod; (4) knife lines corrosion; (5) thermal stress caused by low thermal conductivity; (6) local accelerated erosion due to anticorrosive film peeling by the rapid flow of strong corrosive solution.

Table 8.1 lists some composition of stainless steel possible for reactor material. Nickel alloys, such as Nichrone, Monel metal, Inconel and hastelloys, etc., is also important, especially because they are quite suitable for molten salt and alkali metal hydroxide.

As far as we can see, zirconium is a very good structure and fuel cladding material in a thermal neutron reactor, so here is a discussion of some of its important properties.

Main mineral of zirconium is silicate zircon ($ZrSiO_4$). Zirconium is not rare elements; it ranks seventh in the common elements within the crust of the earth. Therefore, it is richer than copper, lead and zinc. However, from the point of view of nuclear engineering, one important fact is that zirconium ore contains hafnium about 0.5% to 3.0%, while the latter has a high thermal neutron absorption cross section (115 bar). Therefore, zirconium in the thermal reactor applications must think of a way to remove most of the hafnium. This is not a simple thing, because the chemical properties of hafnium and zirconium are very similar.

Zirconium is a silvery white metal, between room temperature and the melting point (1845 °C) are two allotropes: One is FCC lattice, has been stable to 863 °C;

Another is BCC lattice at higher temperature. Zirconium physical properties, such as thermal expansion and thermal conductivity, etc., change with the way to deal with. Under normal temperature, the line expansion coefficient along the hexagonal axis direction is $1.03 \times 10^{-5}/°C$ and in the vertical direction is $4.5 \times 10^{-6}/°C$. Thermal conductivity decreases from 20.9 W/(m°C) at 25 °C to 18.8 W/(m°C) at 300 °C.

In temperature up to 400 °C, a pure zirconium and its alloy (especially containing a small amount of tin) has a very high resistance to corrosion of the air and water (or steam). The metal surface will generate a layer of thin, stable, adhesion of the oxide film, to protect it from further erosion. In more than 400 °C, it is relatively action with oxygen and water even with nitrogen and lost most of its mechanical strength, so for zirconium as structure material may only be limited to 400 °C below.

Zirconium and many other metals can form alloys, and some of the alloys are more strength than pure zirconium, and are more resistant to corrosion.

The mechanical properties of zirconium by metal treatment and the presence of trace impurities influence Table 8.8 in some mechanical properties of zirconium, which can be seen when the temperature increased, tensile strength and yield strength drop. If with other elements such as aluminum, tin, molybdenum and niobium alloy, the situation can be improved.

The mechanical properties of zirconium are influenced greatly by metal treatment and the presence of impurities. Table 8.8 shows some mechanical properties of zirconium. It can be seen that the tensile strength and the yield strength decrease when the temperature rises. If the alloy is made with other elements such as aluminum, tin, molybdenum and niobium, the situation can be improved.

It is worth noting that a small amount of oxygen and nitrogen will have a significant impact on the mechanical properties of zirconium metals. For example: 0.1% weight of oxygen can increase the tensile strength to about 286 MPa, and make the yield strength of 0.2% offset increased to 177 MPa, while the hardness increased to A 38. However, the elongation decreases to about 14% at the same time. Thus, the extension of this metal becomes worse. The hardening effect of oxygen and nitrogen (and carbon) decreases with the increase of temperature, and the effect is too small to be ignored at 400 °C, so these elements do not improve the properties of Zr at elevated temperatures.

Zirconium metal is easy to be hydrogen embrittlement. At room temperature, ten ppm hydrogen content is enough to cause the Charpy V-score values significantly

Table 8.8 Mechanical properties of zirconium

Properties	At 20 °C	At 210 °C	At 320 °C
Tensile strength/MPa	218	136	109
Yield strength (0.2% deviation)/MPa	122	68	54
Elongation/%	25	50	60
Elastic modulus/10^3 MPa	95	82	75
Poisson ratio	0.35		
Hardness/Rockwell A	30		

reduced. However, at a high temperature of 300 °C or so this effect is too small to be ignored. Zirconium may absorb hydrogen in the process of production and subsequent treatment, but it can be removed by annealing at 800–900 °C in good vacuum.

Zirconium processing can be used for all ordinary processing methods, such as cutting, hot rolling, cold rolling, forging, extrusion and drawing, as long as the reasonable measures to prevent oxidation. For the last part of the forming process by cutting, and even the above measures cannot be required. Large ingots can be hot forging and rolling with ordinary industrial rolling equipment, without air isolation. At this time, a little surface contamination can be dealt with the method of removing the sand blasting or impregnation in stainless steel. On the other hand, if it is necessary to process the parts with the exact size, it should be protected with rare gases or steel sleeve. In the extrusion process, steel sleeve is also required for zirconium or will be scratched by hard mold. Another way is to use high temperature glass film. It can protect and smooth zirconium.

Because the heat treatment method has the advantages of easy to reduce the size of the zirconium ingot, generally forging and then hot rolling as the first process. After cleaning the surface, the metal is easily cooled again, and then tempered in a vacuum or rare gas to avoid the erosion and contamination of oxygen and nitrogen in the air. After tempering, the cooling of the zirconium is very ductilitable (assuming that oxygen, nitrogen and carbon are removed), and that it is easy to cut.

Zirconium metal has superior ability to be welded, but this operation must be carried out in high purity rare gas (such as helium or argon) to avoid contamination. The molten zirconium has the ability to dissolve the oxide and nitride surface films, which is an important factor in producing highly solid joints. If we take appropriate protection measures, the common method of welding and brazing can be used.

One of the outstanding properties of zirconium is that it can resist the corrosion of most acid, alkali and salt-water solution under all reasonable conditions. Solutions can corrode this metal are hydrofluoric acid, concentrated hydrochloric acid, phosphoric acid, ferric chloride and copper chloride.

From the viewpoint of reactor engineering, it is very important to resist the corrosion and erosion of water at high temperature. The corrosion properties of zirconium in water under high temperature are very sensitive to the presence of impurities. Among all possible impurity elements, nitrogen, carbon, aluminum and titanium are particularly harmful. Other harmful elements are oxygen, magnesium, silicon and calcium, but the presence of these elements is usually a very small amount, it will not produce any effect. Copper, tungsten, iron, chromium and nickel at low concentrations (up to about 0.1%) seem to have little effect on the corrosion of zirconium in high temperature water.

Due to the high purity of the production of zirconium, metal prices are very expensive, have tried to use the method to add a variety of alloy elements to improve the corrosion resistance of industrial products (i.e., arc melting of zirconium). On the one hand, it is very promising to contain tin 1.5%, especially if there are trace amounts of iron, nickel or chromium. These alloys have good mechanical properties at high temperature.

Zirconium is also important in the fight against the erosion of some liquid metals, as the latter may serve as a reactor coolant. If the oxygen content is very low, it is suitable for sodium and sodium potassium alloy at the temperature below 600 °C. According to the information reported, zirconium for lithium, lead bismuth alloy and lead bismuth tin alloy resistance at 300 °C is also very good, but to 600 °C, becomes worse.

8.7.3 Coolant

Early reactors were primarily intended for research or production of plutonium, and therefore did not want to cause the reactor core temperature to rise too high because of the fission reactions. Therefore, the fluid flows through the reactor core to circulate, and the heat is removed, and the fluid is called a coolant. This name has been in use ever since. However, for today's reactors, the role of the coolant is to transport heat from reactor core to the place where it is used (heat exchanger or turbine). It cools the reactor fuel rods, and transport the heat from chain fission reaction to the outside of the reactor. In this sense, the more accurate name should be the heat transfer agent. However, due to the name of the coolant has been widely used, we use the word coolant.

A nuclear reactor must have an appropriate cooling system to prevent the reactor from reaching an elevated temperature in the form of heat, as most of the energy produced by fission occurs. The temperature may depending on the properties of the fuel, such as uranium metal at 662 °C occurred allotropic phase change and volume decrease of considerable; Or on to the coolant, for example in pressurized water reactor do not want water boiling occurs. The medium used for cooling can be a gas or liquid, and the specific requirements for this material are mainly determined by the heat release rate (i.e., the power density of the reactor) and the operating temperature. The cooling system can be designed very simply in the reactor, which is low in power output, and the heat is not needed to be used. On the other hand, the reactor for economic power production must be operated at high temperature, which requires careful consideration of the choice of coolant.

The main technical requirements of reactor coolant are: ① has good thermal physical properties (high specific heat capacity, high density, high thermal conductivity, low melting point, high boiling point) to reduce the heat transfer area needed, from the core to bring out more heat; ② thermal neutron absorption cross section is small, especially for thermal neutron reactors, radioactivity is weak; ③ low viscosity, in order to make the power consumption of reactor coolant pump small; ④ has good thermal stability and irradiation stability in reactor; ⑤ nuclear fuels and structural materials have good biocompatibility; ⑥ cheap and easy to obtain. The coolant material commonly used in thermal neutron reactor are light water, heavy water, carbon dioxide and helium. Liquid metals such as sodium or lead bismuth alloys are commonly used in fast neutron reactors. In practical applications, the choice of coolant used in any particular case is a compromise between these conflicting requirements.

Since the primary purpose of coolant is to transport heat, it must have good heat transfer properties. In a certain flow geometry and dynamic conditions, to meet the above requirements, thermal conductivity and high specific heat are very important. In a nonmetallic liquid, low viscosity is also good for this aspect. The coolant flows through the reactor and heat exchanger, the pumping power required should be small. This requires a high density and low viscosity of the coolant. In order to be used at high temperatures, the vapor pressure of the liquid coolant should be low so that it can be used without a strong pressure bearing system. At the same time, the coolant should remain in the liquid state at low temperature, in order to avoid solidification when the reactor shutdown. These two kinds of requirements in a certain extent is contradictory, the former requires a high boiling point and the latter has a low melting point. However, it has been known that there are a lot of material such as liquid metal and molten salt, and the liquid phase temperature range is very wide. Naturally, these problems will not occur when using gas as a coolant, such as a high temperature gas cooled reactor.

The coolant should have stability against high temperature and nuclear radiation. At the same time, these fluids must not erode the various structural materials inside and outside the reactor. Because the coolant system is a dynamic system, the phenomenon of material transfer caused by temperature difference may become very important. As far as physical and chemical erosion is concerned, consideration should be given to the selection of coolant and structural materials.

Because the neutron loss must be compensated by the addition of the reactor fuel, the coolant should be composed of one or more elements with small neutron capture cross section. This restriction is not important for fast neutron reactors, because the high energy neutron capture cross sections of various substances are always very small. The induced radioactivity of coolant should very weak and without γ ray, otherwise all pumps, heat exchangers and pipes need shielding. In any case, the coolant is required to be non-toxic and has no other dangers, and it will not be dangerous after exposure to radiation. Of course, the coolant should be made at a low price, and very expensive materials will only be considered in exceptional circumstances.

In thermal neutron reactor, it is also hoped that the coolant contains elements with a lower atomic weight. In other words, if also can be used as a coolant moderator, is more favorable. On the other hand, in the fast neutron reactor, it is necessary to avoid slowing down, so low atomic weight materials should be avoided as far as possible.

There is no single material (or mixture) to meet all the requirements of the ideal reactor coolant. Table 8.9 lists the properties of various materials now or in the future to be used in reactor coolant. Because each coolant must be used within a certain range of temperature, the two temperature values are given in the table as much as possible.

Gas is generally radiation and thermal stability, easy to operate and is not dangerous, it is worth to be recommended as a coolant. Since air is the easiest to

Table 8.9 Physical properties of coolant material

Coolant	Melting point /°C	Boiling point/°C	t/°C	Density/(g/cm³)	Specific heat capacity/(kJ/kg)	Thermal conductivity/(W/cm °C)	σ_a/bar	Induced radioactivity
H (1 atm)	–	Gas	100 300	6.6×10^{-5} 4.3×10^{-5}	14.34 14.63	22.28×10^{-4} 30.76×10^{-4}	0.33	None
He (1 atm)	–	Gas	0 100	1.8×10^{-4} 1.4×10^{-4}	5.23 5.23	13.79×10^{-4} 16.72×10^{-4}	20	None
Air (1 atm)	–	Gas	100 300	9.5×10^{-4} 6.2×10^{-4}	1.00 1.01	31.64×10^{-5} 4.56×10^{-4}	1.5	^{16}N, 7.3 s, 6 meV ^{41}Ar, 1.8 h, 1.4 meV
CO$_2$ (1 atm)	–	Gas	100 300	1.5×10^{-3} 9.5×10^{-4}	0.91 0.97	20.90×10^{-5} 37.62×10^{-5}	0.003	^{16}N, 7.3 s, 6 meV
H$_2$O	0	Pressure dependent	100 250	0.958 0.794	4.22 4.60	7.11×10^{-3} 8.19×10^{-3}	0.22	^{16}N, same as above
Li	179	1317	200 600	0.507 0.474	4.18 4.18	0.38 0.38	0.033	^{8}Li,0.83 s,13 meV bremsstrahlung
Na	98	883	100 400	0.928 0.854	1.38 1.28	0.86 0.71	0.50	^{24}Na, 1.5 h, 1.38 meV, 2.75 meV
Na 22% K 78%	–11	784	100 400	0.774 0.703	0.93 0.88	0.24 0.27	1.7	^{24}Na(same as above), ^{42}K, 12.4 h, 1.5 meV
Pb 44.5% Bi 55.5%	125	1670	200 600	10.46 9.91	0.15 0.15	0.10 0.15	0.1	^{210}Bi, 5d, decay into poisonous ^{210}Po

(continued)

Table 8.9 (continued)

Coolant	Melting point /°C	Boiling point /°C	t/°C	Density/(g/cm³)	Specific heat capacity/(kJ/kg)	Thermal conductivity/(W/cm °C)	σ_a/bar	Induced radioactivity
Pb 97.5% Mg 2.5%	250	~ 1700	300	~ 10	0.15	0.13	0.16	^{27}Mg, 10 min, 0.84 meV, 1.0 meV
NaOH	323	–	350	1.79	2.32	21.74×10^{-3}	0.28	^{24}Na, 1.5 h, 1.38 meV, 2.75 meV
			550	1.67	1.81	11.70×10^{-3}		^{16}N, 7.3 s, 6 meV

achieve, it is certainly the first to be chosen. However, air is not a good heat transmission material, and the general gas, pumping air through the cooling system need high power fan, consume large amounts of electricity. In addition, under the high temperature oxygen (in some cases nitrogen) are prone to erosion such as graphite or beryllium moderator and structural materials and fuel cladding materials. Air is also used as a coolant in some natural uranium graphite reactors, but these are generally large in size and operate at very low power densities. They are very low heat flux, so to keep the fuel temperature at a reasonable level does not require much of air. In this kind of research reactor, the disadvantages of low efficiency of air as coolant can be compensated by the simple design. Some plutonium production reactor in the UK also uses air as coolant in the reactor.

From the point of view of losing heat and pumping power, hydrogen appears to be one of the best gaseous coolant. However, if leakage into air, would cause a serious danger of explosion. In addition, it is also contaminated, especially under high pressure, which requires the use of special materials without hydrogen erosion or embrittlement. Therefore, no one has been suggested to use hydrogen as reactor coolant.

In several ways, helium is worth recommending as a gaseous coolant. Although its thermal transport properties are not as good as that of hydrogen, its thermal neutron capture cross section is small and can be ignored, and it is not dangerous. It is also very stable for heat and radiation. In addition, it is inert in the chemical properties, so it has no effect on the structure and vessel material. Because the price of helium is a little bit high, it is necessary to use a closed system. However, in order to save power by pumping, high pressure is usually needed. There will have problems to prevent leakage.

Historically, it has been suggested that the use of nitrogen as a coolant in the production of plutonium and power reactor at 1 MPa pressure. Nitrogen not only has large neutron absorption cross section, also due to the (n, γ) reaction which will generate a radioactive ^{14}C, if ^{14}C escape will lead to a danger.

The use of carbon dioxide as a coolant has been under consideration for safety. Carbon and oxygen contained in carbon dioxide have very small neutron absorption cross sections, and the gas itself is not toxic or explosive. Carbon dioxide is not easy to erode metal, but it will react with graphite at a very high temperature. In the aspects of heat transmission, pumping power and neutron absorption slightly worse than nitrogen, but of course it is easier to get. Some of the factors that cannot be sure of the use of carbon dioxide as a coolant are the effects of the strong radiation on its stability, as well as the reaction it has with graphite and other materials.

From a cost point of view, the use of water as a coolant is only second to air. Although the thermal conductivity of water is much smaller than that of liquid metal, it has a high specific heat, and its viscosity is very low and therefore easy to pumping. Of course, the biggest advantage is that it can simultaneously as moderator and coolant.

However, there are some disadvantages of using water as a coolant. Some of the more important: (1) neutron absorption cross section is quite large, (2) water will be decomposed by irradiation, (3) corrosion of metal, (4) under normal pressure boiling point is low. (5) The resulting ^{16}N by the ^{16}O fast neutron (n, p) reaction and impurities (such as sodium) captured and the resulting radioactive.

A slightly enriched ^{235}U fuel can be used to compensate for the loss of neutrons. To reduce the degree of irradiation decomposition, the removal of ionic impurities can be as far as possible, and the addition of hydrogen to control, which is described in the Chap. 7 of the reactor water chemistry section.

The corrosion of water at moderate temperature has been able to get enough information, so it can make a satisfactory cooling system design. If the temperature is not high, can be made of aluminum pipe; and at a higher temperature, the use of stainless steel. In order to reduce the pipeline corrosion and scale deposit, we have to use the desalting or distillation processing to make it clean. In some cases, it is necessary to remove gas.

In order to produce electricity efficiently, the reactor must operate at temperatures much above the boiling point of water at atmospheric pressure. In this case, for example in a pressurized water reactor (PWR), high pressure must be used, which means that the reactor must be enclosed in a pressure vessel. In addition, at high temperatures even without the ionic impurities and oxygen, pure water becomes very corrosive. Because of corrosion problems and the need for high pressure, the choice of materials and equipment for power reactor has many restrictions. The necessary strength and corrosion resistance properties of the material are some stainless steel, titanium (or its alloy), Inconel alloy and Monel K alloy. In addition, zirconium can also be used in many places.

Although boiling water (or other liquid) has long been found to be an effective coolant, the latent heat of vaporization can be used. However, at the beginning of the development of nuclear energy, many people think that the formation of bubbles in the reactor may cause instability and thus lead to poor power control. However, the design of the boiling water reactor has been proved that the bubble generation does not increase the power, but can reduce the power. Therefore, the boiling water reactor is a good type of reactor. However, the power reduction is mainly due to the resonance escape probability decreases (in Chap. 11 for details) and moderator is partial discharge caused by neutron leakage.

Because of the self-regulating effect of steam bubble formation, a reactor can run continuously in a stable form under boiling conditions. If the water is boiling more quickly than normal, then many bubbles generated will decreases the reactivity, and fission-generated energy will be reduced. This tends to bring the boiling rate back to the desired value. Great advantages of this form of boiling water reactor is that it can be no steam generator and direct steam generation, so lower reactor temperature and pressure can get the same power production efficiency, but also greatly reduce the pumping power.

As the reactor coolant and moderator, the only meaningful difference of ordinary water and heavy water is the thermal neutron absorption cross section. This means that the fissile material needed in heavy water reactor to make the reactor critical may be less, so it is possible to use natural uranium as fuel. For example, a CANDU type reactor design on the use of heavy water as coolant and moderator, natural uranium as fuel. However, in order to improve the economy, CANDU also committed to use light water as coolant, heavy water moderator and slightly enriched uranium as fuel.

If heavy water can be obtained at a reasonable price, it will use heavy water cooling; the moderator will be especially advantageous for an application of the method in the boiling water reactor.

In a reactor with high thermal neutron flux and high temperature, it seems attractive to use liquid metal as coolant. Liquid metals have excellent heat transfer properties, such as high heat conductivity, low vapor pressure. Low atomic weight metals such as lithium and sodium while also having a high specific heat and volumetric heat capacity. In addition, the liquid metal is stable at high temperature and strong radiation field. Their main disadvantage is that the operation is relatively difficult and some metals are corrosive at high temperatures.

In an element having a smaller thermal neutron absorption cross-section and a melting point lower than 500 °C, some metals suitable for a thermal neutron reactor can be selected and included in Table 8.10. These elements can be divided into two categories, which are cross sections of less than 1b and cross sections in the 1–10b. Mercury thermal neutron absorption cross sections are large (340b) and are therefore not included in the table. However, for the fast neutron, the absorption cross section is much smaller, so in the Los Alamos a fast neutron reactor on the use of mercury as a coolant. However, the boiling point of mercury is only 357 °C, while its vapor pressure is quite large at low temperatures, so it is generally believed that it is not as good as the other possible metal coolant.

Among the various elements of Table 8.10, potassium, gallium and thallium showed no significant advantages, especially these two elements were also very expensive. In addition, liquid gallium at high temperatures it is difficult to have a container to hold, and tin is the same. Although bismuth and lead have relatively low cross-section, but their melting point is quite high. The two elements of the low melting point alloy, containing 44.5% of the weight of lead, melting point of 125 °C, it is possible as a reactor coolant. Another possibility is to use a low melting point of lead magnesium alloy, containing 97.5% of the lead, the melting point of 250 °C.

One of the most important disadvantage of bismuth in reactor is that it will happen (n, γ) reaction and generates ^{210}Bi. This is a half-life of 5 days of the beta particle emission and decay generates ^{210}Po. ^{210}Po is very toxic and difficult to deal with, so it will pose a particularly serious potential hazard. However, this problem can be solved; the United States has been designed to be a successful use of uranium

Table 8.10 Some metals suitable for thermal neutron reactors

Low neutron cross section			Medium neutron cross section		
Metal	σ_a/B	Melting point/°C	Metal	σ_a/B	Melting point/°C
Bismuth	0.032	271	Potassium	2.0	62
Lead	0.17	327	Gallium	2.7	30
Sodium	0.50	98	Thallium	3.3	302
Tin	0.65	232			

dissolved in bismuth as a circulating fuel reactor. Liquid bismuth container material is chrome steel, such as 2.25%Cr-1%Mo and 5%Cr-1.5%Si steel. At a high temperature of about 500 °C mass transfer phenomenon becomes more serious, you can add a small amount of zirconium and magnesium in the liquid metal to control.

Chinese use liquid sodium as a coolant in experimental fast reactor. Equally striking is an alloy of 22% (wt%) sodium and 78% potassium (NaK), which is a liquid at room temperature. At present, it is generally believed that sodium is the most suitable for use as coolant in liquid metals at high temperatures. If the oxygen is removed, the liquid sodium does not erode stainless steel, nickel, and many nickel alloys, beryllium, or graphite at temperatures below 600 °C. At higher temperatures, the mass transfer in sodium may occur, but it can also be controlled using additives. Due to its high melting point (98 °C), sodium is likely to solidify in the reactor cooling system, which can be avoided by using an electric heater to wrap the sodium channel. However, when reactor has been in high power for a period of time, after the shutdown, the residual heat is enough within a few days to maintain sodium in liquid. In the experimental breeding reactor (EBR) there is also a choice of NaK as a coolant, because NaK is a liquid at room temperature, so the risk of no solidification inside and outside the reactor. At the same time, in the fast reactor, like EBR, with a high atomic weight elements instead of sodium is also advantageous.

One disadvantage of using sodium or sodium alloy as a reactor coolant is the result of neutron capture, which produces ^{24}Na, which has a half-life of about 15 h. In addition to the release of beta particles, also released two very high energy (1.38 and 2.75 meV) of the gamma ray. Therefore, the coolant tank, pipe, pump and heat exchanger need some shielding, and the maintenance of the problem has increased. In the EBR and Chinese experimental fast reactor uses a secondary cooling system to reduce the volume of shielding. The heat exchanger to transfer heat from the primary loop to the secondary loop is shielded, but the third loop, including steam generator, are non-radioactive, so there is no need to shield.

One of the main reasons for the opposition to the use of liquid sodium (or its alloys) as a coolant is the risk of fire and explosion when the sodium is exposed to air or water. This risk can be greatly reduced after taking appropriate measures. Due to that the confidence of the operation of hot liquid sodium is very large, so in the United States submarine intermediate neutron reactor is also take sodium as primary coolant and mercury is used as an intermediate fluid transport heat to steam generator; therefore even heat exchanger leaks, sodium does not directly contact with water, and the existence of this leakage can be observed and measured easily use of mercury.

8.7.4   Moderator

In thermal neutron reactors, fast fission neutron should be reduced to thermal neutron energy levels using moderator (about neutron slowing down in detail please refer to Chap. 11). Moderator in thermal neutron reactor is used to transform the fast neutron of fission reaction to thermal neutron to maintain the fission chain. The slowing down of the neutron is the most favorable to light mass number close to the neutron. In addition, demand rebound performance of moderator material is good and less neutron absorption. Moderator absorption during slowdown is very unfavorable, so to use neutron materials with small cross-section as a moderator. Light water, heavy water and graphite is good moderator. In addition, the moderator must be better compatibility with other materials, better radiation stability and low cost, easy to get.

Ordinary water, heavy water (deuterium oxide), beryllium (whether is metal, oxide or carbide) and carbon (as graphite) can be used as moderator material. The properties of these materials will be discussed in turn below [10].

Because of the good thermal neutron reflector is similar to moderator. Therefore, described below can be seen as equally applicable to moderator and reflector. The important properties of five kinds of moderators are summarized in Table 8.11. None of these includes beryllium carbide, as it is associated with the chemical activity of air and water and has limited for its use. Density of graphite and beryllium oxide dependent with production method; the values in the table are less than the theoretical density, but these values can be used as a typical value. SDP and MR respectively

Table 8.11 Several main properties of moderator

Properties	Ordinary water	Heavy water (99.75%D_2O)	Beryllium	Oxide beryllium	Graphite
Atomic (molecular) weight	18.0	20.0	9.01	25.0	12.0
Density/(g/cm^3)	1.00	1.10	1.84	2.80	1.62
N/(10^{22}/cm^3)	3.3	3.3	12.0	6.7	8.1
σ_f/b (super thermal neutron)	49	10.5	6.0	9.8	4.8
σ_a/b (thermal neutron)	0.66	0.0026	0.009	0.0092	0.0045
Σ_f cm^{-1} (super thermal neutron)	1.64	0.35	0.74	0.66	0.39
Σ_a cm^{-1} (thermal neutron)	0.023	0.000037	0.0012	0.00061	0.00038
ζ	0.927	0.510	0.207	0.17	0.158
$\zeta\Sigma_s$ cm^{-1} (SDP)	1.425	0.177	0.154	0.11	0.083
$\zeta\Sigma_s/\Sigma_a$ (MR)	62	4830	126	180	216
Diffusion coefficient D/cm	0.18	0.85	0.61	0.56	0.92
Diffusion length L/cm	2.88	100	23.6	30	50
Fermi age /cm^2	33	120	98	110	350
Migration length/cm	6.4	101	26	32	54

represent the slow ability and slow ratio (see Chap. 11). In calculating these quantities, the cross sections of the super thermal neutron (i.e., the energy exceeding the thermal energy value) are used, because they determine the slow rate of neutrons. All other properties in the table are the numerical values of thermal neutrons at room temperature.

Water is a kind of attractive moderator material. Because of its low price and the ability to slow down. On the other hand, the neutron absorption cross section is a little bit high, so water moderator reactor has to use enriched uranium as fuel to get critical system. However, as long as one get enriched uranium, water as moderator makes a small size of reactor. Another advantage is the water can be also used as a coolant and moderator. Then the water must be removed impurities, because the latter will not only neutron capture, and due to (n, γ) reaction, may become radioactive. Water of high purity to reduce the formation of corrosion and scale is necessary; and the presence of impurity ions often will promote the water by nuclear radiation, the role of decomposition, release of hydrogen and oxygen.

One of the main disadvantages of using water as a moderator is its relatively low boiling point. This means that when it is used in high temperature reactors, the pressure must be high. Moreover, it will cause some problems in equipment manufacturing.

Because of the high slow ability and slow ratio of heavy water, it is a good moderator. When the other conditions are the same, resonance escape probability and thermal neutron utilization factor of heavy water moderator are better than graphite. Therefore, using heavy water as moderator in non-uniform natural uranium reactor, reactor size and fuel needed are both less than using graphite as moderator. In fact, the uniform system of natural uranium and heavy water can achieve critical while a homogeneous mixture of natural uranium and graphite may not reach the critical. With the improvement and development of heavy water production, heavy water is being applied more and more.

Pure deuterium oxide in 3.82 °C melting in 101.42 °C boiling, in two cases these values are a little bit higher than the corresponding temperature of the ordinary water. Therefore, as the system with ordinary water at high temperature needs high pressure, heavy water also needs high pressure. The density of heavy water at room temperature is about 1.10 g/cm^3.

All natural water contains about 1/6500 (about 0.015%) of deuterium oxide, while the heavy water is obtained from this tiny concentration enriched to the desired purity. Several methods have been used to deuterium enrichment in water, of which there are three methods for the large-scale industrial application, this is electrolytic method, distillation and chemical exchange method. In addition, there is a possibility that the separation of deuterium by cryogenic distillation of liquid hydrogen.

Production of heavy water electrolysis principle is when the aqueous solution electrolysis, namely when the current through the aqueous solution, cathode emit hydrogen contained isotopically light weight proportion of isotope predominate. Therefore, stay in the electrolytic tank water deuterium is a partial concentration of the components. The electrolyte used in the business is sodium or potassium hydroxide or carbonate aqueous solution; anode is made of nickel or nickel-plated steel, and the cathode is made of nickel or steel. A steel can be used as electrolytic tank container and cathode. It is possible to obtain a continuous electrolysis of heavy water with high purity. However, in order to obtain high concentration degree of heavy water needs a lot of power, and thus to the electrolytic method adopted by ordinary water began to extract the heavy water is not feasible. However, we will see below, when the feed water has some concentration by other methods after electrolysis method is particularly valuable.

Distillation is based on the small difference of the vapour pressure of light water and heavy water. This makes a tiny difference in the boiling point. As a result, when the water is distilled, the liquid is distilled contains a relatively high proportion of light isotopes, while the residual liquid contains more heavy isotopes. Therefore, by distillation, it is likely to concentrate heavy water. Because of the difference of vapor pressure is very small, the distillation column must contain a large number of distillation tray or equivalent level. The highest separation efficiency can be obtained at low pressure. No matter the huge size needed, high purity deuterium has been obtained by distillation. After more than 8 level, the content of deuterium in water is increased from 0.015% to about 90%.

Chemical exchange method using small differences in the chemical activity of isotope. Although the isotope has exactly the same chemical properties, but the same element of various isotopic composition or its compounds often have a slightly different reaction rate. Therefore, the equilibrium constant of the isotope exchange reaction, that is, the ratio of positive and negative reactions, is slightly different from number 1. This makes it possible to separate the isotopes.

Several hydrogen deuterium exchange reaction have been studied, and some of them have been used to produce heavy water. One of these reactions is the reaction between water and hydrogen. When the content of deuterium is not too high, the isotope exists in the form of HD and HDO, so the exchange reaction equilibrium equations of these two isotopes can be written as follows.

$$H_2O + HD \leftrightarrow HDO + H_2 \qquad (8.17)$$

The equilibrium constant is

$$K = \frac{[HDO][H_2]}{[H_2O][HD]} \qquad (8.18)$$

If all of the substances in the reaction are in the vapor phase, the numerical range of the equilibrium constant is between 2.6 at room temperature and 3.4 at 100 °C. This means that if the hydrogen containing some HD and water (H_2O) vapor to reach a balance, then the reaction results will be generated HDO. You can use ordinary HDO by water vapor phase and water washout, the deuterium is concentrated.

Because the isotope exchange equilibrium between water and hydrogen is very slow, even in the case of 100 °C, it is necessary to use the catalyst. The catalyst can be spread on the basis of platinum powder charcoal; can also be a mixture of nickel and chromium oxide. The mixture of water vapor and hydrogen gas through the catalyst above, and then washed out the liquid with a liquid generated HDO. The latter will be evaporated and then sent over the catalyst with hydrogen, which will generate more HDO, so continue to go. If in a convective system repeated catalytic exchange reaction and water vapor with the washing water of the equilibrium reaction, it can be a considerable concentration of deuterium in water.

Hydrogen—water exchange method requires a huge volume of hydrogen, so the most economical way is to make it linked to another process with hydrogen as an important reactant. The production of ammonia nitrogen and hydrogen synthesis to meet this requirement. Hydrogen was first introduced the deuterium concentration workshop, a certain proportion of deuterium content in a lighter isotope exchange is removed; the rest of the gas so communicated with the ammonia synthesis of the workshop, for the latter gas composition slightly changed had little effect.

Previously used a modified chemical exchange principle, called "double temperature method". It is the use of the difference of two kinds of temperature exchange reaction equilibrium constant, this method does not require a large amount of hydrogen (or other gases), because the hydrogen used only as a deuterium carrier. Figure 8.19 shows a simple form of double temperature method using the above principles for the catalytic water hydrogen exchange reaction.

The feed water fed by the left is divided into two parts: the first part evaporates under rather low temperature T_1 is then fed into the reactor I with hydrogen. Due to the results of isotope exchange reaction, the water will be deuterium concentration, which is then reduced in dilute hydrogen isotope. Another portion of the feed water is evaporated at a temperature higher than T_1 and is associated with a reduced hydrogen temperature in the reaction chamber I. When the mixture of gases through the heat exchanger to increase the temperature to T_2 (T_2 is much higher than T_1), it entered the reaction chamber II. The equilibrium constant of the water hydrogen exchange reaction at T_2 is smaller than that of the T_1 temperature, and thus the deuterium is reduced by water. After removing the water with the condensation method, the

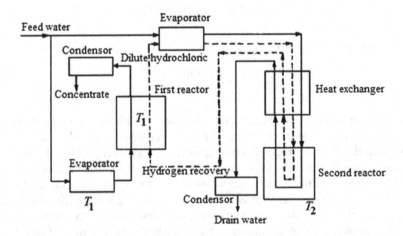

Fig. 8.19 Flow chart of double temperature method

hydrogen containing deuterium has been restored to the original state of the hydrogen and returned to the reaction chamber I, where it is once again diluted, so endless cycle.

If we consider the heavy water from 0.015% concentration to 99.75% (this is used in reactor), it seems that none of the above methods can cover the whole concentration range well than the others. Most of the costs of concentration are spent at the beginning of a few level. This is mainly because of the need to have many factories to deal with massive amounts of water. However, when heavy water ratio increases, this situation has changed. The result makes the electrolytic method usually cannot be started at all levels, but in the last few steps it can be used. Therefore, the most economical solution is the use of distillation and chemical exchange method, or the combination, to concentrate a certain extent and then the use of electrolysis method to desired concentrations.

Moreover, there is a need to mention the possibility of at low temperatures with liquid hydrogen fractionation of deuterium separation, according to certain data shows this method should produce more cheap heavy water than the above method. Although the boiling point of liquid hydrogen and liquid deuterium at atmospheric pressure is only 3.12 °C, the ratio of their vapor pressure at a certain temperature is particularly high. Thus, at about 22 K temperature, the temperature between the boiling point of two kinds of pure liquid, vapor pressure ratio in H_2/HD is 1.6, while in the H_2/D_2 is almost 2.5. Such a high relative volatility means that it is possible to obtain a high separation efficiency by using the fractionation method.

The main problem of this approach lies in: (1) requires large-scale plant running under the low temperature of -250 °C, which is necessary for hydrogen liquefaction and distillation; (2) under this low temperature, trace impurities in hydrogen will generated solid deposits in the pipe; (3) a lot of liquid hydrogen operation brought risk.

Nuclear properties of graphite is not as good as heavy water, but very high purity graphite can be at a reasonable price. It has good mechanical properties and thermal stability. However, it reacts with oxygen and water vapor at high temperatures and can react with certain metals or metal oxides to form carbides. Although graphite is a kind of non-metal, but it is a good conductor of heat, it is necessary for moderator. The main disadvantages of graphite under normal operating conditions are as following: (1) it can be oxidized when exposed to air; (2) the resistance intensity is low; (3) it will be affected by nuclear radiation.

Graphite exists in large quantities in nature. However, due to the state of graphite is fairly impure, so the reactor grade graphite with artificially made from petroleum coke. Firstly, the coke heating to expel volatile substances, and then ground and with coal tar pitch binder blended, then the temperature shall keep in the range of appropriate liquidity within. The mixture is extruded into a rod shape and is melted in a gas furnace to 1500 °C, which causes the asphalt to be burnt and hardened. In order to increase the total density of the product, it will be used in the vacuum infiltration and then melt the asphalt. So the gas roast carbon, and finally by resistance heating to graphite. In the first few days the temperature rose to about 2700–3000 °C, then let it cool gradually in three or four weeks. The physical properties of the final product in a certain extent depending on coke grinding, leaching infiltration with asphalt types and quantity and graphite treatment temperature and time.

Exercises

1. Fill 1 m^3 volume of the cube space with table tennis ball, BCC, FCC and HCP which kind of arrangement with the most? Why?
2. A cross-sectional area of 8 mm^2 carbon steel wire, how much force is needed to pull off?
3. What are the main differences between military and industrial plutonium?
4. Calculate the integral thermal conductivity of UO_2 in the temperature range of 600–1400 °C.
5. What is the temperature of nil ductile transition? What effect does it have on the life of the pressure vessel in nuclear power plants?
6. What are the elements that can be used for reactor core materials with low thermal neutron cross sections?
7. What is the coefficient of linear expansion of the material? How to use it to calculate thermal stress?
8. What are the methods of producing heavy water? What are the advantages and disadvantages of each?
9. What factors should be considered in the selection of coolant material for the reactor?

References

1. U. S. Department of Energy: Material science. Washington, D.C.: Lulu.com; 2016.
2. Berry WE. Corrosion problems in light water nuclear reactors. Speller award lecture, presented during CORROSION/84, New Orleans, Louisiana; 1984.
3. Makansi. Solving power plant corrosion problems, power special report; 1983.
4. Metcalfe HC, Williams JE, Castka JF. Modern chemistry. New York, NY: Holt, Rinehart, and Winston; 1982.
5. Owens. Stress corrosion cracking, presented during CORROSION/85, Paper No. 93, NACE, Houston, Texas; 1985.
6. Foster AR, Wright RL. Basic nuclear engineering. 4th ed. Allyn and Bacon, Inc.; 1983.
7. Weisman J. Elements of nuclear reactor design. Elsevier Scientific Publishing Company; 1983.
8. Glastones S, Sesonske A. Nuclear reactor engineering. 3rd ed. Bel Air, CA: Van Nostand Reinthold Company; 1981.
9. Tweeddale JG. The mechanical properties of metals assessment and significance. American Elsevier Publishing Company; 1964.
10. Glasstone S, Sesonske A. Nuclear reactor engineering. 3rd ed. Van Nostrand Reinhold Company; 1981.

Chapter 9
Mechanical Science

In the field of nuclear engineering using a large number of general machinery equipment. We here introduce some major general machinery, such as internal combustion engines, heat exchangers, pumps, valves, air compressors, hydraulic machine, an evaporator, a steam generator, a cooling tower, a voltage regulator, a diffusion separator.

9.1 Diesel Engine

One of the most common prime movers is the diesel engine. Before gaining an understanding of how the engine operates a basic understanding of the engine's components must be gained. This chapter reviews the major components of a generic diesel engine.

Internal combustion engine is a mechanical power; it is through the fuel burning inside the machine, and the release of heat energy direct conversion for the mechanical power of the heat engine. General internal combustion engine includes not only the reciprocating piston type internal combustion engine, rotation piston engine and free piston type engine, including rotary vane type gas turbine and jet engines. However, usually said internal combustion engine is point to a reciprocating piston type internal combustion engine [1].

Most nuclear facilities require some type of prime mover to supply mechanical power for pumping, electrical power generation, operation of heavy equipment, and to act as a backup electrical generator for emergency use during the loss of the normal power source. Although several types of prime movers are available (gasoline engines, steam and gas turbines), the diesel engine is the most commonly used. Diesel engines provide a self-reliant energy source that is available in sizes from a few horsepower to 10,000 hp. Relatively speaking, diesel engines are small, inexpensive, powerful, fuel efficient, and extremely reliable if maintained properly.

© Tsinghua University Press 2022
J. Yu, *Fundamental Principles of Nuclear Engineering*,
https://doi.org/10.1007/978-981-16-0839-1_9

Because of the widespread use of diesel engines at nuclear facilities, a basic understanding of the operation of a diesel engine will help ensure they are operated and maintained properly. Due to the large variety of sizes, brands, and types of engines in service, this chapter is intended to provide the fundamentals and theory of operation of a diesel engine. Specific information on a particular engine should be obtained from the vendor's manual.

Reciprocating piston internal combustion engines are the most common. Piston type internal combustion engine fuel and air mixture in the cylinder combustion, the release of heat energy to produce high temperature and pressure in the cylinder gas. Gas expansion drives the piston to do work, then the output mechanical power through the crank and connecting rod mechanism or other institutions, drive driven mechanical work. Diesel and gasoline engines are often seen.

The production of modern diesel engines is the result of the internal combustion principle first proposed by Sadie Carnot in the early nineteenth century. Dr. Rudolf Diesel applies sadie Carnot's principles to patented cycle or combustion methods, known as "diesel" cycles. The heat generated by his patented engine during the compressed air fuel charge causes the mixture to ignite and the mixture expands at constant pressure during the engine's full power stroke.

Dr Diesel's first engine ran on coal dust and used a compression pressure of 1500 psi to improve theoretical efficiency. In addition, his first engine did not specify any type of cooling system. Thus, between extreme pressure and lack of cooling, the engine explodes, almost killing its inventor. After recovering from the injury, Diesel again tried to use oil as fuel, adding cooling jackets around the cylinder and reducing the compression pressure to approximately 550 psi. This combination turned out to be a success. The production rights were sold to Adolphus Bush, who built the first diesel engine for commercial use and installed it at his St. Louis brewery to drive various pumps.

Diesel engines are similar to gasoline engines used in most cars. Both engines are internal combustion engines, which means they burn a mixture of fuel air in the cylinder. Both are reciprocating engines, driven by pistons moving laterally in two directions. Most of them are similar. Although diesel engines have similar components to gasoline engines, diesel engines are heavier than gasoline engines with the same horsepower because of the stronger, heavier materials used to withstand higher combustion pressures in diesel engines.

The higher the combustion pressure, the higher the compression ratio of the diesel engine. The compression ratio is a measure of the engine compressed cylinder gas. In gasoline engines, the air pressure ratio (controlling compression temperature) is limited by the air fuel mixture entering the cylinder. The low ignition temperature of the gasoline causes the gasoline to ignite (burn) at a compression ratio of less than 10:1. The compression ratio of a normal car is 7:1. In diesel engines, compression ratios from 14:1 to up to 24:1 are typically used. A higher compression ratio is possible because only air is compressed and then injected into the fuel. This is one of the factors that makes diesel engines so efficient. Compression is discussed in more detail later in this chapter.

Another difference between a gasoline engine and a diesel engine is the way the engine speed is controlled. In any engine, speed (or power) is a direct function of cylinder fuel consumption. Gasoline engines are self-limiting because they use methods to control the amount of air entering the engine. Engine speed is indirectly controlled by the butterfly valve in the oil carburetor. The butterfly valve in the oil carburetor limits the amount of air entering the engine. In the carburetor, the air flow rate determines the amount of gasoline mixed with the air. Limiting the amount of air entering the engine limits the amount of fuel entering the engine, thereby limiting the speed of the engine. By limiting the amount of air entering the engine, adding more fuel does not increase the engine speed beyond the point at which the fuel burns 100% available air (oxygen).

Diesel engines are not self-limiting because the air (oxygen) entering the engine is always the most. As a result, engine speed is limited only by the amount of fuel injected into the engine cylinder. As a result, the engine always has enough oxygen to burn and the engine will attempt to accelerate to meet the new fuel injection rate. As a result, manual fuel control is not possible because these engines can accelerate at more than 2000 revolutions per second in unloading. Diesel engines require a speed limiter, commonly known as a governor, to control the amount of fuel injected into the engine.

Unlike petrol engines, diesel engines do not require an ignition system because in diesel engines, fuel is injected into the cylinder when the piston reaches the top of the compression stroke. When fuel is injected, the fuel evaporates and ignites due to the heat generated by the compression of air in the cylinder.

Figure 9.1 is a large internal combustion engine.

Fig. 9.1 An example of a large-sized diesel engine

9.1.1 Major Components of a Diesel Engine

To understand how a diesel engine operates, an understanding of the major components and how they work together is necessary [2]. Figure 9.1 is an example of a large-sized, four-stroke, supercharged, diesel engine with inlet ports and exhaust valves. Figure 9.2 provides a cross section of a similarly sized V-type diesel engine.

Figure 9.3 provides a cross section of a similarly sized V-type diesel engine.

The cylinder block, as shown in Fig. 9.4, is generally a single unit made from cast iron. In a liquid-cooled diesel, the block also provides the structure and rigid frame for the engine's cylinders, water coolant and oil passages, and support for the crankshaft and camshaft bearings.

The crankcase is usually located on the bottom of the cylinder block. The crankcase is defined as the area around the crankshaft and crankshaft bearings. This area encloses the rotating crankshaft and crankshaft counter weights and directs returning oil into the oil pan. The oil pan is located at the bottom of the crankcase as shown in Figs. 9.2 and 9.3. The oil pan collects and stores the engine's supply of lubricating oil. Large diesel engines may have the oil pan divided into several separate pans.

Diesel engines use one of two types of cylinders. In one type, each cylinder only needs to be machined or drilled into a block casting, making the block and cylinder an integral part. In the second type, the processed steel sleeve is pressed into the block casting to form the cylinder. Figures 9.2 and 9.3 provide examples of sleeve diesel engines. Using both methods, the cylinder sleeve or bore provides the engine

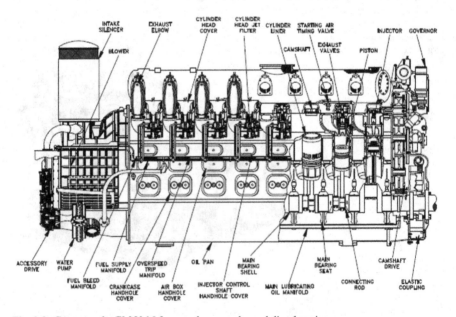

Fig. 9.2 Cutaway of a GM V-16 four-stroke supercharged diesel engine

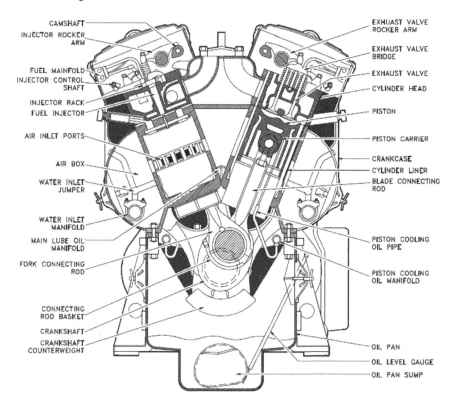

Fig. 9.3   Cross section of a V-type four stroke diesel engine

Fig. 9.4   The cylinder block

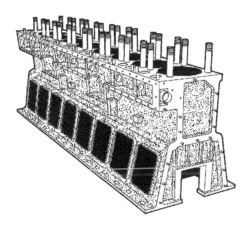

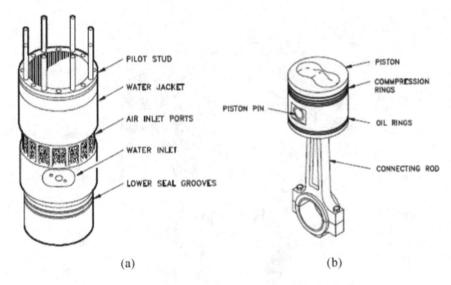

Fig. 9.5 Diesel engine wet cylinder sleeve

with the cylindrical structure needed to limit combustion gas and acts as a guide to the engine piston.

In engines that use sleeves, there are two types of sleeves, wet and dry. The dry sleeve is surrounded by block of metal and does not come into direct contact with the coolant (water) of the engine. The wet sleeve comes into direct contact with the coolant of the engine. Figure 9.5a provides an example of a wet sleeve. The volume surrounding the sleeve or bore is called the combustion chamber and is the space to burn fuel.

In either type of cylinder, sleeved or bored, the diameter of the cylinder is called the engine bore and stated in inches. For example, a 350 cubic inch Chevrolet gasoline engine has a bore of 4 in.

Most diesel engines are multi-cylinder engines, usually arranged in one of two ways, an in-line or "V", but the other combinations exit. In in-line engines, as the name suggests, all cylinders are in a row. In "V" engines, cylinders are arranged in two exhaust cylinders, which are aligned with each other at angles aligned with normal crankshafts. Each set of cylinders that make up the "V" side is called as a bank of cylinders.

The piston transforms the energy of the expanding gasses into mechanical energy. The piston rides in the cylinder liner or sleeve as shown in Figs. 9.2 and 9.3. Pistons are commonly made of aluminum or cast iron alloys.

To prevent the combustion gasses from bypassing the piston and to keep friction to a minimum, each piston has several metal rings around it, as illustrated by Fig. 9.5b.

These rings act as seals between the piston and cylinder walls and reduce friction by minimizing the contact area between the pistons and cylinder walls. Rings are usually made of cast iron and coated with chrome or molybdenum. Most diesel engine pistons have several rings, typically 2 to 5, each performing a different function. The top ring acts primarily as a pressure seal. The intermediate ring acts as a wiper ring to remove and control the amount of oil film on the cylinder wall. The bottom ring is an oiler ring that ensures that the supply of lubricant is deposited evenly on the cylinder wall.

The connecting rod connects the piston to the crankshaft. For information on the position of the connecting rods in the engine, see Figs. 9.2 and 9.3. The rods are made of dropforged, heat-treated steel to provide the desired strength. Both ends of the rod are bored, and the smaller top bore are attached to the piston pin (wrist pin) of the piston, as shown in Fig. 9.5a. The large bore end of the rod is split in half and bolted so that the rod is attached to the crankshaft. Some diesel engine connection rods drill down the center, allowing oil to travel upward from the crankshaft and into the piston pin and piston lubrication.

The variant of the V-engine that affects the connecting rod is to place the cylinders on the left and right sides that are opposite to each other, rather than staggered (the most common configuration). This arrangement requires that the connecting rods of the two opposing cylinders have the same main journal bearing on the crankshaft. To allow this configuration, one of the connecting rods must be disassembled or forked around the other.

Crankshaft

The crankshaft converts the linear motion of the piston into the rotational motion transmitted to the load. The crankshaft is made of forged steel. Forged crankshafts produce crankshaft bearings and connecting rod bearing surfaces by processing. The rod bearings are eccentric, or offset, from the center of the crankshaft as shown in Fig. 9.5b. This offset converts the reciprocating (up and down) motion of the piston into the rotational motion of the crankshaft. The offset determines the engine's stroke (piston travel distance).

The crankshaft does not ride directly on the cast iron block crankshaft bracket, but on the special bearing material shown in Fig. 9.6. The connecting rod is also inserted into the bearing between the crankshaft and the connecting rod. Bearing material is a soft metal alloy that provides a replaceable wear surface and prevents galling between two similar metals, the crankshaft and the connecting rod. Each bearing is divided into two halves in order to assemble the engine. The crankshaft is drilled with an oil channel that allows the engine to deliver oil to each crankshaft bearing and the connecting rod bearing and into the connecting rod itself.

The crankshaft has a large weight, called the counter weight that balances the weight of the connecting rod. These weights ensure balanced forces during the rotation of moving parts.

The flywheel is located at one end of the crankshaft and has three uses. First, through inertia, it reduces vibration by smoothing the power stroke of each cylinder. Second, it is used to secure the engine bolts to the mounting surface of the load. Third, on some diesel engines, the flywheel is surrounded by gear teeth, allowing the starting motor to engage and crank diesel.

The cylinder head of the diesel engine performs a variety of functions. First, they provide a top seal for cylinder bore or sleeves. Second, they provide structural

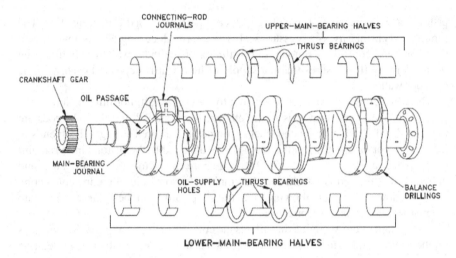

Fig. 9.6 Diesel engine crankshaft and bearings

hold exhaust valves (and intake valves if applicable), fuel injectors and necessary connections. The engine head of a diesel engine is manufactured in one of two ways. In one method, each cylinder has its own head casting, which is bolted to the block. This method is mainly used in larger diesel engines. In the second method, for smaller engines, the head of the engine is cast into a piece (multi-cylinder head).

Diesel engines have two ways to absorb and exhaust. They can use ports or valves or a combination of both. The port is located in the cylinder wall of the lower 1/3 of the bore. For an example of an intake port, see Figs. 9.2 and 9.3 and note their relative position to the rest of the engine. Depending on the type of port, the port is "opened" when the piston is below the port level and fresh air or exhaust gas can enter or leave.

Then, when the piston returns above the port level, the port is "closed". Valves (Fig. 9.7) are mechanically opened and closed to admit or exhaust the gas needs. The valve is located in the head casting of the engine. The point at which the valve seals the head is called the seat. Most medium diesel engines have air intakes ports or exhaust valves, or both.

In order to reduce the centrifugal force generated during motion, the crankshaft journal is often made hollow. Oil holes are provided on each axle neck surface to introduce the oil into or out of the lubricating journal surface. In order to reduce the stress concentration, the main shaft neck, the crank pin and the connecting point of the crank arm are connected with transition arc.

The crankshaft counterweight is to balance the centrifugal force and torque, sometimes also can balance the reciprocating inertia force and torque. When the force and torque self-balance, balance weight can also be used to reduce the load of the main bearing. The number of balance weight, size and placement according to the engine number of cylinders, cylinder arrangement and crankshaft shape factors to consider.

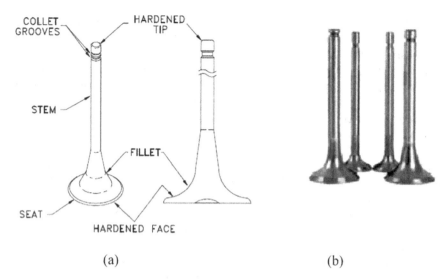

COLLET GROOVES
HARDENED TIP
STEM
FILLET
SEAT
HARDENED FACE

(a)   (b)

Fig. 9.7 Diesel engine valve

Balance weight and crankshaft casting or forging into a whole. The counterweight and crankshaft of high power diesel engine manufactured separately and bolt connection together.

One end of the crankshaft is connected to the transmission gear so that the rotating machinery can be transferred to the components needed to move (such as a car's tire), and the other end is generally connected to a flywheel. The flywheel is a very large part of disc, which acts as an energy storage device. For the four-stroke engine, each of the four-piston stroke, namely acting stroke, while the exhaust, air intake and compression stroke to power consumption. Therefore, the torque of the output of the crankshaft will change periodically, so that the speed of the crankshaft is not stable. In order to improve this condition, the flywheel in the rear-end of the crankshaft. The main function of the flywheel is stored outside the engine power stroke and inertial energy. The flywheel can be used to reduce the speed fluctuation of the engine running process.

Cylinder head, its role is to seal the cylinder, and the piston together to form a combustion space of high temperature and high pressure gas. The cylinder head bears the gas force and the mechanical load caused by the tightening of the cylinder bolt, and is subjected to a high thermal load due to the high temperature gas contact. In order to ensure the good seal of the cylinder, the cylinder head cannot be damaged, and cannot be deformed. To this end, the cylinder head should have sufficient strength and rigidity. Cylinder head is generally made of high quality gray cast iron or alloy cast iron, car gasoline engine is more use of aluminum alloy cylinder head.

The cylinder head is a box shaped part with complex structure. It is processed with inlet and outlet valve seat hole, valve guide hole, spark plug mounting hole (gasoline engine) or injector installation hole. Cylinder cover is cast with a water jacket, inlet and exhaust ports and combustion chamber or a part of combustion chamber. If the camshaft is mounted on the cylinder head, the cylinder cover is also provided with a cam-bearing hole or a cam bearing seat and a lubricating oil path.

Cylinder head of a water-cooled diesel engine has three types: integral, block type and single type. In a multi cylinder diesel engine, all cylinders sharing a cylinder cover is called the cylinder cover is an integral type cylinder cover; if every two cylinders of a cover or cylinder cover, the cylinder cover is divided into block type; if each cylinder cover, single type.

Before burning the fresh air to enter the cylinder, after burning exhaust gas to discharge the cylinder, usually installed in the cylinder head with the intake valve and exhaust valve (as shown in Fig. 9.7). Air intake valve and exhaust valve structure is almost same, can be understood as a sliding one-way valve. Also some diesel engine with chamfer of inner wall of cylinder liner of the intake and exhaust. When the piston moves to the trench, gas can be in and out of the cylinder through the slot. There is also a mixed design and dredging valve.

In order for the diesel engine to operate, all its components must perform piston motion-related functions at very precise intervals. To do this, use a component called a camshaft. Figure 9.8 illustrates the camshaft and camshaft drive gears.

The camshaft is a long bar with egg-shaped eccentric lobes, one lobe for each valve and one lobe for each fuel injector. Each lobe has a follower, as shown in Fig. 9.9. When the camshaft rotates, the follower forces up and down the profile of the cam lobe. Followers connect to the engine's valves and fuel injectors through various types of linkages called pushrods and rocker arms. The pushrod and rocker arm transfer the reciprocating movement generated by the camshaft lobes to the valves and injectors and open and close them as needed. The valves are maintained closed by the springs.

When the valve is opened by the camshaft, it compresses the valve spring. Then, when the camshaft lobe rotates under the follower, the energy stored in the valve

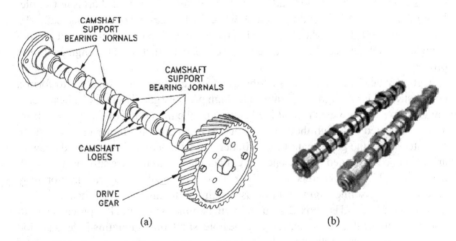

Fig. 9.8 Diesel engine camshaft and drive gear

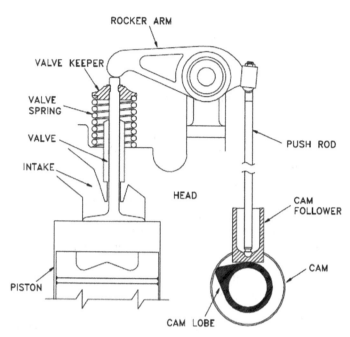

Fig. 9.9 Diesel engine valve train

spring is used to close the valve. Because the temperature of the engine varies considerably, its components must be designed to allow thermal expansion. Therefore, valves, valve pushrods and rocker arms must have some methods that allow expansion. This is achieved by using valve lash. Valve lash is the term given to the "slop" or "give" in a valve train before the cam actually starts opening the valve.

The camshaft is driven by the crankshaft of the engine through a series of gears called idler gears and timing gears. Gears allow the rotation of the camshaft to correspond to or be timely to the rotation of the crankshaft, allowing valve opening, valve closure and fuel injection to occur at precise intervals during the piston's travel. To increase valve opening, valve closure, and fuel injection flexibility, as well as increase power or reduce costs, the engine may have one or more camshafts. Typically, in medium- and large V-engines, each bank will have one or more camshafts. In larger engines, the intake valve, exhaust valve, and fuel injectors may share a common camshaft or have independent camshafts.

Depending on the type of engine and the manufacture, the position of the camshaft or shaft will vary. The camshafts in the inline engine are usually located at the head of the engine or at the top of the block running down one side of the cylinder bank. Figure 9.9 provides an example of an engine with camshafts located on the side of the engine.

The blower of a diesel engine is part of an intake system that compresses incoming fresh air for delivery to the cylinder for combustion. The blower can be part of a turbocharged or supercharged intake system. Additional information about these two types of blowers will be provided later in this chapter.

9.1.2   Diesel Engine Support Systems

A diesel engine requires five supporting systems in order to operate: cooling, lubrication, fuel injection, air intake, and exhaust. Depending on the size, power, and application of the diesel, these systems vary in size and complexity.

Diesel engine cooling system is a system, which uses heat absorbing medium to cool high temperature parts to keep the diesel engine operating at the best temperature. In a diesel engine, due to the cylinder liner, cylinder head, piston and valve etc. parts directly and high-temperature gas contact by intense heating, temperature of parts is very high. This will not only lead to lower mechanical strength, and may induce thermal stress to mechanical damage. High temperature will also destroy the oil film on the cylinder wall, so that the lubrication oil oxidation, so that the piston, piston ring and cylinder liner serious wear and tear, bite or stick. In addition, the high temperature will also make the air into the cylinder density decreased, causing knock, early combustion and other abnormal combustion. In order to ensure the normal and reliable operation of diesel engine, it is necessary to cool the high temperature parts through the cooling system. When the diesel engine works, the heat released by the combustion of the fuel is 20–35%, which is dispersed by the cooling system. Almost all of the diesel engine with liquid cooling, carrying out heat inside the cylinder. Diesel engine cooling system as shown in Fig. 9.10, including radiator, and coolant pump and thermostat and closed loop structure (water jacket) [3].

Water cooling system, heat transferred from high-temperature parts to the water firstly, and then the heat dissipated. Water cooling system consists of thermostat, radiator, fan, water pump and water jacket. In the cylinder head and water jacket are cast in parts of water, so that water can touch the heated parts.

The centrifugal water pump with compact structure and large water delivery is adopted in general, and the water pump is driven by a crankshaft through a belt. The cooling water pump into the water jacket and from the cylinder wall and cylinder cover after absorbing heat into the radiator on the water tank. A plurality of flat straight pipe, which is made of metal material, is arranged between the upper and the lower water tanks of the radiator, and the surrounding of the pipe is provided with a plurality of thin heat radiating fins. The fan installed in the pump pulley, fan the air flowing from the radiating fin radiator. The hot water in the water tank of the radiator is cooled by transfer the heat to the air when the hot water flows down the water tank through the flat tube of the radiator.

Diesel engine speed and load change frequently [4]. According to the design of the cooling system of high load of diesel engine, in the low load operation, the cooling water scattered is reduced, so the cooling capacity of cooling system should also be reduced accordingly, so it is necessary to regulate the heat transfer ability of the cooling system. Adjustment method is using the shutters to change airflow rate to

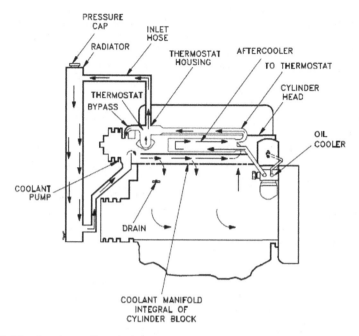

Fig. 9.10   Diesel engine cooling system

the radiator, or using temperature control by loading silicone oil or magneto electric clutch in the fan. When the temperature is raised, the clutch is gradually engaged, and the fan is driven to increase the amount of air flowing through the radiator; conversely, the amount of air is reduced. The adjustable fan can make the structure complex, but it can reduce the power consumption of the cooling system and save fuel. Another way is to change the cooling water flow rate to adjust the cooling intensity. The volatile liquid is filled in the corrugated tube of thermostat on the outlet of cylinder head. When the water temperature is about 80 °C, the vapor pressure of the volatile liquid is enlarged, and the bellows is expanded to drive the main valve to open the main valve. The auxiliary valve closed, all water from the jacket in cylinder head flows into the radiator for cooling. When the water temperature is below 70 °C, the main valve is closed, and the auxiliary valve is opened. At this time, the cooling water from the jacket of cylinder all flow to the auxiliary valve and flows into the pump, to be pumped into the circulation, cooling water is no longer flows into radiator. In this way, the water temperature is quickly controlled at the normal operating temperature, that is 80–90 °C.

When diesel engine works, there is a friction surface (such as the crankshaft journal and bearing, the shaft journal and the bearing, the piston ring and the cylinder wall, timing gear, etc.) at a high speed for relative motion. If there is no lubricating oil, direct metal and metal contact between parts metal surface friction generated huge heat in a few minutes will make the work surface melting, lead to the abnormal operation of diesel engine. Therefore, in order to ensure the normal operation of the diesel engine, relative moving parts on diesel engine must be surface lubricated. The friction surface should be covered with a layer of lubricant (oil or grease), between the metal surface interval a layer of thin oil film, to reduce friction, lower power loss and reduce wear and prolong engine service life. Lubricating oil can to not only lubricate the moving parts of the surface, the friction and heat less as far as possible, and the lubricating oil system can also cool the components. The lubricating system of the diesel engine is shown in Fig. 9.11. Oil is stored in the engine oil tank, the oil from the bottom of the oil pump to pump up, before the pump through a filter to filter-out impurities. After filtering, the lubricating oil through the oil pump into the top of the oil distribution tank began to drop oil droplets to different parts of the lubricating oil in the role of gravity automatic return to the engine oil tank. In order to prevent the oil pressure is too high, the outlet of the oil pump has a pressure relief bypass valve, the pressure is greater than the set pressure (for example, the diesel engine is 3.5 atm) when the pressure relief valve will automatically open, so as to bypass off the lubricating oil.

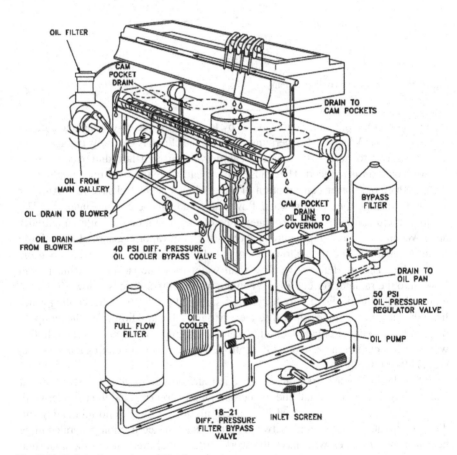

Fig. 9.11 Lubricating system of diesel engine

Fig. 9.12 Diesel engine fuel
supply system

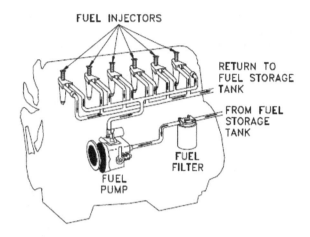

Fuel supply system is a system used to supply the fuel mixture. Of course, because the fuel is different, the supply will be different. Figure 9.12 is a schematic diagram of the diesel fuel supply system. Oil pump from the oil tank, through a filter to filter out the possible presence of impurities in the fuel. Oil pump into the fuel injector to the various cylinders of the fuel, the excess fuel will return to the fuel tank. Not all of engines are designed to have a return flow; some of the gasoline engines do not have a return device. The design of the return flow is mainly to provide the necessary cooling nozzle; large diesel engine will have this necessary.

Diesel engine air intake system has two major categories: wet and dry. The suction system of the wet diesel engine is shown in Fig. 9.13. The air will pass through the bottom of the oil chamber to remove impurities in the air and reduce the temperature of the air. Dry air system will be dry filter, paper, cloth or metal filter is commonly used as a filter material. Lowering the temperature of the air intake is beneficial to improve the efficiency of the diesel engine. This is because the lower the temperature, the greater the density, the mass of oxygen in the air will be more, making the burning more fully.

In order to improve the efficiency of the diesel engine, the inhaled air is pressurized. Turbo is often used. Turbocharger is a kind of technology, which uses the diesel engine to drive the exhaust gas to drive the air compressor. Turbocharger is actually an air compressor, through the compressed air to increase the intake of diesel engine. In general, exhaust heat is used to drive the turbocharger. The turbine drives the coaxial impeller, and the air from the air filter pipe is compressed to make it enter the cylinder, as shown in Fig. 9.13.

When the engine speed increases, the exhaust gas discharge speed and the turbine speed also increases rapidly. Therefore, the impeller will compress more air into the cylinder. The increase of air pressure and density can cause more fuel to be fully burnt, and accordingly, the fuel supply should be increased, and the output power of the diesel engine can be increased (Fig. 9.14).

Fig. 9.13 Schematic
diagram of diesel engine
inspiration system

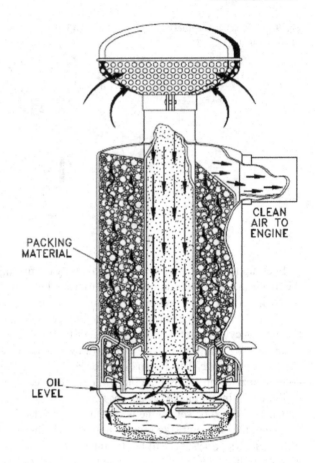

CLEAN
AIR TO
ENGINE

PACKING
MATERIAL

OIL
LEVEL

Fig. 9.14 Turbo

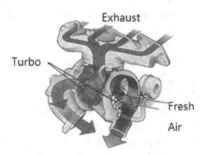

Exhaust

Turbo

Fresh

Air

Supercharging an engine performs the same function as turbocharging an engine. The difference is the source of power used to drive the device that compresses the incoming fresh air. In a supercharged engine, the air is commonly compressed in a device called a blower. The blower is driven through gears directly from the engines crankshaft. The most common type of blower uses two rotating rotors to compress the air. Supercharging is more commonly found on two-stroke engines where the higher pressures that a supercharger is capable of generating are needed.

Diesel engine exhaust system to achieve three functions, one is the exhaust gas, the second is to turbocharged gas (if turbocharged), the third is the silencing.

9.1.3 Principle of Diesel Engine

Before a detailed operation of a diesel engine can be explained [3], several terms must be defined.

Bore and Stroke
Bore and stroke are terms used to define engine size. As mentioned earlier, a bore refers to the diameter of the engine cylinder and a stroke refers to the distance of the piston from the top of the cylinder to the bottom. The highest point of travel for pistons is called the Top Death Center (TDC), and the lowest point of travel is called the Bottom Death Center (BDC). There are 180° trips between BDC and TDC, or a stroke.

Engine Displacement
Engine displacement is one of the terms used to compare one engine with another. Displacement refers to the total capacity of all pistons to shift in one stroke. Displacement is usually used in cubic inches or liters. To calculate the displacement of the engine, the volume of a cylinder must be determined. The volume of one cylinder is multiplied by the number of cylinders to obtain the total engine displacement.

Degree of Crankshaft Rotation
All events in the engine are related to the position of the piston. Since the piston is connected to the crankshaft, any position of the piston corresponds directly to the specific degree of crankshaft rotation.

The position of the crank, which can then be said to be XX degrees before or XX degrees after, the top or bottom of the dead center.

Firing Order
The ignition order is the order in which each cylinder fires in a multi-cylinder engine (power stroke). For example, the ignition order for a four-cylinder engine might be 1-4-3-2. This means that cylinder 1 catches fire, then cylinder 4 catches fire, then cylinder 3 catches fire, and so on. The engine is designed to make the power stroke as uniform as possible, i.e. when the crankshaft rotates a certain number of degrees, one of the cylinders will experience a power stroke. This reduces vibration and allows the power generated by the engine to be applied to the load in a smoother manner, rather than when they fire at the same time or at odd multiples.

Compression Ratio and Clearance Volume

Clearance volume is the volume remaining in the cylinder when the piston is at TDC. Because of the irregular shape of the combustion chamber (volume in the head), the clearance volume is calculated empirically by filling the chamber with a measured amount of fluid while the piston is at TDC. This volume is then added to the displacement volume in the cylinder to obtain the cylinders total volume.

An engine's compression ratio is determined by taking the volume of the cylinder with piston at TDC (highest point of travel) and dividing the volume of the cylinder when the piston is at BDC (lowest point of travel), as shown in Fig. 9.15.

Compression ratio is one of the most important parameters of diesel engine. Compression ratio of modern automobile engine, gasoline engine due to the limit of knock, compression ratio is generally 8–11. Diesel engine without knocking limit, compression ratio is generally 12–22. The compression ratio of the diesel engine in Figs. 9.15 and 9.16.

Diesel engines operate under the principle of the internal combustion engine. There are two basic types of diesel engines, two-cycle and four-cycle. An understanding of how each cycle operates is required to understand how to correctly operate and maintain a diesel engine.

Fig. 9.15 Compression ratio and clearance volume

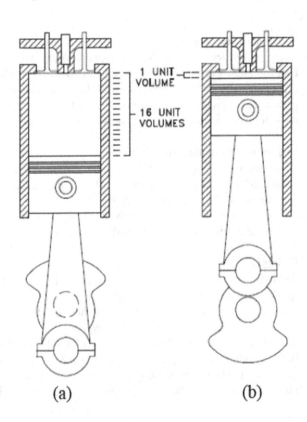

1 UNIT VOLUME

16 UNIT VOLUMES

(a)   (b)

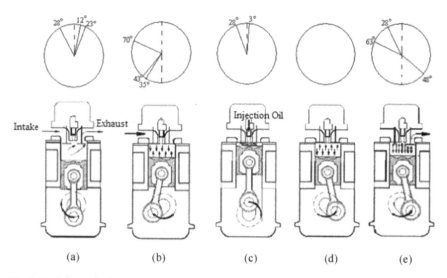

Fig. 9.16 Schematic diagram of four-stroke diesel engine cycle

A diesel engine is a thermal engine that uses an internal combustion process to convert the energy stored in the fuel chemistry into useful mechanical energy. This is done in two steps. First, the fuel reacts (burns) and releases energy in the form of heat. Second, heat causes the gas trapped in the cylinder to expand, the expanding gas is limited by the cylinder, must move the piston in order to expand. The reciprocating motion of the piston is then converted from crankshaft to rotational motion.

To convert the chemical energy of fuel into useful mechanical energy, all internal combustion engines must experience four events: intake, compression, power and exhaust. The time and manner in which these events occur distinguish between various types of engines.

All diesel engines are divided into two categories, two-stroke or four-stroke cycle engines. A word cycle is any action or series of events that are repeated. In the case of a four-stroke cycle engine, the engine requires a full cycle of four strokes (intake, compression, power, and exhaust) of the piston. Therefore, it requires two crankshaft rotations, or 720° crankshaft rotations, to complete a cycle. In a two-stroke cycle engine, events occur only in one rotation of the crankshaft (i.e. 360°).

When discussing the diesel cycle below, it is important to remember the time frame within which each action needs to occur. It takes time to remove the exhaust gas from the cylinder, deliver fresh air to the cylinder, compressed air, injected fuel, and burned fuel. If the four-stroke diesel engine is running at a constant speed of 2100 rpm, the crankshaft will rotate at 35 revolutions per second. One stroke is completed in approximately 0.01429 s.

In a four-stroke engine, the camshaft's gears rotate at half the speed of the crankshaft (1:2). This means that the crankshaft must undergo two complete revolutions before the camshaft can complete a revolution. The next section will describe a four-stroke, normally aspirated, diesel engine with intake and exhaust valves with a 3.5-inch bore and a 4-inch stroke with a 16:1 compression ratio as it passes through a full cycle. We'll start with a stroke. All timing marks given are generic and will vary depending on the engine. Refer to Fig. 9.16 in the following discussion.

Intake

As the piston moves upward, approaching 28° before the top death center (BTDC), by the measurement of the crankshaft rotation, the camshaft lobe begin to lift the cam follower. This causes the pushrod to move upward and pivots the rocker arm on the rocker shaft. When the valve lash is taken up, the rocker arm pushes the air valve down and the valve opens. The intake stroke now starts and the exhaust valve is still open. The flow of exhaust gas will create a low-pressure condition in the cylinder and help to draw in fresh air charge, as shown in Fig. 9.16a.

The piston continues upward through the top death center (TDC), while fresh air enters and exhaust gases leave. At approximately 12° after the top death center (ATDC), the camshaft exhaust lobe rotates so that the exhaust valve starts to close. The valve is completely closed at the 23° ATDC. This is done through a valve spring, which was compressed when the valve is opened, forcing the rocker arm and cam follower back against the cam lobe as it rotates. The time range for both the intake and exhaust valves to open is called valve overlap (51° overlapping in this example) and is necessary to allow fresh air to help scavenge (remove) the exhaust gas and cooling cylinders. In most engines, 30–50 times the cylinder volume is scavenged by cylinder during overlap. This excess cold air also provides the necessary cooling effect on engine components.

As the piston passes through the TDC and begins to move along the cylinder bore, the movement of the piston creates suction and continues to absorb fresh air into the cylinder.

Compression

At 35° after the bottom death center (ABDC), the intake valve starts to close. At 43° ABDC (or 137° BTDC), the intake valve is located on the seat and is completely closed. At this point, the air charge is at normal pressure (1 atm) and ambient temperature (25 °C), as shown in Fig. 9.16b.

At approximately 70° BTDC, the piston travels approximately 2.125 inches, or half of its stroke, reducing the volume in the cylinder by half. The temperature has now doubled to 70 °C and the pressure is ~2.3 atm.

The stroke and volume of the stroke has been halved again at about 43° BTDC pistons that have been up 3.062 in. As a result, the temperature is doubled again to about 160 °C, with a pressure of ~5.78 atm. When the piston reaches 3.530 in, it halves again, reaching a temperature of 337 °C and a pressure of 15.4 atm. When the piston reaches 3.757 in, or halves again, the temperature climbs to 600 °C and the pressure reaches 50 atm. The piston area is 9.616 in^2 and the pressure in the cylinder exerts about 30,000 N force.

The above numbers are ideal and provide a good example of what happens to the engine during compression. In the actual engine, the pressure only reaches about 47 atm. This is mainly due to the loss of heat from the surrounding engine components.

Fuel Injection

Fuel in liquid state is injected into the cylinder at precise time and rate to ensure that combustion pressure forces on pistons neither too early nor too late, as shown in Fig. 9.16c. The fuel enters the cylinder where compressed air is heated; However, it only burns when it is in an evaporative state and is closely mixed with the oxygen supply. In the first minute, the fuel droplets enter the combustion chamber and evaporate rapidly. The evaporation of the fuel cools the air around the fuel, which takes time to reheat to ignite the evaporated fuel. However, once ignition begins, the additional heat generated by combustion helps to further evaporate the new fuel entering the chamber, as long as oxygen is present. The fuel injection starts at 28° BTDC and ends at 3° ATDC; As a result, the fuel injection duration is 31°.

Power

Both valves are closed and the fresh air charge is compressed. Fuel has been injected and started to burn. When the piston passes through the TDC, the fuel ignites quickly, causing the cylinder pressure to increase. The combustion temperature is around 1280 °C. The rise in pressure forces the piston down and increases the force on the crankshaft for the power stroke of the crankshaft, as shown in Fig. 9.16d.

Not all of the energy generated by the combustion process is utilized. In two-stroke diesel engines, only about 38% of the power generated is used to do work, about 30% of the power is wasted on the cooling system in the form of heat, and about 32% of the heat is exhausted. In contrast, a four-stroke diesel engine with a thermal distribution of 42% converted to useful work, 28% of the heat rejected to the cooling system, and 30% of the heat is discharged from the exhaust.

Exhaust

As the piston approaches the 48° BBDC, the cam in the exhaust lobe begins to force the follower upward, causing the exhaust valve to lift up from the seat. As shown in Fig. 9.16e, due to cylinder pressure, exhaust gas begins to flow out of the exhaust valve and into the exhaust manifold. After passing through the BDC, the piston moves upward and accelerates to maximum speed at 63° BTDC. From this the piston is slowing down. When the piston slows down, the speed of the gas flowing out of the cylinder creates pressure slightly lower than atmospheric pressure. At the 28° BTDC, the intake valve opens and the cycle starts again.

9.2 Heat Exchanger

In almost any nuclear, chemical, or mechanical system, heat must be transferred from one place to another or from one fluid to another. Heat exchangers are used to transfer heat from one fluid to another. A basic understanding of the mechanical components of a heat exchanger is important to understanding how they function and operate.

A heat exchanger is a component that allows the transfer of heat from one fluid (liquid or gas) to another fluid. Reasons for heat transfer include the following:

1. To heat a cooler fluid by means of a hotter fluid
2. To reduce the temperature of a hot fluid by means of a cooler fluid
3. To boil a liquid by means of a hotter fluid
4. To condense a gaseous fluid by means of a cooler fluid
5. To boil a liquid while condensing a hotter gaseous fluid.

Regardless of the function the heat exchanger fulfills, in order to transfer heat the fluids involved must be at different temperatures and they must come into thermal contact. Heat can flow only from the hotter to the cooler fluid.

In a heat exchanger there is no direct contact between the two fluids. The heat is transferred from the hot fluid to the metal isolating the two fluids and then to the cooler fluid.

Although heat exchangers come in every shape and size imaginable, the construction of most heat exchangers fall into one of two categories: tube and shell, or plate. As in all mechanical devices, each type has its advantages and disadvantages.

The most basic and common types of heat exchanger structures are tubes and shell, as shown in Fig. 9.17. This type of heat exchanger consists of a set of tubes in a container called a shell. The fluid flowing through the tube is called the tube side fluid, and the fluid flowing through the outside of the tube is called the shell side fluid. At the end of the tube, the pipe side fluid is separated from the shell side fluid by the tube sheet. The pipe is rolled, pressed or welded into the sheet to provide a tight seal of leakage. In systems where the two fluids are at very different pressures, high-pressure fluids are usually directed through tubes and low-pressure fluids circulate on the shell side. This is due to economic reasons, as heat exchangers tubes can withstand higher pressures and are much less expensive than heat exchanger shell. The support plate shown on Fig. 9.17 also acts as a baffles, guiding fluid from the shell back and forth through the tube.

As shown in Fig. 9.18, the plate heat exchanger consists of a plate instead of a tube to separate hot and cold liquids. Hot and cold liquids alternate between each plate. The baffles guide the flow of fluid between plates. Each plate provides a very large heat transfer area for each fluid. As a result, plate heat exchangers can transfer more heat than similar-sized tubes and shell heat exchangers. This is due to the larger area provided by the plates. Because of the high heat transfer efficiency, the plate heat exchanger is usually very small. Plate heat exchangers are not widely used because large gaskets between each plate cannot be reliably sealed. Because of this problem, plate heat exchangers are only used in small low-pressure applications, such as the engine's oil cooler. However, new improvements in gasket design and overall heat exchanger design have led to some large-scale applications of plate heat exchangers. With the upgrading of old facilities or the completion of newly designed facilities, large plate heat exchangers are replacing tubes and shell heat exchangers and are becoming more common [5].

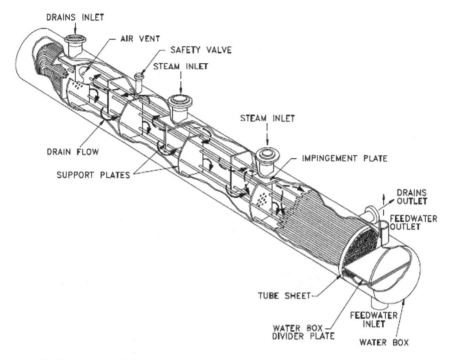

DRAINS INLET

AIR VENT

SAFETY VALVE

STEAM INLET

STEAM INLET

DRAIN FLOW

IMPINGEMENT PLATE

SUPPORT PLATES

DRAINS OUTLET

FEEDWATER OUTLET

TUBE SHEET

FEEDWATER INLET

WATER BOX DIVIDER PLATE

WATER BOX

Fig. 9.17 Tube and shell heat exchanger

9.3 Pump

A pump is a mechanical device that pumps fluid or presses fluid. It transfers the mechanical energy or other external energy of the prime mover to the fluid, which increases the energy of the fluid. The general pump is mainly used to transport liquids for transporting gas are usually called fan. According to the nature of the delivery of liquid can be divided into water pumps, oil pumps and mud pumps, etc.

The pump lift or head refers to the height, usually expressed in H, the unit is m or mH_2O. It is proportional to the pressure difference between the outlet and the inlet of the pump under the rated condition.

The work done by a pump in unit time, is called power, use the symbol P to represent. Power from pump shaft, called the shaft power, can be understood as the input power of pump. Generally speaking, the pump power is the power of the shaft. Due to the friction resistance between filler and bearing, friction during rotation of the impeller and the water, flow of the water pump in the whirlpool, refluxing clearance, import and export impact and other reasons, is bound to consume a part of power, so the pump is not possible transfer shaft power completely to fluid. There must be a power loss. That is to say, the effective power of the pump plus the loss of power within the pump equals the pump shaft power. The ratio of effective power and shaft power is pump efficiency.

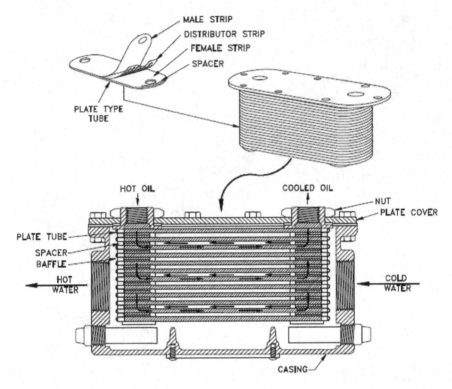

Fig. 9.18 Plate type heat exchanger

There is a certain relationship between head and power; we give an example to illustrate.

Example 9.1 The inlet pressure of a pump is 1 bar, the outlet pressure is 8 bar, the flow rate is 5000 kg/s, the fluid density is 1000 kg/m³, the efficiency of the pump is 85%, assuming that the inlet and outlet height is the same; the flow rate is the same. Calculate pump power.

Solution: According to the principle of conservation of energy, the pump shaft power is

$$P = \frac{Q(p_2 - p_1)\frac{1}{\rho}}{\eta} = \frac{5000 \times \left(8 \times 10^5 - 1 \times 10^5\right) \times \frac{1}{1000}}{0.85} = 4.1 \text{ MW}$$

There are a variety of pumps, below we first introduced the centrifugal pump.

9.3.1 Centrifugal Pump

Centrifugal pumps are the most common type of pumps found in DOE facilities. Centrifugal pumps enjoy widespread application partly due to their ability to operate over a wide range of flow rates and pump heads.

Centrifugal pump is the most commonly used pump, centrifugal pump is rotating impeller and the water have centrifugal motion, as shown in Fig. 9.19. A pump before starting must make the shell and the water pump tube filled with water. And then start the motor, the pump shaft driven impeller and water do high-speed rotation movement. Movement of the centrifugal water thrown to the outer edge of the impeller and the cochlear shape in the flow passage of the pump shell into the outlet pipe of the pump [6].

When the fluid enters the centrifugal pump, the first to enter the impeller inlet. With the impeller rotation, transfer energy to the entering fluid, so that the fluid is accelerated, increasing the kinetic energy of the fluid. The high-speed fluid subsequently pushed into the volute. Volute area is expanding a channel section area, in order to convert kinetic energy to fluid flow energy, pV, the velocity of the fluid decreased and elevated pressure. Design of volute district has two types, as shown in Fig. 9.20, respectively single volute design and double volute design. Double volute design can reduce centrifugal force on the pump shaft and bearing force.

Some centrifugal pumps contain diffusers. A diffuser is a set of fixed blades that surround the impeller. The purpose of the diffuser is to reduce the speed by allowing the liquid to expand more gradually and have fewer turbulent areas to increase the efficiency of the centrifugal pump. Diffuser blades are designed to encounter increasing

Fig. 9.19 Centrifugal pump

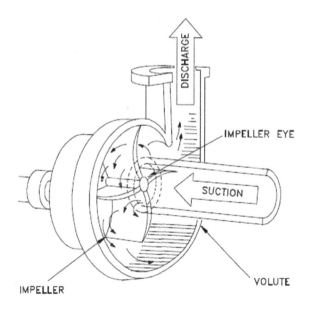

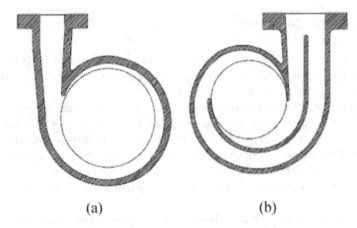

(a)   (b)

Fig. 9.20 Single and double volutes

flow areas as liquid passes through the diffuser. The increase of flow area leads to a decrease in flow rate, which converts kinetic energy into flow pressure (Fig. 9.21).

The impeller of the pump is classified according to the number of points that the liquid can enter the impeller and the amount of webbing between the impeller blades.

Impellers can be single or double suction. Single-suction impellers allow liquid to enter the center of the blade in only one direction. The double-suction impeller allows liquid to enter the center of the impeller blade from both sides at the same time. Figure 9.22 shows a simplified diagram of single-suction and double-sucking impellers.

Impellers can be opened, half opened, or enclosed. The open impeller consists only of blades connected to the hub. The construction of the semi-open impeller is connected to a circular plate (the web) on one side of the blade. Round plates are attached to the blades of the closed impeller. Enclosed impellers are also known

Fig. 9.21 Centrifugal pump diffuser

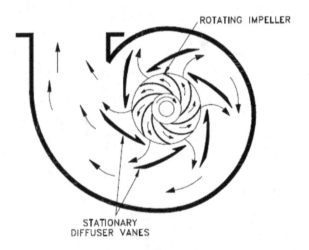

ROTATING IMPELLER

STATIONARY
DIFFUSER VANES

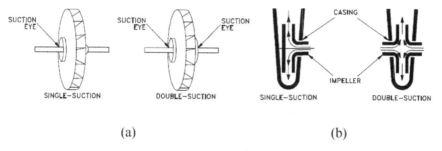

Fig. 9.22 Single-Suction and double-suction impellers

as shrouded impellers. Figure 9.23 illustrates examples of open, semi-open, and enclosed impellers.

The impeller sometimes contains balancing holes that connect the space around the hub to the suction side of the impeller. The balancing holes have a total cross-sectional area that is considerably greater than the cross-sectional area of the annular space between the wearing ring and the hub. The result is suction pressure on both sides of the impeller hub, which maintains a hydraulic balance of axial thrust.

Centrifugal pumps with a single impeller can produce a difference of more than 10 atm between suction and discharge, having been designed and built at a difficult and costly cost. A more economical way to develop high pressure using a single centrifugal pump is to install multiple impellers on a normal shaft in the same pump housing. The internal passages in the pump housing transmit the discharge of one impeller to the suction of the other impeller. Figure 9.24 shows the arrangement of the four-stage pump impellers. Water enters the pump from the upper left corner and passes continuously through four impellers, left to right. The water goes from the volute surrounding the discharge of one impeller to the suction of the next impeller.

A pump stage is defined as that portion of a centrifugal pump consisting of one impeller and its associated components. Most centrifugal pumps are single-stage pumps, containing only one impeller. A pump containing seven impellers within

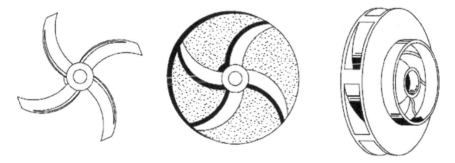

Fig. 9.23 Open, semi-open, and enclosed impellers

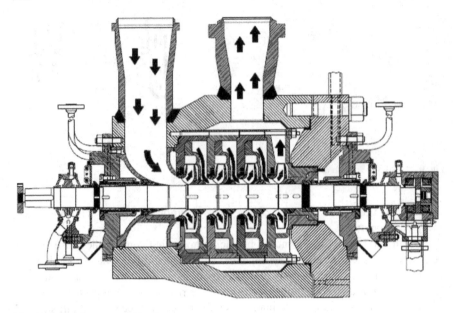

Fig. 9.24 Multi stage centrifugal pump

a single casing would be referred to as a seven-stage pump or, or generally, as a multi-stage pump.

Many centrifugal pumps are designed to allow the pump to run continuously for months or even years. These centrifugal pumps typically rely on pumped liquids to cool and lubricate pump bearings and other internal components of the pump. If the flow through the pump is stopped while the pump is still running, the pump will no longer cool sufficiently and the pump may be damaged quickly. Pump damage can also result from pumping a liquid whose temperature is close to saturated conditions.

The flow area in the eye of the pump impeller is usually smaller than the flow area of the pump suction piping or through the impeller vanes. When the pumped fluid enters the eyes of the centrifugal pump, a decrease in flow area leads to an increase in flow rate and a decrease in pressure. The greater the pump flow rate, the greater the pressure drop between the pump suction and the impeller eye. If the pressure drops large enough, or if the temperature is high enough, the pressure drops below the saturation pressure of the pumped fluid, which may be enough to cause the liquid to flash to steam. Any steam bubbles formed by the pressure drop in impeller eye are swept along the impeller blades by the flow of fluid. Bubbles burst when they enter areas where local pressure is greater than saturation pressure. The process of formation and subsequent collapse of steam bubbles in the pump is called cavitation.

Cavitation in centrifugal pumps have a significant effect on pump performance. Cavitation degrades pump performance, causing fluctuations in flow and discharge pressure. Cavitation can also be disruptive to the internal components of the pump. When the pump cavity and steam bubbles are formed directly in the low-pressure area behind the rotating impeller blades. These steam bubbles then move toward the oncoming impeller blades, where they collapse, causing a physical impact on the front of the impeller blades. This physical shock creates small pits at the forefront of the impeller blades. Each pit is microscopic in size, but the cumulative effect of millions of pits formed over hours or days can actually destroy pump impellers. Cavitation can also cause excessive pump vibration, which can damage pump bearings, wearing rings and seals.

A small number of centrifugal pumps are designed to operate under unavoidable cavity conditions. These pumps must be specifically designed and maintained to withstand the small number of cavities that occur during operation. Most centrifugal pumps are not designed to withstand sustained cavitation.

Noise is one of the signs of cavitating. The cavitation pump sounds like a can of marbles being shaken. Other signs observed from the remote operating station are fluctuations in discharge pressure, flow rate and pump motor current. Methods to block or prevent cavities are described in the following paragraphs.

It is difficult to completely eliminate cavitation erosion in the long run. The effective measure to reduce cavitation is to prevent the generation of bubbles as much as possible. First, the liquid contact surface has a very good streamline, to avoid the vortex in the local area, because the vortex area of low pressure, easy to produce bubbles. In addition, it should reduce the dissolved gas content in the liquid and the disturbance in the liquid flow, and also can restrict the formation of bubbles.

The main hazards of cavitation are:

(1) To produce vibration and noise. Bubble collapse, liquid particle colliding, also hit the metal surface, the noise of various frequencies. Can be serious when the pump has heard "crackling" sound of the explosion, and vibration.
(2) Reduce the pump performance. Cavitation produced a large number of bubbles, and even full of flow channels, the destruction of the continuous flow of liquid in the pump. This will make the pump flow, head and efficiency decreased significantly.
(3) Damage to internal mechanical components. Due to mechanical erosion and electrochemical corrosion, the metal material is damaged, usually by the cavitation damage of the parts in the vicinity of the impeller exit and near the entrance of the liquid discharge chamber. Early cavitation performance for metal surface pitting, then present the cavernous, groove, honeycomb, scaly, traces, serious when can cause blade or front and back cover of perforation, or impeller rupture, causing a serious accident.

To avoid cavitation in centrifugal pumps, the pressure of the fluid at all points within the pump must remain above saturation pressure. The quantity used to determine if the pressure of the liquid being pumped is adequate to avoid cavitation is the net positive suction head (NPSH). The net positive suction head available ($NPSH_A$) is the difference between the pressure at the suction of the pump and the saturation pressure for the liquid being pumped. The net positive suction head required ($NPSH_R$) is the minimum net positive suction head necessary to avoid cavitation.

Fig. 9.25 Pump head flow characteristics and operating points

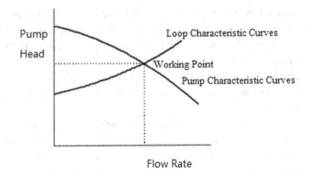

$$\text{NPSH} = p_{su} - p_{sa} \tag{9.1}$$

where, p_{su} is the pressure at the inlet of the impeller, mH_2O; p_{sa} is the saturation pressure at the corresponding temperature, mH_2O.

The necessary cavitation margin of large pump is generally 50–75 mH_2O.

There is a certain relationship between the head and the flow of centrifugal pump. The "pump characteristic" curve of Fig. 9.25 is the relationship between the flow rate and the head of a typical centrifugal pump.

According to this feature, we can use two methods to adjust the pump flow. The first is by changing the actual characteristics of the pipeline (for example, at the outlet of the pump to install a control valve); the second way is to change the speed of the pump (such as governor). One of the first methods is low cost, but when the valve opening changes the pressure head loss increases, the power supply to the pump is greater than the actual needs of power. The cost of the second methods is higher, but the pressure head loss does not change, the power provided by the power is close to the actual needs of power.

Centrifugal pumps also follow some of the inherent laws. For example, flow and pump speed is proportional, the head and the square of pump speed is proportional, the pump power and the three square of pump speed is proportional. These are known as the pump's laws, and are sorted as follows:

$$Q \propto n \tag{9.2}$$

$$H_p \propto n^2 \tag{9.3}$$

$$P \propto n^3 \tag{9.4}$$

其中, n is the rotational speed, rpm; Q is the volume flow, m^3/s; H_p is the head, mH_2O; P is the power, W. An example is given to illustrate the application of these laws to the pump.

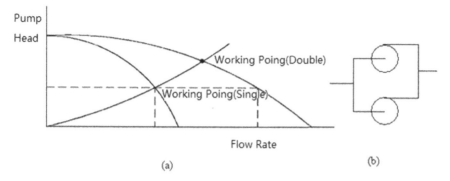

Fig. 9.26 Working point of double pump parallel connection

Example 9.2 A coolant pump operating at 1500 rpm, the volume flow is 10 m³/s, head is 15 mH$_2$O, the pump power is 45 kW, determine the flow rate, head and power after the speed is adjusted to 1800 rpm.

Solution:

$$\text{flow rate}: Q_2 = Q_1 \times (n_2/n_1) = 10 \times (1800/1500) = 12 \text{ m}^3/\text{s}$$

$$\text{head}: H_2 = H_1 \times (n_2/n_1)^2 = 15 \times (1800/1500)^2 = 21.6 \text{ mH}_2\text{O}$$

$$\text{power}: P_2 = P_1 \times (n_2/n_1)^3 = 45 \times (1800/1500)^3 = 77.76 \text{ kW}$$

Centrifugal pump due to fewer moving parts, can be adapted to different drives, including DC or AC motors, diesel engines, steam turbines or air engines, etc.. Centrifugal pump usually has the characteristics of small size, low cost, low head, large flow rate.

If you want to increase the flow, you can achieve through the parallel multiple pumps; if you want to increase the head, you can use a series of ways. Figure 9.26 is the flow head characteristic of two parallel pumps. When considering the characteristics of the pipeline, the head of two parallel pumps is also increased, and the flow rate is not two times the flow. Two pumps in series as shown in the case of Fig. 9.27.

9.3.2 Positive Displacement Pump

Positive displacement pump is a pump that uses the change of pump cylinder content to transport liquid. Positive displacement pump can be divided into three types: reciprocating pump, rotor pump, diaphragm pump.

Reciprocating pump is the use of reciprocating motion of the piston to transport liquid pump, by the reciprocating motion of the piston to the energy directly in the form of static pressure to pump liquid. Because liquid is almost incompressible, so when the piston pressure to send liquid, you can make the liquid to withstand high

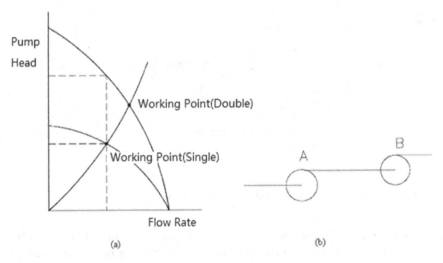

Fig. 9.27 Working point of double pumps in series

pressure, thereby obtaining a high head. Bicycle manual inflator is a typical example of a reciprocating pump.

The rotor pump is composed of a stationary pump shell and a rotating rotor. It has no suction valve and discharge valve, by the pump body of the rotor and the liquid contact side of the energy in the form of static pressure directly to the liquid, and the liquid through the rotating rotor of the extrusion. At the same time, in the other side to set aside space, the formation of low pressure, so that liquid continuous inhalation. Rotor pump with higher-pressure head, flow rate is usually smaller, uniform liquid discharge. Applicable to transport high viscosity, with lubrication, but does not contain solid particles of liquid. Rotor pumps are commonly used in the type of gear pump, screw pump, vane pump, flexible impeller pump, rotary pump, rotary piston pumps, etc. Which gear pump and screw pump is the most common rotor pump.

Diaphragm pump, also known as control pump, is the main type of actuator. By receiving a control signal output from the modulation unit. Diaphragm pump with the power to change the flow of fluid. Diaphragm pump according to the different liquid media were used to nitrile rubber, chloroprene rubber, fluorine rubber, poly partial fluorine ethylene, polyethylene as a membrane material. The diaphragm can be placed in a variety of special occasions, for pumping various medium that conventional pumps cannot.

Reciprocating Positive Displacement Pump

The principle of reciprocating pump as shown in Fig. 9.28. Piston in the cylinder for reciprocating motion, when the piston moves to the left (a), the suction valve opens, and the discharge valve closed, from the water source to the cylinder suction water. When the piston moves to the right (b), the valve of the suction pipe is closed,

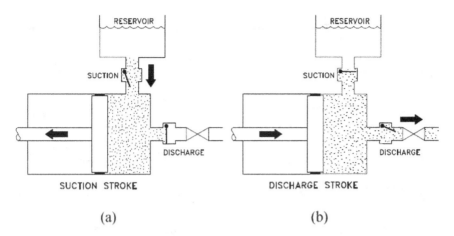

Fig. 9.28 Working principle of reciprocating pump

the valve of the discharge pipe is opened, and the water is discharged to complete a cycle. Continue to go back and forth on the water supply to the destination.

Reciprocating displacement pump is the key to two one-way check valve, control fluid can flow only to one direction. Figure 9.28 in the reciprocating pump outlet fluid is not continuous. Because the check valve works a bit like a diode inside the circuit, so Fig. 9.28 reciprocating pump is a bit similar to the half wave rectifier circuit shown in Fig. 5.31. From here we can get inspiration: if you can design the Fig. 5.33 as shown in the bridge full wave rectifier, you can make a continuous flow. This is not difficult to achieve, such as the design of Fig. 9.29. Among them, Fig. 9.29a is a single-acting reciprocating pump, Fig. 9.29b is a double-acting pumps useing four check valve.

Rotary Pump

Gear pump is a typical rotor pump, its working principle as shown in Fig. 9.30. The motor drives the gear to rotate, and the mechanical energy can be directly translated into the flow energy of the conveying fluid through the pump. Pump flow depends only on the size and speed of the gear, in theory and the export of pressure is not independent. Therefore, the gear type rotor pump head can be very high.

The gear pump is a rotary pump which is formed by the working volume change and movement of the pump cylinder and the meshing gear. Consists of two gears, the pump body and the front and back cover to form two closed spaces. When the gear rotaries, the space volume of gear disengaged side varies from small to large, negative pressure is formed, the inhalation of the liquid; the space volume of the gear meshing side varies from larger to small and liquid to squeeze into the pipeline flow. The suction chamber and the discharge chamber are separated by the meshing lines of the two gears. The pressure of the discharge port of the gear pump is completely determined by the resistance at the pump outlet.

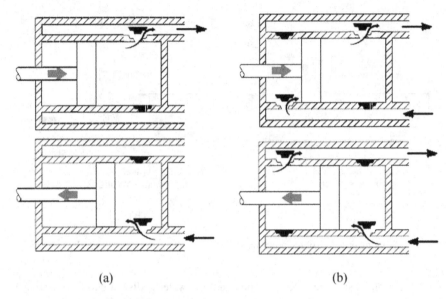

Fig. 9.29 Single-Acting and double-acting pumps

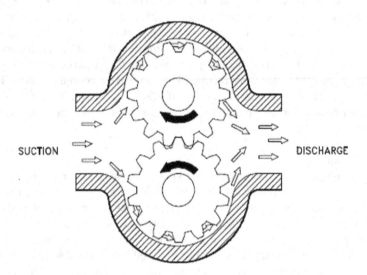

Fig. 9.30 Gear pump

The concept of gear pump is very simple, that is, the basic form of it is that the two gears of the same size in a tightly matched shell are engaged in one another. The inside of the shell is similar to the "8" shape, the two gears are mounted inside, the outer diameter of the gear and the two sides of the gear are closely matched with the casing. The material from the extruder enters the middle of the two gears in the intake port, and is filled with the space, along with the rotation of the gear, the gear is rotated along the housing, and finally, the fluid is discharged at the meshing point of the two gears.

In each turn, the amount of discharge is the same. With the continuous rotation of the drive shaft, the pump will not be interrupted to discharge fluid. The flow rate of the pump is directly related to the speed of the pump. In fact, because both sides of the seal is not possible to 100%, so there will be a small amount of loss in the pump, which makes the pump running efficiency can not reach 100%. These leaks can be used to lubricate the bearings and gears on both sides of the gear so that the pump can run well. For most of the extruded materials, the efficiency can reach 93–98%.

When the viscosity or density of the fluid changes in the process, this type of pump will not be affected too much. If there is a damper, such as a filter or a limiter on the side of the discharge port, the pump will push the fluid through them. If the damper changes at work, i.e., if the mesh become dirty, blocked or limiting back pressure increased, the pump will maintain a constant flow until it reaches the mechanical limit in the weakest parts of the device (usually equipped with a torque limiter).

The lobe type pump shown in Fig. 9.31 is another variation of the simple gear pump. It is considered as a simple gear pump having only two or three teeth per rotor; otherwise, its operation or the explanation of the function of its parts is no different. Some designs of lobe pumps are fitted with replaceable gibs, that is, thin plates carried in grooves at the extremity of each lobe where they make contact with the casing. The gib promotes tightness and absorbs radial wear.

Screw pump is a volume type rotor pump, it is to rely on the screw and bushing formed by the sealing chamber volume change to suck and discharge liquid. Screw pump according to the number of screw is divided into single screw pump, double screw pump, three screw pump and five screw pump. Screw pump is characterized

Fig. 9.31 Lobe type pump

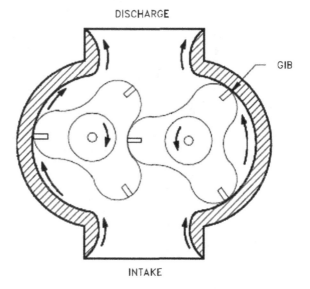

by smooth flow, the pressure fluctuation is small, self-priming ability, low noise, high efficiency, long service life, reliable work, and its outstanding advantage is the transmission medium does not form a vortex, insensitive to viscosity of medium, the transport of high viscosity medium. Its schematic diagram, as shown in Fig. 9.32.

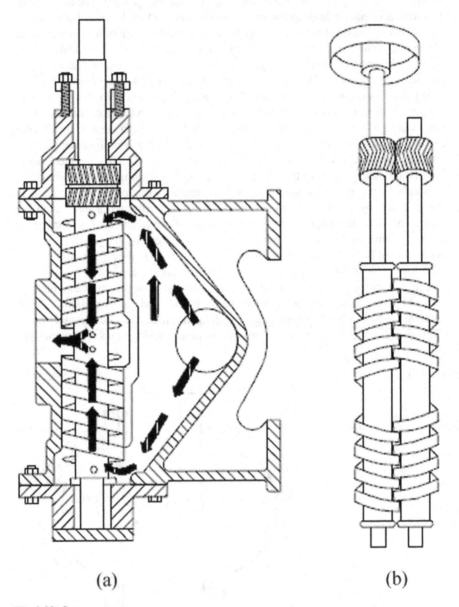

(a)   (b)

Fig. 9.32 Screw pump

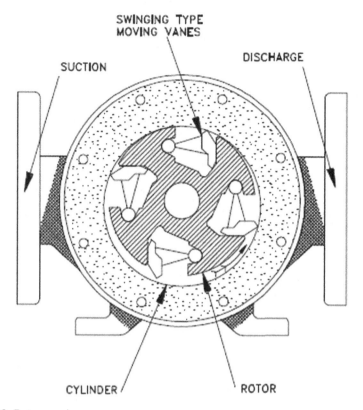

Fig. 9.33 Rotary moving vane pump

The rotating moving vane pump shown in Fig. 9.33 is another positive displacement pump used. The pump consists of a cylindrical, boring housing with an inlet on
one side and an discharge port on the other. A cylindrical rotor less than the diameter
of cylindrical is driven around an axis placed on the centerline of the cylinder. The
gap between the rotor and the cylinder is small at the top, but increases at the bottom.
The rotor carries vanes that move in and out and maintains a sealed space between
the rotor and the cylinder wall as it rotates. The vanes trap the liquid or gas on the
suction side and delivers it to the discharge side, where space shrinks to drain it from
the discharge line. The vanes may swing on pivots, or they may slide in slots in the
rotor.

Diaphragm pumps are also classified as positive displacement pumps because the
diaphragm acts as a finite displacement piston. The pump acts when the diaphragm
is forced to return through mechanical linkage, compressed air, or liquid from a
pulsating external source. The pump structure eliminates any contact between the
pumped liquid and the energy source. This eliminates the possibility of leakage,
which is important when dealing with toxic or very expensive liquids. Disadvantages

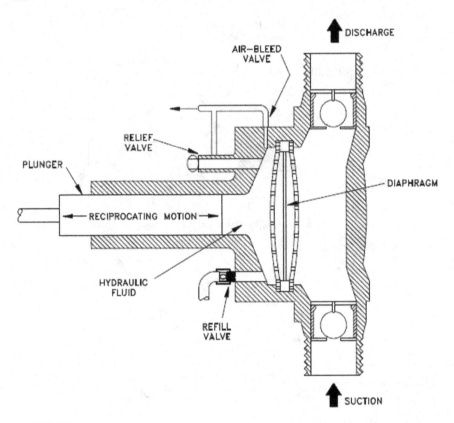

Fig. 9.34 Diaphragm pump

include limited head and capacity ranges and the need to check valves in suction and discharge nozzles. An example of a diaphragm pump is shown in Fig. 9.34.

The diaphragm pump mainly consists of two parts: the transmission part and the cylinder head. Driving part is driven by the driving mechanism of the diaphragm back and forth. Its transmission form has mechanical transmission, hydraulic transmission and pneumatic transmission, etc. Which is widely used in hydraulic transmission. When the diaphragm pump works, the crank connecting rod mechanism is driven by the motor to drive the piston to move back and forth. Movement of the piston through the liquid cylinder of the working fluid (generally for oil) and spread to the diaphragm, so that the diaphragm back and forth.

The cylinder head part of the diaphragm pump mainly consists of a diaphragm piece, which is separated from the liquid and the working liquid. When the diaphragm is moved to the side of the transmission mechanism, the pump cylinder is negative pressure to suck liquid; when the diaphragm moves to the other side, the liquid is discharged. The liquid in the pump cylinder is separated by diaphragm and working liquid, only contact with pump cylinder, a suction valve and discharge valve and pump side diaphragm and without touching the piston and sealing device, which makes the piston important parts in oil medium work in good state.

Diaphragm must have good flexibility, but also has a better corrosion resistance. Diaphragm is usually made of PTFE, rubber and other materials. The sealing performance of the diaphragm pump is good, and can be easily reached without leakage, and can be used for transporting acid, alkali, salt and other corrosive liquids and high viscosity liquids.

9.3.3 Coolant Pump for Pressurized Water Reactor Nuclear Power Plant

The coolant pump is used to circulate the reactor coolant in a loop, which transfers the heat from the reactor core to the steam generator, also known as the main pump. Usually install one or two main pumps in each loop.

Because the reactor coolant has a strong radioactive, so the main technical difficulties of the main pump is to be controlled by leakage. Also due to the loss of power in the main pump, it is required to maintain a certain flow at the best time to cool the core, so the main pump generally with heavy flywheel, to increase the moment of inertia of the rotor.

The main pump is usually vertical, single-stage, centrifugal pump, driven by AC motor. The main parts of the main pump including shaft seal, flywheel, thrust bearings, impeller and guide vane, pump casing, shaft and motor, etc., as shown in Fig. 9.35.

Shaft seal is the most important part of the main pump, its design and manufacture is difficult. At present, the commonly used seal form is to control the leakage and non-contact mechanical seal. The basic principle is that the sealing medium is introduced between the two end faces of the moving ring and the static ring, and the liquid film, which is formed by a layer of a few microns thick, is formed by the action of lubricating and cooling. According to the details of the structure can be divided into static pressure and dynamic pressure seal. The former is usually for groove type; the latter lubrication groove is provided on the end surface of the friction surface, medium into the slot, reuse generated by the rotation of the fluid wedge pressure action, squeeze into the surface, the formation of liquid film..

In order to establish the liquid friction between the moving ring and the static ring, the contact end surface requires the material with very low roughness and strict control of the tolerance, and the proper choice of the material with good physical and mechanical properties. Commonly used materials are graphite, tungsten carbide, silicon carbide, silicon nitride, alumina, etc. In addition, the operation to control the shaft seal water quality, especially the particle size of impurities containing in the water, usually in the injection pipeline to set the ultra-fine filter, can filter out 5 μm particles of impurities.

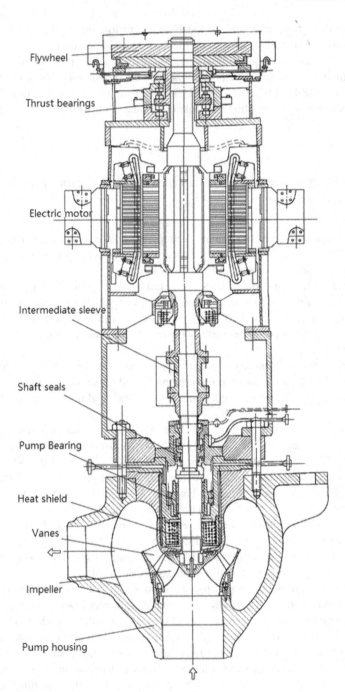

Flywheel

Thrust bearings

Electric motor

Intermediate sleeve

Shaft seals

Pump Bearing

Heat shield

Vanes

Impeller

Pump housing

Fig. 9.35 Three bearing pump group structure profile primary coolant pump

A shaft seal usually has three stages. Typical assembly with a first main seal, bear all or most of the pressure, the large amount of leakage; and the second stage is the security seal, in the first level failure but short bearing the system full running pressure; third level is also safe sealing, low working pressure difference, a small amount of leakage, sometimes use contact seal.

The structure design of shaft seal must be verified by experiment. Usually with full size test pieces (such as dynamic and static ring effective diameter 254 mm), in the test bench to work under normal operating pressure, temperature and water injection conditions for a long time, determine the amount of leakage, vibration and other data. In the case of changing the temperature and pressure of the injection water, the experiment was carried out. Test time to exceed the design life of shaft seal. After the test, the test piece is disintegrated, and the contact surface and the deformation of the moving ring are checked.

During normal operation, the seal water in most of the nuclear power plant is supplied by the pump of the chemical and volume control system. After it enters a sealed shell divided into approximately equal of two streams: One stream shares down into the pump body, and coolant mixing; another shares on the first level and second level seal were introduced volume control tank and coolant drain tank. The third seal has a separate water injection, also into two shares, one mixes with the secondary sealing leakage; another is lead to coolant drain tank or containment. When a fault of normal injection water occurs, the coolant in the pump is cooled by a heat shield heat exchanger installed on the pump cover or cooled by a high-pressure cooler outside the pump, and the cooling water system is continuously supply water.

When the main pump lose power, the inertia of the flywheel can make the pump run for a long time. The rotational inertia of components can provide the appropriate flow in a certain period of time and usually require in the 30 s flow is not less than 50%. Taking the main pump of 1000 W nuclear power plant as an example, the flywheel mass is about 5 tons, and the moment of inertia is about 1800 kg m^2. Flywheel material for low alloy steel requires a high strength and impact toughness. When the pump is over speeding, the flywheel will fall off from the shaft to avoid the cracking caused by the super stress and produce fly objects.

Table 9.1 are some of the main parameters of the typical pump for the reader's reference.

9.4 Valve

Valve is fluid transport control system components, with cut off, regulation, diversion, to prevent the counter flow, pressure regulator, and diversion or overflow relief pressure. Valves for fluid control systems, from the simplest to the most complex automatic control system used in a variety of valves, the variety and specifications are quite a number of. Valve can be used to control the air, water, steam, all kinds of corrosive media, mud, oil, liquid metal and radioactive media, and other types of fluid flow.

Table 9.1 Typical main pump parameters

Parameters	Qinshan (50 Hz)	93D type (50 Hz)	100 type (60 Hz)	100 type (50 Hz)
Rated flow /(m³/h)	16,800	21,350	22,620	22,620
Rated head /m	75	86.31	100	100
Rated efficiency /%	79	82	81	87
Casting weight /t	88	31.8	29	29
Casting diameter /m	–	2.65	2.44	2.44
Water outlet position	Tangential	Tangential	Radial	Tangential
Critical speed /rpm	2500	2600	1610	1800
Standard moment of inertia /(kg·m²)	1750	2318	4638	2967
Motor rated power /kW	4000	5147	5882	5882
Synchronous speed /rpm	1500	1500	1200	1500

9.4.1   Valve Type

According to the role of the valve and the use of classification, can be divided into.

(1)   Cut off valve: such as gate valve, globe valve, plug valve, ball valve, butterfly valve, needle valve, diaphragm valve, etc. The main function of the cut-off valve is to switch on or cut off the medium in the pipeline.

(2)   One way valve: such as check valve (also known as a one-way valve). Check valve is an automatic valve, its main role is to prevent the flow of media in the pipeline back to prevent the pump and drive motor reverse, to prevent the leakage of the medium inside the container.

(3)   Safety valve: such as explosion-proof valve, safety valve, etc. The function of safety valve is to prevent the pipeline or the device in the medium pressure exceeds the specified value, so as to achieve the purpose of safety protection.

(4)   Regulating valve: such as throttle valve and pressure reducing valve. Its role is to regulate the media pressure, flow and other parameters.

(5)   Vacuum valve: such as vacuum ball valve, vacuum baffle valve, vacuum charging valve, pneumatic vacuum valve, etc. Its function is in the vacuum system, is used to change the direction of airflow, regulate the flow of gas, cut off or switch on the vacuum system components.

(6)   Special valve: such as pigging valve, vent valve, drain valve, exhaust valve, filter valve etc. The exhaust valve is an essential auxiliary component in the pipeline system, which is widely used in boiler, air conditioning, oil and gas, water supply and drainage pipeline. Exhaust valve is often installed in the commanding heights or elbow, the exclusion of excess gas pipeline, improve the efficiency of the use of pipes and reduce energy consumption.

According to the main parameters of the valve, can be classified as follows.

(a) **According to the rated work pressure**
(1) Vacuum valve: refers to the work pressure is lower than the standard atmospheric pressure valve.
(2) refers to the low pressure valve rated working pressure p_n of the valve is less than or equal to 2.5 MPa.
(3) Medium pressure valve: refer to the rated working pressure p_n in 2.5–6.5 MPa.
(4) High pressure valve: refers to the rated working pressure p_n in 6.5–80.0 MPa.
(5) Ultra-high pressure valve: refers to rated working pressure p_n greater than 100.0 MPa.
(b) According to the working temperature
(1) Ultra-low temperature valve: the valve used for medium operating temperature is lower than -100 °C.
(2) Normal temperature valve: the valve used for medium operating temperature at -29–120 °C.
(3) Medium temperature valve: valve used for medium operating temperature at 120–425 °C.
(4) The high temperature valve: the valve used for medium operating temperature is higher than 425 °C.
(c) According to the nominal diameter
(1) Small diameter valve: nominal diameter $D_n \leq 40$ mm.
(2) Medium diameter valve: nominal diameter $D_n = 50$–300 mm.
(3) Large diameter valve: nominal diameter $D_n = 350$–1200 mm.
(4) Ultra large diameter valve: nominal diameter $D_n \geq 1400$ mm.

According to the structural characteristics of the valve can be divided into:

(1) The door shape: off pieces along the center of the seat to move. Such as cut-off valve.
(2) The plug and the ball: the closing part is a plunger or ball, rotating around the center of itself. Such as plug valves, ball valves, etc.
(3) Gate shape: the closure member moves along the center of the vertical valve seat. Such as gate valve, gate, etc.
(4) Swing: the closure of a shaft around the valve seat rotation. Such as swing check valve.
(5) Butterfly: the disc of the closure, rotating around the axis of the valve seat. Such as butterfly valve, butterfly valve and so on.
(6) Sliding type: closure in the direction perpendicular to the sliding channel.

According to the valve connection, can be divided into:

(1) The threaded connection valve: the valve body with an internal thread or external thread, and pipe thread connection.
(2) Flange connection valve: the valve body with a flange, and pipe flange connection.

(3)   Welding connection valve: valve with welding groove, welding and piping connections.
(4)   Clamp connection valve: the valve body with a clip, and pipe clamp connection.
(5)   The cutting sleeve connection valve: with the pipe uses the cutting sleeve connection.
(6)   The clamp connection valve: the valve and the two ends of the pipe will be connected together by the bolt directly.

According to body material, can be divided into:

(1)   Metal material valve: most of the parts are made of metal materials. Such as cast iron valve, cast steel valve, alloy steel valves, copper alloy valves, aluminum alloy valve, lead alloy valve, titanium alloy valve, and Mongolian Monel alloy valve.
(2)   Nonmetallic material valves: most of the parts are made of non-metallic materials. Such as plastic valve, ceramic valve, ceramic valve, glass steel valve, etc.

9.4.2   Basic Structure of Valve

No matter what type of valve, have the following basic components: body, valve cover, valve components (including the valve, valve seat, sealing ring, valve stem, etc.), valve actuators, packing, etc., as shown in Fig. 9.36.

The valve body is sometimes called the valve shell, which is the main pressure boundary of the valve and the main body of the valve. The rest of the parts are attached to the body. The valve body is also connected with the fluid pipe, which can be connected with the screw thread and the flange.

The valve cover is used to encapsulate the inner part of the valve. Usually designed to be removable in order to facilitate the maintenance of the valve. In addition, some valves designed to be divided into two parts together in a way, this time may not need valve cover. Due to the valve cover and the valve is connected, so the valve cover is also a pressure boundary. Special attention should be paid to the valve cover due to the leakage of the valve.

According to different design, valve have different components; generally have the valve clack, valve seat, stem, sealing, valve stem sleeve and other components. The valve clack is generally a disc shaped component, and the gap between the valve seat is allowed to flow. By controlling the size of the gap, the flow of the fluid can be controlled. Many valves are designed to be named according to the design of the valve clack.

The valve seat is used for engaging with the disc in order to achieve sealing, some of the valve has no seat, and the valve body is directly used as the occlusal surface.

Valve stem is a rod connecting the valve actuator and disk, transfer the movement of the actuator to the valve clack, so that the corresponding movement of the valve to achieve the purpose of opening or closing the valve.

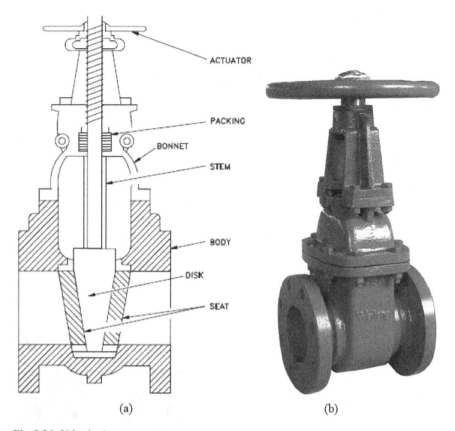

(a)   (b)

Fig. 9.36 Valve basic components

Valve actuator is to control the valve stem up and down movement. The valve actuator can be either manual or electric, or pneumatic, etc. The valve actuator in Fig. 9.36 is manual.

Packing is generally used for sealing. Between the valve cover and the valve stem generally need to fill the packing, can be used as a porous linen or PTFE. The filler cannot fill too loose or too tight. Too loose will leak, too tight will damage the stem.

9.4.3   Typical Valves

Here we mainly introduce some of the nuclear engineering field commonly used valves, such as gate valve, globe valve, ball valve, plug valve, diaphragm valve, etc..

Gate Valve

Gate valve is a gate opening and closing parts of the gate. The movement direction of the gate is generally perpendicular to the direction of the fluid flow, as shown in Fig. 9.37. Gate valve is generally used only for fully open or fully closed, is not used to adjust the flow rate. This is mainly because the flow through the valve and the position of the valve stem does not form a linear proportional relationship.

Gate valve in general with the stem for a straight-line movement, known as the hand wheel type gate valve. Usually on the lifting rod with trapezoidal thread, through the nut on the top of the valve and the guide slot on the body, change the rotating motion into a linear motion, that is, the operating torque into the operating thrust. When the valve is opened, if the gate height is equal to 100% of the diameter of the valve, the passage of the fluid is completely free. However, this position is not

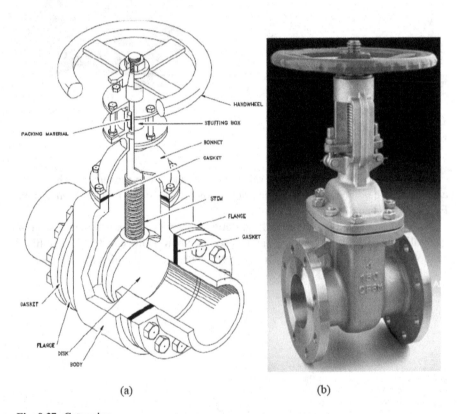

(a)   (b)

Fig. 9.37 Gate valve

monitored. Actual use is the apex of the stem as a sign, that is, the open position, as its fully open position. In order to avoid locked because of the temperature change, usually in open to the vertex position, and then rewind 1/2–1 circle, as a fully open valve position. Therefore, the valve fully open position, according to the location of the gate (that is, stroke) to determine.

Some gate valve stem nut is located on the gate, the hand wheel rotation to drive the valve stem rotation, and so that the gate to lift, this valve is called the rotating rod gate valve or called the dark bar gate valve.

Full closing of the gate, the gate and valve seat between the upper reaches of the pressure between the liquid seal, known as self-sealing. Stem and valve cover sealing by packing and gasket.

Gate valve has the following characteristics:

(1) Flow resistance is small. Internal media channel is straight, flow resistance is small.
(2) Not too hard to turn on or off. Is compared with the cut-off valve, because whether it is open or closed, the direction of movement of the gate perpendicular to the direction of flow.
(3) The height is relatively large, open and close for a long time. Gate opening and closing stroke larger. Lifting is carried out through the screw, so water hammer phenomenon is not easy to happen.
(4) Medium can flow in any direction on both sides, easy to install. Gate valves are symmetrical on both sides of the channel.
(5) The length of the structure (the distance between the two connecting end faces of the shell) is small.
(6) The form is simple, the manufacture craft is good, and the scope of application is wide.
(7) Compact structure, good rigidity of the valve, the sealing surface of stainless steel and hard alloy, long life, the use of PTFE filler, reliable sealing, flexible and flexible operation.

The main drawback is that the sealing surface easily lead to erosion and abrasion, maintenance is more difficult.

Globe Valve

Globe valve is one of the most commonly used valves. Globe valve and ball valve is sometimes referred to as ball valve, easy to confuse. Globe valve is the body's shape looks like a ball, the shape of the valve can be a variety of. The ball valve is a valve with a ball valve clack; the appearance is not necessarily spherical. Schematic diagram of the globe valve as shown in Fig. 9.38.

Globe valve can be used to close, open and adjust the flow. The flow area between the valve clack and valve seat can be adjusted continuously and will not occur flow induced vibration, this is because it is consistent with the direction of movement of the valve clack, rather than like a gate that is vertical. Therefore, the ball valve can be used to adjust the flow. When the valve is open, the direction of movement can be the same as the movement direction of the fluid. When the direction is the same,

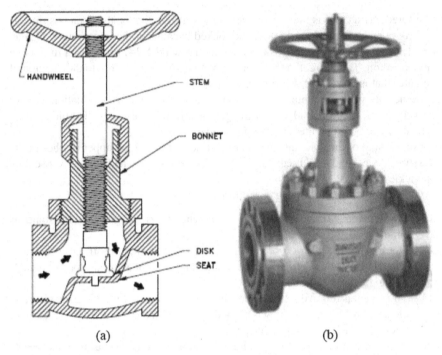

Fig. 9.38 Z type globe valve

the kinetic energy of the fluid is beneficial to the valve to open; when the direction is opposite, it is not conducive to open but it is conducive to close.

Of course, there are shortcomings of the globe valve; the main drawback is the greater flow resistance. Fluid in the flow through the ball valve, in order to adjust the direction of the flow of fluid and the direction of movement of the valve, the fluid needs to turn a few bends. As shown in Fig. 9.38 Z type globe valve has two bends.

In order to reduce the flow resistance, Y type globe valve is designed, as shown in Fig. 9.39. Y type bend globe valve can also replace the elbow in pipeline, to reduce the flow resistance (because the elbow would have a flow resistance).

There are three kinds of design of the globe valve: ball, plug and composite clack. Ball valve and valve seat depends on the thrust of the valve stem tightly combined, commonly used in low temperature and low pressure system. Plug valve clack has a better throttle performance than the ball valve, the general design into a longer conical valve plug. The composite clack can be inserted into a ball or plug valve, which is commonly used in the system of steam and hot water. The composite clack has good sealing performance, and the contact surface between the valve clack and the valve seat will not be destroyed when the solid particles are in the fluid.

Ball Valve

Ball valve refers to the valve clack is a ball, usually through the method of rotating the ball clack to achieve open and close, as shown in Fig. 9.40.

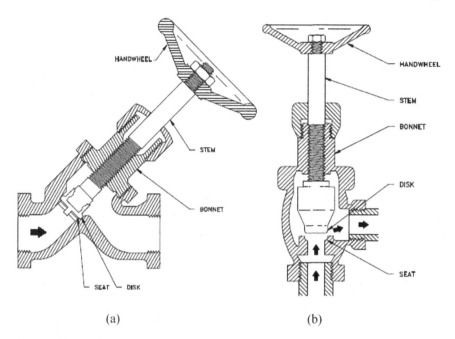

Fig. 9.39 Y type globe valve

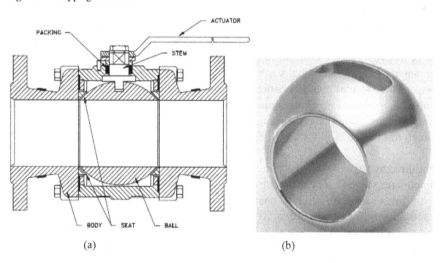

Fig. 9.40 Ball valve

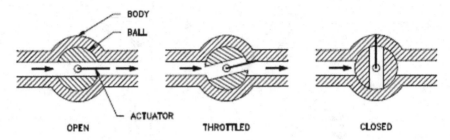

Fig. 9.41 Schematic diagram of the opening and closing of ball valve

The ball valve depends on the position of the top handle to control the angle of the inner ball clack, so as to realize the function of opening, closing and throttling. If you look down from the top, turn the handle to adjust the opening and closing as shown in Fig. 9.41.

Ball valve is a fast action valve, only need to turn 90° on the transition from fully open to fully close. Ball valve is the cheapest of all the valves, but also easy to maintain. Ball valve has the advantages of small opening and closing torque, no need of lubrication, good sealing performance, etc. The disadvantage is that the throttle adjustment ability is poor, and a long period of time in the throttle, the valve seat is easy to be washed by the fluid and corrosion.

A variant of the ball clack is plug valve, as shown in Fig. 9.42. Plug valve is a plunger shaped rotary valve, by turn 90° to make the channel on valve plug and the channel on valve body to close or separate, to achieve open or closed. Plug has the shape of cylindrical or conical. In a cylindrical valve plug, the channel is generally rectangular, and in a conical valve plug, the channel is formed into a trapezoidal shape. These shapes make the structure of the plug valve light, but at the same time, it also has a certain loss of pressure. Plug valve is very suitable as a cut and switch on the media and shunt, but according to the sealing surface of the erosion resistance, sometimes can be used for throttling.

Diaphragm Valve

Diaphragm valve is very different in the form of the general valve. It is to rely on a soft rubber or plastic film to control the movement of fluid. Diaphragm valve is a special form of cut-off valve, its opening and closing pieces are made of a soft material, the diaphragm, the valve body cavity and the valve cover cavity and drive components, such as Fig. 9.43. Commonly used diaphragm valves have rubber lined diaphragm valves, fluorine lined diaphragm valves, none lined diaphragm valves, plastic diaphragm valves, etc.

Diaphragm valve with corrosion-resistant lining of the body and corrosion-resistant diaphragm instead of the valve components, the use of the movement of the diaphragm to play a regulatory role. Diaphragm valve body material using cast iron, cast steel or cast stainless steel, and lined with a variety of corrosion resistant or wear resistant materials, rubber and PTFE. Corrosion resistant lining of the diaphragm, adjusting suitable for acid, alkali and other strong corrosive media.

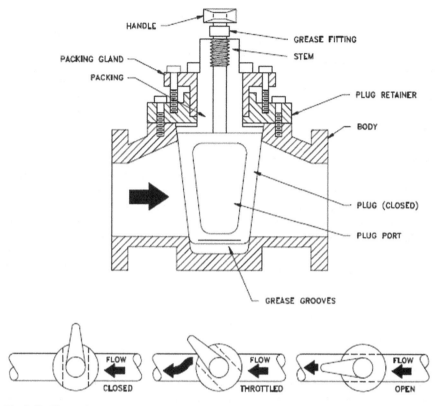

Fig. 9.42 Plug valve

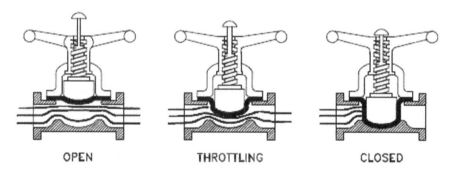

Fig. 9.43 Diaphragm valve

The diaphragm valve has simple structure, small fluid resistance, and other types of flow capacity is larger than other types of specifications, no leakage, can be used for high viscosity and suspension of granular media regulation. Diaphragm to the isolation of the medium and the valve stem, so there is no media fill and the media will not leak. However, due to the limitations of the diaphragm and lining materials, pressure resistance, temperature resistance is poor, generally only apply to the 1.6 MPa pressure and temperature of 150 °C.

The flow characteristics of diaphragm valve close to fast open characteristics, before the 60% stroke is approximation linear, after 60% the change of flow rate is small. Pneumatic form of the diaphragm valves can still be attached to the feedback signal, limit position device and positioning device, to meet the needs of self-control, program control or regulate the flow.

Reducing Valve

Reducing valve is through the adjustment, the import pressure will be reduced to a need of the export pressure, and rely on the energy of the medium itself, so that the export pressure to maintain the stability of the valve automatically. From the point of view of fluid mechanics, the pressure-reducing valve is a local resistance can change the throttling element, that is, through changing the throttling area, the velocity of the fluid and kinetic energy changes, resulting in different pressure loss, so as to achieve the purpose of decompression. Then rely on the control and adjustment of the system, so that the valve after the pressure and the spring force balance, so that the pressure after the valve in a certain range of error remain constant.

Pressure reducing valve used to control valve opening and closing parts of the valve to adjust medium flow and reduce the pressure of the medium, and at the same time, using the valve pressure regulating opening and closing parts of the valve, the pressure after the valve was maintained at a certain range. Pressure relief valve is characterized by changing the inlet pressure in the case, to maintain the export pressure within a certain range. The schematic diagram of the pressure relief valve is shown in Fig. 9.44.

If the main valve is closed, the pressure of the upstream fluid will be directed to the auxiliary valve through the high-pressure side of the drainage hole. As the main valve at this time downstream pressure is low, the auxiliary valve is in the open state. After the high pressure fluid is passed through the auxiliary valve, the piston is pressed through the piston to function at the top of the piston. Due to the area at the top of the piston to the ratio of the area of the main valve flap, so under the same pressure, the downward force greater than the upward force, so that under the impetus of the piston, the main valve of the valve flap downward movement, open the main valve. Once the main valve is opened, the main valve downstream pressure will rise. Downstream pressure will be increased through the low pressure side of the drainage hole to control the diaphragm, and the auxiliary valve is in the control of the diaphragm and the role of the auxiliary valve spring action. Control of the diaphragm in the lower reaches of the pressure will be pushed upward movement, so that the auxiliary valve opening becomes smaller, the pressure on the top of the piston is small, and the main valve will move upward, the opening is small.

Therefore, the principle of the valve is a mechanical feedback principle; the valve will be automatically adjusted according to the downstream pressure to open the

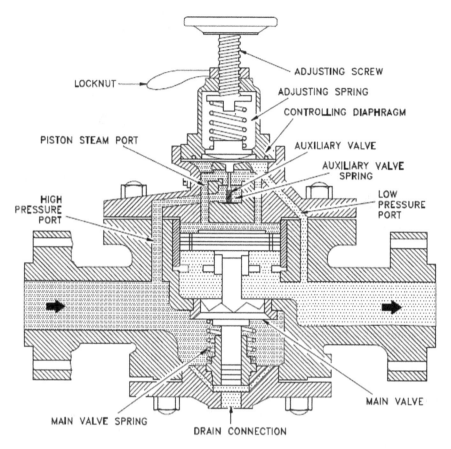

LOCKNUT

ADJUSTING SCREW

ADJUSTING SPRING

CONTROLLING DIAPHRAGM

PISTON STEAM PORT

AUXILIARY VALVE

AUXILIARY VALVE
SPRING

HIGH
PRESSURE
PORT

LOW
PRESSURE
PORT

MAIN VALVE

MAIN VALVE SPRING

DRAIN CONNECTION

Fig. 9.44 Reducing valve

main valve in order to control the downstream pressure in a certain range. Pressure range can be adjusted by adjusting wheel.

Pinch Valve

Pinch valve also known as the balloon valve, pinch off valve, etc. Casing is the most important part of any pinch valve, and it is the core of the pinch valve. Casing pipe needs corrosion resistance, wear resistance and bearing capacity. The quality of the pinch valve depends on the quality of the casing. Pneumatic, electric, manual or hydraulic to squeeze the casing, to achieve the role of switch or regulation, as shown in Fig. 9.45.

Media only through the hose, the other parts of the body does not need to be in contact with the chemical corrosive media, just replace the hose can be. This means

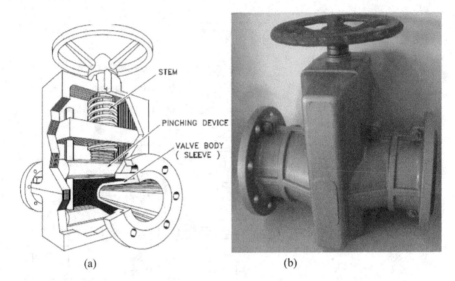

STEM

PINCHING DEVICE

VALVE BODY
(SLEEVE)

(a)   (b)

Fig. 9.45 Pinch valves schematic

that, compared with other types of valves, the use of pinch valve in the corrosive pipeline is more economical.

Butterfly Valve

Butterfly valve is a commonly used throttle valve. Butterfly valve structure is according to the principle of pull the tube block, the flow control element is tilted plate (the material can be metal or plastic covered metal, Teflon and other), disc fixed on the valve stem, by rotating the valve stem to control opening and closing (opening, closing only rotated 90°). The seat can be made of metal, rubber, Teflon and other materials, fixed in the wall. As shown in Fig. 9.46.

Butterfly valve is simple in structure, its body weight is light, the space is small, and it is suitable for the throttle and the opening and closing of the valve. Especially for large flow of control (not suitable for small flow).

Needle Valve

Needle valve is a precise adjustment valve. The valve needle is a very sharp cone, as if a needle into the valve seat, hence the name. Its principles as shown in Fig. 9.47.

Needle valve can withstand greater pressure than other types of valves, good sealing performance; it is generally used for small flow, high pressure gas or liquid media. Needle valve with pressure gauge is the most appropriate use, the general needle valve are made of thread connection.

Check Valve

The reverse check valve is only allowed to flow in one direction of the media, and to prevent the direction of the reverse flow. Also known as no return valve, counter flow

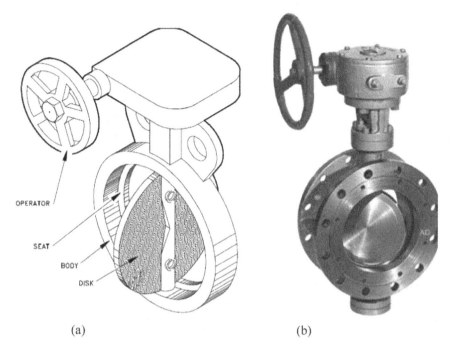

OPERATOR

SEAT

BODY

DISK

(a)   (b)

Fig. 9.46 Butterfly valve

valve or back pressure valve. This valve usually works automatically, in a direction of flow of fluid under pressure, the valve flap to open; fluid flow in the opposite direction, by the self-weight of the valve flap and the fluid pressure, valve flap closure, thereby cutting off flow. As shown in Fig. 9.48.

Valve clack be all made of metal, can also be embedded in the metal leather, rubber, or the use of synthetic coverage, depending on the requirements of the use of performance. Swing check valve in the fully open position, the fluid pressure is almost unimpeded, so the pressure drop through the valve is relatively small.

9.4.4   Pressure Relief Valve and Safety Valve

Pressure relief valve is based on the working pressure of the system can automatically open and close, generally installed in the closed system of equipment or piping to protect the safety of the valve system. When the equipment or pipeline pressure exceeds the setpoint of pressure relief valve, will automatically open the relief valve, to ensure that the pressure in the pipeline under the setpoint of pressure, the protection of equipment and pipelines to prevent accidents.

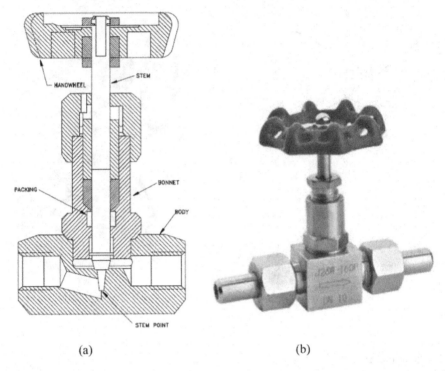

(a)   (b)

Fig. 9.47 Needle valve

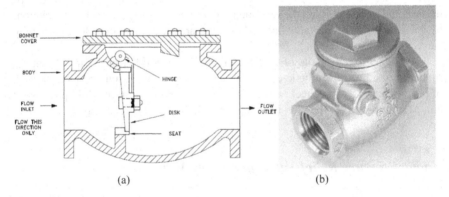

(a)   (b)

Fig. 9.48 Check valve

The structure of pressure relief valve mainly has two major categories: spring type and lever type. Spring refers to the seal is maintained by the spring force of valve and valve seat, as shown in Fig. 9.49. The lever type force is on the lever and weight.

Pressure relief valves are generally used for incompressible fluids, such as oil or water. Safety valve, as shown in Fig. 9.50, is usually used for compressible fluids,

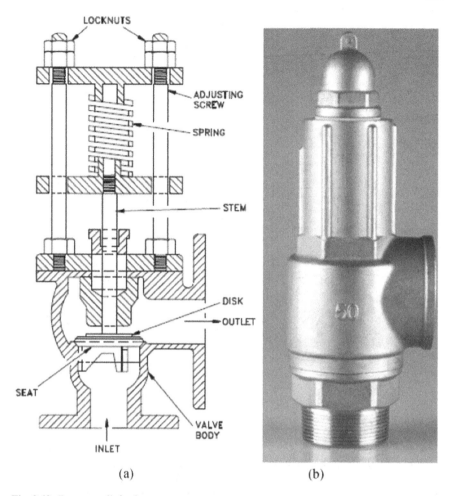

Fig. 9.49 Pressure relief valve

such as steam or other gases. Safety valve and pressure relief valve of the appearance is very easy to distinguish, as long as there is no handle to see. The safety valve has a pressure relief handle, and the valve can be opened when the handle is lifted, and the on-line detection of the safety valve is convenient for running. The opening pressure of the valve can be adjusted by adjusting the pressing force of the spring through the adjusting nut on the top.

With the need of large capacity pressure relief, there is a pulse pressure relief valve, also known as the pilot pressure relief valve. It is composed of the main pressure relief valve and the auxiliary valve. When the pressure in pipeline exceeds the required pressure value, auxiliary valve opens first, medium along the catheter into the main relief pressure valve, and the main relief pressure valve is opened, the higher medium pressure reduced. Its working principle is similar to the relief valve.

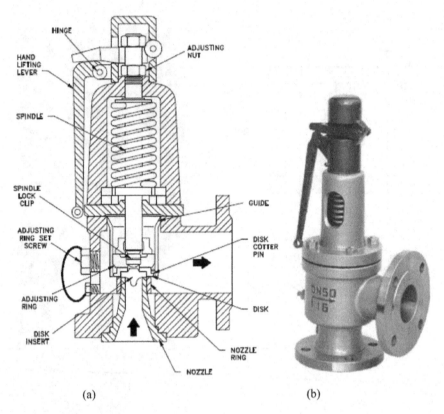

Fig. 9.50 Safety valve

Here we talk about the function of the pressure relief valve and the safety valve in a nuclear power plant. The pressure relief valve or safety valve will have a specific meaning. For example, some nuclear power plants to the normal operation of the system pressure control valve known as the pressure relief valve or release valve, and the safety system valve known as the safety valve.

9.5 Miscellaneous Mechanical Components

Here we introduce some of the nuclear engineering field commonly used mechanical equipment, such as air compressors, hydraulic machines, evaporator, steam generator, cooling tower, desalination, regulator, diffusion separator, etc..

9.5.1 Air Compressor

Air compressor is a device that can continuously provide compressed air. Compressed air is very useful in the industry as a power source, such as pneumatic valve actuator is to rely on the power of compressed air to act. According to the compression principle, air compressor can be divided into three types: reciprocating, rotary and centrifugal.

Reciprocating air compressor schematic diagram as shown in Fig. 9.51 is the piston in reciprocating motion of the cylinder body. The change of the chamber volume and mechanical pumped in and press out the air [7].

The crankshaft drives the connecting rod and the connecting rod drives the piston to move back and forth. Movement of the piston to cylinder volume changes, when the downward movement of the piston, cylinder volume increased, intake valve open, close the exhaust valve, air is sucked into to complete the intake process; when the upward movement of the piston, cylinder volume decrease, outlet valve is opened, air inlet valve is closed, the compression process. Piston rings are usually used to seal the gap between the cylinder and the piston.

Reciprocating compressors have cylinders, pistons and valves. The working process of the compressed air can be divided into four strokes, such as expansion, suction, compression and exhaust, as shown in Fig. 9.52.

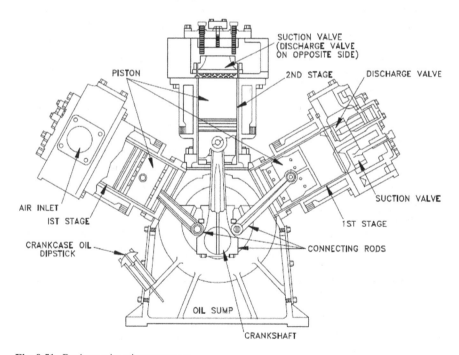

Fig. 9.51 Reciprocating air compressor

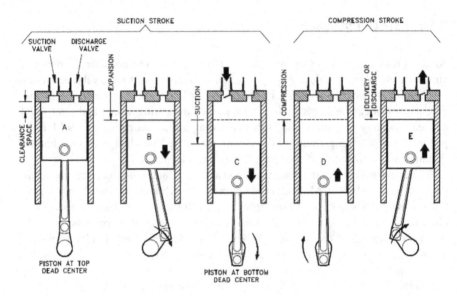

Fig. 9.52 Suction and compression strokes of a reciprocating air compressor

(1) Expansion: when the piston moves downward, the volume of the cylinder increases, the pressure drop, the residual gas in the cylinder is expanding.

(2) Inhalation: when the pressure is reduced to a little less than the gas pressure in the intake pipe, the air in the intake pipe will push the suction valve into the cylinder. As the piston moves down, continue to enter the gas, until the piston move to the end (also called BDC).

(3) Compression: when the piston moves upward, cylinder volume gradually reduced thus began the process of compressed gas. Because of the suction check valve, the gas in cylinder cannot flow back to the inlet pipe. The gas pressure in the outlet pipe is higher than the gas pressure in the cylinder, and the gas in the cylinder cannot flow from the exhaust valve to the cylinder. The gas in the outlet pipe cannot be discharged into the cylinder due to the reverse action of the discharge valve. Therefore, the mass of gases in the cylinder to maintain a certain value. The piston continues to move upward, reducing the volume of air space in the cylinder, increasing the pressure of the gas.

(4) discharge: as the piston moves up, the compressed gas pressure increases to slightly larger than the gas pressure of outlet pipe, exhaust valve opens, gas flows into the outlet pipe, and is continuously discharged out, until the piston to move to the end (also known as dead) so far.

Then, the piston begins to repeat the action. Piston in the cylinder reciprocating motion, so that the cylinder reciprocating cycle of suction and discharge of gas, each time the piston is referred to as a work cycle.

Due to the design principle of the compressor, it determines the characteristics of piston compressor. First, a number of moving parts, an inlet valve, exhaust valve, a piston, a piston ring, connecting rod, crankshaft, etc.; the second is the unbalanced stress, there is no way to control the reciprocating inertia force, and usually requires multistage compression, complex structure. Finally, because it is a reciprocating motion, compression air is not continuous discharged, it is pulsed. The advantages are that the heat efficiency is high, the unit electricity consumption is small; the processing is convenient, the material requirement is low, the cost is low; the design, the production is early, the manufacture technology is mature.

The principle of the rotary air compressor is similar to that of a positive displacement pump, as shown in Fig. 9.53. Rotary compressor with excellent performance, compact structure, less parts, long service life, widely used in room air conditioning, refrigeration equipment, automobile air conditioning and compressed gas device.

Centrifugal air compressor is driven by the impeller of the gas for high-speed rotation, so that the centrifugal force of gas. Due to the expansion of gas in the impeller flow, so that the gas flow through the impeller and the pressure has been enlarged, continuous production of compressed air, the principle of which is shown in Fig. 9.54.

Centrifugal air compressor mainly consists of two parts: rotor and stator. Rotor including impeller and shaft. With the impeller blades, in addition to the balance disc and seal. The main body is the stator shell, stator and diffuser, bend, refluxing device, air inlet pipe, the exhaust pipe and seal etc. The working principle of centrifugal compressor is almost same as centrifugal pump, when the impeller rotates at high speed, the gas will rotate, under the centrifugal force; the gas is thrown to the back of the diffuser. A vacuum zone formed in the impeller, fresh gas enters into the impeller. As the impeller rotates, the gas continues to breathe in and out, so as to keep the gas flowing continuously.

Centrifugal air compressor based on the change of the kinetic energy to improve the gas pressure. When the rotor with blade rotates, the blade drives the gas to rotate, and the power is passed to the gas, so that the gas obtains the kinetic energy. After entering the stator part, the kinetic energy is converted into the required pressure, the speed is lower and the pressure is increased. At the same time, using the guiding

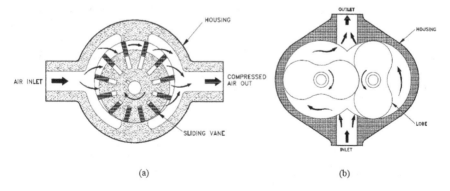

(a)   (b)

Fig. 9.53 Rotary air compressor

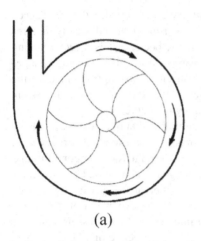

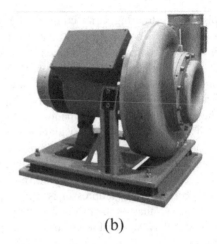

(a)   (b)

Fig. 9.54 Centrifugal air compressor

role of the stator part enters into the next level of the impeller to continue to increase pressure, and finally discharged by the spiral case.

The centrifugal air compressor is a speed type compressor with large flow rate and low pressure difference. It is stable and reliable in the centrifugal air compressor when the load is stable. Centrifugal air compressor has a compact structure, light weight, large displacement range; less wearing parts, reliable operation, long service life; without exhaust by lubricating oil pollution, high quality of air supply; large displacement high efficiency and is conducive to energy saving and other advantages.

9.5.2 Hydraulic Press

Hydraulic machine is a kind of machine that uses liquid as working medium, according to Pascal principle to transfer energy in order to achieve a variety of processes. As shown in Fig. 9.55.

The principle of the hydraulic press is Pascal principle. Assuming the area of the piston is A, the pressure is p, and then the force output of the piston is pA. For example: the pressure is 10 MPa, the area is 10 cm^2, and then the force can reach 10,000 N [8].

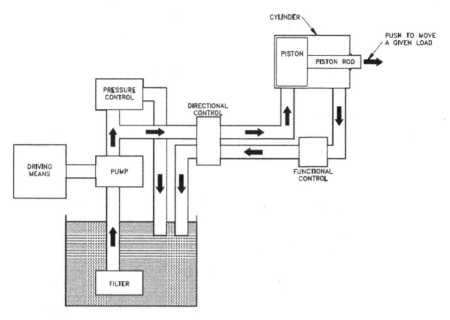

Fig. 9.55 Principle diagram of hydraulic press

9.5.3 Evaporator

Evaporator is also known as the boiling device, steam boiler, etc. According to the different heat source, can be divided into fuel, gas and electric heating evaporator. Figure 9.56 shows a schematic diagram of a gas fired evaporator.

Heating equipment (combustion chamber) to release heat, the heat is absorbed by the water cooled wall, vaporization and produce steam to the drum separator, separation of saturated steam into the gas pipeline (also by radiation and convection to absorb the heat of flue gas in the furnace chamber top and flue, and make it become superheated steam). Feed water can also be preheated by the heat of flue gas in order to improve efficiency.

9.5.4 Steam Generator

Steam generator (SG) is a special equipment for generating steam in a pressurized water reactor nuclear power plant, which passes the coolant heat from the primary side to the second side. It is the boundaries between the primary loop and the secondary loop, the generation of steam used to drive steam turbine.

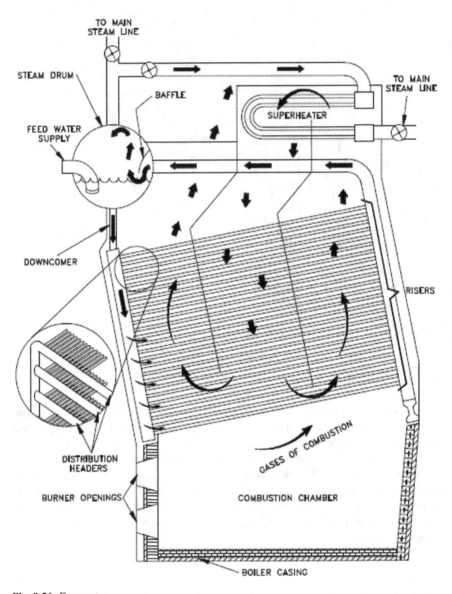

Fig. 9.56 Evaporator

According to the structure, steam generator can be divided into three categories, that is, the horizontal type U tube steam generator, vertical once through steam generator (OTSG) and vertical type U tube steam generator.

Figure 9.57 shows a horizontal U-tube steam generator structure. It has the advantage of good flow properties of water and steam, no sludge deposition around the heat transfer tube to cause corrosion, no tube plate, processing is convenient; small

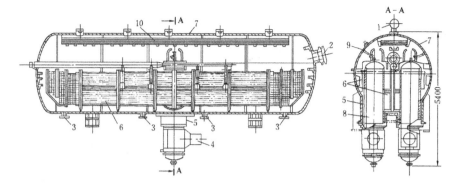

1 - Steam outlet header; 2 - manhole. 3 - sewage and drainage; 4 - inlet and outlet nozzle; 5 - coolant inlet header; 6 - heat transfer tube; 7 – cylinder; 8 - coolant outlet header; 9 - exhaust pipe; 10 steam water separator

Fig. 9.57 Horizontal U type tube steam generator. 1—Steam outlet header; 2—manhole. 3—sewage and drainage; 4—inlet and outlet nozzle; 5—coolant inlet header; 6—heat transfer tube; 7—cylinder; 8—coolant outlet header; 9—exhaust pipe; 10—steam water separator

steam load at steam water interface, simple steam and water separation device. The disadvantage is that the area is large, so that the diameter of the containment is large. Some of Russia's pressurized water reactor nuclear power plants use this form [9].

A vertical straight tube steam generator structure [10] is shown in Fig. 9.58. It has the advantage to produce 25–30 °C of overheat steam, so that the thermal efficiency of the power plant to improve 1.5–2%. The disadvantage is that the requirements for the heat transfer tube material, secondary loop water quality and automatic control of feed water are high. This form of steam generator in the Three Mile Island nuclear power plant accident has also been exposed the shortcomings as a result of the secondary side water capacity is small and easy to burn.

The vertical type U tube steam generator is the most widely used in the pressurized water reactor nuclear power plant. According to the capacity and structure can be divided into two types:

(1)   the combustion engineering company development type, heat transfer tube for the inverted U-shaped, with the feed water preheater and single heat rate of about 2000 MW, weighing 800t, as the world's largest steam generator.
(2)   the U.S. Westinghouse representative type, single thermal power is about 1000 MW, about 20 m, the weight of about 300t (see Fig. 9.59). France, Germany and Japan imported this Westinghouse's technology to develop, currently about more than 70% of the total capacity of the steam generator.

Here we introduce the basic principle of vertical type U tube steam generator. The coolant from reactor core enters tubes from the inlet water chamber, and the heat is released to the secondary side, leaves from the outlet water chamber, which is then returned to the reactor after being boosted by the main pump. The secondary

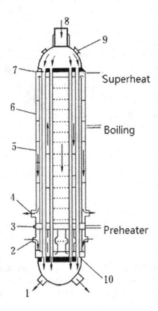

1 - Coolant outlet; 2 - feed water inlet; 3 - emergency feed water inlet;
4 - Superheated steam outlet; 5 - heat transfer tube; 6 - sleeve; 7 - tube sheet;
8 - Coolant inlet; 9 - manhole; 10 - tube plate

Fig. 9.58 Vertical straight tube steam generator. 1—Coolant outlet; 2—feed water inlet; 3—emergency feed water inlet; 4—Superheated steam outlet; 5—heat transfer tube; 6—sleeve; 7—tube sheet; 8—Coolant inlet; 9—manhole; 10—tube plate

side feed water is fed into an annular feed water pipe, and a plurality of inverted J shaped tubes are arranged on the annular tube, the function of which is to prevent the water hammer when the water supply is interrupted and the annular tube will not be emptied. Inverted J shaped tubes are arranged at different intervals, so that the water flow in the cold and hot sections of the tube bundle is properly allocated to improve the heat transfer performance of the tube bundle. The feed water is ejected from the inverted J shaped pipe and flows into the lower channel between the lower cylinder and the tube bundle sleeve. The lower end of the sleeve is folded into the tube bundle, the outer side of the heat transfer pipe is upward flow, and the heat is obtained by the heat transfer tube. When reaching the top of the tube bundle, the quality of 20–30% (quality of steam). After the steam and water separator and steam dryer, steam drying degree of 99.75% or more. The outlet pipe at the top of the dry saturated steam is taken over along the main steam pipeline to the steam turbine. Steam flow limiter is usually located in the outlet pipe of the Venturi tube, in the main steam line break accident (One of nuclear power plant design basis accident), can limit the steam flow at around 200% of the rated flow, reactor coolant temperature will not drop too much, to prevent more serious consequences of the accident.

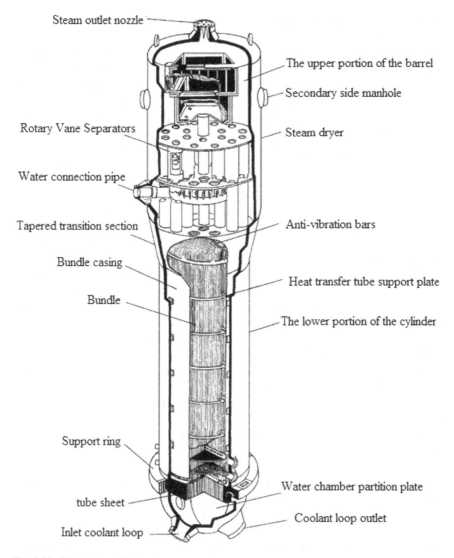

Fig. 9.59 Vertical type U tube steam generator

The structure of the vertical U type tube steam generator is mainly composed of a cylinder body, a heat transfer tube bundle, a steam water separating device and a supporting component.

The cylinder body is divided into two sections of different diameter cylinders, which are connected by a conical tube. The upper barrel body holds steam water separating device and water supply pipe component. The upper end of the top of the head is usually a standard ellipsoid, and the main steam outlet takes over the position in the center. The lower cylinder diameter is smaller, and the heat transfer tube bundle and the related components are arranged in the lower end and the lower end of the tube plate is connected with the pipe plate to be thickened to reduce the stress at the connecting point. Expansion joint and welding between two ends of U shaped heat transfer tube and tube plate. The thickness of the tube sheet is 500–550 mm, the material is low alloy steel, and the surface of reactor coolant is exposed to 6 mm thick nickel base alloy. Lower head usually hemispherical, inner surface of the welding of austenitic stainless steel, separated by a flat or curved plate for import and export of the two water chambers, each with coolant pipe and manhole of maintenance. The lower end of the head and the tube plate is sometimes provided with a height of 300–400 mm transition tube section, the lower header weld for local heat treatment of pipe and pipe plate joint temperature is not too high, so as to avoid tube sensitization and expansion joint relaxation.

The heat transfer tube is thin wall U tube, the common tube diameter is 22 mm, 19 mm, 15.8 mm, and the corresponding wall thickness is 1.27 mm, 1.09 mm, 0.86 mm. The steam generator with a single heat power 1000 MW has 4000 or so heat transfer tubes, and is arranged in a semi-circle tube bundle according to a triangle or a square array, and the two ends of the steam generator are connected with the tube plate at both ends. 6–8 supporting plates with about 20 mm thickness is arranged along the height of the straight section of the tube bundle to prevent the flow induced vibration of the heat transfer tube. An anti-vibration strip is embedded in the tube of the U type bending section to prevent the transverse vibration of the bending section, and the structure is usually two columns or three rows of V shaped strips. The material is usually nickel-based alloy, and the surface is plated with hard chromium to improve the abrasion resistance.

The heat transfer tube leakage or rupture may cause by corrosion or abrasion mechanism of micro vibration. It will significantly affect the availability of power plants, due to excessive plugging and cannot be full power operation even in less than ten years of operation, the need to replace the whole. Improvement direction of this kind of fault are: correct selection of heat transfer tube material; maintain good water quality; improved structure design; increase circulating ratio of the steam generator and reduce the secondary side stagnation as far as possible to avoid the local concentration of impurities in the water or sludge accumulation. For example, in the early, the tube hole of the supporting plate is a circular hole, the material is carbon steel, and the heat transfer tube is depressed due to the accumulation of corrosion products in the crevice. Later the tube holes are changed with three or four leaf shape (see Fig. 9.60), and tube material is well matched with the anti-abrasion materials.

Grid shaped supporting plate was developed by the combustion engineering company (see Fig. 9.61), it has larger flow area and smaller contact area than multi hole plate.

A flow distribution plate is arranged above the tube plate surface about 400 mm, it is a whole plate with a large circle hole or some polygon holes, to rinse the tube bundle, at a higher flow rate across the tube on the surface of slab, can prevent sludge deposition. In addition, the expansion and welding process of pipe and pipe plate connection is improved to avoid corrosion; the development of eddy current testing

Fig. 9.60 Support plate with
three-leaf water hole

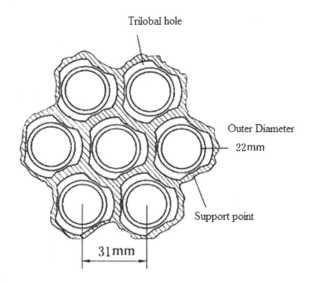

Trilobal hole

Outer Diameter
22 mm

Support point

31 mm

Fig. 9.61 Grid supporting
plate

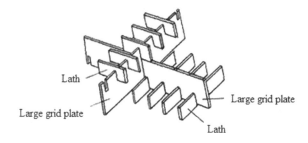

Lath

Large grid plate

Large grid plate

Lath

technology, inspection in the service to detect the thinning of the wall, and timely
blocking. These measures have been taken to reduce the pipe leakage and break.

The steam water separating device is generally two stages: (1) the steam water
separation (or the primary separator). Usually consists of three or more, up to more
than 100 cylinder shaped primary separators and a swirl blade is positioned in each
cylinder. When the steam water mixture rises into the cylinder, swirl flow happens
through blades. The water close to the cylinder wall is discharged by the centrifugal
force, and the steam rises and flows out of the center of the cylinder. Before reaching
the dryer, the steam is separated by about 800 mm gravity space, and the larger drops
of water separated by gravity. (2) Steam dryer (or secondary separator). Usually as
a combination of corrugated plate with multilayer parallel to the vertical. Wet steam
flow in the plate waves, entrained droplets attached to the wall, captured by the hook
plate and discharged (see Fig. 9.62). A plurality of assemblies are composed of a
double layer square, a hexagonal shape or a human shaped structure, which is fixed
on the upper end of the steam generator.

Requirements of the steam water separation device has a good separation capacity
and hydrophobic capacity. The combined effect of the steam water separator, the

Fig. 9.62 Chevron separator

gravity separation space and the steam dryer can make the outlet steam dry up to more than 99.75%. Steam water separator outlet steam is usually more than 90%, the steam load of the unit cross-sectional area of the separator is up to 400–500 t/h per m². The steam dryer is required when the inlet quality more than 70%, the outlet of the steam dryer can reach more than 99.75%. The structure and size of the separation device are usually determined by test.

Steam generator has its own weight of several hundred tons, and must be supported carefully. Usually on the head or tube plate arranges some support lugs, with a hinged vertical pillar fixed on the bottom plate of the house structure, steam generator is free to move along the hot leg direction; and in the building structure design block device, and the displacement is limited in a certain range. The guide rail is arranged on both sides of the tube plate and is parallel with the hot leg to limit lateral movement. In the upper part of the lower cylinder body, at the height of gravity center of entire equipment, set with dampers lateral support, to prevent the seismic loads or pipeline rupture load makes the steam generator roll.

Due to the deposition of a large amount of sludge on the steam generator tube sheet will affect the service life of the heat transfer tube of the steam generator. Therefore, in the operation of the steam generator to carry out continuous sewage, and set a sewage system of steam generator.

The main factors that affect the performance of steam generator include heat transfer area, circulation ratio and steady state characteristic curve.

In order to ensure the steam output of the steam generator under the rated power of the nuclear power plant, the steam generator must maintain a large enough heat transfer area. The heat transfer area of the general steam generator should be considered in the design of a certain dirt factor and blocking factor, so that the heat transfer area of the steam generator has a certain margin. Margin usually takes 10–15%. The heat transfer area of a steam generator with a thermal power of 1000 MW is about 5000 m^2, that is, the heat transfer efficiency per square meter is about 0.2 MW.

The ratio of the circulating water flow rate and the outlet steam flow rate is called the circulating ratio, which is expressed by the symbol K. The physical significance of the circulating ratio is: to generate 1 kg steam in the rise of the tube bundle, should enter the circulating water of Kkg; or to make 1 kg water to all become steam, need to circulate in the loop K times. Its value is the ratio of the steam water two-phase mixture flow rate in the arbitrary section of the tube bundle to the steam flow at outlet. It is also equal to the reciprocal of the steam rate of the steam water mixture at the outlet of the tube bundle. For example, a steam generator steam output for G, if steam water mixture containing 25% steam and 75% water at the outlet of the tube bundles, two-phase flow rate is 4G, and so the circulating ratio is 4G/G = 4. Low circulating ratio will lead to local impurities concentration in the secondary side; or lead to stagnation in the secondary side of bundle. Stagnation and local concentration will cause corrosion of the heat transfer tube. And may also cause part of the pipe wall thinning due to the velocity of secondary loop water passing tube plate surface is low and the sludge in the tube plate surface accumulated. High circulating ratio will cause the high load of the steam water separator, which may affect the separation ability, and make the flow rate of flushing heat transfer tube is too high, which may cause the vibration of heat transfer tube. Usually requires a circulating ratio of 3–5, there are some of the steam generator circulating ratio reached 8–10.

When the steam generator load is reduced, the average temperature difference between the primary side and the secondary side decreases, pressure of steam increased. To zero load, the highest pressure, approximately equal to the saturation pressure of reactor coolant temperature at this time. The steam pressure curve with different load is the steady state characteristic curve of steam generator. Zero load with a certain margin as the design pressure of steam generator. Proper selection of reactor coolant temperature with load, the steam pressure difference between full load and zero load will not be too big, so that the secondary design pressure is not too high.

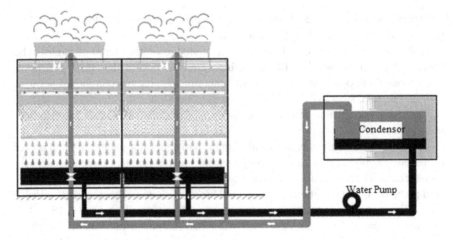

Fig. 9.63 Schematic diagram of cooling tower

9.5.5 Cooling Tower

The cooling tower is a device that absorbs heat from the system to the atmosphere to reduce the temperature, using water as a circulating coolant. The principle is that the heat exchange between water and air is used to carry out heat. It is the use of evaporative cooling, convective heat transfer and radiation heat transfer and other principles to disperse the heat of industrial or refrigeration air conditioning. The general design of the barrel shaped device [11], named for the cooling tower, as shown in Fig. 9.63.

With the development of thermal power and inland nuclear power, the design of the air-cooling tower is getting bigger and bigger, as shown in Fig. 9.64.

9.5.6 Pressurizers

Pressurizer is used for pressure control of the reactor system. Pressurizers are components that allow water systems, such as reactor coolant systems in PWR facilities, to remain hot and not boiling [12].

There are two types of pressurizer: static and dynamic. A static pressurizer is a partially filled tank with a gas pressure trapped in a gap area. A dynamic pressurizer is a tank that controls its saturation and environment by using heaters (controlling temperature) and sprays (controlling pressure).

Here, we focus on dynamic pressurizer. Dynamic pressurizer utilizes controlled pressure control to prevent high temperature fluid from boiling, even if the system fluctuates abnormally.

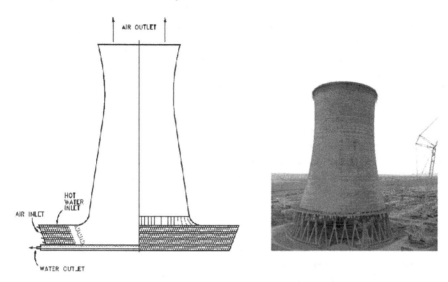

Fig. 9.64 Air cooling tower for large power plant

Some preliminary information about fluids will prove helpful before discussing the purpose, construction and operation of the pressure.

Evaporation is the process by which a liquid is converted into steam at a temperature below boiling point. All molecules in the liquid are constantly moving. The fastest moving molecules have the most energy. This energy occasionally escapes from the liquid surface and enters the atmosphere. When a molecule enters the atmosphere, it is gaseous or steamy.

Liquids at high temperatures have more molecules escaping to steam because they can only escape at higher speeds. If the liquid is in an enclosed container, the space above the liquid is filled with steam molecules, although some return to the liquid as they decelerate. The return of steam to liquid form is called condensation. There is a dynamic balance between liquid and steam when the molecular weight of condensation is equal to the molecular weight of evaporation.

The pressure that steam exerts on the surface of a liquid is called steam pressure. The steam pressure increases as the temperature of the liquid increases until the saturation pressure is reached, at which point the liquid boils. When a liquid evaporates, it loses its most dynamic molecule, and the average energy of each molecule in the system decreases. This causes the liquid temperature to decrease.

Boiling is an observed activity in a liquid when it changes from a liquid phase to a steam phase by increasing heat. The term saturated liquid is used for liquids at boiling points. Water and standard atmospheric pressure at 100 °C are an example of saturated liquids.

Saturated steam is steam whose temperature and pressure are the same as the water that forms the water. It is water, in the form of saturated liquids, that increases the potential heat of steaming. When heat is added to saturated steam that does not come into contact with liquids, its temperature increases and the steam over-heats. The temperature of an overheated steam, expressed above saturation, is called overheating.

The pressurizer provides a point in the reactor system for control purposes, where liquids and steam can be balanced under saturation conditions.

The role of dynamic pressurizer is to:

- Keep system pressure above saturation point,
- Provides a method for controlling the expansion and contraction of fluid,
- Provides the means to control the pressure of the control system,
- The steam space of the ventilation press provides a way to remove dissolved gas from the system.

Dynamic pressurizer are made from tanks equipped with heat sources such as electric heaters, cold water sources and nozzles for the base. The nozzle is a device located on top of the pressurizer to atomize the incoming water.

Dynamic pressurizer must be connected to the system in order to have differential pressure throughout the system. The bottom connection, also known as the surge line, is the lower part of the two pressure lines. The top connection, called the spray line, is a higher pressure line. Differential pressure is obtained by connecting the pressurizer to the suction and discharge side of the pump serving a particular system. Specifically, the surge (bottom connection) is connected to the suction side of the pump: the spray line (top connection) is connected to the discharge side of the pump. Figure 9.65 shows a basic press.

The top and bottom of the hemisphere are usually made of carbon steel, with all surfaces exposed to water from the reactor system and stainless steel cladding.

The pressurizer can be activated in two ways. The press for partially filled system water is the first. When the water reaches its intended level, the heater will be involved in raising the water temperature. When water reaches saturation temperature, it starts to boil. Boiling water fills the gap above the water level, creating a saturated water and steam environment. Another option is to completely fill the press, heat the water to the desired temperature, and then partially drain the mixture of water and steam, creating a steam gap at the top of the container.

The temperature of the water determines the amount of pressure that is formed in the steam space, and the longer the heater is involved, the hotter the environment. The hotter the environment, the greater the pressure.

Install the control valve in the spray tube and absorb cooling water from the top of the sprayer through the nozzle. Adding cooling water compresses steam bubbles, lowers the temperature of existing water, and reduces system pressure.

The level of water in the pressurizer depends directly on the temperature in the system to which the pressurizer is connected, and thus on the density of the water. The increase in system temperature causes the density of water to decrease. This decrease in density causes the water to expand, causing the water level in the container to increase. The rise in water levels in pressurizer is known as a boom. The surge compresses the steam space, which in turn causes the system pressure to rise. This causes the steam to overheat slightly when in contact with the subthermal booster.

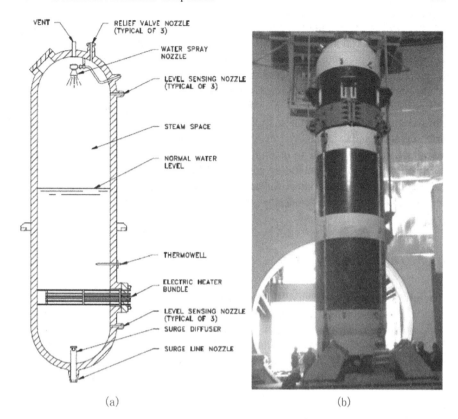

(a)   (b)

Fig. 9.65 Basic pressurizer

Overheated steam carries heat to the walls of liquids and presses. This re-establishes and maintains saturation.

A decrease in the system temperature leads to an increase in density and a contraction in the amount of water in the system. The contraction (decrease) of the pressurizer water level and the increase in steam space are called extroversions. Increased steam space causes pressure to drop, flashing heating water and producing more steam. The increase in steam has re-established saturation. The flicker continues until the water level drops and the saturation condition returns at a slightly lower pressure.

In each case, the final condition gives new value to the stress level. If the level change is not too extreme, the system pressure remains at the previous value and the pressure change is relatively small during the level change.

In practice, it is unrealistic to rely on saturation to deal with all changes in stress. For example, additional control is gained by activating the spray when the system water flows into the booster faster than the booster can hold. This spray condenses steam faster, controlling the magnitude of the pressure rise.

When a large eruption occurs, the water level drops rapidly and the water does not steam quickly. This can lead to a decrease in stress. The installed heater increases the energy of the water, allowing it to steam quickly, thereby reducing the pressure drop. The heater can also continue to heat up to re-establish the original saturation temperature and pressure. In some designs, pressure heaters are constantly energized to compensate for the loss of heat to the environment.

The pressurizer's heater and spray functions are designed to compensate for expected surges. Surge volume is a volume adapted to system expansion and contraction, designed as typical normal supercharged performance. Plant transients can cause larger orgasms and external orgasms than normal. When the surge volume exceeds, the booster may not be able to maintain pressure under normal operating pressure.

Pressurizer operation, including spray and heater operation, is usually automatically controlled. If the control function fails, monitoring is required because the impact on the system can be catastrophic without operator action.

9.5.7 Diffusion Separator

The diffusion separator is a gas diffusion separation equipment of $^{235}UF_6$, also known as the diffuser, such as Fig. 9.66 shows. The S shaped coil inside the separator is made of a micro porous separation membrane. Because the UF_6 gas is fed into the separator by the compressor, the temperature of the gas will be higher after the compressor. Therefore, after the inflow of the inlet, the first to go through the cooling of evaporative cooler. The cooler is arranged inside the separator, and is a

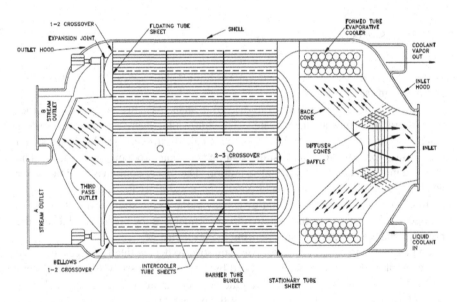

Fig. 9.66 Diffusion separator

Table 9.2 Parameters of separation plant

Type of diffuser	Length/in	Diameter/in	Numbers
33	291	155	640
31	221	105	500
29	185	90	600
27	129	47	720
25	129	38	1560

tube bundle evaporator. The cooling medium is liquid water, and the heat is taken away by evaporating the liquid water.

After cooling, the UF_6 gas then enters the S—shaped coil. In the tube will go through 3 stages of processes. The near shell is from right to left flow process 1, after 1-2 turn into from left to right the process 2. To the right of the tube plate and after 2-3 turning into the process 3. After the process 3, flow out through outlet B, entrance to the next level. Moreover, the enriched flow is flowed out from outlet A, and enters the upper stage.

In the gas diffusion plant, there will be a cascade of thousands of separators; Table 9.2 lists some of the parameters of all stages of diffuser of a separation plant in the United States.

Exercises

1. What are the functions of the crankshaft and the camshaft of a diesel engine?
2. What is the support system of diesel engine? What are the functions?
3. Compare the advantages and disadvantages of direct turbo and the use of exhaust gas turbo of diesel engine.
4. What is the compression ratio of a diesel engine? Why the compression ratio of vehicle gasoline engine is not too large?
5. What are the main characteristics of plate type heat exchanger and shell—tube heat exchanger?
6. What is the main function of a pump? How to achieve a continuous flow of a pump?
7. A large inertia of the flywheel is needed for nuclear power plant main pump, why?
8. In a nuclear power plant, the first stage of the main pump seal water provided by what system? The direction of water flow should be designed toward the inside or outside.
9. What type of valve is not suitable for use as a throttle valve?
10. Describe the principle of pressure control of pressure reducing valve.
11. What are the main factors affecting the performance of a vertical natural circulation steam generator? How to consider these factors in design?
12. If mixture reaches the top of a vertical U type natural circulation steam generator at quality of 15%, determine the circulation ratio of this steam generator?

13. What are the main factors affecting the performance of a pressurizer?
14. If you need to control the system at a pressure of 15.5 MPa, determine the temperature inside the pressurizer should be controlled?

References

1. U.S. Department of Energy. Mechanical science. Washington, D.C.: Lulu.com; 2016.
2. Li F. Internal combustion mechanism construction and principles of the . . . Version 2 . Beijing: China Railway Press; 2006.
3. Benson & Whitehouse, Internal Combustion Engines, Pergamon.
4. Stinson KW. Diesel Engineering Handbook, Diesel Publications Incorporated.
5. Heat Transfer, Thermodynamics and Fluid Flow Fundamentals, Columbia, MD, General Physics Corporation, Library of Congress Card #A 326517.
6. Cheremisinoff NP. Fluid Flow, Pumps, Pipes and Channels, Ann Arbor Science.
7. Scheel, Gas and Air Compression Machinery, McGraw/Hill.
8. Stewart HL. Pneumatics & Hydraulics, Theodore Audel & Company; 1984.
9. Skrotzki and Vopat, Steam and Gas Turbines, McGraw/Hill.
10. Babcock & Wilcox, Steam, Its Generations and Use, Babcock & Wilcox Co.
11. Marley, Cooling Tower Fundamentals and Applications, The Marley Company.
12. Westinghouse Technical Manual 1440-C307, SNUPPS, Pressurizer Instructions, Westinghouse.

Chapter 10
Nuclear Physics

Nuclear physics known as atomic physics is a branch established in the twentieth century. It studies the structure of nuclei and variation; generating radiation beam, detection and analysis techniques; and with nuclear energy, nuclear technology issues. It is both a profound theoretical significance, but also of great practical significance discipline [1].

In 1896, Becquerel discovered natural radioactivity, this is the first time observed nuclear transformation. Usually we put this important discovery as the beginning of nuclear physics. The next 40 years [2], people are mainly engaged in law of radioactive decay and radiation properties, and the use of radioactive rays on nuclei do a preliminary study, which is an early stage of development of nuclear physics. During this period, people in order to detect a variety of rays, and determine the identification of the types of energy, initially created a series of detection methods and measuring instruments. Most of the principles and methods of detection in the future has been developed and applied. Some basic equipment, such as counters, ionization chambers, etc., are still in use. Detecting, recording radiation and determine its nature, it has always been a central part of nuclear physics and nuclear technology applications. Studies have shown that the radioactive decay of an element can decay into another element, overthrow the view of element cannot be changed, established statistical decay law. Statistics is a microcosm of the important features of motion of matter, with the laws of classical mechanics and electromagnetism are different in principle.

Radioactive elements can emit a lot of energy rays, which provides an unprecedented tool for exploring atoms and nuclei. In 1911, Rutherford, who use a variety of α-rays bombarding atoms, observing α-rays deflection occurred, thus establishing the nuclear structure of the atom, proposed the planetary model of atomic structure, this achievement for the study of atomic structure laid the foundation. Shortly thereafter, people will find out the initial shell structure and the movement of electrons in atoms, the establishment and development of the physical description of the microscopic world.

© Tsinghua University Press 2022

J. Yu, *Fundamental Principles of Nuclear Engineering*,

https://doi.org/10.1007/978-981-16-0839-1_10

In 1919, Rutherford, etc. also found with α particle bombardment of nitrogen nuclei will release protons, this is the first artificial nuclear transmutation reaction implemented. Thereafter nuclei bombarded with radiation to induce nuclear reaction method is becoming the primary means of nuclear research [3].

In the initial study of nuclear reactions, the most important achievement is the discovery of the neutron in 1932 and synthesis of artificial radionuclides in 1934. Nucleus is composed of neutrons and protons, neutrons for the study of nuclear structure was found to provide the necessary precondition. Uncharged neutrons, nuclear charge is not exclusive, easy to enter the nucleus and cause a nuclear reaction. Thus, the neutron reaction become an important means of nuclear research. In the 1930s, it is also through the study of cosmic rays discovered positrons and mesons, which are found in particle physics precedent.

The late 1920s, it has been exploring the principles of accelerated charged particles. By the early 1930s, static electricity, and other types of linear accelerator and cyclotron has been developed, people took preliminary nuclear reaction experiment in high-voltage doubler. Use beam accelerator can get more powerful, more energy and a wider range of the beam, thus greatly expanding the study of nuclear reactions. Thereafter, the accelerator is becoming necessary equipment research and application of nuclear technology. In the 1930s, most people can only accelerate protons to 1 meV magnitude, and by the 1970s, it has been able to accelerate protons to 400 GeV, and can produce a variety of very small energy spread based on operational need, or a particularly high degree of collimation strong flow particularly large beam.

In the initial stages of the development of nuclear physics, people noticed rays may apply, and soon discovered the therapeutic effects of radioactive emissions of certain diseases. This is an important reason for it at that time on the subject of social importance, until today, nuclear medicine is still an important field of nuclear technology applications.

Before and after the 1940s, with the discovery of nuclear physics and application of nuclear fission has entered a stage of great development. In 1939, Hahn and Strassmann discovered nuclear fission phenomenon; in 1942, Fermi built the first chain fission reactor, which is the milestone of peaceful utilization of nuclear energy. Particle detection technology has been greatly developed. Semiconductor detector application greatly improves the measurement of ray energy resolution. The rapid development of nuclear electronics and computing technology to improve the acquisition and processing of experimental data capacity fundamentally, but also greatly expands the range of theoretical calculations. All this has opened up a range of nuclear phenomena observable improve the accuracy of observation and theoretical analysis capability, which greatly promoted the application of nuclear physics and nuclear technology.

Through a large number of experimental and theoretical studies, people on the basic structure and changes of the nucleus with a deeper understanding. Understand the basic nature of the interaction of a variety of nucleons (protons and neutrons collectively) between the stability of long-lived radionuclides or radionuclide ground state and excited states of nature has accumulated more systematic experimental data. In addition, theoretical analysis, the establishment of a variety of suitable models.

By nuclear reactions, we have been synthesized 17 different atomic number greater than 92 transuranic elements, and thousands of new radionuclides. These studies further showed that the element is only under certain conditions, a relatively stable substance structural unit, is not eternal.

Through interaction of high energy and ultra-high energy beam, it was discovered that hundreds of short-lived particles such as baryons, mesons, leptons and various resonances particles. The emergence of a large family of particles, the study of the physical world forward to a new stage—the establishment of a new science—particle physics, sometimes also referred to as high-energy physics. A variety of high-energy beam is a new way, and they can provide some knowledge of nuclear structure cannot be achieved by other means.

In the past, through the study of macroscopic objects, people know that there are two kinds of electromagnetic interaction and the gravitational interaction between matters—long-range interaction. Depth study of the atomic nucleus found there are two short-range interaction between substance, namely strong interaction and weak interaction. The phenomenon of parity violation in the weak interaction was found, it is a major breakthrough to the traditional physics (time and space). Research on the interaction between the law and the four possible link between them; explore new interactions possible, it has become an important topic in particle physics. Undoubtedly, nuclear physics research will also make new and important contribution in this regard.

Nuclear physics development, continue to provide for the design of nuclear devices increasingly accurate data, thereby improving the efficiency of the use of nuclear energy and economic indicators, and to prepare the conditions for larger-scale use of nuclear energy. Application of various isotopes have been artificially prepared throughout the engineering, agriculture and health sectors. New nuclear technologies, such as nuclear magnetic resonance, Mössbauer spectroscopy, crystal channeling effect and blocking effect, and perturbed angular correlation technique so rapidly applied. Extensive application of nuclear technology has become a symbol of modern science and technology.

10.1 Atomic Nucleus

Nuclear reactor applications depends on various neutron reactions with nuclei. In order to understand the nature and characteristics of these reactions, the best first discuss some of the basics of atomic physics and nuclear physics. Previous chapter has been initially introduced, the atom consists of positively charged atomic nucleus and some negatively charged electrons, the entire atom is electrically neutral. Nuclei consists of two nucleons (i.e. protons and neutrons). A proton is positively charged unit, which value is equal to the electron charge. In fact it is the hydrogen nucleus, is a hydrogen atom loses the only one of its extra nuclear.

Since the nuclear is very small, for convenience, a measure of nuclear usually use specialized units. In Chap. 8, we have introduced a measure of the mass of the

Nuclei	Nuclei radius/10^{-13} cm
$_{1}^{1}$H	1.25
$_{5}^{10}$B	2.69
$_{26}^{56}$Fe	4.78
$_{72}^{178}$Hf	7.01
$_{92}^{238}$U	7.74
$_{98}^{252}$Cf	7.89

Table 10.1 Some nuclei radius

nuclei usually amu, namely atomic mass units. It is defined as 1/12 of the mass of the ^{12}C atoms, 1.66×10^{-24} g. The commonly used measure of energy is electron volts (eV), 1 eV = 1.602×10^{-19} J.

Mass of the proton is equal to 1.007277 amu. Neutrons are electrically neutral particles with no charge; mass of the neutron is 1.008665 amu. Electron with a negative charge, the mass of electron is 0.0005486amu.

The radius of a nucleus is about 10^{-13} cm of the order, as shown in Table 10.1 lists the atomic radius of some nuclides.

10.1.1 Atomic Number and Mass Number

The number of protons in the nucleus of any element contained in the nucleus is equal to the number of positive charges carried; it is called the atomic number of the element, usually by the symbol Z represents. It is the same as the number of elements in the periodic table of the sequence.

Therefore, atomic number 1 hydrogen, helium is 2, lithium is 3, and so on, until the uranium is 92, which is the highest atomic weight elements in nature may have sensed naturally occurring. It has been artificially produced many heavier elements [4].

The total number of protons and neutrons in the nucleus is called the mass number of the element, with the A representatives. Above have said, is the number of protons Z, so that the number of neutrons in the nucleus is equal to A-Z. Since the mass of protons and neutrons in atomic mass units calculations nearly 1 amu, so the mass is closest to an integer number of atomic matter in question. Representation of nuclides using convention shown in Fig. 10.1.

10.1.2 Isotope

Having the same number of protons, the different elements of different nuclides of the same number of neutrons each other isotopes (Isotope). There is difference between

Fig. 10.1 Representation
nuclides

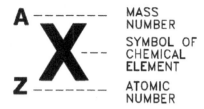

the word nuclide and element, "element" is the chemical term and "nuclide" is the nuclear physics terms.

We know that the atomic number (the number of protons in the nucleus) of an element decides its chemical properties. This is due to the chemical nature depends on the number of orbital electrons, which is equal to the number of protons in the nucleus. Thus, the same number of protons in the nucleus (i.e. the same atomic number) but different mass numbers of atoms in the chemical properties are almost exactly the same (some isotopes, such as hydrogen and deuterium, the equilibrium constant of hydrogen–deuterium exchange can have nuances). Although they have significant differences in the properties of nuclei often, these same atomic number but different mass number of substances, each other isotopes. Here the meaning of "same position" refers to their position in the periodic table, which is the same. They are generally chemically indistinguishable, but with different atomic weight. Isotope separation methods, we will introduce in the next chapter.

On the release of nuclear energy the most important elements uranium, in nature there are at least three isotopes, whose mass numbers were 234, 235 and 238. In the upper left corner in order to distinguish various isotopes, generally written in the mass element symbol, or the mass number behind the name, this radionuclide ^{238}U can also be referred to as uranium-238, U238 and the like. Ratio of the natural uranium in a variety of isotopes listed in Table 10.2.

You can see, ^{238}U is the most abundant isotope, in the natural uranium contains ^{235}U only slightly more than 0.7%. These two isotopes in the field of nuclear energy occupies an important position. ^{234}U ratio is very small, almost negligible.

From the point of view of nuclear energy, there is a very important element, and that is thorium. Thorium atomic number 90, it is almost exclusively only single nuclides exist in nature, i.e., the nuclear mass number 232. Although there are trace amounts of other isotopes, but the proportion is very small and can be ignored.

Table 10.2 Isotopic composition of natural uranium

Isotopes	Abundance/%	Atomic weight/amu
^{234}U	0.006	234.11
^{235}U	0.712	235.12
^{238}U	99.282	238.12

10.1.3 Chart of Nuclides

Chart of nuclides with atomic nuclei and the number of neutrons produced as an independent coordinate graph shown in Fig. 10.2.

In the chart of nuclides, each grid, which is a nuclide usually, indicate a half-life of the radionuclide inside. For example Be7 half-life of 53.28 days. A complete chart of nuclides in which each nuclide information is shown in Fig. 10.3, in which Fig. 10.3a is a stable nuclide, Fig. 10.3b is unstable nuclides.

Stable nuclide is not much among all nuclides in the nuclide chart. Their distribution is shown in Fig. 10.4. Can be found in Fig. 10.4, with the increase of the number of protons, the ratio of the number of neutrons and protons of stable isotope changes. The larger the mass number nuclides, to stabilize the number of neutrons

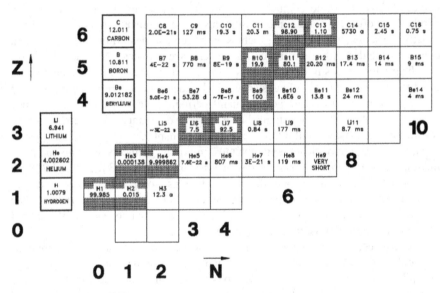

Fig. 10.2 Chart of nuclides

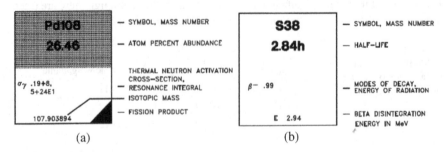

Fig. 10.3 Nuclide information

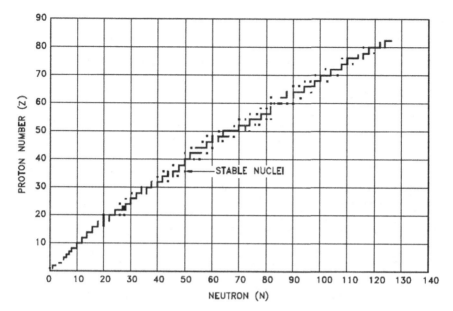

Fig. 10.4 Stable nuclide distribution

needed more. This law is crucial for fission nuclear energy, because it means that the process of a heavy nucleus into two-daughter nuclei fission releases excess neutrons.

10.2   Mass Defect and Binding Energy

The so-called mass defect, mass is not disappear, but rather refers to the difference between the sum of all protons and neutrons in the nucleus and the mass of the composition of the entire nuclear mass.

At the level of the nucleus, macro mass and energy conservation is no longer accurate each. It may occur between mass and energy into each other. The sum of the energy and mass conservation preached one kind of law, that mass-energy conservation. The relationship between energy and mass is the famous Einstein's mass-energy equation:

$$E = mc^2 \tag{10.1}$$

where, c is the speed of light in vacuum, 2.998×10^8 m/s.

10.2.1 Mass Loss

When the number of neutrons and protons together to form a number of atomic time, accurate measurement display the mass is slightly smaller than the atomic mass of the atom consisting of protons and neutrons, and a phenomenon known as mass loss [5] (Δm).

$$\Delta m = \left[Z(m_p + m_e) + (A - Z)m_n \right] - m_a \tag{10.2}$$

wherein, Z is the atomic number, A is the mass number, the subscript p indicates a proton, subscript e is an electron, the subscript n represents neutron, and atom represents a subscript.

Example 10.1 If known ^{7}Li atomic mass of 7.016003amu, calculated ^{7}Li atomic mass loss.

Solution: $\Delta m = \left[Z(m_p + m_e) + (A - Z)m_n \right] - m_a$

$$\Delta m = [3(1.007277 + 0.0005486) + (7 - 3)1.008665] - 7.016003$$
$$= 0.0421335 amu$$

10.2.2 Binding Energy

When an object (system) is composed of two or more parts, among the components of the force due to the interaction, so that they are joined together. If the components separately, of course, requires a certain energy to overcome related attractiveness of the need to do work. The magnitude of the work required to do, indicating that part of the combination of various components of the tightness; the object is called binding energy. The energy molecule split into free atoms needed is called chemical binding energy, and the energy of the atom split into nuclear required is called nuclear binding energy.

Nuclear binding can also be understood as the respective nuclear nucleus can together form an energy released. Therefore, according to the conservation of mass and energy, the binding energy is equal to mass loss correspond to the energy, that is

$$E_b = \Delta mc^2 \tag{10.3}$$

Since $1 amu = 1.6606 \times 10^{-27}$ kg, then the mass loss of 1amu is

$$E_{amu} = 1.6606 \times 10^{-27} \times \left(2.998 \times 10^8 \right)^2 = 1.4924 \times 10^{-10} J \tag{10.4}$$

Usually we use eV or MeV to represent energy, according to $1 \text{ eV} = 1.6022 \times 10^{-19}$ J, so

$$E_{\text{amu}} = \frac{1.4924 \times 10^{-10}}{1.6022 \times 10^{-19}} = 9.315 \times 10^8 \text{eV} = 931.5\text{MeV} \tag{10.5}$$

Example 10.2 Calculate the binding energy of ^{235}U. ^{235}U atomic mass is known as 235.043924amu。

Solution: $\Delta m = [Z(m_p + m_e) + (A - Z)m_n] - m_a$

$$\Delta m = [92(1.007277 + 0.0005486) + (235 - 92)1.008665] - 235.043924$$
$$= 1.91517\text{amu}$$

Then the binding energy of ^{235}U is

$$E_b = \Delta m \cdot E_{\text{amu}} = 1.91517 \times 931.5 = 1784\text{MeV}$$

10.2.3 Energy Level Theory

Energy level theory was originally a theory to explain the movement of extra nuclear atomic orbits and developed. It believed that the electron only on a specific, discrete orbital motion of each electron has a separate orbit of energy; the energy value is the energy level. Electronic transitions can occur between different tracks. Electrons can absorb energy from low energy level transition to the high energy level, transition from a high energy level to a lower energy level is radiated photons, often referred to as the X-ray photon.

The nuclear levels are based on extra nuclear level theory that the nucleus also has a variety of quantized energy states. These states reflect the law of interaction and multi-body nuclear systems of nucleons between nuclei. That in addition to stabilizing the nuclear ground state, all of the nuclear levels are unstable. They can be excited by the strong nuclear interaction, nuclear or other strong sub-group to be γ photons emitted by electromagnetic effects or by weakly interacting electron and neutrino emission, and decays to a lower energy state of the excited state or neighboring nuclides or ground state.

Therefore, X-rays and γ-rays are different. X-ray photons is due to the transition of atoms between two energy levels of electrons, γ-rays is due to the nuclei level transition. X-rays and γ-ray, although their essence is photon, but because of the different sources have different names.

Figure 10.5 shows the energy level diagram of ^{60}Ni nuclei. Usually painted on the bottom of a horizontal line represents the ground state and the excited states are all represented as a horizontal line above the ground state. Energy is proportional

Fig. 10.5 ^{60}Ni nuclear
energy level diagram

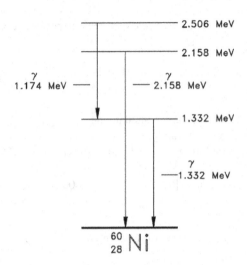

to the difference between the two horizontal lines and the distance between energy levels. Energy and a level difference between the ground state is called the level of excitation energy. Excited energy of ground state is zero. Next to the usual horizontal line will be marked on the level of excitation energy. For example, the first level of ^{60}Ni nucleus excitation energy is 1.332 meV, the second is 2.158 meV, and the third is 2.506 meV. Down arrow between two horizontal lines represent transitions between energy levels; γ-ray is released when the excitation energy of the transition depends on the difference between two energy levels. Such as γ rays ^{60}Ni from the second stage (2.158 meV) transition to the ground state is released 2.158 meV, γ-ray transitions from the third stage (2.506 meV) to the first stage (1.332 meV) when released to 1.174 meV.

10.3   Radioactive Decay

Most of the nucleus are naturally occurring in the ground state, the excitation energy is zero. Therefore, they are stable, and not the spontaneous release of energy or particles. However, there are also some of the atoms in the excited state of a wide variety, will release α or β-ray radiation through nuclear decay spontaneously (sometimes emit γ-rays), we call radioactivity.

Therefore, according to the nucleus is stable or not, the radionuclide can be divided into two types of stable isotope and radioactive nuclides. One species of nuclei spontaneously emit some rays into a lower energy level of the state or into other nuclides phenomenon, known as radioactive decay of nuclei.

Thus, radioactivity refers to certain radionuclide spontaneously emit particles or γ-rays, or electron capture orbit after the release of X-rays, or the nature of the occurrence of spontaneous fission.

10.3.1 Discovery of Radioactive Decay

In 1896, French physicist Becquerel in research experiments of uranium salts first discovered the natural radioactivity of uranium nuclei. In a further study, he found that uranium salts emitted rays and X-rays are very similar: make the air ionization, can also penetrate the black paper to make photographic film sensitive to light. He also found that changes in the external factors such as pressure and temperature experiments would not have any impact. Becquerel's discovery of this far-reaching, it makes people on the microstructure of the material has been updated knowledge, and thus opened the door to nuclear physics. In 1898, Marie Curie discovered radium and polonium radioactivity stronger. Because of this epoch-making discovery of natural radioactivity, Marie Curie and Becquerel were jointly awarded the 1903 Nobel Prize in Physics. Thereafter, the Marie Curie continue to study the application of radium in chemistry and medicine, and in 1902 isolated high-purity metal radium. Thus, Marie Curie was awarded the 1911 Nobel Prize in Chemistry. On the basis of Becquerel and Marie Curie, who study, and later gradually discovered many radionuclides other elements. It has been found in more than 100 kinds of elements, about more than 2600 kinds of nuclides. Which is just over 280 kinds of stable isotope (belonging to 81 kinds of elements), there are more than 2300 kinds of radionuclides. Radionuclides can be divided into natural radionuclides and artificial radionuclides two categories. Atomic number 83 (bismuth) or more elements are radioactive, but some elements of atomic number less than 83 (such as technetium) is also radioactive.

Since the nucleus is a system composed by a variety of nuclear, when the system can be at a lower energy state, it will by releasing particles (energy) ways to reduce their energy state.

10.3.2 Category Decay

In the process of radioactive decay of nuclei, through careful research and found the following to be compliance with the laws.

The first is the total number of charge conservation. The total charge of the system becomes neither more nor fewer. It allows the occurrence of positive and negative charge and allows generating a pair of positive and negative charges, but the overall charge of algebra conservation.

Followed by the mass number is conserved. Mass number is the number of neutrons and protons in the nuclear reaction process remains unchanged. Allowing protons into neutrons, also allows neutrons into protons.

The third is the mass-energy conservation. The total kinetic energy means that the system changes and mass losses corresponding to the energy conservation to maintain. Energy can be converted to mass, mass can be converted into energy.

Finally, conservation of momentum, the total momentum before and after the nuclear reaction system is conserved. Conservation of momentum of the decay products in the kinetic energy distribution play a decisive role.

There are several forms of radioactive decay α, β, γ, spontaneous fission, capture and other electronics. α decay, the decay refers to after the release of a decay process α particles, for example:

$$^{234}_{92}U \rightarrow {}^{230}_{90}Th + {}^4_2\alpha + \gamma + E_k \tag{10.6}$$

wherein, Ek is the kinetic energy released from the decay process.

In this reaction, radioactive decay, ^{234}U and ^{230}Th decay of α particle, maintaining the mass conservation. Since the binding energy of ^{234U} is lower than the binding energy of ^{230Th} with α particle, so the process will release energy, (lower binding energy indicates the higher energy state). Part of the energy released by γ-rays (this reaction is 0.068 meV), into the rest of sub-nuclear kinetic energy. Because to satisfy conservation of momentum, ^{230}Th mass is much greater than α particles, so the main kinetic energy (about 98%) is carried by the α particles.

β decay, the decay refers to after the release of a β particle decay process, for example:

$$^{239}_{93}Np \rightarrow {}^{239}_{94}Pu + {}^0_{-1}\beta + {}^0_0\overline{\nu} \tag{10.7}$$

$$^{13}_{7}N \rightarrow {}^{13}_{6}C + {}^0_{+1}\beta + {}^0_0\nu \tag{10.8}$$

In order to maintain mass conservation, energy conservation and momentum conservation, this process must release a neutrino or antineutrino. We called positron decay into neutrinos, negative electron antineutrino decay.

Neutrinos are one of nature's fundamental particles composed of, common symbols ν representation. Neutrinos are uncharged spin 1/2, the mass is very light (less than a millionth of the electron), close to the speed of light. Small size neutrino, uncharged, freely through the earth, and very weak interaction with other substances, known as the universe of "Invisible Man". An analogy: If the sun is a uranium nucleus, the Earth is like the electron; if the sun is an electron, the Earth is like a neutrino. Neutrino is too small; the scientific community predicted its existence from indirect to justify its existence by several decades.

The discovery of neutrino come from radioactivity research in the late 19th and early twentieth century. Some researchers found that in the quantum world, the energy absorption and emission is not continuous. Not only atomic spectrum is not continuous, and the nucleus emitted α ray and γ-rays is discontinuous. This is because nuclear energy transitions between different energy levels when released, is in line with the laws of the quantum world. The strange thing is, the β-ray energy spectrum during β decay process is continuous and only took part of the total electron energy, as well as some energy disappeared.

In 1930, Austrian physicist Pauli proposed a hypothesis: that the β decay in the process, in addition to electronics, the same time there is a zero rest mass, electrically neutral, with different photon radiation out of new particles, take away another part of the energy. This interaction of particles with matter very weak, difficult to detect. Unknown particles, the sum of the electron and nuclear recoil energy of a certain value, the conservation of mass and energy is still valid, but this unknown energy distribution ratio can vary between particles and electrons only. Pauli was such particles would be named "neutron", he initially thought that such particles originally present in the nucleus of an atom. In 1931, Pauli at the American Physical Society raised in a discussion, such particles are not originally present in the nucleus, but the decay. Pauli predicted the stolen energy "thief" is the neutrino. 1932 real neutron was discovered, the Italian physicist Fermi change Pauli's "neutron" to "neutrino".

In 1933, Italian physicist Fei Miti presented a quantitative theory of β decay, noting that in addition to the known nature of gravitational and electromagnetic forces, there is a third interaction, that is, the weak interaction. β decay is a neutron in the nucleus by the weak interaction decays into an electron, a proton, and a neutrino. His theory quantitatively describe the β and β-ray spectrum of continuous decay half-life of the law, continuous spectrum β mystery finally solved.

American physicist Cowan and Reines for the first time experimentally confirmed the existence of neutrinos in 1957. Their experiment is actually detected electrons and anti-neutrinos emitted by nuclear reactors β decay. The electron and antineutrino with hydrogen nuclei (i.e. protons) occur β decay; there is a specific correlation between the signal at the detector and the reactor power, enabling the observation of neutrinos. Their discovery won the Nobel Prize in Physics in 1995.

Electron capture reaction also belongs to the β decay, sometimes also called inverse β decay. If the number of protons in the nucleus is too much, an inner track electron may be trapped by a protons inside the nucleus (usually from the K shells, called K capture, if from the L layer, trapping is called L capture), the formation of a neutron and a neutrino, such as Eq. (10.9) below. When a proton transformed into a neutron, the number of neutrons plus 1, the number of protons reduce 1, atomic mass number remains unchanged. Since the reaction minus one proton, an element is transformed into another element in the electron capture reaction. Since for the new nuclei, missing an electron in K layer, and therefore inevitable in an excited state. Other layers electrons will transition to K layer to release an X-ray, as shown in Fig. 10.6.

$$^7_4\text{Be} + ^0_{-1}\text{e} \rightarrow ^7_3\text{Li} + ^0_0\nu \tag{10.9}$$

γ-ray is released from the nucleus energy transition process, it is a kind of high-energy photons. Nuclei in the excited state can be entered to the ground state by way of γ-ray decay. The γ-rays released must also pass through the extra nuclear layer; there may be an internal conversion. Here the so-called internal conversion means one of the nuclear external electron absorbs a γ-ray energy and free. Then it occurred extra nuclear level transitions of atoms to the ground state. In this process, γ-ray energy is converted into an electron and an X-ray.

Fig. 10.6 Electronic
schematic capture

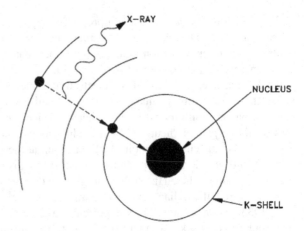

    Nuclei in the decay process, possibly in the isomeric state (having an average life
expectancy that is long enough to be observed nuclei excited state) of γ transition. It
is a form of radioactive decay. Long-lived isomeric state is usually in the back upper
left corner of radionuclide "m" symbol to represent. For example, charge number and
mass of ^{60m}Co and ^{60}Co are the same, but different half-lives, the former 10.5 min,
and the latter 5.27a. Typically, the same mass number and atomic number, but in
different energy states nuclides, called isomer prime.
    Spontaneous fission is a type of nuclear decay. Refers to spontaneous fission,
fission phenomenon in the ground state or isomeric states of nuclei occur in the
absence of additional particles or energy. Spontaneous fission decay and α are two
different ways of heavy nuclei decay, there is competition between the two. Spontaneous fission decay of Uranium nuclei is much smaller than α decay and just can be
detected. However, for some overweight artificial radionuclides, such as ^{252}Cf spontaneous fission decay is the main form. ^{252}Cf is an important spontaneous fission
neutron source.

10.3.3   Decay Chain

When there is a nuclide undergoes radioactive decay, the nuclide generated are not
necessarily stable. We will continue to radioactive decay, thereby forming a chain
decay. There are a series of nuclide decay chain in which one radionuclide through
radioactive decay is converted to the next nuclide until a stable nuclide. Before
converting nuclear called a parent nuclide, nuclide transformed called daughters,
along with continuous, sub-body decay can have many generations.
    E.g.,

$$^{91}_{37}\text{Rb} \xrightarrow[58.0s]{\beta^-} {}^{91}_{38}\text{Sr} \xrightarrow[9.5h]{\beta^-} {}^{91}_{39}\text{Y} \xrightarrow[58.5d]{\beta^-} {}^{91}_{40}\text{Zr} \qquad (10.10)$$

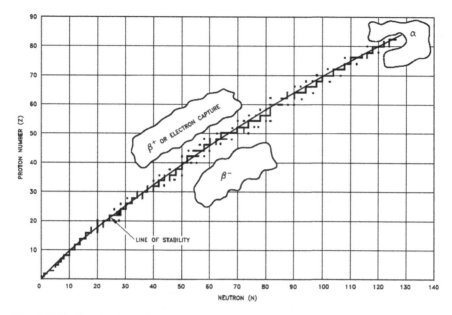

Fig. 10.7 Radioactive decay forecast map

$$\ce{^{215}_{85}At} \xrightarrow[0.10\text{ms}]{\alpha} \ce{^{211}_{83}Bi} \xrightarrow[2.14\text{min}]{\alpha} \ce{^{207}_{81}Tl} \xrightarrow[4.77\text{min}]{\beta^-} \ce{^{207}_{82}Pb} \qquad (10.11)$$

Continuous decay series known as radioactive series. There are three natural decay in the earth's crust. For example, the series of thorium from ^{232}Th, through 10 consecutive decay, and finally to the stable nuclide ^{208}Pb. Fission products inside the reactor are often continuous decay, until far into a stable nuclide. For example, ^{140}Xe four times to go through β- decay, the transition to stable nuclide ^{140}Ce.

According to a large number of observations found in the nuclei decay chain, which will be a radionuclide decay occurs can be predicted. As shown in Fig. 10.7 prediction method.

According to Fig. 10.7, α decay generally occurs only in the total number of nucleons large area, β$^-$ decay, electron capture occurs in the number of protons in the high side of the area, β + decay, or electron capture occurs in the number of neutrons in the high side region.

10.3.4 Half-Life

Often with the "half-life" to represent the decay speed of radionuclides. Half-life is the amount of nuclear decay half time off required. The half-life range from 10^{10}a ~

10^{-9} s. In addition to the half-life is too long or too short, difficult to measure, and the rest are available by different methods from 10^{-9} s to years.

Any kind of radioactive nuclei in the presence of separate decays exponentially with time. The number of nuclei at time t is.

$$N(t) = N_0 e^{-\lambda t} \tag{10.12}$$

Where λ is the decay constant, N_0 represents time zero moment nucleus number. Decay constant λ determines the size of the speed of decay. It is only with the type of radionuclide related. Decay constant λ and half-life $t_{1/2}$ is inversely proportional. The greater λ, it indicates radioactive decay faster, apparently to decay to half the time required is shorter. Their relationship is

$$\lambda = \frac{\ln 2}{t_{1/2}} = \frac{0.693}{t_{1/2}} \tag{10.13}$$

Decay constant is a certain probability that a nucleus of radionuclides in unit time for the spontaneous decay. Since λ is a constant, in unit of 1/s. Therefore, whenever each nucleus decays the probability per unit time decay are the same. This means that the individual nuclei decay is independent unrelated, each nucleus decays completely random event. Define the half-life after radionuclide decay law as shown in Fig. 10.8. Each through a half-life, reduced by half the number of the nucleus.

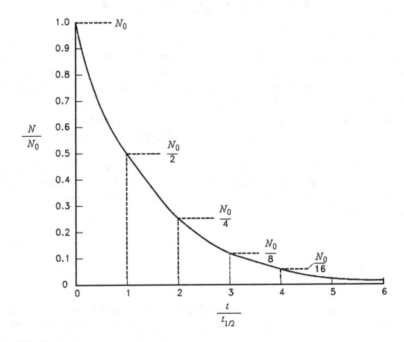

Fig. 10.8 Nuclear decay rates schematic

τ is the average life expectancy of an average survival time under a certain condition or atomic nuclei of the system. A large number of radioactive nuclei is concerned, some nuclear decay first, some nuclear decay later. Each nuclear of longevity in general is different from $t = 0$ to $t \rightarrow \infty$ is possible. However, for a certain type of radionuclide, the average lifetime τ is a constant. Average life and the decay constant reciprocal, that is,

$$\tau = \frac{1}{\lambda} \tag{10.14}$$

10.3.5 Radioactivity

Since the measurements of the number of radionuclides extremely inconvenient, and often is not necessary, and people are interested in and easy to measure: a certain amount of radioactive material, the number of spontaneous decay in a reasonably short time intervals that occur divided by the interval quotient, that the decay rate $-\frac{dN}{dt}$, also known as radioactivity. Its expression is

$$A \equiv -\frac{dN}{dt} = \lambda N = \lambda N_0 e^{-\lambda t} = A_0 e^{-\lambda t} \tag{10.15}$$

where $A_0 = \lambda N_0$, is radioactivity when $t = 0$. Radioactivity and radioactive nuclei number have the same exponential decay law.

Thus, the radioactivity is the number of atoms of radioactive elements or isotopes decay per second. Currently international units of radioactivity is Bq, 1 Bq equals to an atom decays per second. Historically Curie (Ci) as a unit of radioactivity. 1 Curie refers to a gram of radium's radioactivity, equal to 3.7×10^{10} Bq.

Example 10.3 A radioactive source has $20\mu g$ ^{252}Cf, known ^{252}Cf half-life of 2.638y, calculate the radioactivity of initial time and radioactivity after 12 years.

Solution: First, you need to calculate the number of atoms at the initial time

$$N_0 = \frac{20 \times 10^{-6}}{252.08} \times 6.022 \times 10^{23} = 4.78 \times 10^{16}$$

Decay constant

$$\lambda = \frac{\ln 2}{t_{1/2}} = 0.263\frac{1}{a} = \frac{0.263}{3600 \times 24 \times 365.25s} = 8.334 \times 10^{-9}\frac{1}{s}$$

Then you can get

$$A_0 = \lambda N_0 = \frac{\ln 2}{t_{1/2}} N_0 = 8.334 \times 10^{-9} \times \frac{4.78 \times 10^{16}}{3.7 \times 10^{10}} = 0.0108\text{Ci}$$

Radioactivity after 12 years

$$A_{12} = A_0 e^{-0.263 \times 12} = 0.04268 A_0 = 0.000461\text{Ci}$$

10.3.6   Radioactive Equilibrium

In the decay chain, there may be radioactive equilibrium. It refers to a certain kind of balance in the decay chain, according to the radioactivity of each chain precursor nuclides average life expectancy at any time intercropping exponential change. Therefore, the average life expectancy in the average length of life than this precursor radionuclide decay chains in any other generation daughter nuclides, it is possible to equilibrium phenomenon that the specific radioactivity of each radiator does not change over time.

Radioactive equilibrium, there are two cases; one is if the average expectancy life of precursor nuclides is not very long, but longer than any other generation in the chain daughter nuclides. When the time is long enough, the whole decay series will reach a temporary equilibrium, with each child decay with the half-life of the parent (or average expectancy life) and attenuation, as Fig. 10.9a. The curves represent a child radioactivity over time; parent curve b shows changes with time of radioactivity; curve c represents the total activity of the parent, the children over time; curve d represents changes of the child alone.

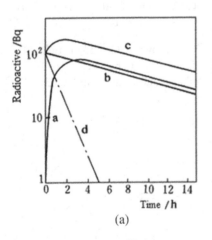

(a)

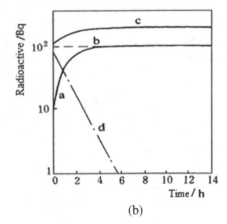

(b)

Fig. 10.9 Radioactive equilibrium

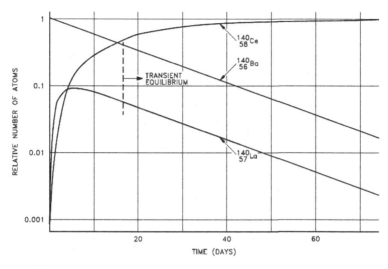

Fig. 10.10 ^{140}Ba Balancing process of Decay Chain

There is another situation, if the average expectancy life for long precursor nuclides, the overall change in the precursor nuclides can be ignored. So after quite a long time (usually five to seven times the longest continuous series decay half-life of the child above) can achieve long-term series of radiation balance. That radioactivity of each child are equal to the mother's activity, such as Fig. 10.9b. Before equilibrium is not reached, the child's activity increases with time, until it reaches the radioactive equilibrium reached.

The growth of artificial radionuclides similar long-term equilibrium process, so be produced in reactors or accelerators radionuclide, about five half-life of irradiation, radioactivity can be considered reached saturation point.

For example, consider the production ^{140}Ce following decay chain:

$$^{140}_{56}\text{Ba} \xrightarrow[12.75d]{\beta^-} {}^{140}_{57}\text{La} \xrightarrow[1.678d]{\beta^-} {}^{140}_{58}\text{Ce} \tag{10.16}$$

Its decay chain balancing process as shown in Fig. 10.10.

10.4 Neutron Interactions with Matter

Neutrons and nuclei can occur a variety of interactions, including scattering, absorption or fission process [6].

10.4.1   Scattering Process

Interactions between neutrons and very few nuclear (One of which is ^{235}U), can make the practical application of nuclear energy. If somehow you can get free state of neutrons, neutrons outside atoms. Since neutron no charge, when it approaches a nucleus (positively charged), unlike the positively charged protons will be as huge electric repulsion. Charged particles (such as protons) must have a lot of energy in order to overcome the electrical repulsion to touch a nuclear; but for neutrons, as long as there is a small energy neutron, which can cause nuclear reactions.

So where are free neutrons come from? Generally, we need a nuclear reaction to produce free neutrons. Neutrons emitted from the nuclear reaction, without exception, have a lot of energy, its magnitude is generally 1 ~ 10 meV. Since the mass of the neutron is about 1.660566×10^{-24} g, is easy to calculate the movement speed of these high energy neutrons is about 10^7 m/s t, i.e., about one-tenth the speed of light. Such fast neutron often called "fast" refers to the meaning of the high neutron velocity [3].

As neutrons pass through matter, they collide with the nuclei and scattered, then put the fast neutron energy to slow motion nuclear. Such scattering collision of two types: elastic and inelastic.

In elastic scattering, the momentum and kinetic energy are conserved between the two, as shown in Fig. 10.11. After the collision, the kinetic energy of the neutron part (or all) into a kinetic energy of the target. This process may be the laws of classical mechanics, with "Pinball" type of collision to calculate. In each collision of neutron and substantially stationary nucleus, the former will be a part (or all) of the kinetic energy transferred to the nucleus, and thus the speed is reduced. For a certain nuclei, the neutrons pass out how much energy will depend on the scattering angle. For a given scattering angle, the smaller the scattering kernel mass, the greater the proportion of outgoing neutron energy, and thus the greater the degree of moderation.

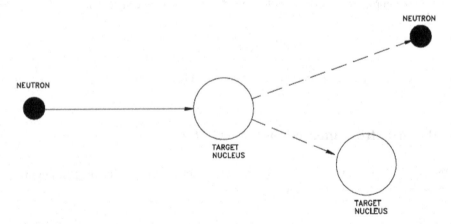

Fig. 10.11 Elastic scattering schematic

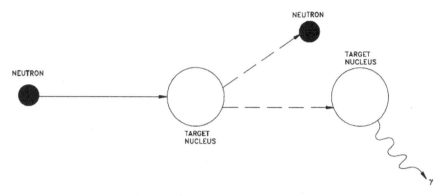

Fig. 10.12 Inelastic scattering schematic

Here appears one called "moderation" of the word, is a process of slowing down the velocity of neutron. The average energy produced by nuclear reactions is about 2 meV. We call it fast neutrons. When fast neutrons continuous loss of energy during the collision, so that their energy and speed decreases to become thermal neutron son, this process is called as moderation. The word "thermal neutron", usually refers to the kinetic energy of thermal neutrons is about 0.0253 eV (a rate of about 2200 m/s) of free neutrons. This also corresponds to the Maxwell-Boltzmann most probable velocity distribution at room temperature. Therefore, "thermal" refers to the exact meaning of the surrounding medium to reach thermal equilibrium.

In an inelastic collision, the momentum is conserved, and kinetic energy is not conserved, as shown in Fig. 10.12.

Part of the kinetic energy of the neutrons transformed into the nucleus internal energy (potential energy, sometimes called the excitation energy) during inelastic scattering. Inelastic neutron scattering and the specific properties of the scattering kernel are related and generally occurs only in the case of very high neutron kinetic energy. For elements of moderate or high mass number, over 0.1 meV neutrons can be involved in inelastic collisions. If the mass number of the scattering material is low, the neutron energy is required even higher.

10.4.2 Thermal Neutron

Within about 0.1 meV or less energy range, it is no longer inelastic collision. However, elastic collisions between neutron and nuclear will still be effective in slowing down neutrons, until the average kinetic energy of neutrons is equal to the average kinetic energy of the scattering medium atoms or molecules (thermal equilibrium). This energy depends on the temperature of the medium, the neutron energy has been reduced to within this range of values is called thermal neutrons. A mean energy of thermal neutrons is given by:

Table 10.3 Average speed
and energy of thermal neutron

Temperature/°C	Energy/eV	Speed/(m/s)
25	0.026	2.2×10^3
200	0.041	2.8×10^3
400	0.058	3.4×10^3
600	0.075	3.8×10^3
800	0.092	4.2×10^3

$$\text{Thermal neutron average energy} = 8.6 \times 10^{-5} t_k \text{eV} \qquad (10.17)$$

Here t_k is the absolute temperature of the scattering medium, K. Thus, at room temperature (i.e., about 295 K), the average energy of thermal neutrons is about 0.0253 eV.

The average speed of thermal neutrons by the following formula:

$$\text{Average speed of thermal neutrons} = 1.3 \times 10^4 \sqrt{t_k} \text{cm/s} \qquad (10.18)$$

Therefore, at room temperature, which is about 2.2×10^3 m/s. At various temperatures, and the speed of the thermal neutrons are listed in Table 10.3.

In order to draw a line between fast and thermal neutrons in nuclear engineering, the fast neutron energy refers 0.1 meV (i.e. 10^5 eV) or more. Energy from 10^5 to 1 eV neutrons are called medium-speed neutrons. The energy of about 1 eV or less is called slow neutrons. The thermal neutrons are neutrons with the surrounding environment in the thermal equilibrium state. Slightly higher than the thermal energy (i.e., in the lower speed range) neutrons sometimes called epithermal neutrons.

Neutron moderate problem occupy a very important position in many reactors, and the material used for this purpose is called the moderator. A good moderator is able to greatly reduce the speed of neutron in several collision. Thus, the material of low mass number of atoms is a good moderator. Ordinary water (H_2O), heavy water (D_2O), beryllium, beryllium oxide and carbon have been used in a variety of reactors as moderator. As good moderator material must not significantly absorb neutrons, so the light elements lithium and boron was excluded.

10.4.3 Radiative Capture Effect

It can occur several different types of neutron capture reactions with nuclei; here only discuss the two reactions more important for a nuclear reactor.

In radiative capture, the incident neutron enters the target nucleus forming a compound nucleus. The compound nucleus then decays to its ground state by gamma emission. An example of a radiative capture reaction is shown below.

$$_0^1 \text{n} + _{92}^{238}\text{U} \rightarrow \left(_{92}^{239}\text{U}\right)^* \rightarrow _{92}^{239}\text{U} + _0^0 \gamma \qquad (10.19)$$

The upper right corner marked with an asterisk means a compound nuclear. We can see that the radiative capture reaction product is an isotope of the original reactants, because they have the same atomic number, but one more mass number. In many cases, the compound nucleus are a bit unstable than the latter, and showed radioactivity through decay became another nuclear, some kind of identification radiation emitted. In this case, as discussed, nuclear radiation is almost always composed of electrons, then generally known electronic β particles, often accompanied with the occurrence of γ-rays to take away excess energy.

Radiation capture reaction by the symbol (n, γ), means in the process of a neutron is captured, while a γ-rays is emitted. The resulting formula (10.19) ^{239}U nucleus is radioactive; it gives out a negative β decay of particles, namely

$$_0^1 n + _{92}^{238}U \rightarrow _{93}^{239}Np + _0^0\gamma + _{-1}^0\beta \tag{10.20}$$

The product is the atomic number of an isotope of element 93, called neptunium (Np), it is not naturally exist. Neptunium 239 is β radioactive decay and become very rapid decay process is as follows:

$$_{93}^{239}Np \rightarrow _{94}^{239}Pu + _{-1}^0\beta \tag{10.21}$$

At this time, the atomic number of 94 elements of the isotope ^{239}Pu, called plutonium. Elemental plutonium exists only in a very small amount in nature. ^{239}Pu is radioactive, although it decays very slow, alpha particles. However, the importance of ^{239}Pu is that it can be used both in the atomic bomb and in the nuclear reactor to release nuclear energy.

From the natural existence of the ^{232}Th (n, γ) reaction, it will occur in a series of similar processes decay. At this case,

$$_{90}^{232}Th + _0^1 n \rightarrow _{90}^{233}Th + \gamma \tag{10.22}$$

The product is an isotope of thorium ^{233}Th. It has two stages of decay, the first stage is

$$_{90}^{233}Th \rightarrow _{91}^{233}Pa + _{-1}^0\beta \tag{10.23}$$

The second stage is

$$_{91}^{233}Pa \rightarrow _{92}^{233}U + _{-1}^0\beta \tag{10.24}$$

The product here is ^{233}U, which is essentially a fissile isotope that does not exist in nature. It is very slow to emit alpha particles, and, although it has not yet produced a large number of production, but in the future it may occupy an important position in the use of nuclear energy.

The example of the radiation capture process is exactly two elements of high atomic number, but in fact, all of the elements from hydrogen to uranium, are more or less the occurrence of this reaction. The process for hydrogen is

$$_1^1\text{H} + _0^1\text{n} \rightarrow _1^2\text{H} + \gamma \tag{10.25}$$

The product of the symbol ^2H is the stable hydrogen isotope deuterium of hydrogen. Although this reaction is not used to produce deuterium, it is important for some aspects of the design of the reactor.

10.4.4   Particle Emission

In a particle ejection reaction the incident particle enters the target nucleus forming a compound nucleus. The newly formed compound nucleus has been excited to a high enough energy level to cause it to eject a new particle while the incident neutron remains in the nucleus. After the new particle is ejected, the remaining nucleus may or may not exist in an excited state depending upon the mass-energy balance of the reaction. An example of a particle ejection reaction is shown below.

$$_5^{10}\text{B} + _0^1\text{n} \rightarrow \left(_5^{11}\text{B}\right)^* \rightarrow _3^7\text{Li} + _2^4\alpha \tag{10.26}$$

This reaction is somewhat similar to the fission reaction described below, which is the fission of the ^{10}B nucleus under the action of the neutron.

10.4.5   Fission

There is also a very important type of neutron reaction. In fission process, the compound nucleus, which is formed by the neutron absorbing, is very unstable, so that it is immediately split into two or more fission fragments. There are many kinds of specific ways of nuclear fission and only a few of the fission nuclei are divided in a symmetrical form. As a result, when a nuclear fission occurs, many kinds of fission fragments are generated. Most of them are radioactive, decay at different rates, and emit beta particles and gamma rays, and the resulting product itself is often radioactive. For example, in ^{235}U fission, will generate a mass number by more than and 80–160 of the 72 primary products. Each product is converted to a stable nucleus by an average of three stages of radioactive decay. Results, in a very short time after fission there are more than 30 kinds of different elements of more than 200 kinds of radioactive isotopes.

Nuclear fission [7], which is caused by neutron capture, exists only in the most important element, and the most important is thorium, uranium and plutonium. Some of the isotopes, most notably ^{233}U, ^{235}U and ^{239}Pu, are fissionable by thermal neutrons and fast neutrons, and other isotopes such as ^{232}Th and ^{238}U, which require fast neutrons to cause fission.

For ^{232}Th and ^{238}U, there is a fairly clear threshold energy limit at about 1 meV, and the neutron energy below this value is not likely to cause any appreciable fission. However, it is important that these two substances (especially the latter) will be spontaneous fission, such as 1 g ^{238}U per hour has 24 nuclear spontaneous fission. This fact is of great importance for the start-up of some reactors.

It is believed that only the high mass nuclei with odd number of neutrons can be quite stable under ordinary conditions, but also can react with slow neutron fission reactions. Only ^{235}U can meet the above conditions in the natural. Because of its mass number is equal to 235, while the atomic number is equal to 92, so the number of neutrons contained within it (i.e., 235–92 = 143) is an odd number. Artificial ^{233}U and ^{239}Pu also contain odd numbers of neutrons, which can be caused by thermal neutrons to cause fission. As for other stable material with this kind of nature, it seems that there is no practical significance because it is impossible to get enough weight.

From the point of view of the use of nuclear energy, there are two aspects of the importance of fission. First, this process releases a lot of energy; second, the reaction caused by the neutron itself will release the neutron. Therefore, it is possible to make this fission reaction in the appropriate conditions once the start, on their own continued, and continuously produce energy.

10.5 Nuclear Fission

Nuclear fission refers to fissile nuclear into two, in a few cases, can be split into three or more same order of mass nuclei and release of nuclear energy. Fission reactions including chain fission and spontaneous fission induced by neutron bombardment. In addition to spontaneous fission neutron source is used as external (e.g. ^{252}Cf), others such as ^{240}Pu and other spontaneous fission in the reactor, generally not be considered. So in a sense it refers to the reactor by neutron bombardment of some fissile nuclei, the cause fission of heavy nuclei a reaction to occur. Often in such a large amount of energy released fission process, and accompanied by the release of a number of secondary neutrons.

Equation (10.27) shows in the formula describing nuclear reaction.

$$U + n \rightarrow X_1 + X_2 + \nu n + E \tag{10.27}$$

Where, U is fissile nuclear, n is the neutron, X_1 and X_2, respectively, represent two nuclear fission fragments, ν represents the number of secondary neutrons for each fission average emitted, E represents each fission process released energy.

10.5.1   The Liquid Drop Model of Nuclear Fission

Droplet approximation model [8] is a model from the nature of the nucleon and nucleon strong interaction. This model can be clarified to some extent the phenomenon of nuclear fission, as shown in Fig. 10.13.

In the Fig. 10.13a state of the nuclei in the ground state, since the nuclear force inside the nucleus of neutrons and protons together to form a stable nucleus. When a neutron enters, the original balance is broken, the compound nucleus formed in (b) is a high energy level, the internal nuclear force is not sufficient to put all nuclear bound together, spherical nuclei are deformed. If the excitation energy is greater than a threshold energy (the minimum required to undergo fission excitation energy, the critical energy of different materials are shown in Table 10.4), deformed into a dumbbell shape of the nucleus. Due to the short-range nuclear forces, this time in the middle part of the nuclear force is weakened, and the electrical repulsion between protons is long-range force, unchanged, so the repulsion is greater than nuclear force, fission occurs (c).

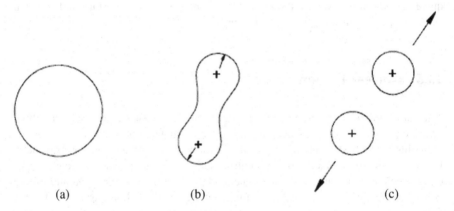

| (a)   (b)   (c) |

Fig. 10.13   The liquid drop model of nuclear fission

Table 10.4   Critical energy and the last neutron binding energy

Nuclide	Critical energy E_c/MeV	The last neutron binding energy E_{bn}/MeV	$(E_{bn} - E_c)$/MeV
^{232}Th	7.5	5.4	−2.1
^{238}U	7.0	5.5	−1.5
^{235}U	6.5	6.8	+ 0.3
^{233}U	6.0	7.0	+ 1.0
^{239}Pu	5.0	6.6	+ 1.6

10.5.2 Fissile Material

Capable of fission (whether caused by what process) called fissile nuclide, which nuclei are generally large mass of heavy nuclei. The most important fissile nuclide ^{233}U, ^{235}U, ^{239}Pu and ^{232}Th, ^{238}U. By their ease of nuclear fission into two categories. When using any energy neutron bombardment, which can cause nuclear fission, called fissile material. Such as ^{233}U, ^{235}U, ^{239}Pu these three nuclides, according to Table 10.4, their $(E_{bn} - E_c)$ are greater than zero, so when using any energy neutron bombardment will cause fission.

In nature, fissile material only ^{235}U, it's a typical fission reaction is

$$_0^1n + _{92}^{235}U \rightarrow \left(_{92}^{236}U\right)^* \rightarrow _{55}^{140}Cs + _{37}^{93}Rb + 3\left(_0^1n\right) \qquad (10.28)$$

In this reaction, ^{235}U hit by a neutron after the formation of a compound nucleus ^{236}U, ^{236}U are excited to the excited state and the nucleus split occurred, and the release of three neutrons.

The other is only when the energy is greater than a threshold neutron bombardment to its nucleus, will cause radionuclide fission reaction, known as fissionable material. For example, ^{238}U, only with an energy greater than 1.5 meV neutrons to bombard their nuclei, will have a fission reaction.

Some nuclides after neutron capture, through radioactive decay will generate new artificial fissile nuclides; such materials may be referred to as fertile material. For example, ^{238}U capture a neutron, after two β decay and eventually become fissile nuclides ^{239}Pu, such as Eqs. (10.19), (10.20), (10.21). Similarly, ^{232}Th nucleus to capture a neutron ultimately generate new artificial fissile nuclides ^{233}U, such as Eqs. (10.22), (10.23), (10.24). The two main fertile materials is shown in Fig. 10.14.

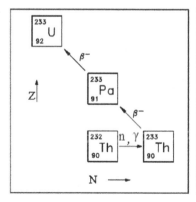 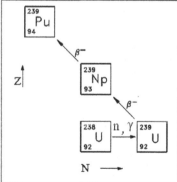

Fig. 10.14 Fertile process

10.5.3   Specific Binding Energy

In the previous, we introduce binding energy. The calculated binding energy of ^{235}U
is 1784 meV, which is the formation of the total binding energy of ^{235}U. If the average
per nucleon, it is the specific binding energy. For ^{235}U, the specific binding energy
is 1784 meV/235 = 7.59 meV. If the specific binding energy of each stable nuclide
can be calculated, it can draw a curve as shown in Fig. 10.15, specific binding energy
curve. Table 10.5 lists binding energy and specific binding energy of some nuclides
[9].

As can be seen from Fig. 10.15, starts from the mass number of relatively small,
with increasing mass number, specific binding energy shows an increasing trend.
When the mass number reaches 56, specific binding energy reaches a maximum of
8.79 meV. In the mass number more than about 60, showing a downward trend. For
^{235}U, mass number is 235; the specific binding energy is 7.59 meV. Therefore, from
the viewpoint of binding energy, low mass, if the number of nuclei can be combined
into a larger mass number of nuclei, which are capable of specific binding energy
increases, and therefore such a process should be able to release energy, which is
basically a fusion reaction principle. Conversely, if a large mass number of nuclei,
split into two medium-mass nuclei, its specific binding energy is increased, which is
the basic principle of the fission reaction.

The specific binding energy curve can also explain some of the other reactions.
For example due to α particles are no extra nuclear He nucleus, its mass number is
4, specific binding energy is 7.07 meV, higher than the specific binding energy of
^{10}B, it is possible to occur Eq. (10.26) type of reaction. This reaction can be used for
neutron measurement.

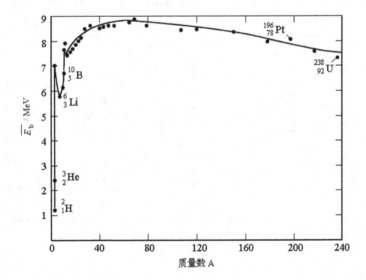

Fig. 10.15   Specific binding energy curve

Table 10.5 Binding energy and specific binding energy

Nuclide	Binding energy E_b/ MeV	Specific binding energy $\overline{E}_b$/(MeV/nucleon)
^{2}H	2.224	1.112
^{2}He	8.481	2.827
^{4}He	28.30	7.07
^{6}Li	11.99	5.33
^{7}Li	39.24	5.61
^{12}C	92.16	7.68
^{14}N	104.66	7.48
^{15}N	115.49	7.70
^{15}O	119.95	7.46
^{16}O	127.61	7.98
^{17}O	131.76	7.75
^{17}F	128.22	7.54
^{19}F	147.80	7.78
^{40}Ca	342.05	8.55
^{56}Fe	492.3	8.79
^{107}Ag	915.2	8.55
^{129}Xe	1087.6	8.43
^{131}Xe	1103.5	8.42
^{132}Xe	1112.4	8.43
^{208}Pb	1636.4	7.87
^{235}U	1783.8	7.59
^{238}U	1801.6	7.57

10.5.4 The Energy Released from Nuclear Fission

If a large mass number of nuclei split into two medium-mass nuclei, the total binding energy is increased, thereby releasing energy. So how to calculate the energy released fission reaction out of it? Look at the following fission reaction [10]:

$$_0^1n + {}_{92}^{235}U \rightarrow \left({}_{92}^{236}U\right)^* \rightarrow {}_{55}^{140}Cs + {}_{37}^{93}Rb + 3\left({}_0^1n\right) \qquad (10.29)$$

This fission reaction of ^{235}U become Cs (mass number 140) and Rb (mass number 93), and the release of three neutrons. Cs and Rb will be β decay. If the specific binding energy of Cs and Rb can be determined, we can calculate the binding energy changes. As shown in Fig. 10.16.

By calculation, ^{93}Rb binding energy is 809 meV, ^{140}Cs binding energy is 1176 meV, so you can get

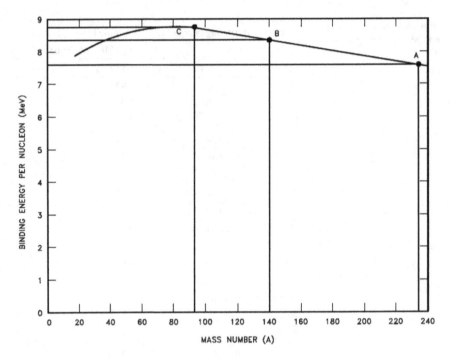

Fig. 10.16 Specific binding energy of fission reaction

$$\Delta E_b = (809 + 1176) - 1786 = 199 \text{MeV} \tag{10.30}$$

In addition to the use of calculated binding energy fission energy, the method can also be used to calculate the mass loss of fission energy, this method is often more accurate.

According to the data in Table 10.6, you can get mass loss of Eq. (10.29) fission reaction is 0.200509amu, whereby fission energy is 186.8 meV.

In the actual process of ^{235}U fissile nuclei, not all of the fission reactions are accordance with Eq. (10.29). Results fission fragments generated several medium mass numbers. There are many possible ways of nuclear fission, most of which are split into two fission fragments nuclei. For fission induced by thermal neutrons of ^{235}U, it has been found that about 30 different ways fission, i.e., about 60 kinds of fission fragments. Most mass distribution of fission fragments between 72 and 158.

Table 10.6 ^{235}U fission nuclides and mass

Nuclide	Mass/amu
^{235}U	235.043924
^{93}Rb	92.91699
^{140}Cs	139.90910
^{1}n	1.008665

Almost all of the fission fragments are unstable; they go through a series of β and γ decay. Thus in the final fission products may include more than 300 kinds of various radioactive and stable isotopes of different nuclides. Figure 10.17 shows a number of different mass in thermal neutron and 14 meV neutron effects of fission products yield distribution. Therefore, the most probable yield is 95 and 140. The actual measurements show, ^{235}U fission released every available energy around 200 meV, of which more than 80% is in the form of kinetic energy of fission fragments. Fission energy distribution as shown in Table 10.7.

In the reactor, fission fragments are very short range in the fuel pellets is approximately 0.0127 mm. It is considered that most of the kinetic energy of fission fragments are converted into heat in the fuel. Most of the kinetic energy of fission neutrons

Fig. 10.17 Yield of fission products distribution

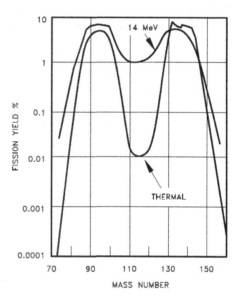

Table 10.7 ^{235}U fission energy release distribution

Energy form	Energy/MeV	Time
Kinetic energy of prompt fission fragments	167	Prompt
Prompt fission neutron kinetic energy	5	Prompt
Prompt γ energy	5	Prompt
(n, γ) reaction	10	Prompt
Fission product γ decay energy	6	Delayed
Fission product β decay energy	7	Delayed
Neutrino energy	10	Delayed
Total	210	

emitted is converted into heat absorbing materials. Fission neutrons itself will be absorbed within the reactor materials, the occurrence of (n, γ) reaction, and the release of energy of about 10 meV. Although it is not part of the energy released in the nuclear fission directly, but it is also the direct consequence of fission, and is instantaneous form released. A considerable part of γ-rays will be absorbed and converted into heat in the reactor; the reactor therefore generally in the calculation of them also included the release of the fission reaction in the energy available.

Since neutrinos is not charged, and its mass is very small, hardly reaction with any material in reactor. Therefore, 10 meV about neutrino energy cannot be utilized.

Rather, the available energy after each fission reaction will be released along with the type of reactor makes a difference. However, when calculated as a general, you can think approximately of each of ^{235}U nuclear fission, the energy can be utilized in the reactor is about 200 meV. Other nuclear fissionable nuclides emitted per fission available energy is also approximately this value. PWR fission energy available, about 97% are distributed within the fuel, less than 1% (as γ-rays) of energy in the reactor shield, the rest of the energy distribution in the moderator, coolant and within the structure of the material.

Available energy also includes fission products released during the decay of γ and β rays, but this part of the release of energy is a time delay. We estimate β decay energy of fission products, consider the following two decay chains,

$$\,^{93}_{37}\text{Rb} \xrightarrow{\beta} \,^{93}_{38}\text{Sr} \xrightarrow{\beta} \,^{93}_{39}\text{Y} \xrightarrow{\beta} \,^{93}_{40}\text{Zr} \xrightarrow{\beta} \,^{93}_{41}\text{Nb} \tag{10.31}$$

$$\,^{140}_{55}\text{Cs} \xrightarrow{\beta} \,^{140}_{56}\text{Ba} \xrightarrow{\beta} \,^{140}_{57}\text{La} \xrightarrow{\beta} \,^{140}_{58}\text{Ce} \tag{10.32}$$

For the Eq. (10.31), we have

$$\begin{aligned}\Delta E &= [m_{\text{Rb-93}} - (m_{\text{Nb-93}} + 4m_e)] \times 931.5 \\ &= [92.91699 - (92.90638 + 4 \times 0.0005486)] \times 931.5 \\ &= 7.84\text{MeV}\end{aligned}$$

For the Eq. (10.32), we have

$$\begin{aligned}\Delta E &= [m_{\text{Cs-140}} - (m_{\text{Ce-140}} + 3m_e)] \times 931.5 \\ &= [139.90910 - (139.90543 + 3 \times 0.0005486)] \times 931.5 \\ &= 1.89\text{MeV}\end{aligned}$$

You can see, this part of the decay energy is about 9.73 meV. Taking into account the amount of fission products yield distribution, usually taken 7 meV. In addition, the γ decay energy of fission products takes 6 meV, see Table 10.7. They accounted for a total available energy of about 6.5%. When the reactor is shun down, most of the fission energy due to the termination of the fission reaction is no longer released. However, before the shutdown fission products formed, this time still exist, and in a constant process of decay. Therefore, the fission products decay, β and γ rays' emitted energy, remained after the shutdown within a period, therefore still a need for cooling and shielding. To remove these decay heat from the reactor core after shutdown, the reactor safety research has become one of the important issues.

Some fission product nuclide has a longer half-life or a strong radioactivity, which will bring a number of special problems on transport and ultimately secure storage. This is one of the important issues that also must be considered in the use of fission energy. Some fission products such as ^{135}Xe and ^{149}Sm have a large thermal neutron absorption cross section; they will absorb thermal neutrons within the reactor, thus affecting the reactor's neutron balance. These issues we will discuss in the next chapter.

Exercises

1. Why a heavy nucleus split into two daughter nuclei fission process will release neutrons?
2. What is the isomer prime? What is the difference between isotope and isomer prime?
3. A neutron source has 1 mg ^{252}Cf; try to calculate the radioactivity of the initial time.
4. Archaeologists use ^{14}C radioactive decay to determine the age of paleontological. Because the content of ^{14}C within a living organism is same as the content of the ^{14}C in atmosphere. Moreover, after metabolism once stopped, ^{14}C content began to change. It is possible to determine the age of living in accordance with existing paleontological paleontology amount of ^{14}C. Known ^{14}C decay constant λ is 0.00012097 / year, if a paleontology fossil sensed ^{14}C remaining amount of 5% of the content of the atmosphere. How many years ago this creature died?
5. What kind of neutron interaction with matter?
6. Why after reactor shutdown will continue to release heat?

References

1. Fujia Y, Yansen W. Lu Fuquan Atomic Nucleus Physics (M). Shanghai: Fudan University Press; 2006.
2. Knief RA. Nuclear energy technology: theory and practice of commercial nuclear power. McGraw-Hill; 1981.
3. Kaplan I. Nuclear physics. 2nd ed. Addison-Wesley Company; 1962.
4. Foster AR, Wright RL Jr., Basic nuclear engineering. 3rd ed. Allyn and Bacon, Inc.; 1977.
5. Jacobs AM, Kline DE, Remick FJ. Basic principles of nuclear science and reactors. Van Nostrand Company, Inc.; 1960.
6. Academic Program for Nuclear Power Plant Personnel, Volume III, Columbia, MD, General Physics Corporation, Library of Congress Card #A 326517; 1982
7. Glasstone S. Sourcebook on atomic energy. Robert F: Krieger Publishing Company Inc.; 1979.

8. Glasstone S, Sesonske A. Nuclear reactor engineering. 3rd ed. Van Nostrand Reinhold Company; 1981.
9. Lamarsh JR. Introduction to nuclear engineering. Addison-Wesley Company; 1977.
10. Lamarsh JR. Introduction to nuclear reactor theory. Addison-Wesley Company; 1972.

Chapter 11
Reactor Theory

Nuclear reactor, or simply the reactor, is able to maintain a controlled, self-sustaining nuclear fission chain reaction in order to achieve the use of nuclear energy. The principles of a nuclear reactor theory discussed in this chapter include nuclear reactor physics, operation of nuclear reactors, nuclear fuel cycle and so on.

Nuclear reactor physics is the study of time, space, energy distribution and the direction of motion of neutron. On the physical nature, reactor physics is basically to establish a discipline in the following two aspects of knowledge on the basis of:

1. Some basic results of neutron nuclear reactions, in particular the various radionuclide neutron absorption, scattering and fission cross section with the variation of neutron energy. Reactor physics use these achievements in nuclear physics, microscopic cross section of neutron nuclear reactions collated, edited and evaluated, as the basic data of reactor physics.
2. describe the mathematical model of spatial neutron population movement and proliferation process. Reactor physics research method is to use a known cross-sectional data, mathematical models or quantitative description of the experimental method and the proliferation of neutron group time and space transport, and calculate some of the main physical parameters to illustrate the physical properties of the neutron multiplication system. These include critical properties, reactivity, power distribution, dynamic parameters, fuel consumption, control characteristics, growth characteristics.

Security features of reactor depends largely on the physical characteristics of the reactor. For example, for pressurized water reactor nuclear power plant, due to an external disturbance caused when the power rises, the temperature of the coolant rises; due to the negative effect of coolant temperature reactivity, resulting in decreased reaction, forcing power back down. Therefore, it has a negative feedback effect, capable of stable operation of the reactor protection. Reactor physics design in addition to ensure safe operation of the reactor has a good feature, but also consider a variety of physical state reactor accident conditions, to prevent significant radioactive release accident could endanger the public and the environment.

© Tsinghua University Press 2022
J. Yu, *Fundamental Principles of Nuclear Engineering*,
https://doi.org/10.1007/978-981-16-0839-1_11

Reactor physics calculations by theoretical calculation, researches the law of the interaction of a large number of neutron and matter in reactor, multiplication of neutrons in reactor. Reactor physics calculations include critical calculations, fuel consumption calculations, power distribution control, reactivity control, reactor stability and security and so on; they are to meet the design requirements and safety guidelines.

In terms of a single neutron, it is in the media has been motion until it is absorbed or escape from the reactor. Its trajectory is chaotic polyline, which is a random stochastic process. But, in fact, to be discussed is the statistical behavior of a large number of neutrons, macroscopic behavior caused by them can be described. Noting the movement of neutrons relates not only position in space, but also its speed (i.e., energy) and direction. Neutron transport equations are established.

A basic principle to establish neutron transport equation is the conservation of the number of neutrons or neutrons balance. In a certain volume, the number of neutron density with time rate of change should be equal to its production rate minus the rate of disappearance. This gives the transport equation can be accurately expressed in neutron space, energy distribution and direction of motion. But in general, difficult to obtain analytical solution of the transport equation. Even on a computer using numerical methods to solve, it is very complex and difficult task. Therefore, in the early reactor physics calculations, it is often only used in some local area actually requires accurate calculation of, or as a reference for comparison.

In a large reactor core, the spatial distribution of neutrons is nearly isotropic. This can be approximated that the neutron distribution is independent of the direction of movement, the problem is greatly simplified. By this approximation to simplify the resulting equation is called neutron diffusion equation. After the clustering method is applied to the diffusion equation, so that the final equation obtained in the spatial distribution of the neutron reactor is a set of simultaneous multi-group diffusion equations. Appears only in spatial variables in each equation, neutron cross-section parameters related to energy will appear as a constant in the equation. Multigroup diffusion equation is most commonly used in reactor physics calculation.

Since the reactor core components, the complexity of the geometry, multi-group diffusion equation is impossible analytically. With the development of computers and computing technology, the current numerical method by means of a computer has become almost the main method of reactor physics calculations commonly used.

In summary, the principles of a nuclear reactor theory is the study of neutron behavior, we first discuss the source of free neutrons.

11.1 Neutron Source

Neutron source refers to a device capable of releasing neutrons. Neutron source could be hand-held radiation source, research reactor facility and fission source. Neutron sources are generally used for general industrial applications such as oil well logging, neutron therapy. Here we mainly discuss neutron source inside nuclear reactor. Starting a nuclear reactor must have a neutron source, the reactor; there are the main source of neutrons in two ways: First, the natural source of spontaneous fission, secondly, artificial neutron source.

11.1.1 Natural Neutron Source

Some materials in the reactor core will be spontaneous fission (spontaneous fission) or α decay, resulting in a certain amount of neutrons. Table 11.1 shows some of the spontaneous fission nuclides. Spontaneous nuclear fission is occurring in the absence of external energy or external particle bombardment, it is one of the radioactive decay mode. In 1940 K. A. Peter Zach and G. N. Frio Rove first observed ^{238}U nuclei self-fission phenomenon, and estimated half-life of about 10^{16}–10^{17} years. The latest studies suggest that physical, spontaneous fission is the result of quantum mechanical tunneling effect. Nuclear fission has a border barrier, atomic number lower than thorium nuclides, fission barrier is too high, almost impossible to spontaneous fission; With the continuous generation of transuranic nuclides, spontaneous fission rate increases, becoming some of the transuranic nuclear the main decay mode pigment. For example, ^{254}Cf α decay of only 0.31%, accounting for 99.69% of spontaneous fission.

Possible intrinsic neutron source in reactor is another boron nuclear reaction, as follows:

$$\,^{11}_{5}B + \,^{4}_{2}\alpha \rightarrow \,^{14}_{7}N + \,^{1}_{0}n \tag{11.1}$$

Boron in natural abundance ^{11}B was 80.1% (see Fig. 10.2). To take advantage of this reaction as a neutron source some difficulty, this is because the range of α particle is very short, usually cannot penetrate the fuel rod cladding and therefore require special structural design.

For a nuclear reactor has been run, such as refueling after the restart, there is an important neutron source is a deuterium following reaction:

$$\,^{2}_{1}H + \,^{0}_{0}\gamma \rightarrow \left(\,^{2}_{1}H\right)^{*} \rightarrow \,^{1}_{1}H + \,^{1}_{0}n \tag{11.2}$$

Table 11.1 Neutron source of spontaneous fission and α decay

Nuclide	Half-life of fission/a	Half-life of α decay/a	Neutron/(g·s)
^{235}U	1.8×10^{17}	6.8×10^{8}	8.0×10^{-4}
^{238}U	8.0×10^{15}	4.5×10^{9}	1.6×10^{-2}
^{239}Pu	5.5×10^{5}	2.4×10^{4}	3.0×10^{-2}
^{240}Pu	1.2×10^{11}	6.6×10^{3}	1.0×10^{3}
^{252}Cf	66.0	2.65	2.3×10^{12}

5
80 11 Reactor Theory

Has been run since the inner reactor has enough high energy γ rays, and all
the water is used as the reactor coolant are the presence of deuterium (deuterium
abundance in natural water is 0.015%), so this is a very important in neutron source.

11.1.2 Artificial Neutron Source

Due to the above-described natural weak or dependent on the neutron source reactor
operating history, so sometimes need to start reactor by artificial neutron source.

The strongest artificial neutron source installed in reactor is ^{252}Cf, the only disadvantages of this kind of neutron source is the very expensive price. The main reason
is that californium very rare, extremely difficult and synthetic. Until 1975, the world
was about 1 g californium, so it is one of the most expensive elements. United States
with a nuclear reactor plant manufacturing californium 252, 1980 cumulative production of 2 g. In 1980, the price was as high as $1 million/mg. By 1994, the price was
dropped to $10 million/g, a million times of the price of gold. Therefore, some of
the nuclear reactors use beryllium neutron source.

$$^{9}_{4}\text{Be} + ^{4}_{2}\alpha \rightarrow \left(^{13}_{6}\text{C}\right)^* \rightarrow ^{12}_{6}\text{C} + ^{1}_{0}\text{n} \tag{11.3}$$

Such neutron source need to provide a number of radionuclides decay α particles
to dope into beryllium metal, such as doped radium, polonium or plutonium, and then
wrap it up with a stainless steel or other alloys. In addition, the binding energy of the
last neutron of beryllium is only 1.66 MeV, so if γ-ray energy exceeding 1.66 MeV,
the reaction can also take place as follows:

$$^{9}_{4}\text{Be} + ^{0}_{0}\gamma \rightarrow \left(^{9}_{4}\text{Be}\right)^* \rightarrow ^{8}_{4}\text{Be} + ^{1}_{0}\text{n} \tag{11.4}$$

In this case, you need to be filled in an appropriate amount of beryllium neutron
source with strong γ radiation materials. Irradiated antimony (Sb) is a good strong γ
radiation. Irradiation can be installed in good antimony beryllium foil package into
a small cylinder, and then wrapped sheath outside.

11.1.3 PWR Neutron Source Assembly

PWR neutron source components including primary and secondary neutron source
for improving the level of reactor neutron fluence rate at start, so that the source-scale
nuclear testing instrument can be reliably measured neutron flux level, thus ensuring
the safety of the reactor started.

The structure of primary neutron source is basically the same as burnable poisons
by connecting the handle and the primary source rod components. 2 groups of primary
neutron source assembly, each consisting of a primary neutron source rod. Neutron

source material is Cf or Po–Be, it will spontaneously emit neutrons. The source strength of each primary source rod is not less than 3.6×10^8 n/s. Primary neutron source is double-coated by stainless steel. It is used for the first time starting of reactor.

The structure of secondary neutron source assemblies is basically the same as control rod assembly by connecting the handle and a secondary source rod component. Secondary source rod made of stainless steel cladding, Sb–Be pellets source and composition of the upper and lower end plug. Sb–Be source is a stable source material, antimony absorb neutrons during reactor operation, antimony γ-ray bombardment of neutrons released beryllium. Secondary neutron source assembly starts reactor for refueling, the minimum strength requirements are still not less than 3.6×10^8 n/s neutron after three to four months refueled.

Primary and secondary neutron source components are fixed without moving parts. Neutron source is mainly used to start, once the start-up, the reactor is capable of producing neutron own and maintain a certain number of neutrons. Because the reactor is able to maintain control, means self-sustaining chain reaction of nuclear fission. Trough reasonable arrangement of nuclear fuel, a self-sustaining chain reaction of nuclear fission can occur without additional neutron source. Below we discuss how to achieve "maintain controlled". Because the main implication of "maintain controlled" is to maintain the level of a certain number of neutrons in the reactor, and therefore need to discuss the interaction between neutrons and matter.

11.2 Nuclear Cross-Section

In order to discuss the interaction of neutrons with matter, we have once introduced σ in Eq. (6.14), where σ is the thermal neutron absorption cross section of the emitter material, cm^2. At that time, we did not introduce the physical meaning of the neutron reaction cross section in detail. Now we have to focus on at these very basic physical quantities.

11.2.1 Neutron Reaction Cross Section

Neutron reaction cross section [1] is a kind of measurement of the probability of the incident neutron particles and nuclear material with various nuclear reactions. Therefore, neutron reaction cross section is used to describe a nuclear reaction probability. People defined microscopic cross-sectional and macroscopic cross-section, we first look at microscopic cross section.

The probability of a particular reaction occurring between a neutron and a nucleus is called the microscopic cross section of the nucleus for the particular reaction. This cross section will vary with the energy of the neutron. The microscopic cross section may also be regarded as the effective area the nucleus presents to the neutron for

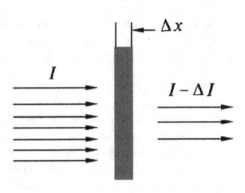

the particular reaction. The larger the effective area, the greater the probability for
reaction.

In nuclear and particle physics, the concept of a microscopic neutron cross section
σ is used to express the likelihood of interaction between an incident neutron and
a target nucleus [2]. The physical meaning is when a single neutron incident on the
unit area with only a target nucleus, the probability of nuclear reaction of this neutron
and target nuclei. If there is a one-way parallel uniform neutron beam, the intensity
is I (i.e., the unit area per unit time, go through a plane perpendicular to the direction
of flight of the neutron, have I neutrons), normal incidence on a thin target unit area,
thin target thickness Δx, within the target chip the nucleon number density is N,
some nuclear reactions resorted to exit neutron beam intensity diminished ΔI, see
Fig. 11.1.

The microscopic cross section is defined as

$$\sigma = \frac{-\Delta I / I}{N \Delta x} \tag{11.5}$$

where, $-\Delta I/I$ is the fraction of neutron reactions to target nuclei; $N \Delta x$ is target
nuclei per unit area on the thin target. Equation (11.5) was developed for measuring
microscopic cross section, it is a little bit difficult to understand, let us use a little
common way to introduce the concept of microscopic cross section.

First think about a hypothetical question: If the football field sprinkled with soccer
uniform (as shown in Fig. 11.2), an average of one football per square meter, then
dropped from a height of a small grain of rice. What is the probability of a small
grain of rice can hit football is?

Hold on, we will first analyze the probability and what factors related. First, is
there any relationship between the probability and the size of football field? Just
make sure that the small grain can fall into the football stadium, the probability
does not change with the size of football field, only dependent with the number
of footballs per unit area. Secondly, the size of football matters or not? Obviously
it matters, precisely, should be cross-section of football (the projected area of the
parallel beam) matters. Therefore, the cross-section of football is a measure of the
probability of a small grain hits a football.

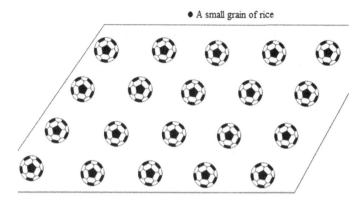

Fig. 11.2 Small grain shooting football schematic

If a football has a cross-section of 0.03 m², one football per square meter, then the probability of a small grain of rice hit football is 0.03. If there are three footballs per square meter, and each football position is random, the probability of 1 − 0.97³ = 0.087; if not overlap for each football, the probability is equal to 3 × 0.03 = 0.09.

Therefore, microscopic cross section having an area dimension is a measure of the probability of a neutron and the size of a nucleus-target interaction occurs. Or that its value is equal to the probability of the interaction of a neutron and a target nucleus per unit area occurs. It should be noted that "probability" is dimensionless, and "the measure of probability" may has a dimension of area.

According to Table 10.1, the radius of ^{238}U nuclei is 7.74×10^{-13} cm, the radius of ^{235}U nuclei and it almost. Thus, a uranium nucleus in a parallel beam projection area is about 2×10^{-28} m². Measure microscopic cross section of nuclei generally used bar (b) as a unit, $1b = 10^{-28}$ m², or $1b = 10^{-24}$ cm².

Due to a variety of nuclear reactions can occur between neutrons and the target nuclei, nuclear reactions have different microscopic cross section. σ_a represents microscopic absorption cross section; σ_s represents microscopic scattering cross section and so on. σ_t microscopic total cross section for a variety of microscopic cross section and that $\sigma_t = \sigma_a + \sigma_s + \dots$. Some of commonly used nuclear material's neutron fission cross section and absorption cross section at 0.0253 eV energy neutron are shown in Table 11.2.

Table 11.2 Some of material's fission cross section and absorption cross section at 0.0253 eV energy neutron (2200 m/s)

Material	Fission cross section σ_f/b	Absorption cross section σ_a/b
^{233}U	531	579
^{235}U	582	681
^{238}U	–	2.70
^{239}Pu	743	1012

Example 11.1 Suppose the 1 cm^2 area of uniformly distributed random 10^5 ^{235}U atoms, flew into a neutron at a speed of 2200 m/s inside the area. Determine the probability of reaction between this neutron and ^{235}U atomic fission reaction. If the area is 1 nm^2, and the probability?

Solution: According to Table 11.2, to 2200 m/s neutron, ^{235}U atomic fission cross section is 582 b. If the atoms cannot overlap each other, the probability is

$$p = n\frac{\sigma}{A} = 10^5 \times \frac{582 \times 10^{-28}}{1 \times 10^{-18}} = 5.82 \times 10^{-17}$$

If an area of 1 nm^2, assume that atoms cannot overlap each other, the probability is

$$p = n\frac{\sigma}{A} = 10^5 \times \frac{582 \times 10^{-28}}{1 \times 10^{-18}} = 5.82 \times 10^{-3}$$

However, if atoms can overlap each other, then the probability is

$$p = 1 - \left(1 - \frac{\sigma}{A}\right)^n = 1 - \left(1 - \frac{582 \times 10^{-28}}{1 \times 10^{-18}}\right)^{100000} = 5.803 \times 10^{-3}$$

As can be seen from Example 11.1, when the probability of the order of 10^{-3} or less, irrespective of the overlap is acceptable. Of course, the phenomenon of micro-world examples and above compared to neutron small grain of rice, compared to the target nucleus football there is a little difference. In football, only a small grain of rice within the projected area to hit the soccer and football collide or interact. In the microscopic world actually is not necessarily, as long as it is possible to fly into neutrons within a certain sphere of influence of the target nucleus and the target nucleus interactions can occur. That is, in the microscopic world, ^{235}U nuclei and neutron interactions may occur sectional cross-section much greater than the projection of nuclei about 2 b. For example, 0.0253 eV thermal neutron fission cross section of ^{235}U nuclei can reach 582 b.

For a considerable number of nuclides of moderately high (or high) mass numbers, an examination of the variation of the absorption cross section with the energy of the incident neutron reveals the existence of three regions on a curve of absorption cross section versus neutron energy. This cross section is illustrated in Fig. 11.3. First, the cross section decreases steadily with increasing neutron energy in a low energy region, which includes the thermal range ($E < 1$ eV). In this region the absorption cross section, which is often high, is inversely proportional to the velocity (v). This region is frequently referred to as the "$1/v$ region," because the absorption cross section is proportional to $1/v$, which is the reciprocal of neutron velocity. Following the $1/v$ region, there occurs the "resonance region" in which the cross sections rise sharply to high values called "resonance peaks" for neutrons of certain energies, and then fall again. These energies are called resonance energies and are a result of the affinity of the nucleus for neutrons whose energies closely match its discrete,

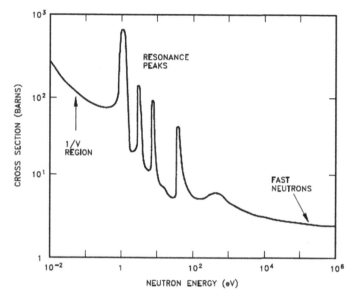

Fig. 11.3 Relations of typical absorption cross-sections and neutron energy

quantum energy levels. That is, when the binding energy of a neutron plus the kinetic energy of the neutron is exactly equal to the amount required to raise a compound nucleus from its ground state to a quantum level, resonance absorption occurs. The following example problem further illustrates this point.

Let us look at an example of a formant. ^{236}U has a 6.8 MeV quantum level, since the binding energy released from a ^{235}U atom absorbs a neutron can be calculated from the mass loss for

$$(235.043925 + 1.008665 - 236.045563) \times 931 = 6.54 \, \text{MeV}$$

Therefore, for ^{235}U atom, energy of $6.8 - 6.54 = 0.26$ MeV neutron resonance absorption will occur.

Once you have the concept of microscopic section, let us establish the concept of the macro section. The macroscopic cross section Σ is a measurement of the probability of a neutron reaction with nuclei in unit volume. Look at an example, if uniformly distributed within 1 nm^3 has 10^5 ^{235}U atom, a neutron speed of 2200 m/s fly into this region, and what is the probability of ^{235}U atomic fission reaction?

It is different with the previous example. After passing through the first layer of the target nucleus, neutrons may also reaction with the behind target nucleus. The original two-dimensional problem changes into a three-dimensional probability. To this end, we define a physical quantity called the macroscopic cross section. Macroscopic cross section is the product of microscopic cross section and the number of the target nucleus per unit volume (nucleon number density).

$$\Sigma = N\sigma \qquad (11.6)$$

For a uniform distribution within the case, here we have 10^5 have ^{235}U atoms in 1 nm^3, nuclear number density is $10^5/10^{-27} = 10^{32}/m^3$, then the macroscopic cross section $\Sigma = N\sigma = 5.82 \times 10^6$ m^{-1}. In order to determine the probability of fission with ^{235}U atoms in this case, we also need to know the thickness of target. If the incident is a two-dimensional area 1 nm^2, (neutrons incident randomly in this area), the target thickness of 1 nm, the probability is about 5.82×10^{-3}. This is equivalent to the probability that there are 10^5 ^{235}U atoms within 1nm^2 (Example 11.1). So, with the same volume, the probability will change with the incident area varies. Thus, the macroscopic cross-section describes a measure of the probability of reaction of neutrons with atomic nuclei in a unit volume with a unit surface area.

Here we add some mathematical knowledge. In a problem of calculating probability of an event, the first important thing is to determine what amount of randomly distributed within the range. Such as the previously mentioned "incident neutrons randomly on a unit surface area". We give an example to illustrate.

Example 11.2 Within a unit circle randomly draw a chord, determine the probability of the chord length greater than the side of an equilateral triangle?

Solution 1: Given any circle and a chord, there must be two intersections. Assuming these two intersections on the circumference of the uniform random distribution on circle. Since the point is random, may assume a fixed point in the A position (Fig. 11.4), you can calculate the probability of the length of the string is greater than the inscribed equilateral triangle side is 1/3. (F point falls BDC meet the conditions on the arc.).

Fig. 11.4 Randomly chord within a unit circle

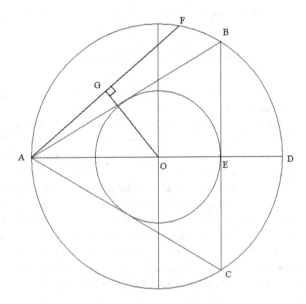

Solution 2: Consider the distance between a chord to center (OG), if the length OG is assumed randomly distributed between 0 and 1 (uniformly), you can calculate the probability of the length of the string is greater than the inscribed equilateral triangle of side is 1/2.

Solution 3: Given an arbitrary string randomly, it has a midpoint (G), assuming the position of the midpoint G random distribution in the area of the unit circle (uniformly), you can calculate the probability of the length of the string is greater than the inscribed equilateral triangle of side is 1/4 (the area of half radius to the unit circle area).

Three solutions above are all correct, except that of different understanding of the word of "randomly draw a chord" and "uniformly distributed".

Therefore, macroscopic neutron cross section is a measure of the interaction probability per unit distance of the target nucleus, its units is m^{-1}. What is needed is the conditions in the neutron incident per unit area is random incident neutron, through a unit distance, so it is the probability per unit volume. Macro-sectional is not an area, but a concept extending out from microscopic cross section for the case of three-dimensional probability of neutron and target nucleus interactions. Corresponding to different microscopic cross section has a corresponding macroscopic cross section, for example: $\Sigma_a = N\sigma_a$, showing macroscopic absorption cross section. Similarly, the total macroscopic cross section Σ_t for the sum a variety of macro-section.

Another point to note is that the formula to calculate the macroscopic cross section is $\Sigma = N\sigma$, that is to say the case is used in the projected area of atoms cannot overlap each other. Let us look at Fig. 11.5, in a three-dimensional space without overlapping, when projected onto a two-dimensional plane, overlap may occur. So to calculate the probability of neutron interactions with nucleus of per unit volume should be

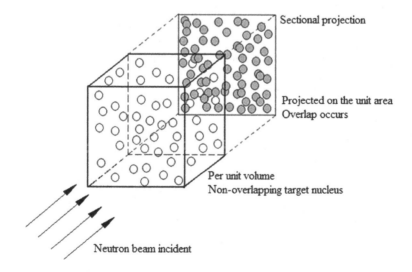

Fig. 11.5 The probability within a unit volume

calculated using the formula:

$$p = 1 - \left(1 - \frac{\sigma}{A}\right)^n = 1 - \left(1 - \frac{\sigma}{A}\right)^{N \cdot V} \tag{11.7}$$

where N is the density of nuclear number, V is the volume. When σ/A is very small, can be approximated as the case do not overlap, that is

$$p \approx (N \cdot V)\frac{\sigma}{A} = N\sigma\frac{V}{A} = N\sigma\Delta x = \Sigma\Delta x \tag{11.8}$$

11.2.2   Mean Free Path

In the statistical mechanics of gas molecules in motion, there is a physical quantity called the mean free path. It is the mean free path of a gas molecule between two successive collisions may pass. The mean free path of microscopic particles refers the average distance to microscopic particles collide with other particles. We can put this concept into the field of neutron science, the definition of the mean free path of neutrons, symbol λ, defined as the average distance of neutrons between two interactions in the nucleus. After analysis, we can get

$$\lambda = \frac{1}{\Sigma} \tag{11.9}$$

The mean free path is the inverse of macroscopic cross section, the unit is m. Similarly, different nuclear reactions have different mean free path, such as λ_a represents absorption mean free path, λ_s represents scattering mean free path.

Introducing the concept of the mean free path is mainly to deepen the understanding of the macroscopic cross section; macroscopic cross section is the reciprocal of the mean free path.

11.2.3   Temperature Effects Cross Section

Thermal neutrons are neutrons in thermal equilibrium with the surrounding medium. In a reactor core, the average temperature reaches 300 °C, while the cross-section data are usually given neutron energy sectional values of 0.0253 eV (295 K, see Table 11.2), for other temperatures thermal neutron the average fission cross section, can be calculated as follows:

$$\sigma_f = \sqrt{\frac{295}{273 + t}} \times \sigma_{f,0.0253} \tag{11.10}$$

where, t is temperature in °C.

Example 11.3 Calculate ^{235}U fission cross-section at temperature of 260 °C.

Solution: $\sigma_f = \sqrt{\frac{295}{273+t}} \times \sigma_{f,0.0253} = \sqrt{\frac{295}{273+260}} \times 582 = 433b$.

11.3 Neutron Flux

When calculating the power of reactor core, the problem we are concerned: in unit volume (for example 1 cm^3) within the fuel, per unit time (1 s), what is the number of nuclear reactions? The physical quantity called nuclear reaction rate or fission rate. Earlier we introduced the concept of cross-section, such as the size of the microscopic cross section reflects the probability of a neutron and a nucleus on the basis of nuclear reactions. Macroscopic cross section and microscopic cross section is foundation for the calculation of nuclear reaction rate. However, we also need a very important physical quantity, the number of neutrons bombard the target nucleus in unit time. Like the previous cited that small grain of rice to hit football example, what is the number of small grain falling from the sky every second? This factor is the physical description of neutron flux, defined as

$$\varphi = nv \qquad (11.11)$$

where in, N is the density of neutrons number, i.e., the number of neutrons per unit volume, 1/m^3; v is neutron velocity, m/s.

The unit of neutron flux is n/(m^{2}s). Therefore, the neutron flux refers to in a point of space; the number of neutrons received whatever direction in unit time, divided by the surface area of the sphere of this point. It is the number of neutrons per unit time through unit area. For the previous example, the grain flux is the number of small grain falls on the football field per unit area and per unit time. Reactor kinetics is devoted to research nuclear reactor neutron flux with time.

Free neutrons per unit volume called neutron number density, indicating the intensity of the free neutrons in the medium. Neutron number density and neutron velocity is termed neutron flux that is the total distance of all free neutrons per unit volume per unit time of flight.

The product of neutron flux and neutron macroscopic cross section called the nuclear reaction rate, which represents the number of nuclear react by free neutrons per unit time, unit volume.

The net number of neutrons flows vertically through unit area of a plane per unit time is called the neutron flux density, which indicates the strength of the case of the free flow of neutrons in the medium. In describing the neutron diffusion phenomenon to use the neutron flux density.

11.3.1 Fick's Law

In the media, by neutron scattering and successive collisions tend to migrate from the high particle number density areas to low-density areas, this phenomenon known as diffusion of neutrons. It is a basic phenomenon of spatial movement of neutrons in nuclear fission reactors and certain neutron experimental device.

Neutrons movement in a medium is a very complex issue. But if the media and neutrons make the following simplifying assumptions: the medium is infinite; medium is uniform (and therefore all cross sections are constant, regardless of the position); no external neutron source; in the laboratory coordinate system scattering is isotropic; neutron flux as a slow change function of position; neutron flux does not change with time. You can get

$$J = -D\,\mathrm{grad}\varphi \tag{11.12}$$

where in, J is the neutron flux density; φ neutron flux; D neutron diffusion coefficient. If we compare the Fourier's law calculate the thermal conduction in Chap. 3, see Eq. (3.3), we can see that the diffusion neutron and heat diffusion have a similar pattern.

From Eq. (11.12) we can find, the negative gradient is proportional to the neutron flux density and neutron flux. This relationship is similar in form to Fick's law is used to describe the same liquid and gas molecular diffusion phenomenon, also known as Fick's law. Although the derivation of Fick's law made a number of assumptions, such that the Law is not a completely accurate description of the movement of neutrons within the reactor. However, due to the law of the basic phenomena it reflected neutron diffusion, coupled with the simplicity of the relationship between the expression neutron flux density and neutron flux, so that became the basis of the laws of neutron diffusion theory.

11.3.2 Neutron Diffusion Equation

The number of neutrons leaking out per unit time per unit volume can be further expressed from Fick's law (if it is a negative value indicates incoming). The physical density also known as leakage rate, it can be proven that it is equal to $-D\nabla^2\varphi$. If you know of neutron production rate and disappearance rate within a small volume, then, the neutron flux changes over time can be obtained to meet the equation

$$\frac{1}{v}\frac{\partial\varphi}{\partial t} = D\nabla^2\varphi - \Sigma_a\varphi + S \tag{11.13}$$

where φ is the neutron flux; v is the neutron velocity; D neutron diffusion coefficient; Σ_a macroscopic absorption cross section; S is the neutron source density; ∇^2

Laplace operator, for the different form of the Laplace operator in different coordinate systems, review Chap. 1.

Equation (11.13) is called diffusion equation. If the neutron flux does not change with time, steady-state neutron diffusion equation can be obtained

$$D\nabla^2\varphi - \Sigma_a\varphi + S = 0 \qquad (11.14)$$

If without a neutron source, you can get the steady state neutron diffusion equation without source,

$$D\nabla^2\varphi - \Sigma_a\varphi = 0 \qquad (11.15)$$

In order to solve this equation, the boundary conditions also needed to determine the neutron diffusion equation. Commonly used boundary conditions are: at the junction with the vacuum, the neutron flux density from vacuum is zero; both neutron flux density and neutron flux continuous at the interface between two different media; neutron flux must be a limited single-valued non-negative real number. Since the neutron diffusion equation is a second-order partial differential equation, generally it requires two boundary conditions.

Depending on the geometry of the medium, it is appropriate to represent different coordinate systems and solving neutron diffusion equation, here we do not discuss specific details of solving the derivation, only lists some conclusions of uniform naked reactor.

Uniform naked reactor is an extremely simplified reactor core models. Although in practice because of the existence of the reactor coolant and the structure of the material within the core, the media cannot be uniformly distributed within the core, but a simplified uniform reactor model, during the time of the theoretical analysis is extremely useful. This is because through the analysis of uniform naked reactor, we can grasp the whole of a reactor. Some analytical solutions for thermal neutron flux of uniform naked reactor is shown in Table 11.3.

Table 11.3 Thermal neutron flux of uniform naked reactor

Geometry	Coordinate	Thermal neutron flux
Infinite thickness flat with thickness of a, a_e is extrapolated thickness	x	$\varphi_0 \cos\left(\frac{\pi x}{a_e}\right)$
Side length a, b, c cuboid, a_e, b_e, c_e extrapolated side	x, y, z	$\varphi_0 \cos\left(\frac{\pi x}{a_e}\right)\cos\left(\frac{\pi y}{b_e}\right)\cos\left(\frac{\pi z}{c_e}\right)$
A sphere of radius R, R_e extrapolated radius, equal to $R + 0.71\lambda$	r	$\varphi_0 \sin\left(\frac{\pi r}{R_e}\right)/\left(\frac{\pi r}{R_e}\right)$
Radius R, height L cylinder, R_e extrapolated radius, L_e extrapolated height, equal to $R + 1.42\lambda$	r, z	$\varphi_0 J_0(2.405r/R_e)\cos\left(\frac{\pi z}{L_e}\right)$

Currently the vast majority of power reactors are used cylindrical core, for a uniform naked cylindrical core, thermal neutron flux density distribution in the height direction is cosine distribution, the radial distribution is zero-order Bessel function.

11.3.3 Self-Shielding

In some locations within the reactor, the flux level may be significantly lower than in other areas due to a phenomenon referred to as neutron shadowing or self-shielding. For example, the interior of a fuel pin or pellet will "see" a lower average flux level than the outer surfaces since an appreciable fraction of the neutrons will have been absorbed and therefore cannot reach the interior of the fuel pin. This is especially important at resonance energies, where the absorption cross sections are large.

11.4 Reactor Power

In order to calculate the fission reactor power or power density, need to make sure of the fission reaction rate, hereinafter referred to as the fission rate.

11.4.1 Fission Rate

In unit time (1 s) per unit volume (1 cm^3) fuel, the number of fission, known as fission reaction rate.

$$R = \Sigma_f \varphi = N_{235} \sigma_f \varphi \tag{11.16}$$

where, the R for fission rate, the unit is $1/(\text{s·cm}^3)$; Σ is macroscopic cross section, the unit is 1/cm; σ is micro section, the unit is cm^2; N_{235} is ^{235}U nuclear density, the unit is 1/cm^3; φ is neutron flux, the unit is n/(cm^2 s).

11.4.2 Volumetric Heat Release Rate

Volumetric heat release rate refers to the heat release rate per unit time per unit volume. Note that the volumetric heat release rate refers to energy converted into heat, not all the energy release per unit volume, because some of the energy is released in other places (such as γ rays) and even some energy can't be converted into heat (such as neutrinos energy).

After homogenization, the volumetric heat release rate in a core is

$$q_V = E_f \Sigma_f \varphi \tag{11.17}$$

where, E_f is heat released of each fission reaction, about 200 MeV, the unit of heat release rate is MeV/(s cm^3).

11.4.3 Nuclear Power of Reactor Core

According to the volumetric heat release rate, we can get the total power of the reactor core, that is

$$P = 1.6021 \times 10^{-7} E_f \Sigma_f \varphi V \tag{11.18}$$

where, P is core total heat output, unit is W; V is volume of the reactor core, the unit is m^3; φ is neutron flux, the unit is 1/(cm^2 s); Σ_f is macro fission cross section, the unit is 1/cm; E_f is heat released of each fission reaction, the unit is MeV.

For a specific reactor, the volume of the reactor core keeps constant. In a relatively short period (a few days or weeks), nuclear fissile material atomic number density is also less change, so the macro fission cross section also basically don't change. Then, according to Eq. (11.18), the reactor power is linear proportional to neutron flux.

In the case of long running, consider burnup, in order to maintain the same power, the neutron flux in the life of the final will be higher.

11.5 Neutron Moderation

Neutron moderation is a neutron energy reducing process caused by neutron scattering. In the thermal neutron reactor (hereinafter referred to as thermal reactor), the neutron energy of the fission reaction is much lower than they just released from fission reaction. Therefore, in a thermal reactor, a moderator material is used to slow down neutrons, usually to slow down as soon as possible.

11.5.1 Neutron Slowing

In a thermal reactor, nuclear fission occurs mainly in the thermal neutron kinetic energy less than 1 eV zone. The neutron released from fission reaction is very fast, the average energy of 2 MeV. Slowing down the velocity of neutron becomes a basic process in a thermal reactor.

Scattering with the nucleus can slow neutron. Reactions include inelastic scattering and elastic scattering. Inelastic scattering can make neutron loss more kinetic energy, but the inelastic scattering occurs only of the order for MeV neutron kinetic energy of the high-energy area, and elastic scattering can happen in all energy area. Therefore, in the thermal reactor neutron slowing down rely mainly on elastic scattering.

After elastic scattering, the ratio of neutron kinetic energy E' and the kinetic energy E before scattering is

$$\frac{E'}{E} = \frac{1}{2}[(1 + \alpha) + (1 - \alpha)\cos\theta_c] \tag{11.19}$$

where, the θ_c is the scattering angle in the centroid coordinates (see Fig. 11.6); α is associated with the mass number of nuclei,

$$\alpha = \left(\frac{A-1}{A+1}\right)^2 \tag{11.20}$$

where, A is the mass number of nuclei scattering with neutrons.

When $\theta_c = 0$, $E' = E$. If the scattering neutrons do not change direction, the neutron has no loss of kinetic energy. In fact, in this case, neutron has no collision with nucleus. As long as the collision happened, will be changing direction.

When theta $\theta_c = \pi$, $E' = \alpha E$. If the movement direction of the neutron after scattering on the contrary, most the neutron kinetic energy is reduced. However, only portion of the kinetic energy loss in this case, the size of the loss of the kinetic energy associated with the mass number of target nuclei. The mass number of nucleus is smaller, the greater the kinetic energy lost by. Therefore, tell from scattering point of view, the reactor moderator material choose the nuclear mass number small elements. In addition, the moderator of the macro scattering cross section Σ_s should be larger, macroscopic absorption cross section Σ_a should be smaller.

The average natural logarithm of neutron energy can reduce in each collision is the logarithmic average loss factor, as ξ (the greater the value of ξ, the greater the

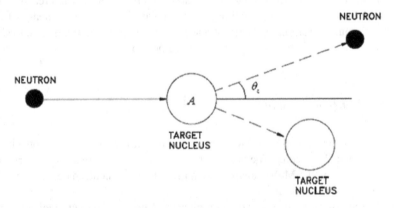

Fig. 11.6 Elastic scattering of neutrons and the nucleus

fraction of the average loss of neutron kinetic energy in each scattering).

$$\xi = \ln E - \ln E' = \ln\left(\frac{E}{E'}\right) \tag{11.21}$$

ξ is constant for each material, different material has different average. So it is used to compare the material ability in slowing down neutron. It is usually listed in a table in order to refer to, in the absence of table can also use the estimate equation:

$$\xi = \frac{2}{A + \frac{2}{3}} \tag{11.22}$$

This estimation formula is more accurate when the mass number of nuclei is greater than 10. There will be more than 3% of the error if $A < 10$.

With logarithmic average loss factor, can calculate the collision numbers needed for neutron energy from E_1 moderated to E_2, namely

$$N = \frac{\ln E_1 - \ln E_2}{\xi} \tag{11.23}$$

We usually use the product of logarithmic average loss factor and macro cross section of thermal neutron scattering, namely factor $\xi \Sigma_s$ to show the ability of moderator, called macroscopic slowing down power. $\xi \Sigma_s / \Sigma_a$ is the ratio of moderation, called moderating ratio. Slowing down than is the moderator moderated the comprehensive index of performance. The greater the ratio of moderation, the better comprehensive performance. Some commonly used moderator material parameters as shown in Table 11.4.

Example 11.4 From 2 MeV neutron energy slowing down to 1 eV, using water as a moderator, what is the number of collision needed.

Solution: according to the Table 11.4, the logarithmic average loss factor of water is 0.927, then

Table 11.4 Moderator material parameters from 2 MeV slowing down to 0.0253 eV

Material	ξ	Number of collision N	Macroscopic slowing down power $\Xi \sigma_S$	Moderating ratio $\xi \sigma_S / \Sigma_A$
H_2O	0.927	20	1.425	62
D_2O	0.510	36	0.177	4830
He	0.427	43	9×10^{-6}	51
Be	0.207	88	0.154	126
B	0.171	106	0.092	0.00086
C	0.158	115	0.083	216

$$N = \frac{\ln E_1 - \ln E_2}{\xi} = \frac{\ln \frac{2 \times 10^6}{1}}{0.927} = 16$$

11.5.2   The Release of Fission Neutron

Neutrons released in the process of fission reaction known as fission neutron. The average amount of neutrons released in each fission reaction is v. The value of v relates with fission type and neutron energy. The greater the neutron energy, the greater the value of v. Thermal neutron bombardment, for example, ^{235}U, v value is 2.43 (i.e., each fission reaction, release 2.43 neutrons averagely); If use thermal neutron bombardment ^{239}Pu, the value is 2.84, but if use fast (for example, the energy is 1 MeV) to bombard ^{239}Pu, then the value is 2.98. Because of there are neutrons released in the fission reaction, and the v value is greater than 1, so it is possible to sustain a chain reaction. v value is a critical value.

Daily life tells us that it is difficult to use a match to light coal directly. This is because although coal and oxygen reaction is exothermic reaction, but the reaction need to absorb heat. When use a match to light a bunch of coal, the reaction cannot "self-sustain", namely heat released from coal is not enough to heat up the surrounding coal. Only after the burning of coal releases enough heat, the reaction is sustainable, coal will continue to burn, until it is burned out. This kind of phenomenon is somewhat similar to the self-sustaining chain fission reaction in a nuclear reactor, only this time the conditions of the self-sustaining is not energy, but the number of neutrons.

In a nuclear reactor, the role of the secondary neutrons can be summarized as the following aspects. Firstly, to maintain chain reaction, at least have one secondary neutron to bombardment fissile nuclide and induce fission. Secondary, there is a part of the secondary neutrons due to motion and to leak out of the reactor. Finally, there is absorption of other materials, which can be to convert to new fissile nuclide (such as ^{238}U). So try to improve the v-value, and try to reduce leakage and useless absorption, it is possible to make fissile nuclides are consumed in the reactor generated new fissile nuclide, in order to achieve conversion of fissile nuclide, may even cause proliferation of fissile nuclide. This is the basic principle of fast breeder reactor.

The average energy of the secondary neutrons released by fission is about 2 MeV. Therefore, if the thermal neutron is used to fission reaction of ^{235}U, the energy of the secondary neutron must be reduced to the thermal energy in order to keep the chain reaction. The requirements in a thermal neutron reactor core placed moderator, the high-energy secondary neutron with nuclei collide; reduce the neutron energy, resulting in thermal neutron.

Secondary neutrons released in the fission reaction in most (99%) are in fission within a very short period of time (probably less than 10^{-13} s). This part of the neutron is often referred to as the prompt neutron. The neutron energy contains a very wide range, from 10 MeV to a very small number of values. However, most of the energy is between 1 and 2 MeV (see Fig. 11.7). In addition, a small fraction (less than 1%) is due to the release of fission fragments in the process of decay, called delayed neutrons. Such as the following reaction:

$$\ce{^{87}_{35}Br} \xrightarrow[55.9s]{\beta^-} \ce{^{87}_{36}Kr} \xrightarrow[\tilde{0}s]{} \ce{^{86}_{36}Kr} + \ce{^{1}_{0}n} \tag{11.24}$$

For ^{235}U fission, the total number of delayed neutrons accounts for about 0.6% of the total number of fission neutrons. It is actually released by the decay of several different fission fragments.

The precursor nucleus of the ^{235}U thermal fission can be divided into 6 groups. Table 11.5 gives the six groups of data of ^{235}U delayed neutron.

The fraction of delayed neutrons in all fission neutrons is expressed in β, which is called the delayed neutron fraction. Thermal neutron fission of ^{235}U, $\beta = 0.0065$.

The average energy of the delayed neutron is lower than that of the prompt neutron.

Fig. 11.7 ^{235}U energy spectrum of fission neutrons

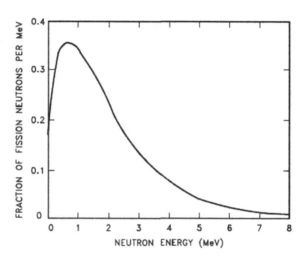

Table 11.5 Delayed neutrons in ^{235}U thermal neutron fission

Groups	Half-life t_i/s	Energy/keV	Fraction β_i	Mean time l_i/s
1	55.7	250	0.00021	78.64
2	22.7	560	0.00142	31.51
3	6.22	430	0.00127	8.66
4	2.3	620	0.00257	3.22
5	0.61	420	0.00075	0.716
6	0.23	430	0.00027	0.258
Total	–	–	0.0065	–

Although the proportion of delayed neutrons in the total number of fission neutrons is small (less than 1%), it has a very important influence on the dynamic process of the reactor. Because of the existence of delayed neutrons, the chain fission reaction can be controlled.

11.5.3   Neutron Generation Time

The average time between the two generations of neutrons, called the neutron generation time. Three factors determine the generation time of the prompt neutron:

1.   The time required for fast neutron slowing down.
2.   The time required before the nuclear fuel absorbs the thermal neutron.
3.   The time required to absorb and release the next generation of neutrons.

Fast neutron slowing down time is about 10^{-4}–10^{-6} s, depending on the moderator material. In a water reactor, the thermal neutron is about 10^{-4} s before being absorbed by the nuclear fuel. Absorbed and release of the next generation of fission neutrons need time is about 10^{-13} s. Therefore, the transient neutron generation in the thermal reactor is about 10^{-4} s.

For the delayed neutrons, the average time of the six groups is the average time of the delayed neutrons. Also contains three factors:

1.   The time required for fast neutron slowing down.
2.   The time required before the nuclear fuel absorb the thermal neutron.
3.   The time required to release the next generation of neutrons into the delayed neutron precursor.

The main effect is the third one. Therefore, the generation of the delay time is about 12.5 s in thermal reactor.

If the time between two generation of neutrons is only 10^{-4} s, the reactor will be very difficult to control. Because of the delayed neutrons, the average generation time is

$$\bar{t} = \bar{t}_p(1 - \beta) + \bar{t}_s\beta = 10^{-4} \times 0.9935 + 12.5 \times 0.0065 = 0.0813\,\text{s} \quad (11.25)$$

In Eq. (11.25), the generation time of fast neutron can be ignored. Because of delayed neutrons, the generation time of the neutron increases, which makes the chain fission reaction controlled. Otherwise, with the current level of control, it can only be detonated as a nuclear weapon, unable to control the rate at which the energy is released.

11.5.4 Neutron Energy Spectrum

Neutrons in the core are in a variety of energy states. The law of the distribution of the quantity or the fraction of the neutron flux as the energy distribution is the neutron energy spectrum.

The neutrons produced by fission are high energy neutrons, and almost all fission neutrons have energies between 0.1 and 10 MeV. The neutron energy distribution, or spectrum, may best be described by plotting the fraction of neutrons per MeV as a function of neutron energy, as shown in Fig. 11.7. From this figure, it can be seen that the most probable neutron energy is about 0.7 MeV. In addition, from this data it can be shown that the average energy of fission neutrons is about 2 MeV. Figure 11.7 is the neutron energy spectrum for thermal fission in uranium-235. The values will vary slightly for other nuclides.

The spectrum of neutron energies produced by fission varies significantly from the energy spectrum, or flux, existing in a reactor at a given time. Figure 11.8 illustrates the difference in neutron flux spectra between a thermal reactor and a fast breeder reactor. The energy distribution of neutrons from fission is essentially the same for both reactors, so the differences in the curve shapes may be attributed to the neutron moderation or slowing down effects.

No attempt is made to thermalize or slow down neutrons in the fast breeder reactor (liquid metal cooled); therefore, an insignificant number of neutrons exist in the thermal range. For the thermal reactor (water moderated), the spectrum of neutrons in the fast region (>0.1 MeV) has a shape similar to that for the spectrum of neutrons emitted by the fission process.

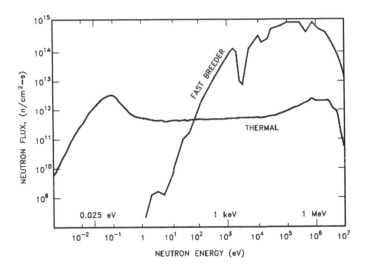

Fig. 11.8 Comparison of neutron flux spectra for thermal and fast reactor

In the thermal reactor, the flux in the intermediate energy region (1 eV–0.1 MeV) has approximately a $1/E$ dependence. That is, if the energy (E) is halved, the flux doubles. This $1/E$ dependence is caused by the slowing down process, where elastic collisions remove a constant fraction of the neutron energy per collision (on the average), independent of energy; thus, the neutron loses larger amounts of energy per collision at higher energies than at lower energies. The fact that the neutrons lose a constant fraction of energy per collision causes the neutrons to tend to "pile up" at lower energies, that is, a greater number of neutrons exist at the lower energies because of this behavior.

In the thermal region, the neutrons achieve a thermal equilibrium with the atoms of the moderator material. In any given collision, they may gain or lose energy, and over successive collisions will gain as much energy as they lose. These thermal neutrons, even at a specific temperature, do not all have the same energy or velocity; there is a distribution of energies, usually referred to as the Maxwell distribution (e.g., Fig. 11.8). The energies of most thermal neutrons lie close to the most probable energy, but there is a spread of neutrons above and below this value.

If the moderator is hydrogen, absorbent has infinite mass, in which two nuclei of infinite mixed media, neutron moderation can satisfies the spectrum:

$$\Sigma_t(E)\varphi(E) = \frac{S_0}{E_0} + \int_E^{E_0} \frac{\Sigma_s^H(E')\varphi(E')}{E'} dE' \tag{11.26}$$

where, $\Sigma_t(E)$ is the total macroscopic cross section at energy E; $\Sigma_s^H(E)$ is macroscopic scattering cross section of hydrogen; $\varphi(E)$ is the moderate spectrum in infinite mediumis; S_0 is neutron source; E_0 is the kinetic energy of neutron source; E is neutron kinetic energy. If the absorption is small, the solution of the equation is

$$\varphi(E) = \frac{S_0}{\Sigma_s(E)E} \propto \frac{1}{E} \tag{11.27}$$

In the process of slowing down, the moderate spectrum of neutron is directly proportional to the $1/E$, and the moderate spectrum of this form is called the $1/E$ spectrum or the Fermi spectrum.

If it is assumed that the neutron is scattered with the nucleus, the kinetic energy is loss a little per collision, and the scattering of a large number of neutrons and nuclei forms a continuous change of energy. This model is continuous slowing model. The Fermi age equation can be obtained by combining the continuous slowing model and the neutron diffusion model, which can be used to solve the neutron energy spectrum.

11.5.5 Fermi Age Model

In order to describe the Fermi age model, we first introduce two physical variables.

Collision Density: The total number of interactions of neutron and nucleus in unit time, unit volume and unit energy intervals.

Moderate Density: The total number of neutrons slowed down under a particular energy in unit time and unit volume.

The Fermi age equation is

$$\nabla^2 q(r, \tau) = \frac{\partial q(r, \tau)}{\partial \tau} \tag{11.28}$$

where, q is moderate density, τ is Fermi age, defined as

$$\tau(E) = \int_E^{E_0} \frac{D(E')}{\xi \Sigma_s(E') E'} dE' \tag{11.29}$$

where, E, D, ξ and Σ_s are neutron kinetic energy, neutron diffusion coefficient, average logarithmic energy loss and the macroscopic scattering cross section respectively; E_0 is the kinetic energy of neutron source.

Equation (11.28) is an equation that is related to the neutron moderation density and space position distribution. Equation (11.28) describes the variation of the spatial distribution of the neutron with the reduction of the neutron kinetic energy.

We look at a special case, if in the infinite medium of $r = 0$ there is a neutron source S, you can solve the moderate neutron density to meet the following relationship:

$$\nabla q(r, \tau) = S \frac{e^{-r^2/4\tau}}{(4\pi\tau)^{3/2}} \tag{11.30}$$

From Eq. (11.30), smaller Fermi age neutrons (i.e. neutron kinetic energy is close to the neutron source energy) in closer to the source in space distribution, larger Fermi age neutrons is far away from the source (i.e. neutron kinetic energy is low).

The unit of Fermi age is m^2. Its physical significance is that for isotropic monoenergetic neutron source, neutron energy E_0 moderated to energy E, one sixth of the mean square value of linear displacement of neutron in the medium. The mean square value of linear displacement can use experimental method to measure, which means the Fermi age can also experimentally measured indirectly.

11.5.6 Most Probable Neutron Velocities

The most probable velocity (v_p) of a thermal neutron is determined by the temperature of the medium and can be determined by Eq. (11.31).

$$v_p = \sqrt{\frac{2kT}{m_n}} \qquad (11.31)$$

where:

v_p most probable velocity of neutron (m/s)
k Boltzmann's constant (1.38×10^{-23} J/K)
T absolute temperature in degrees Kelvin (K)
m mass of neutron (1.66×10^{-27} kg).

Example 11.5 Calculate the most probable velocities for neutrons in thermal equilibrium with their surroundings at the following temperatures. (a) 20 °C, (b) 260 °C.

   Solution: 20 °C,

$$v_p = \sqrt{\frac{2kT}{m_n}} = \sqrt{\frac{2 \times \left(1.38 \times 10^{-23}\right) \times (20 + 273)}{1.66 \times 10^{-27}}} = 2207 \text{ m/s}$$

260 °C,

$$v_p = \sqrt{\frac{2kT}{m_n}} = \sqrt{\frac{2 \times \left(1.38 \times 10^{-23}\right) \times (260 + 273)}{1.66 \times 10^{-27}}} = 2977 \text{ m/s}$$

   From these calculations, it is evident that the most probable velocity of a thermal neutron increases as temperature increases. The most probable velocity at 20 °C is of particular importance since reference data, such as nuclear cross sections, are tabulated for a neutron velocity of 2200 m/s.

11.6 Neutron Life Cycle and Critical

The neutron cycle, also known as the neutron life cycle, is from the beginning of fission in reactor until the new fission and the next generation of fission neutrons are produced. The cycle includes scattering, slowing, leakage, absorption, fission and other factors. Fission chain reaction is based on the continuity of the neutron cycle that is sustainable controlled. The neutron cycle of fast reactor and thermal reactor are very different, mainly discussed here is the thermal reactor neutron cycle.

   We examine the total number of ants in a region, the new ants continue to hatch out, and the ant will experience the growth, natural enemies to destroy, hunger, natural disasters, natural death and other factors. Therefore, if the rate of the newly hatched and lost is basically the same, the total amount of the ant colony will reach a balance. If a number of factors change, so that the balance is broken, then the total number of colonies will change.

The critical reactor is a state that maintains a balance between the production rate and disappearance rate of neutrons, so that the chain reaction can be carried out continuously.

The minimum required size of a core or device with a given geometry and material composition to reach critical is critical dimension or critical size. The fuel loading of a critical reactor core, it is to maintain self-sustaining fission chain reaction required minimum mass of fissile material, known as the critical mass.

The critical mass of a reactor generally refers to the critical mass of the reactor core without the control rod and chemical compensation poison. The critical mass of reactor depends on the type of reactor, material composition, geometry and other conditions, but for any particular reactor system, it is a determined value. However, for different designs, the critical mass can vary greatly. For example, using ^{235}U as the fuel of reactor, the critical mass can be less than 1 kg, but also can be as large as 200 kg. The former is the critical mass of a uranium salt solution system with ^{235}U enrichment of about 90%; the latter is the mass of ^{235}U in a natural uranium graphite reactor. The critical condition of a reactor can be expressed by the multiplication factor.

11.6.1 Multiplication Factor

As described before, and neutron-induced fission reaction is able to reproduce neutrons, this process clearly has the potential for a chain reaction. However, if you want to be used in the reactors, fission chain must be self-sustaining. If each of generated by fission can initiate another nuclear fission, then two or more neutrons produced by fission at a time average, can without a doubt make fission chain sustainable. However, for various reasons, only part of the fission neutrons causing further fission.

Secondly, it should be noted that fission neutrons would participate in a variety of non-fission reactions. Even in the purest of fissionable material, and not all absorbed neutrons can cause fission. This is because there is always some other process of probability of occurrence (depends on the neutron energy), especially the radiative capture process. For example, in ^{235}U absorbs thermal neutrons, only 84% can cause fission. For ^{239}Pu, this ratio is only 65%. In addition, all kinds of non-fissile nuclear reactor materials also absorb some of the neutrons and therefore can no longer be used to cause fission, neutrons is absorbed.

Finally, because of the geometry of the system size is usually limited, neutron will inevitably leak out system, and these neutrons are lost.

A nuclear reactor is considered as consisting of the following materials: containing fissile material (for example, ^{233}U, ^{235}U or ^{239}Pu) fuel, moderator to slow neutrons, coolant to transport the heat generated by fission as well as the need to maintain a fixed geometry structure of the material. Therefore achieving a self-sustaining chain reaction depends on neutron relative magnitude involved in the four main factors:

1. Neutron leakage in the system, typically referred to as leakage.
2. Non-fission capture of the fissile nuclear fuel and other nuclear fissile (^{238}U and ^{232}Th). This process is sometimes also known as a resonant capture because it is most likely to happen on the resonance energy.
3. Non-fission capture of coolant, moderator, and various other substances (such as structural materials, fission products and possible impurities), it is called a parasitic captured.
4. Fission by fissile material.

These four factors summarized above require consume of neutrons, only the last one (that is, the fission reaction) has new neutron produced. If the neutrons produced by fission process is equal to (or exceeded) the number of neutrons by leak, parasitic capture, fission and total number of non-fissile neutron capture losses, then it should be possible to produce a self-sustaining reaction. Figure 11.9 shows a possible neutron cycle of a reactor takes ^{235}U as fuel. In this case, the total number of neutrons absorbed or leaks is 1000, and fission neutrons is 1000, which stands for a steady-state condition of self-sustained chain fission reaction.

If the escaping and captured neutrons in a non-fission process over 587, it can cause thermal neutrons to ^{235}U fission would be less than 413 and then produce a new generation of neutrons must be less than 1000. This fission chain is convergence, due to the short time interval between the two generations, so chain reaction will stop in a very short time. To the other hand, if the non-fission capture is less than the number of 587, then has more than 413 neutrons to cause fission, the result of fission-neutron number will be greater than 1000. Fission chain is spread, more and more number of neutrons generated.

The chain reaction conditions described above can be represented by a quantity called the multiplication factor. Multiplication factor is ratio of the generation of neutrons in the reactor to their directly parents (the previous generation of neutrons),

Fig. 11.9 Neutron balance diagram

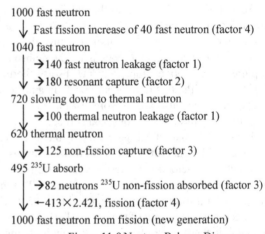

1000 fast neutron

↓ Fast fission increase of 40 fast neutron (factor 4)

1040 fast neutron

→140 fast neutron leakage (factor 1)

↓ →180 resonant capture (factor 2)

720 slowing down to thermal neutron

↓ →100 thermal neutron leakage (factor 1)

620 thermal neutron

↓ →125 non-fission capture (factor 3)

495 ^{235}U absorb

→82 neutrons ^{235}U non-fission absorbed (factor 3)

↓ ←413×2.421, fission (factor 4)

1000 fast neutron from fission (new generation)

Figure 11-9 Neutron Balance Diagram

or the ratio of neutron generation rate to neutron loss rate, usually represented by the symbol k. In a reactor system, neutron is mainly produced by fission reactions.

Neutron disappearing in two ways, which is absorbed in reactor and leak from the reactor surface. Depending on whether or not to consider leakage, the multiplication factor can be divided into an infinite multiplication factor and the effective multiplication factor.

1. The infinite multiplication factor: imaginary infinite proliferation of media multiplication factor, usually k_∞ (pronounced as k-infinite). For infinity systems, no neutron leakage, neutrons produced by nuclear fission, and only as a result of absorption of various materials in the system.
2. The effective multiplication factor: finite system proliferation factor, usually represented by k_{eff} (pronounced as k-effective).

Infinite multiplication factor

$$k_\infty = \text{neutron number of new generation/number of neutrons absorbed} \quad (11.32)$$

Taking into account leakage, the effective multiplication factor

$$k_{\text{eff}} = \text{neutron number of new generation/neutron number disappeared} \quad (11.33)$$

For limited size systems must consider neutron leakage. Non-leakage probability not only associated with the material properties of the system, also associated with the size and geometry of the system. Thus, when there is no external neutron sources, the critical condition of limited size reactor is $k_{\text{eff}} = 1$, reactor at steady state, the reactor neutrons have a stable distribution. If $k_{\text{eff}} < 1$, system is subcritical. If there is no external neutron source, the neutron flux will continue to decay to zero. If $k_{\text{eff}} = > 1$, system is supercritical, neutron flux will constantly grow exponentially over time. Mention "no external neutron source" is a prerequisite, in case of external neutron source, the conclusions above will change.

11.6.2 Four Factor Formula

Four-factor formula is the infinite multiplication factor analysis.

$$k_\infty = \varepsilon p f \eta \quad (11.34)$$

where, ε is fast fission factor, p is resonance escape probability, f is the thermal utilization factor, η is the reproduction factor.

Fast fission factor. The fast fission factor is defined as the ratio of the net number of fast neutrons produced by all fissions to the number of fast neutrons produced by thermal fissions. The mathematical expression of this ratio is shown below.

Table 11.6 average number of neutrons liberated in fission

Fissile nucleus	Thermal neutrons		Fast neutrons	
	v	η	v	η
^{233}U	2.49	2.29	2.58	2.40
^{235}U	2.42	2.07	2.51	2.35
^{239}Pu	2.93	2.15	3.04	2.90

$$\varepsilon = \frac{\text{number of fast neutrons produced by all fissions}}{\text{number of fast neutrons produced by thermal fissions}} \tag{11.35}$$

In the example in Fig. 11.9, $\varepsilon = 1040/1000 = 1.04$.

Resonance Escape Probability. The resonance escape probability is defined as the ratio of the number of neutrons that reach thermal energies to the number of fast neutrons that start to slow down. This ratio is shown below.

$$p = \frac{\text{number of neutrons that reach thermal energy}}{\text{number of fast neutrons that start to slow down}} \tag{11.36}$$

In the example in Fig. 11.9, $p = 720/900 = 0.8$.

Thermal utilization factor. The thermal utilization factor is defined as the ratio of the number of thermal neutrons absorbed in the fuel to the number of thermal neutrons absorbed in any reactor material. This ratio is shown below.

$$f = \frac{\text{number of thermal neutrons absorbed in the fuel}}{\text{number of thermal neutrons absorbed in all reactor materials}} \tag{11.37}$$

In the example in Fig. 11.9, $f = 495/620 = 0.799$.

The reproduction factor. The reproduction factor is defined as the ratio of the number of fast neutrons produces by thermal fission to the number of thermal neutrons absorbed in the fuel. The reproduction factor is shown below.

$$\eta = \frac{\text{number of fast neutrons produced by thermal fission}}{\text{number of thermal neutrons absorbed in the fuel}} \tag{11.38}$$

In the example in Fig. 11.9, $\eta = 1000/495 = 2.02$.

Thus, to obtain $k_\infty = \varepsilon p f \eta = 1.04 \times 0.8 \times 0.799 \times 2.02 = 1.343$.

Table 11.6 lists values of v and η for fission of several different materials by thermal neutrons and fast neutrons.

11.6.3 Effective Multiplication Factor

The infinite multiplication factor can fully represent only a reactor that is infinitely large, because it assumes that no neutrons leak out of the reactor. To completely describe the neutron life cycle in a real, finite reactor, it is necessary to account for neutrons that leak out. The multiplication factor that takes leakage into account is the effective multiplication factor (k_{eff}), which is defined as the ratio of the neutrons produced by fission in one generation to the number of neutrons lost through absorption and leakage in the preceding generation.

The effective multiplication factor for a finite reactor may be expressed mathematically in terms of the infinite multiplication factor and two additional factors, which account for neutron leakage as shown below. With the inclusion of these last two factors, it is possible to determine the fraction of neutrons that remain after every possible process in a nuclear reactor. The effective multiplication factor can then be determined by the product of six terms.

$$k_{eff} = \varepsilon L_f p L_t f \eta \qquad (11.39)$$

where, L_f is fast non-leakage probability, is L_t thermal non-leakage probability.

In a realistic reactor of finite size, some of the fast neutrons leak out of the boundaries of the reactor core before they begin the slowing down process. The fast non-leakage probability (L_f) is defined as the ratio of the number of fast neutrons that do not leak from the reactor core to the number of fast neutrons produced by all fissions. This ratio is stated as follows.

Neutrons can also leak out of a finite reactor core after they reach thermal energies. The thermal non-leakage probability (L_t) is defined as the ratio of the number of thermal neutrons that do not leak from the reactor core to the number of neutrons that reach thermal energies. The thermal non-leakage probability is represented by the following.

In the example in Fig. 11.9, $L_f = (1040\text{--}140)/1040 = 0.865$, $L_t = (720\text{--}100)/720 = 0.861$.

Figure 11.10 illustrates a neutron life cycle with nominal values provided for each of the six factors. Refer to Fig. 11.10 for the remainder of the discussion on the neutron life cycle and sample calculations. The generation begins with 1000 neutrons. The first process is fast fission and the population has been increased by the neutrons from this fast fission process. From the definition of the fast fission factor, it is possible to calculate its value based on the number of neutrons before and after fast fission occur.

Effective multiplication factor can be considered to consist of two parts. The first part depends on the composition and arrangement of materials exist within the system, and another part depends on the size of the system. Former known as infinite multiplication factor k_∞, it is equal to the ratio of the average number of neutrons produced in each generation and the average number of neutrons absorbed. In other words, it is in fact an infinite no neutron leakage in the system. The latter related to the geometry of the system, generally speaking the larger the reactor, the smaller the leaks out proportion of neutrons.

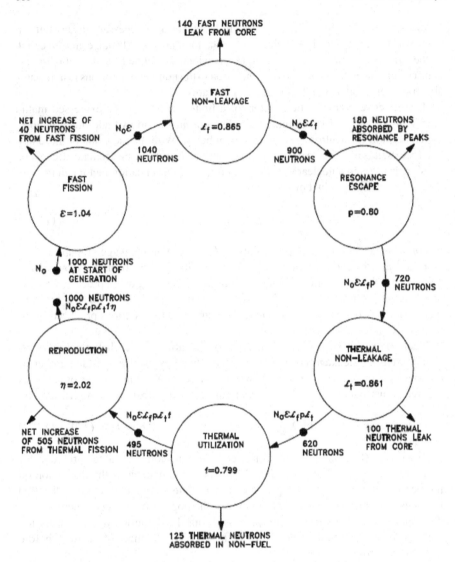

Fig. 11.10 Neutron cycle schematic

11.6.4 Critical Size

For specific material of components and layout, if infinite multiplication factor k_∞ is greater than 1, so this system must will has a size, makes k_{eff} can be equal to 1. In other words, such of reactor must has a species size, makes by fissile produced of neutron number and leak and absorption of neutron number just mutual balance, this is critical size, is makes chain reaction just can self-sustaining of size. Here we define a quantity called the leakage probability, as follows:

$$P = k_{\text{eff}}/k_\infty = L_f L_t \qquad (11.40)$$

Because of the neutrons generated in the whole reactor, while the leak only occurs on the outer surface, it is clear that when the system volume/area ratio increases, neutron leakage probability increases. For example, we discuss a sphere of radius r, volume/area ratio is proportional to r, so smaller radius, greater leakage probability. Therefore, if a system is less than the critical size, the p values is less than the p-value of critical reactor, thus $k_\infty p$ will be less than 1. From the above it can be concluded that chain reactions will then converge, and the system is subcritical. The other hand, if the dimensions of the reactor exceeds reactor critical size, so k_{eff} would be greater than 1, and the system will become supercritical, it can produce a divergent fission chain.

One of the factors that determine the critical size is obviously proliferation factor of infinite medium. If the latter is much larger than 1, then p is small you can meet the requirement of critical system $k_{\text{eff}} = 1$. If k_∞ is only slightly bigger than 1, such as reactors using natural uranium as fuel, so the system becomes critical only in the size is very large, then p is only slightly smaller than the 1.

If a system makes $k_\infty < 1$, so it never became critical and self-sustaining chain reaction is impossible. For example, natural uranium metal fuel and graphite moderator homogeneous mixture is the case. But, will fuel and moderator layout into lattice array of form, makes natural metal uranium block or rod appropriate distribution in middle of graphite lattice, k_∞ value will be slightly greater than 1. If reactor is big enough, makes neutron of relative leak reduced, so not leak probability on full close to 1, and can make k_{eff} equal to 1. So, natural uranium-graphite reactor are is big size. For example, the United States of the first reactor in Chicago (CP-1) and its extension (CP-2), the X-10 reactor in Oak Ridge, Brookhaven reactor and Belinda Hanford belongs to this design.

Various factors in determining the critical size, k_∞ is the most basic one, and there are other factors. Including reactor of shape (or geometric shapes). Because in all of equal volume geometric shapes, the surface area of a sphere is the smallest one. Obviously, for the specific fuel-moderator system, not leak probability is the least when the reactor is spherical and thus critical volume (or mass) the smallest. For this reason, the world's first successful chain reaction system (CP-1) is designed as a spherical, but spherical has not made it through when it has reached a critical mass, so it is shaped a bit like an Oblate spheroid.

With a material wrapped a reactor to reflect neutrons back into the reactor core, can also make the critical size smaller in a specific shape and composition. This substance is the reflective layer. In certain circumstances of core size, reflectors reduce the leak of neutrons, thus increasing the probability of not leakage. Therefore, in the case of equal infinite multiplication factor, the critical size of a reactor with a reflective layer is smaller than the naked one (the reactors without reflector). Reflector materials depends mainly on neutron energy, for thermal neutrons, the best reflector material should contain elements of low atomic weight and is not significantly thermal slow neutrons, such as heavy water, beryllium (or its oxide) and graphite (carbon).

11.6.5 Critical Calculation

Reactor critical computing tasks can be grouped into the following categories of problems:

1. Given the reactor material composition, determine the critical dimensions of it.
2. Given the shape and size of the reactor (for power reactors, these factors are determined by calculations in other areas, such as material, thermal hydraulics) for the determination of critical reactor material components. Is to determine the enrichment of ^{235}U fuel, or the number of control rods and layout of the poison. This is commonly encountered in engineering design.
3. In design and calculation of burnup reactor physics, often encounter a problem, to be called the criticality calculation. Reactor criticality calculation of geometric shapes, sizes and core components of fuel and other materials have been given; you need to calculate the effective multiplication factor or reactivity of the reactor.

Critical calculation is important part of reactor physics and design, in addition to find out the reactor critical size and fuel composition and fuel load, but there is one important task is to determine the critical state of neutron flux distribution in the system. The heat release rate or the distribution of power in the reactor core are proportional to neutron flux distribution.

Commonly used Reactor critical calculation methods are: continuous moderate diffusion theory and energy group diffusion theory. Theory of continuous moderate is the age equation that described earlier. Study on the energy group method is an approximate method for neutron diffusion. It is effective and commonly used method to calculate a reactor with reflective layer.

In the theory of energy group diffusion, the neutron energies from the ceiling to thermal energy into energy group, or "group". The same group neutrons diffusion, scattering and absorption using appropriate averaging group parameters and cross section (known as "group constant") to describe. The neutron transport for energy group diffusion equation. Group theory in this, the simplest is "single group" theory, but it can only show the approximate results. In calculation of thermal neutron reactor, especially for graphite or heavy water as a moderator of the reactor, two groups of diffusion theory is used. At this time, as long as the group constant properly selected, will be able to give better results. With the development of technology and new reactor designs (such as fast breeder reactors) appears, and the calculation of reactor put forward higher requirements, many design use groups (2–4 group) or multiple groups calculations.

When calculating a non-uniform reactor, you must consider the effect of non-uniform grids. Usually done in two steps:

1. Calculation of physical parameters of non-uniform grids must be considered in the calculation to the influence of non-uniform effect. Equivalent into equivalent uniform reactor, this is called a uniform process.

2. Apply the theory of homogeneous reactor to calculate the equivalent uniform reactor.

11.7 Reactivity

When we analyze neutron cycling by k_{eff}, if in every generation of neutrons k_{eff} remains a constant, you can predict the changes over time in the number of neutrons. If the initial number of neutrons is N_0, then the neutron number is

$$N_n = N_0(k_{eff})^n \qquad (11.41)$$

Example 11.6 If the initial 1000 neutrons, $k_{eff} = 1.002$, calculate how many neutrons after 50 generations?

Solution: $N_{50} = N_0(k_{eff})^{50} = 1000 \times 1.002^{50} = 1105$.

If previous generations have N_0 neutrons, then $N_0 k_{eff}$ neutrons in present generation. Changes in the number of neutrons is $(N_0 k_{eff} - N_0)$, the relative rate of change of the number of neutrons between the two generations is

$$\rho = \frac{N_0 k_{eff} - N_0}{N_0 k_{eff}} = \frac{k_{eff} - 1}{k_{eff}} \qquad (11.42)$$

We define this as the reactivity. According to different k_{eff}, the reactivity may be greater than zero, less than or equal to zero. The absolute value of reactivity means the deviation from critical. Reactivity is sometimes more convenient to represent a departure from the critical level of light water reactors.

Example 11.7 If k_{eff} is 1.002 or 0.998, determine the reactivity respectively

Solution: $\rho = \frac{1.002-1}{1.002} = 0.001996$ or $\rho = \frac{0.998-1}{0.998} = -0.0020$.

We can see, the reactivity is a dimensionless number defined. From the above examples, we can see, the reactivity is usually a very small number. For ease of communication and expression, usually artificially to provided a unit to the reactivity. History with "dollar" as a unit of measure, provision 1 dollar = 0.065, then provisions "cent" 1 cent = 1% dollar. Now out of use. Now more popular unit is $\Delta k/k$. Smaller scale is $\% \Delta k/k$, mk, or pcm.

$$1\% \Delta k/k = 0.01 \Delta k/k \qquad (11.43)$$

$$1 mk = 0.001 \Delta k/k \qquad (11.44)$$

$$1 pcm = 0.00001 \Delta k/k \qquad (11.45)$$

Table 11.7 Various reactivity of several major reactor type $\Delta k/k$

Item	BWR	PWR	CANDU	HTGR	SFR
Excess reactivity of a clean core					
• at 20 °C temperature	0.25	0.293	0.075	0.128	0.050
• operating temperature		0.248	0.065		0.037
• equilibrium Xe and Sm		0.181	0.035	0.073	
Total value of control	0.29	0.32	0.125	0.210	0.074
• value of the control rods	0.17	0.07	0.035	0.16	0.074
• value of burnable poison	0.12	0.08	0.09	0.10	
• chemical compensation		0.17			
Shutdown					
• cold and clean core	0.04	0.03	0.05	0.082	0.024
• hot and equilibrium Xe and Sm		0.14		0.137	0.037

Important reactivity has excess reactivity, control values, shutdown margin of the poison etc. Several major reactor reactivity values shown in Table 11.7.

No poisonous substances control in the reactor (such as compensation for control rods, burnable poison and chemical poisons) conditions, the reactivity of the reactor known as the excess reactivity. The excess reactivity of a reactor associated with the running time and the condition of the reactor. At the time running, the reactor's excess reactivity can be understood as all the amount of positive reactivity that can be put into reactor.

The absolute change of reactivity caused by control poisons is the control value of this kind of poisonous substances, also known as the reactivity of poison control equivalent. While the reactor is running, control values can be understood as the amount of negative reactivity that can be put into reactor.

Shutdown margin. When all poisonous substances are put into reactor, the sub criticality of a reactor can reach is shutdown margin. Shutdown margin is equal to the difference between total control value of poison and the excess reactivity of reactor. It is associated with the operation history and the condition of the reactor.

11.7.1   Reactivity Coefficient

The number of neutrons in the reactor changes over time, and can be determined through the reactivity. Many factors influence the reactivity, such as fuel consumption, temperature, pressure and fission products will affect the reactivity.

We use reactivity coefficient to analyze various factors affecting reactivity. The reactivity coefficient is defined as Eq. (11.46).

$$\alpha_x = \frac{\Delta \rho}{\Delta x} \tag{11.46}$$

where, x means a kind of factor. Considering the effect of temperature, it is the temperature coefficient of reactivity, or temperature coefficient for short; influence of coolant void fraction known as the void coefficient of reactivity, referred to as void coefficient.

The change of these parameters is often the result of reactor neutron flux or power change. While the change of neutron flux is caused by reactivity change. This forms a feedback effect. The strength of feedback characterized by reactivity coefficient. Positive or negative feedback will affect the reactor's stability and security. In order to guarantee the safety of the reactor, it is required that the reactivity coefficient should be negative. Commonly used reactivity coefficients are temperature reactivity coefficient, void coefficient and power coefficient, and so on.

Example 11.8 The moderator temperature coefficient for a reactor is -8.2 pcm/°C, calculate the reactivity defect that results from a temperature decrease of 5 °C

Solution: $\Delta\rho = \alpha_T \Delta T = (-8.2) \times (-5) = 41$ pcm.

11.7.2 Temperature Reactivity Coefficient

The change in reactivity per degree change in temperature is called the temperature coefficient of reactivity.

$$\alpha_T = \frac{\Delta\rho}{\Delta T} \tag{11.47}$$

Because different materials in the reactor have different reactivity changes with temperature and the various materials are at different temperatures during reactor operation, several different temperature coefficients are used. The overall temperature coefficient equal to the temperature coefficient of the sum of the respective components, i.e.

$$\alpha_T = \sum_i \frac{\Delta\rho}{\Delta T_i} = \sum_i \alpha_{Ti} \tag{11.48}$$

where, T_i is the temperature of material, α_{Ti} is the temperature coefficients of this kind of material. Usually, the two dominant temperature coefficients are the moderator temperature coefficient and the fuel temperature coefficient.

1. **Fuel Temperature Coefficient**

The change in reactivity per degree change in fuel temperature is called the fuel temperature coefficient of reactivity.

$$\alpha_{T_f} = \frac{\Delta\rho}{\Delta T_f} \tag{11.49}$$

where, T_f is the temperature of fuel.

This coefficient is also called the "prompt" temperature coefficient because an increase in reactor power causes an immediate change in fuel temperature. A negative fuel temperature coefficient is generally considered to be important because fuel temperature immediately increases following an increase in reactor power. In the event of a large positive reactivity insertion, the fuel temperature coefficient starts adding negative reactivity immediately.

Another name applied to the fuel temperature coefficient of reactivity is the fuel Doppler reactivity coefficient. This name is applied because in typical low enrichment, light water moderated; thermal reactors the fuel temperature coefficient of reactivity is negative and is the result of the Doppler Effect, also called Doppler broadening. The phenomenon of the Doppler Effect is caused by an apparent broadening of the resonances due to thermal motion of nuclei as illustrated in Fig. 11.11. Stationary nuclei absorb only neutrons of energy E_0. If the nucleus is moving away from the neutron, the velocity (and energy) of the neutron must be greater than E_0 to undergo resonance absorption. Likewise, if the nucleus is moving toward the neutron, the neutron needs less energy than E_0 to be absorbed. Raising the temperature causes the nuclei to vibrate more rapidly within their lattice structures, effectively broadening the energy range of neutrons that may be resonantly absorbed in the fuel. Two nuclides present in large amounts in the fuel of some reactors with large resonant peaks that dominate the Doppler Effect are uranium-238 and plutonium-240.

In addition, the nuclear fuel due to temperature changes will cause thermal expansion, resulting in fuel density becomes smaller, which will introduce a change in reactivity, the more important this mechanism of uranium metal fuel.

2.   Moderator Temperature Coefficient

The change in reactivity per degree change in moderator temperature is called the moderator temperature coefficient of reactivity.

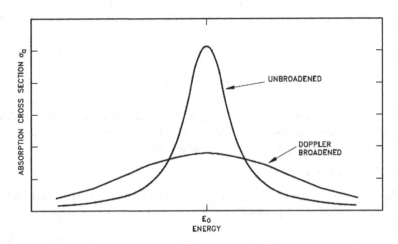

Fig. 11.11 Effect of fuel temperature on resonance absorption peaks

$$\alpha_{T_\mathrm{m}} = \frac{\Delta\rho}{\Delta T_\mathrm{m}} \tag{11.50}$$

where, T_m is the temperature of moderator.

Due to the heat from the fuel to the moderator has a heat transfer process, so temperature variation of moderator typically lags for some time on power. Moderator temperature effect is a lag effect.

As discussed in the previous section, a moderator possesses specific desirable characteristics.

(a)   Large neutron scattering cross section
(b)   Low neutron absorption cross section
(c)   Large neutron energy loss per collision.

A ratio, the moderator-to-fuel ratio ($N_\mathrm{m}/N_\mathrm{u}$), is very important in the discussion of moderators. As the reactor designer increases the amount of moderator in the core (that is, $N_\mathrm{m}/N_\mathrm{u}$ increases), neutron leakage decreases. Neutron absorption in the moderator ($\Sigma_\mathrm{a}^\mathrm{m}$) increases and causes a decrease in the thermal utilization factor. Having insufficient moderator in the core (that is, $N_\mathrm{m}/N_\mathrm{u}$ decreases) causes an increase in slowing down time and results in a greater loss of neutrons by resonance absorption. This also causes an increase in neutron leakage. The effects of varying the moderator-to-fuel ratio on the thermal utilization factor and the resonance probability are shown in Fig. 11.12.

Because the moderator-to-fuel ratio affects the thermal utilization factor and the resonance escape probability, it also affects k_eff. The remaining factors in the six-factor formula are also affected by the moderator-to-fuel ratio, but to a lesser extent than f and p. As illustrated in Fig. 11.12, which is applicable to a large core fueled with

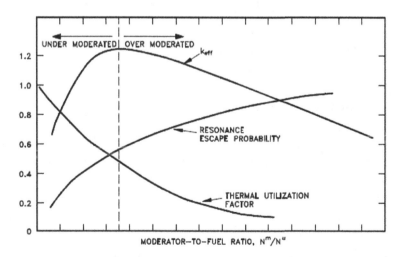

Fig. 11.12 Effects of over and under moderation on k_eff

low-enriched fuel, there is an optimum point above which increasing the moderator-to-fuel ratio decreases k_{eff} due to the dominance of the decreasing thermal utilization factor. Below this point, a decrease in the moderator-to-fuel ratio decreases k_{eff} due to the dominance of the increased resonance absorption in the fuel. If the ratio is above this point, the core is said to be over moderated, and if the ratio is below this point, the core is said to be under moderated.

In practice, water-moderated reactors are designed with a moderator-to-fuel ratio so that the reactor is operated in an under moderated condition. The reason that some reactors are designed to be under moderated is if the reactor were over moderated, an increase in temperature would decrease the N_m/N_u due to the expansion of the water as its density became lower. This decrease in N_m/N_u would be a positive reactivity addition, increasing k_{eff} and further raising power and temperature in a dangerous cycle. If the reactor is under moderated, the same increase in temperature results in the addition of negative reactivity, and the reactor becomes more self-regulating.

After the increase of the temperature of moderator, in addition to water to uranium ratios vary, moderator density (particularly liquid moderator) and micro-neutron cross section changes, this will make decline moderated and neutron energy spectrum hardening. Due to neutron energy spectrum hardening, the probability of nuclear resonance absorption increases, this factor will introduce negative reactivity. As the moderator density decreases, moderator of thermal neutron absorption is reduced accordingly, thus the moderator thermal neutron absorption is reduced, so that the thermal neutron utilization factor increases, this factor will be the introduction of positive reactivity. In addition, due to the hardening of the neutron energy spectrum, after the fuel absorbs a thermal neutron, due to the lower average number of fast neutron fission that introduces negative reactivity. The combination of these factors, the moderator temperature coefficient can be positive or negative, as the case may be. When poison chemical compensation added to the moderator liquid, positive effects likely to increase. Thus in the pressurized water reactor nuclear power plant, in order to give it a negative reactivity coefficient, the content of chemical compensation (B) should be limited (boron concentration limits is bellow $1.3 \times 10^{-3} \sim 1.4 \times 10^{-3}$).

11.7.3 Pressure Coefficient

The reactivity in a reactor core can be affected by the system pressure. The pressure coefficient of reactivity is defined as the change in reactivity per unit change in pressure.

$$\alpha_p = \frac{\Delta\rho}{\Delta p} \tag{11.51}$$

The pressure coefficient of reactivity for the reactor is the result of the effect of pressure on the density of the moderator. For this reason, it is sometimes referred to as the moderator density reactivity coefficient. As pressure increases, density correspondingly increases, which increases the moderator-to-fuel ratio in the core. In the typical under moderated core, the increase in the moderator-to-fuel ratio will result in a positive reactivity addition. In reactors that use water as a moderator, the absolute value of the pressure reactivity coefficient is seldom a major factor because it is very small compared to the moderator temperature coefficient of reactivity.

In boiling water reactor, due to pressure on the density is relatively large, the pressure coefficient will be relatively large, but it is much less than the void coefficient introduced following.

11.7.4 Void Coefficient

In systems with boiling conditions, such as boiling water reactors (BWR), the pressure coefficient becomes an important factor due to the larger density changes that occur when the vapor phase of water undergoes a pressure change. Of prime importance during operation of a BWR, and a factor in some other water-moderated reactors, is the void coefficient. The void coefficient is caused by the formation of steam voids in the moderator.

The void coefficient of reactivity is defined as the change in reactivity per percent change in void volume.

$$\alpha_V = \frac{\Delta\rho}{\Delta x_V} \tag{11.52}$$

where, x_V is volume percentage of bubbles in the coolant.

As the reactor power is raised to the point where the steam voids start to form, voids displace moderator from the coolant channels within the core. This displacement reduces the moderator-to-fuel ratio, and in an under moderated core, results in a negative reactivity addition, thereby limiting reactor power rise. The void coefficient is significant in water-moderated reactors that operate at or near saturated conditions.

To liquid as the moderator and coolant in the reactor, due to coolant boiling (including local boiling) bubbles take the space of liquid moderator, this will cause the following effects: Moderator of neutron absorption by reducing, so that the thermal utilization factor to improve; neutron leakage increases; moderating power decreases; neutron energy spectrum can be hardened. These effects, some positive, some negative. The total net effect is positive or negative associated with the type of reactors and nuclear properties and the position of the relevant void. In General is negative for PWR, positive effects can occur for large fast reactors, particularly if bubbles appear in the center area of the reactor core.

Several typical reactor reactivity coefficient as shown in Table 11.8.

Table 11.8 Typical reactor reactivity coefficient

Item	BWR	PWR	CANDU	HTGR	SFR
Temperature coefficient of fuel (10^{-5}/k)	$-4 \sim -1$	$-4 \sim -1$	$-1 \sim -2$	-7	$-0.1 \sim -0.25$
Temperature coefficient of moderator (10^{-5}/k)	$-50 \sim -8$	$-50 \sim -8$	$-3 \sim -7$	$+1.0$	
Void coefficient (10^{-5}/%FP)	$-200 \sim -100$	0	0	0	$-12 \sim +20$

11.7.5   Power Coefficient

Power coefficient of reactivity is defined as the change in reactivity per unit change in power,

$$\alpha_P = \frac{\Delta \rho}{\Delta P} \tag{11.53}$$

where, P is the reactor's power.

When reactor power change, temperature of the reactor fuel, moderator temperature and void fraction in the coolant is changed, resulting in a total change of reactivity. Therefore, power coefficient α_P is all reactivity changes combined with the power. Throughout the whole life, it must be negative. Usually at the end of life of the reactor core, its negative value will be even greater in absolute terms; this is as a result of increased moderator temperature effect of the negative at the end.

11.8   Neutron Poisons

The poisons here, not the chemical toxicity or ecotoxicity of materials, but especially as a large neutron absorption cross-section of material-neutron poisons. For pressurized water reactor nuclear power plant, for economic considerations, the reactor has greater excess reactivity in order to compensate the effect of burnup (such as fuel, fission product poisons, fuel and coolant temperature effects) caused by the loss of reactivity. In the beginning of life, the excess reactivity using three kinds of poison in the form of compensation, which is the compensation control rods, burnable poison and chemical poisons (instant soluble boron in the coolant). Chemical poison is primarily used to control or compensate for the slow response of reactivity change, compensation burnup and fission products effects. Control rod is mainly used for fast control and compensation of reactive changes [3].

These three poisons are active added, to control the reactivity. Fission products is a passive adding effect, most notably Xenon (^{135}Xe) and samarium (^{149}Sm).

11.8.1 Burnable Poisons

During operation of a reactor the amount of fuel contained in the core constantly decreases. If the reactor is to operate for a long period of time, fuel in excess of that needed for exact criticality must be added when the reactor is built. The positive reactivity due to the excess fuel must be balanced with negative reactivity from neutron-absorbing material. Moveable control rods containing neutron-absorbing material are one method used to offset the excess fuel. Control rods will be discussed in detail in a later. Using control rods alone to balance the excess reactivity may be undesirable or impractical for several reasons. One reason for a particular core design may be that there is physically insufficient room for the control rods and their large mechanisms.

To control large amounts of excess fuel without adding additional control rods, burnable poisons are loaded into the core. Burnable poisons are materials that have a high neutron absorption cross section that are converted into materials of relatively low absorption cross section as the result of neutron absorption. Due to the burnup of the poison material, the negative reactivity of the burnable poison decreases over core life. Ideally, these poisons should decrease their negative reactivity at the same rate the fuel's excess positive reactivity is depleted. Fixed burnable poisons are generally used in the form of compounds of boron or gadolinium that are shaped into separate lattice pins or plates, or introduced as additives to the fuel. Since they can usually be distributed more uniformly than control rods, these poisons are less disruptive to the core power distribution.

Pressurized water reactor in order to reduce the boron concentration required for the initial core excess reactivity, and flattens the neutron flux, avoid posituve moderator temperature coefficient, layout a certain amount of burnable poison rods in reactor core. Burnable neutron-absorbing material (such as boron and gadolinium) package, made of burnable poison rods components. In order to improve fuel consumption and extend fuel-reloading cycle, some pressurized water reactors add Gd_2O_3, Er_2O_3, or zirconium diboride into the fuel pellets. The economics of nuclear power plants has been greatly improved.

11.8.2 Soluble Poisons

Soluble poisons, also called chemical shim, produce a spatially uniform neutron absorption when dissolved in the water coolant. The most common soluble poison in commercial pressurized water reactors (PWR) is boric acid, which is often referred to as "soluble boron", or simply "solbor". The boric acid in the coolant decreases the thermal utilization factor, causing a decrease in reactivity. By varying the concentration of boric acid in the coolant (a process referred to as boration and dilution), the reactivity of the core can be easily varied. If the boron concentration is increased, the coolant/moderator absorbs more neutrons, adding negative reactivity. If the boron concentration is reduced (dilution), positive reactivity is added. The changing of boron concentration in a PWR is a slow process and is used primarily to compensate for fuel burnout or poison buildup. The variation in boron concentration allows control rod use to be minimized, which results in a flatter flux profile over the core than can be produced by rod insertion. The flatter flux profile is because there are no regions of depressed flux like those that would be produced in the vicinity of inserted control rods.

Emergency shutdown systems inject solutions containing neutron poisons into the system that circulates reactor coolant. Various solutions, including sodium polyborate and gadolinium nitrate, are used.

Fixed burnable poisons possess some advantages over chemical shim. Fixed burnable poisons may be discretely loaded in specific locations in order to shape or control flux profiles in the core. Also, fixed burnable poisons do not make the moderator temperature reactivity coefficient less negative as chemical shim does. With chemical shim, as temperature rises and the moderator expands, some moderator is pushed out of the active core area. Boron is also moved out, and this has a positive effect on reactivity. This property of chemical shim limits the allowable boron concentration because any greater concentration makes the moderator temperature coefficient of reactivity positive.

11.8.3   Control Rods

Most reactors contain control rods made of neutron absorbing materials that are used to adjust the reactivity of the core. Control rods can be designed and used for coarse control, fine control, or fast shutdowns.

Control rods are used to control the reactivity of nuclear reactor. In other words, the strength of chain reaction in the reactor can be controlled by control rods. In addition, the control rods can also be used to control the reactor's power distribution, avoid the formation of larger peaking power, and ensure that the temperature does not exceed the design limits of the fuel elements.

Control rods are mainly used to control and compensate reactivity quickly. Main features of control rod are:

1.   To compensate the reactivity changes caused by fuel and moderator temperature changes and power redistribution from hot state zero power to full power;
2.   To increase or decrease reactor power, quick load tracking by moving the control rods, to make nuclear power plant running at variable operating conditions;
3.   Under various operating conditions, quick or emergency shutdown, and keep enough shutdown reactivity margin;
4.   To compensate Xenon effect caused by changes of power, and with the help of the control rods to suppress Xenon oscillating.

Pressurized water reactor control rods are grouped according to different functions, generally divided into shutdown rods, power distribution control rods and power adjustment control rods.

These rods are able to be moved into or out of the reactor core and typically contain elements such as silver, indium, cadmium, boron, or hafnium.

The material used for the control rods varies depending on reactor design. Generally, the material selected should have a good absorption cross section for neutrons and have a long lifetime as an absorber (not burn out rapidly). The ability of a control rod to absorb neutrons can be adjusted during manufacture. A control rod that is referred to as a "black" absorber absorbs essentially all incident neutrons. A "grey" absorber absorbs only a part of them. While it takes more grey rods than black rods for a given reactivity effect, the grey rods are often preferred because they cause smaller depressions in the neutron flux and power near the rod. This leads to a flatter neutron flux profile and more even power distribution in the core.

If grey rods are desired, the amount of material with a high absorption cross section that is loaded in the rod is limited. Material with a very high absorption cross section may not be desired for use in a control rod, because it will burn out rapidly due to its high absorption cross section. The same amount of reactivity worth can be achieved by manufacturing the control rod from material with a slightly lower cross section and by loading more of the material. This also results in a rod that does not burn out as rapidly.

Another factor in control rod material selection is that materials that resonantly absorb neutrons are often preferred to those that merely have high thermal neutron absorption cross sections. Resonance neutron absorbers absorb neutrons in the epithermal energy range. The path length traveled by the epithermal neutrons in a reactor is greater than the path length traveled by thermal neutrons. Therefore, a resonance absorber absorbs neutrons that have their last collision farther (on the average) from the control rod than a thermal absorber. This has the effect of making the area of influence around a resonance absorber larger than around a thermal absorber and is useful in maintaining a flatter flux profile.

Neutron properties of common materials as shown in Table 11.9. Factors to consider selection of a control rod material, in addition to the absorption cross section, also consider the control rod on the distribution of power disturbances and material consumption of the control rod itself. Some of this material has a very large neutron absorption cross-section (black bar), but such rods absorb neutrons almost all around the bar, introducing a large neutron flux distortion (see Fig. 11.13). In addition, because the absorption cross section is large, absorber material consumption will increase. So sometimes, use grey rods. Only some of the neutrons are absorbed around grey rods. By designing the material composition of the gray bar, and can be required to reach the control rods absorb neutrons. take Silver-Indium-Cadmium alloys commonly used in PWR control rod, the mass ratio is 80, 15, 5%, and made of thin rods to reduce distortion of power following the insert control rods.

There are several ways to classify the types of control rods. One classification method is by the purpose of the control rods. Three purposes of control rods are listed below.

Table 11.9 Microscopic absorption cross section and resonance peak of some common used elements

Elements	Nuclear reaction	Abundance/%	Microscopic absorption cross section	
			0.0253 eV	Resonance peak/eV
^{10}B	^{10}B(n, α)^{7}Li	18.8	3837	
(B)*			760	
^{107}Ag	(n, γ)	51.4	35	1225(16.6), 466(42)
^{109}Ag		48.6	92	25,720(5.1), 174(31)
(Ag)			66	
^{113}Cd	(n, γ)	12.3	2000	63,415(0.175)
(Cd)			2550	
^{113}In	(n, γ)	4.3	12	1977(2.7)
^{115}In		95.7	203	2926(1.46), 867(3.8), 1044(9.1)
(In)			196	
^{155}Gd	(n, γ)	14.7	61,000	8500(2.7), 3540(6.4)
^{157}Gd		15.7	240,000	6370(17)
(Gd)			3600	
^{174}Hf	(n, γ)	0.16	400	900
^{176}Hf		5.2	30	8090
^{177}Hf		18.5	370	1610
^{178}Hf		27.1	80	500
^{179}Hf		13.8	65	18
^{180}Hf		35.2	10	

*Element with brackets is natural element

1. Shim rods—used for coarse control and/or to remove reactivity in relatively large amounts.
2. Regulating rods—used for fine adjustments and to maintain desired power or temperature.
3. Safety rods—provide a means for very fast shutdown in the event of an unsafe condition. Addition of a large amount of negative reactivity by rapidly inserting the safety rods is referred to as a "scram" or "trip."

Not all reactors have different control rods to serve the purposes mentioned above. Depending upon the type of reactor and the controls necessary, it is possible to use dual-purpose or even triple-purpose rods. For example, consider a set of control rods that can insert enough reactivity to be used as shim rods. If the same rods can be operated at slow speeds, they will function as regulating rods. Additionally, these same rods can be designed for rapid insertion, or scram. These rods serve a triple function yet meet other specifications such as precise control, range of control, and efficiency.

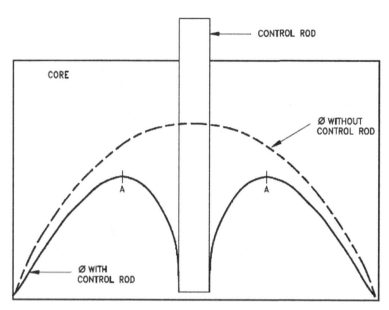

Fig. 11.13 Effect of control rod on radial flux distribution

The exact effect of control rods on reactivity can be determined experimentally. For example, a control rod can be withdrawn in small increments, such as 0.5 inch, and the change in reactivity can be determined following each increment of withdrawal. By plotting the resulting reactivity versus the rod position, a graph similar to Fig. 11.14 is obtained. The graph depicts integral control rod worth over the full range of withdrawal. The integral control rod worth is the total reactivity worth of the rod at that particular degree of withdrawal and is usually defined to be the greatest when the rod is fully withdrawn.

The slope of the curve ($\Delta\rho/\Delta x$), and therefore the amount of reactivity inserted per unit of withdrawal, is greatest when the control rod is midway out of the core. This occurs because the area of greatest neutron flux is near the center of the core; therefore, the amount of change in neutron absorption is greatest in this area. If the slope of the curve for integral rod worth in Fig. 11.14 is taken, the result is a value for rate of change of control rod worth as a function of control rod position. A plot of the slope of the integral rod worth curve, also called the differential control rod worth, is shown in Fig. 11.15. At the bottom of the core, where there are few neutrons, rod movement has little effect so the change in rod worth per inch varies little. As the rod approaches the center of the core its effect becomes greater, and the change in rod worth per inch is greater. At the center of the core, the differential rod worth is greatest and varies little with rod motion. From the center of the core to the top, the rod worth per inch is basically the inverse of the rod worth per inch from the center to the bottom.

Fig. 11.14 Integral control
rod worth

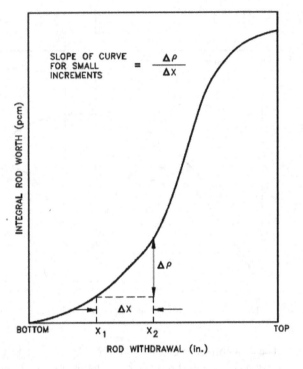

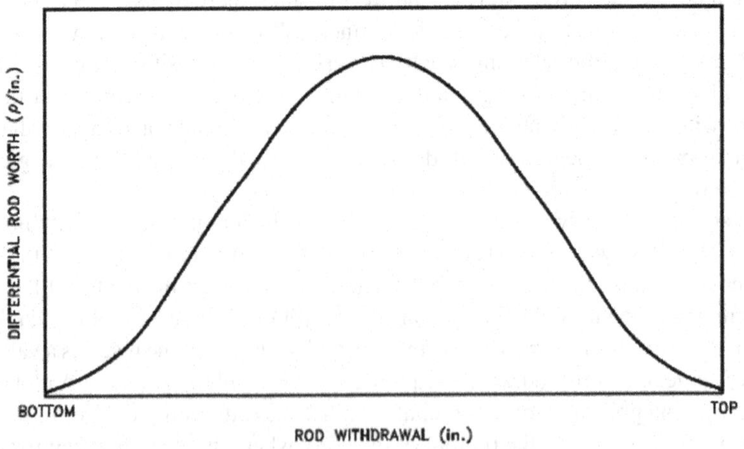

Fig. 11.15 Differential control rod worth

Differential control rod worth is the reactivity change per unit movement of a rod and is normally expressed as ρ/cm, Δk/k per cm, or pcm/cm. The integral rod worth at a given withdrawal is merely the summation of the entire differential rod worth up to that point of withdrawal. It is also the area under the differential rod worth curve at any given withdrawal position.

11.8.4 Xenon

Fission fragments generated at the time of fission decay to produce a variety of fission products. Fission products are of concern in reactors primarily because they become parasitic absorbers of neutrons and result in long-term sources of heat. Although several fission products have significant neutron absorption cross sections, xenon-135 and samarium-149 have the most substantial impact on reactor design and operation. Because these two fission product poisons remove neutrons from the reactor, they will have an impact on the thermal utilization factor and thus k_{eff} and reactivity.

When fission of fissile material, fissile material nucleus is generally split into two medium mass nuclei, known as fission fragments, and other radioactive and emits neutrons. These fission fragments are almost always had a large number of neutron-proton ratio and after a series of beta decay into stable nuclei. Fission fragment and its decay products are collectively called fission products. Fission product radionuclides, such as ^{135}Xe, ^{149}Sm, ^{151}Sm, ^{113}Cd, ^{155}Gd, ^{157}Gd, have considerable thermal neutron absorption cross section, ^{135}Xe and ^{149}Sm is particularly strong. In the reactor, they consume in-core neutron, negative impact on the effective multiplication factor of the reactor as "poisons".

Xenon-135 has a 2.6×10^6 b neutron absorption cross section. It is produced directly by some fissions, but is more commonly a product of the tellurium-135 decay chain shown if Fig. 11.16. The fission yield (γ) for xenon-135 is about 0.3%, while γ for tellurium-135 is about 6%. Understand the source of xenon and balanced in reactor for the safe operation of the reactor is necessary.

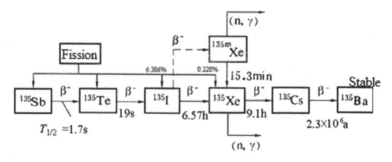

Fig. 11.16 Xenon-135 comes from the decay of iodine-135

The half-life for tellurium-135 is so short compared to the other half-lives that it can be assumed that iodine-135 is produced directly from fission. Iodine-135 is not a strong neutron absorber, but decays to form the neutron poison xenon-135. Ninety-five percent of all the xenon-135 produced comes from the decay of iodine-135. Therefore, the half-life of iodine-135 plays an important role in the amount of xenon-135 present.

After full power running, ^{135}I and ^{135}Xe concentration increase with run time. About 5–6 half-life of the isotope later, reaching equilibrium concentration (saturated concentration). This is equivalent to running 2–3 days in full power, can achieve equilibrium concentration. At this time, the ^{135}I and ^{135}Xe produced rate is exactly equal to its rate of loss, so their concentration remained unchanged. Balancing xenon concentrations causing loss of reactivity is called balance of xenon poison, its size and power density of reactors and nuclear fuel enrichment related.

The rate of change of iodine concentration is equal to the rate of production minus the rate of removal. This can be expressed in the equation below.

$$\frac{dN_I}{dt} = \gamma_I \Sigma_f \varphi - \lambda_I N_I - \sigma_a^I N_I \varphi \tag{11.54}$$

where:

N_I ^{135}I concentration
γ_I fission yield of ^{135}I
Σ_f macroscopic fission cross section fuel
φ thermal neutron flux
λ_I decay constant for ^{135}I
σ_a^I microscopic absorption cross section ^{135}I.

Fission yield ^{135}I can be considered approximately 6.386%.

Since the σ_a^I is very small, the burn up rate term may be ignored, and the expression for the rate change of iodine concentration is modified as shown below.

$$\frac{dN_I}{dt} = \gamma_I \Sigma_f \varphi - \lambda_I N_I \tag{11.55}$$

When the rate of production of iodine equals the rate of removal of iodine, equilibrium exists. The iodine concentration remains constant. The following equation for the equilibrium concentration of iodine can be determined from the preceding equation by setting the two terms equal to each other

$$N_I = \frac{\gamma_I \Sigma_f \varphi}{\lambda_I} \tag{11.56}$$

Since the equilibrium iodine concentration is proportional to the fission reaction rate, it is also proportional to reactor power level. The rate of change of the xenon concentration is equal to the rate of production minus the rate of removal. Recall that

5% of xenon comes directly from fission and 95% comes from the decay of iodine. The rate of change of xenon concentration is expressed by the following equations.

$$\frac{dN_{Xe}}{dt} = \gamma_{Xe}\Sigma_f\varphi + \lambda_I N_I - \lambda_{Xe}N_{Xe} - \sigma_a^{Xe}N_{Xe}\varphi \qquad (11.57)$$

The xenon burnup term above refers to neutron absorption by xenon-135 by the following reaction.

$${}^{135}_{54}\text{Xe} + {}^{1}_{0}\text{n} \rightarrow {}^{136}_{54}\text{Xe} + \gamma \qquad (11.58)$$

Xenon-136 is not a significant neutron absorber; therefore, the neutron absorption by xenon-135 constitutes removal of poison from the reactor. The burnup rate of xenon-135 is dependent upon the neutron flux and the xenon-135 concentration. The equilibrium concentration of xenon-135 is represented as shown below.

$$N_{Xe} = \frac{\gamma_{Xe}\Sigma_f\varphi + \lambda_I N_I}{\lambda_{Xe} + \sigma_a^{Xe}\varphi} \qquad (11.59)$$

For xenon-135 to be in equilibrium, iodine-135 must also be in equilibrium. Substituting the expression for equilibrium iodine-135 concentration into the equation for equilibrium xenon results in the following.

$$N_{Xe} = \frac{(\gamma_{Xe} + \gamma_I)\Sigma_f\varphi}{\lambda_{Xe} + \sigma_a^{Xe}\varphi} \qquad (11.60)$$

From this equation it can be seen that the equilibrium value for xenon-135 increases as power increases, because the numerator is proportional to the fission reaction rate. Thermal flux is also in the denominator; therefore, as the thermal flux exceeds 10^{12} neutrons/cm^2-sec, the term begins to dominate, and at approximately 10^{15} neutrons/cm^2-sec, the xenon-135 concentration approaches a limiting value. The equilibrium iodine-135 and xenon-135 concentrations as a function of neutron flux are illustrated in Fig. 11.17.

The higher the power level, or flux, the higher the equilibrium xenon-135 concentration, but equilibrium xenon-135 is not directly proportional to power level. For example, equilibrium xenon-135 at 25% power is more than half the value for equilibrium xenon-135 at 100% power for many reactors. Because the xenon-135 concentration directly affects the reactivity level in the reactor core, the negative reactivity due to the xenon concentrations for different power levels or conditions are frequently plotted instead of the xenon concentration.

When a reactor is shutdown, the neutron flux is reduced essentially to zero. Therefore, after shutdown, xenon-135 is no longer produced by fission and is no longer removed by burnup. The only remaining production mechanism is the decay of the iodine-135 which was in the core at the time of shutdown. The only removal mechanism for xenon-135 is decay.

Fig. 11.17 Equilibrium
iodine-135 and xenon-135
concentrations versus
neutron flux

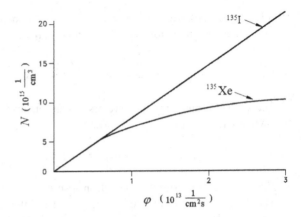

$$\frac{dN_{Xe}}{dt} = \lambda_I N_I - \lambda_{Xe} N_{Xe} \tag{11.61}$$

Because the decay rate of iodine-135 is faster than the decay rate of xenon-135, the xenon concentration builds to a peak. The peak is reached when the product of the terms $\lambda_I N_I$ is equal to $\lambda^{Xe} N^{Xe}$ (in about 10 to 11 h). Subsequently, the production from iodine decay is less than the removal of xenon by decay, and the concentration of xenon-135 decreases. The greater the flux level prior to shutdown, the greater the concentration of iodine-135 at shutdown; therefore, the greater the peak in xenon-135 concentration after shutdown. This phenomenon can be seen in Fig. 11.18, which illustrates the reactivity value of xenon-135 following shutdown from various neutron flux levels.

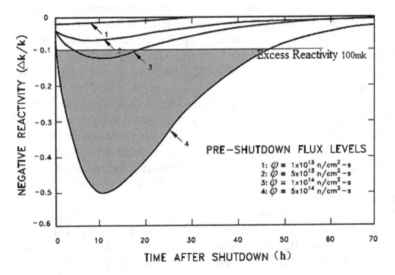

Fig. 11.18 Xenon-135 reactivity after reactor shutdown

Negative xenon reactivity, also called xenon poisoning, may provide sufficient negative reactivity to make the reactor inoperable because there is insufficient positive reactivity available from control rod removal or chemical shim dilution (if used) to counteract it. The inability of the reactor to be started due to the effects of xenon is sometimes referred to as a xenon precluded startup. The period of time where the reactor is unable to "override" the effects of xenon is called xenon dead time. The shaded region in Fig. 11.18 is unable to start region under 100 mk excess reactivity. Because the amount of excess reactivity available to override the negative reactivity of the xenon is usually less than 10% $\Delta k/k$, thermal power reactors are normally limited to flux levels of about 5×10^{13} n/(cm^2 s) so that timely restart can be ensured after shutdown. For reactors with very low thermal flux levels (~5×10^{12} n/(cm^2 s) or less), most xenon is removed by decay as opposed to neutron absorption. For these cases, reactor shutdown does not cause any xenon-135 peaking effect.

Following the peak in xenon-135 concentration about 10 h after shutdown, the xenon-135 concentration will decrease at a rate controlled by the decay of iodine-135 into xenon-135 and the decay rate of xenon-135. For some reactors, the xenon-135 concentration about 20 h after shutdown from full power will be the same as the equilibrium xenon-135 concentration at full power. About 3 days after shutdown, the xenon-135 concentration will have decreased to a small percentage of its pre-shutdown level, and the reactor can be assumed xenon free without a significant error introduced into reactivity calculations.

Xenon also cause power oscillations. Xenon oscillation is a space oscillation phenomenon of reactor power with xenon concentration. Large thermal reactors with little flux coupling between regions may experience spatial power oscillations because of the non-uniform presence of xenon-135. The mechanism is described in the following four steps.

1. An initial lack of symmetry in the core power distribution (for example, individual control rod movement or misalignment) causes an imbalance in fission rates within the reactor core, and therefore, in the iodine-135 buildup and the xenon-135 absorption.
2. In the high-flux region, xenon-135 burnout allows the flux to increase further, while in the low-flux region, the increase in xenon-135 causes a further reduction in flux. The iodine concentration increases where the flux is high and decreases where the flux is low.
3. As soon as the iodine-135 levels build up sufficiently, decay to xenon reverses the initial situation. Flux decreases in this area, and the former low-flux region increases in power.
4. Repetition of these patterns can lead to xenon oscillations moving about the core with periods on the order of about 15 h.

With little change in overall power level, these oscillations can change the local power levels by a factor of three or more. In a reactor system with strongly negative temperature coefficients, the xenon-135 oscillations are damped quite readily. This is one reason for designing reactors to have negative moderator-temperature coefficients.

During periods of steady state operation, at a constant neutron flux level, the xenon-135 concentration builds up to its equilibrium value for that reactor power in about 40–50 h. Figure 11.19 illustrates a typical xenon transient that occurs as a result of a change in reactor power level. At time zero, reactor power is raised from 50% power to 100% power. When the reactor power is increased, xenon concentration initially decreases because the burnup is increased at the new higher power level. Because 95% of the xenon production is from iodine-135 decay, which has a 6–7 h half-life, the production of xenon remains constant for several hours. After a few hours (roughly 4–6 h depending on power levels), the rate of production of xenon from iodine and fission equals the rate of removal of xenon by burnup and decay. At this point, the xenon concentration reaches a minimum. The xenon concentration then increases to the new equilibrium level for the new power level in roughly 40–50 h. It should be noted that the magnitude and the rate of change of xenon concentration during the initial 4–6 h following the power change is dependent upon the initial power level and on the amount of change in power level. The xenon concentration change is greater for a larger change in power level.

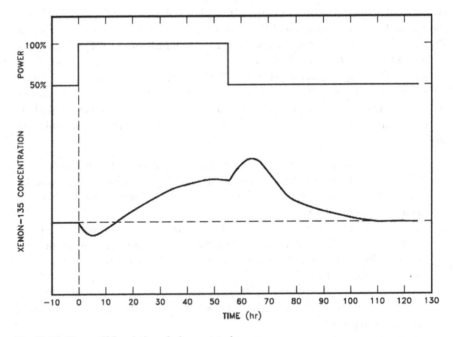

Fig. 11.19 Xenon-135 variations during power changes

11.8.5 Samarium

Samarium-149 is the second most important fission-product poison because of its high thermal neutron absorption cross section of 4.1×10^4b barns. Samarium-149 is produced from the decay of the neodymium-149 fission fragment as shown in the decay chain below.

$$^{149}_{60}\text{Nd} \xrightarrow[1.72\text{h}]{\beta^-} {}^{149}_{61}\text{Pm} \xrightarrow[53.1\text{h}]{\beta^-} {}^{149}_{62}\text{Sm} \tag{11.62}$$

For the purpose of examining the behavior of samarium-149, the 1.72 h half-life of neodymium-149 is sufficiently shorter than the 53.1 h value for promethium-149 that the promethium-149 may be considered as if it were formed directly from fission. This assumption, and neglecting the small amount of promethium burnup, allows the situation to be described as follows.

$$\frac{dN_{\text{Pm}}}{dt} = \gamma_{\text{Pm}} \Sigma_f \varphi - \lambda_{\text{Pm}} N_{\text{Pm}} \tag{11.63}$$

Solving for the equilibrium value of promethium-149 gives the following.

$$N_{\text{Pm}} = \frac{\gamma_{\text{Pm}} \Sigma_f \varphi}{\lambda_{\text{Pm}}} \tag{11.64}$$

The rate of samarium-149 formation is described as follows.

$$\frac{dN_{\text{Sm}}}{dt} = \gamma_{\text{Sm}} \Sigma_f \varphi + \lambda_{\text{Pm}} N_{\text{Pm}} - N_{\text{Sm}} \sigma_a^{\text{Sm}} \varphi \tag{11.65}$$

The fission yield of samarium-149, however, is nearly zero; therefore, the equation becomes the following.

$$\frac{dN_{\text{Sm}}}{dt} = \lambda_{\text{Pm}} N_{\text{Pm}} - N_{\text{Sm}} \sigma_a^{\text{Sm}} \varphi \tag{11.66}$$

Solving this equation for the equilibrium concentration of samarium-149 and substituting Equatoin (11.64) into (11.66) yields the following.

$$N_{\text{Sm}} = \frac{\gamma_{\text{Pm}} \Sigma_f}{\sigma_a^{\text{Sm}}} \tag{11.67}$$

This expression for equilibrium samarium-149 concentration during reactor operation illustrates that equilibrium samarium-149 concentration is independent of neutron flux and power level. The samarium concentration will undergo a transient following a power level change, but it will return to its original value.

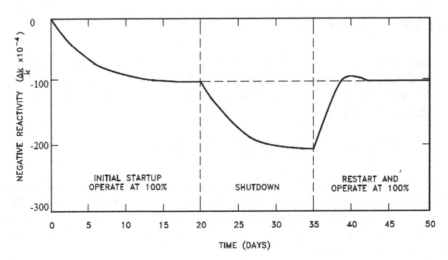

Fig. 11.20 Behavior of samarium-149 in a typical light water reactor

Because samarium-149 is not radioactive and is not removed by decay, it presents problems somewhat different from those encountered with xenon-135, as illustrated in Fig. 11.20. The equilibrium concentration and the poisoning effect build to an equilibrium value during reactor operation. This equilibrium is reached in approximately 20 days (500 h), and since samarium-149 is stable, the concentration remains essentially constant during reactor operation. When the reactor is shutdown, the samarium-149 concentration builds up as a result of the decay of the accumulated promethium-149. The buildup of samarium-149 after shutdown depends upon the power level before shutdown. Samarium-149 does not peak as xenon-135 does, but increases slowly to a maximum value as shown in Fig. 11.20. After shutdown, if the reactor is then operated at power, samarium-149 is burned up and its concentration returns to the equilibrium value. Samarium poisoning is minor when compared to xenon poisoning. Although samarium-149 has a constant poisoning effect during long-term sustained operation, its behavior during initial startup and during post-shutdown and restart periods requires special considerations in reactor design.

The xenon-135 and samarium-149 mechanisms are dependent on their very large thermal neutron cross sections and only affect thermal reactor systems. In fast reactors, neither these nor any other fission products have a major poisoning influence.

11.9 Subcritical Multiplication

Subcritical multiplication is the phenomenon that accounts for the changes in neutron flux that takes place in a subcritical reactor due to reactivity changes. It is important to understand subcritical multiplication in order to understand reactor response to changes in conditions.

When a reactor is in a shutdown condition, neutrons are still present to interact with the fuel. These source neutrons are produced by a variety of methods. If neutrons and fissionable material are present in the reactor, fission will take place. Therefore, a reactor will always be producing a small number of fissions even when it is shutdown.

11.9.1 Subcritical Multiplication Factor

Consider a reactor in which k_{eff} is 0.6. If 100 neutrons are suddenly introduced into the reactor, these 100 neutrons that start the current generation will produce 60 neutrons (100×0.6) from fission to start the next generation. The 60 neutrons that start the second generation will produce 36 neutrons (60×0.6) to start the third generation. The number of neutrons produced by fission in subsequent generations due to the introduction of 100 source neutrons into the reactor is shown in Table 11.10.

Because the reactor is subcritical, neutrons introduced in the reactor will have a decreasing effect on each subsequent generation. The addition of source neutrons to the reactor containing fissionable material has the effect of maintaining a much higher stable neutron level due to the fissions occurring than the neutron level that would result from the source neutrons alone. The effects of adding source neutrons at a rate of 100 neutrons per generation to a reactor with a k_{eff} of 0.6 are shown in Table 11.11.

A neutron source strength of 100 neutrons per generation will result in 250 neutrons per generation being produced from a combination of sources and fission in a shutdown reactor with a k_{eff} of 0.6. If the value of k_{eff} were higher, the source neutrons would produce a greater number of fission neutrons and their effects would be felt for a larger number of subsequent generations after their addition to the reactor.

The effect of fissions in the fuel increasing the effective source strength of a reactor with a k_{eff} of less than one is subcritical multiplication. For a given value of k_{eff} there exists a subcritical multiplication factor (M) that relates the source level to the steady-state neutron level of the core. If the value of k_{eff} is known, the amount that the neutron source strength will be multiplied (M) can easily be determined by Eq. (11.68).

$$M = \frac{1}{1 - k_{eff}} \tag{11.68}$$

Table 11.10 Changes in the number of neutrons per generation

Generation	1	2	3	4	5	6	7	8	9	10	11	12
Number of neutrons	100	60	36	22	13	8	5	3	2	1	0	0

Table 11.11 Changes in the number of neutrons per generation with source neutrons

Generation	1	2	3	4	5	6	7	8	9	10	11	12
	100	60	36	22	13	8	5	3	2	1	0	0
		100	60	36	22	13	8	5	3	2	1	0
			100	60	36	22	13	8	5	3	2	1
				100	60	36	22	13	8	5	3	2
					100	60	36	22	13	8	5	3
						100	60	36	22	13	8	5
							100	60	36	22	13	8
								100	60	36	22	13
									100	60	36	22
										100	60	36
											100	60
												100
Number of neutrons	100	160	196	218	231	239	244	247	249	250	250	250

Example 11.9 Calculate the subcritical multiplication factors for $k_{eff} = 0.986$
Solution: $M = 1/(1-0.986) = 71.4$

The example above illustrates that the subcritical multiplication factor will increase as positive reactivity is added to a shutdown reactor, increasing the value of k_{eff}. If the source strength of this reactor were 1000 n/s, the neutron level would increase to a neutron level of 71,400 n/s.

Example in Table 11.10, we use the number of neutrons per generation to describe, but in fact, each generation neutron lifetime is not 1 s. Is the conclusion of "the 1000 n/s increase to 71,400 n/s" OK or not? Readers can think about yourself.

11.9.2   Effect of Reactivity Changes on Subcritical Multiplication

In a subcritical reactor, the neutron level is related to the source strength by Eq. (11.69).

$$N = S \cdot M \tag{11.69}$$

where:

N neutron level
S neutron source strength
M subcritical multiplication factor.

If the term M in Eq. (11.69) is replaced by the expression $1/1$-keff from Eq. (11.68), the following expression results.

$$N = \frac{S}{1 - k_{\text{eff}}} \quad (11.70)$$

According to the source strength and the k_{eff} can calculate the number of neutrons within the reactor level. To this point, it has been necessary to know the neutron source strength of the reactor in order to use the concept of subcritical multiplication. In most reactors the actual strength of the neutron sources is difficult, if not impossible, to determine. Even though the actual source strength may not be known, it is still possible to relate the change in reactivity to a change in neutron level.

Consider a reactor at two different times when k_{eff} is two different values, k_1 and k_2. The neutron level at each time can be determined based on the neutron source strength and the subcritical multiplication factor using Eq. (11.71).

$$\begin{cases} N_1 = S\left(\frac{1}{1-k_1}\right) \\ N_2 = S\left(\frac{1}{1-k_2}\right) \end{cases} \quad (11.71)$$

Or

$$\frac{N_1}{N_2} = \frac{1 - k_2}{1 - k_1} \quad (11.72)$$

Because the source strength appears in both the numerator and denominator, it cancels out of the equation. Therefore, the neutron level at any time can be determined based on the neutron level present at any other time provided the values of keff or reactivity for both times are known.

The neutron level in a shutdown reactor is typically monitored using instruments that measure the neutron leakage out of the reactor. The neutron leakage is proportional to the neutron level in the reactor. Typical units for displaying the instrument reading are counts per second (cps). Because the instrument count rate is proportional to the neutron level, the above equation can be restated as shown in Eq. (11.73).

$$\frac{C_{R1}}{C_{R2}} = \frac{1 - k_2}{1 - k_1} \quad (11.73)$$

where:

C_{R1} count rate at time 1
C_{R2} count rate at time 2
k_1 k_{eff} at time 1
k_2 k_{eff} at time 2.

Example 11.10 A reactor that has a reactivity of -1000 pcm has a count rate of 42 counts per second (cps) on the neutron monitoring instrumentation. Calculate what the neutron level should be after a positive reactivity insertion of 500 pcm from the withdrawal of control rods.

Solution: $k_1 = \frac{1}{1-\rho_1} = \frac{1}{1-(-0.01000)} = 0.9901$

$$k_2 = \frac{1}{1 - \rho_2} = \frac{1}{1 - (-0.01000 + 0.00500)} = 0.9950$$

$$C_{R2} = C_{R1}\frac{1-k_1}{1-k_2} = 42 \times \frac{1 - 0.9901}{1 - 0.9950} = 83\,(1/\text{s})$$

11.9.3 Use of 1/M Plots

Because the subcritical multiplication factor is related to the value of k_{eff}, it is possible to monitor the approach to criticality through the use of the subcritical multiplication factor. As positive reactivity is added to a subcritical reactor, k_{eff} will get nearer to one. As k_{eff} gets nearer to one, the subcritical multiplication factor (M) gets larger. The closer the reactor is to criticality, the faster M will increase for equal step insertions of positive reactivity. When the reactor becomes critical, M will be infinitely large [4]. For this reason, monitoring and plotting M during an approach to criticality is impractical because there is no value of M at which the reactor clearly becomes critical. Instead of plotting M directly, its inverse ($1/M$) is plotted on a graph of $1/M$ versus rod height.

$$\frac{1}{M} = 1 - k_{eff} \tag{11.74}$$

Or

$$\frac{1}{M} = \frac{C_{R0}(1 - k_0)}{C_R} \tag{11.75}$$

As control rods are withdrawn and k_{eff} approaches one and M approaches infinity, $1/M$ approaches zero. For a critical reactor, $1/M$ is equal to zero. A true $1/M$ plot requires knowledge of the neutron source strength. Because the actual source strength is usually unknown, a reference count rate is substituted, and the calculation of the factor $1/M$ is through the use of Eq. (11.75).

In practice, the reference count rate used is the count rate prior to the beginning of the reactivity change. The startup procedures for many reactors include instructions to insert positive reactivity in incremental steps with delays between the reactivity insertions to allow time for subcritical multiplication to increase the steady-state neutron population to a new, higher level and allow more accurate plotting of $1/M$. The neutron population will typically reach its new steady-state value within 1–2 min,

Table 11.12 Rods distance and count rate

Rods withdraw/inch	Count rate/(1/s)	C_{R0}/C_R
0	50(C_{R0})	1
2	55	0.909
4	67	0.746
6	86	0.581
8	120	0.417
10	192	0.260
12	500	0.100

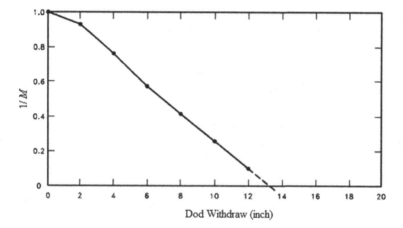

Fig. 11.21 Rods withdraw and $1/M$ curve

but the closer the reactor is to criticality, the longer the time will be to stabilize the neutron population.

Let us look at a specific example. Given the following rod withdrawal data, construct a $1/M$ plot and estimate the rod position when criticality would occur. The initial count rate on the nuclear instrumentation prior to rod withdrawal is 50 cps.

The measured data as shown in Table 11.12. Draw $1/M$ curve as shown in Fig. 11.21. By extrapolation, $1/M = 0$ is the position of the critical point.

11.10 Reactor Kinetics

The response of neutron flux and reactor power to changes in reactivity is much different in a critical reactor than in a subcritical reactor. The reliance of the chain reaction on delayed neutrons makes the rate of change of reactor power controllable.

Reactor kinetics is a branch to research the law of nuclear reactor neutron flux changes over time. Time characteristics of dynamic processes, issues involved in reactor kinetics can be roughly divided into three categories [5]:

1. Transient. Such as the start-up of the reactor, the reactor's power and transients caused by the reactor accident.
2. Slow transients. Fission product Xenon poisoning and oscillations. These processes are slow, often hours or days of orders of magnitude.
3. Long-term changes. Such as reactor fuel, transformation and proliferation of nuclear fuel. These changes are very slow, and tend to be cumulative monthly and yearly basis.

Reactor kinetics focuses on the first type of transient problems. As for the third category of long-term slow process, static analysis method is used step by step.

11.10.1 Reactor Kinetics Equations

Due to changes in reactivity caused by instantaneous change of reactor neutron flux with time, often by a group called the reactor kinetics equations to description. In engineering, in order to describe, often used point reactor model, which holds that the neutron number density changes over time is consistent in the whole reactor, has nothing to do with the spatial position.

Considering the delayed neutron kinetics equation can be expressed as follows.

$$\begin{cases} \frac{dN}{dt} = \frac{\rho-\beta}{\Lambda} N + \sum_{i=1}^{I} \lambda_i C_i + S \\ \frac{dC_i}{dt} = \frac{\beta_i}{\Lambda} N - \lambda_i C_i, i = 1, 2 \ldots, I \end{cases} \tag{11.76}$$

where N is the number density of neutrons in reactor; C_i is the number of nuclear density of the ith group of delayed neutron precursors; λ_i is the decay constant of the ith group of delayed neutron precursor, delayed neutrons are released by β decay process of pioneer nuclei; S is the neutron source strength; Λ is the average neutron generation time; ρ is the reactivity of the reactor; β_i is the fraction of the ith group of delayed neutrons, β is the fraction of the delayed neutron to all fission neutrons.

$$\beta = \sum_{i=1}^{I} \beta_i \tag{11.77}$$

Recall that, the delayed neutron fraction is the fraction of all fission neutrons that are born as delayed neutrons. The value of β depends upon the actual nuclear fuel used. The delayed neutron precursors for a given type of fuel are grouped on the basis of half-life. Table 11.13 lists the fractional neutron yields for each delayed neutron group of three common types of fuel.

Table 11.13 Delayed neutron fraction for various fuels

i	Half-life/s	^{235}U	^{238}U	^{239}Pu
1	55.7	0.00021	0.0002	0.00021
2	22.7	0.00142	0.0022	0.00182
3	6.22	0.00127	0.0025	0.00129
4	2.3	0.00257	0.0061	0.00199
5	0.61	0.00075	0.0035	0.00052
6	0.23	0.00027	0.0012	0.00027
Total	–	0.0065	0.0157	0.00200

In General, dynamic equation is a linear differential equation. It solved using numerical methods. In case of feedback, need to know the reactor's response to changes in various parameters change, characterized by reactivity coefficient of reactivity changes with the effects of the rate of change. While matching some of the parameters varies with time and reactor power change in relationship equations, such as the reactor thermal hydraulic equations, simultaneous solution.

When the spatial distribution of neutron flux occurs local drastic changes, point reactor equations do not apply to solve this kind of problems, take solutions related to the neutron dynamics equation, this equation is generally very complicated. It solved only through numerical methods, or half a numerical approximation methods, approximation methods, common and effective method of factorization method (also called static methods) and synthetic method. In some special cases, the dynamic equations can be simplified to get analytical results.

11.10.2 In-Hour Equation

The point reactor kinetics Eqs. (11.76) represents the prompt neutron number density of neutrons in the reactor over the time rate of change is produced by the fission neutron and delayed neutron and external neutron source. The number of nuclear density of precursors changes over time. Delayed neutrons have an important influence on the dynamic characteristics of a reactor.

In the case of reactivity ρ is constant, the neutron number density in the reactor changes with the reactivity ρ, there is a simple relationship, and this relationship is the reactivity equation or the in-hour equation. We can prove that when ρ in step changes from 0 to a constant ρ_0, solving Eqs. (11.76) we can get

$$N(t) = \sum_{i=1}^{I} A_i e^{\omega_i t} \qquad (11.78)$$

where, A_i is the coefficient of the index entries, as determined by the initial conditions. ω_i satisfies the equation

$$\rho_0 = \Lambda\omega + \sum_{i=1}^{I} \frac{\beta_i \omega}{\omega + \lambda_i} \tag{11.79}$$

That satisfied the reactivity equation or the in-hour equation, which is linked with ω and ρ_0 relationship. This equation has $(I + 1)$ roots, where I roots are negative real numbers, and the other root is a real number with sign same as ρ_0. When ρ_0 is positive, the in-hour equation has a positive root. At this time, in addition to $A_0 e^\omega$, all other items are decay with time. Thus, when the time is long enough, the neutron number density change is mainly determined by the first term, that is,

$$N(t) \approx A_0 e^{\omega_0 t} \tag{11.80}$$

11.10.3 Reactor Period

The reactor period is defined as the time required for reactor power to change by a factor of "e," where "e" is the base of the natural logarithm and is equal to about 2.718. The reactor period is usually expressed in units of seconds. From the definition of reactor period, it is possible to develop the relationship between reactor power and reactor period that is expressed by Eq. (11.81).

$$P(t) = P_0 e^{\omega_0 t} = P_0 e^{t/\tau} \tag{11.81}$$

where:

P transient reactor power
P_0 initial reactor power
τ reactor period (seconds)
t time during the reactor transient (seconds).

The smaller the value of τ, the more rapid the change in reactor power. If the reactor period is positive, reactor power is increasing. If the reactor period is negative, reactor power is decreasing.

When the in-hour Eq. (11.79) is represented in reactor period, one get

$$\rho_0 = \frac{\Lambda}{\tau} + \sum_{i=1}^{I} \frac{\beta_i}{1 + \lambda_i \tau} \tag{11.82}$$

Delayed neutrons do not have the same properties as prompt neutrons released directly from fission. The average energy of prompt neutrons is about 2 MeV. This is much greater than the average energy of delayed neutrons (about 0.5 MeV). The fact that delayed neutrons are born at lower energies has two significant impacts on the way they proceed through the neutron life cycle. First, delayed neutrons have a much lower probability of causing fast fissions than prompt neutrons because their average energy is less than the minimum required for fast fission to occur. Second, delayed neutrons have a lower probability of leaking out of the core while they are at fast energies, because they are born at lower energies and subsequently travel a shorter distance as fast neutrons. These two considerations (lower fast fission factor and higher fast non-leakage probability for delayed neutrons) are taken into account by a term called the importance factor (ζ). The importance factor relates the average delayed neutron fraction to the effective delayed neutron fraction.

The effective delayed neutron fraction is defined as the fraction of neutrons at thermal energies, which were born delayed. The effective delayed neutron fraction is the product of the average delayed neutron fraction and the importance factor.

$$(\beta_i)_{\text{eff}} = \zeta \beta_i \qquad (11.83)$$

$$\beta_{\text{eff}} = \sum_{i=1}^{I} (\beta_i)_{\text{eff}} \qquad (11.84)$$

In a small reactor with highly enriched fuel, the increase in fast non-leakage probability will dominate the decrease in the fast fission factor, and the importance factor will be greater than one. In a large reactor with low enriched fuel, the decrease in the fast fission factor will dominate the increase in the fast non-leakage probability and the importance factor will be less than one (about 0.97 for a commercial PWR).

Another new term has been introduced in the reactor period (τ) equation. That term is λ_{eff}, the effective delayed neutron precursor decay constant. The decay rate for a given delayed neutron precursor can be expressed as the product of precursor concentration and the decay constant (λ) of that precursor. The decay constant of a precursor is simply the fraction of an initial number of the precursor atoms that decays in a given unit time. A decay constant of 0.1 s^{-1}, for example, implies that one-tenth, or ten percent, of a sample of precursor atoms decays within one second. The value for the effective delayed neutron precursor decay constant, λ_{eff}, varies depending upon the balance existing between the concentrations of the precursor groups and the nuclide(s) being used as the fuel.

If the reactor is operating at a constant power, all the precursor groups reach an equilibrium value. During an up-power transient, however, the shorter-lived precursors decaying at any given instant were born at a higher power level (or flux level) than the longer-lived precursors decaying at the same instant. There is, therefore, proportionately more of the shorter-lived and fewer of the longer-lived precursors decaying at that given instant than there are at constant power. The value of λ_{eff} is closer to that of the shorter-lived precursors. During a down-power transient, the longer-lived precursors become more significant. The longer-lived precursors decaying at a given instant were born at a higher power level than the shorter-lived precursors decaying at that instant. Therefore, proportionately more of the longer-lived precursors are decaying at that instant, and the value of λ_{eff} approaches the values of the longer-lived precursors.

Approximate values for λ_{eff} are 0.08 s^{-1} for steady-state operation, 0.1 s^{-1} for a power increase, and 0.05^{-1} for a power decrease. The exact values will depend upon the materials used for fuel and the value of the reactivity of the reactor core.

After getting β_{eff} and λ_{eff}, let us return now to the equation for reactor period.

$$\tau = \frac{l^*}{\rho} + \frac{\beta_{eff} - \rho}{\lambda_{eff}\rho + d\rho/dt} \tag{11.85}$$

where, l^* is the average lifetime of prompt neutrons, prompt neutron-life is very short, for thermal neutron reactors, it is about 10^{-4} s.

If the positive reactivity added is less than the value of λ_{eff}, the emission of prompt fission neutrons alone is not sufficient to overcome losses to non-fission absorption and leakage. If delayed neutrons were not being produced, the neutron population would decrease as long as the reactivity of the core has a value less than the effective delayed neutron fraction. The positive reactivity insertion is followed immediately by a small immediate power increase called the prompt jump. This power increase occurs because the rate of production of prompt neutrons changes abruptly as the reactivity is added. Recall from an earlier module that the generation time for prompt neutrons is on the order of 10^{-13} s. The effect can be seen in Fig. 11.22.

Because the prompt neutrons immediately react to sudden power rise quickly. Soon, because the prompt neutrons alone could not reached critical mass, power rise speed immediately decreases. The power rising rate behind determined largely by delayed neutrons.

Conversely, in the case where negative reactivity is added to the core there will be a prompt drop in reactor power. The prompt drop is the small immediate decrease in reactor power caused by the negative reactivity addition. The prompt drop is

Fig. 11.22 Reactor power response to positive reactivity addition

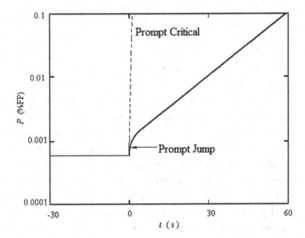

Fig. 11.23 Reactor power response to negative reactivity addition

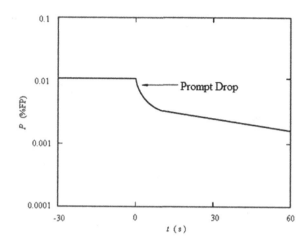

illustrated in Fig. 11.23. After the prompt drop, the rate of change of power slows and approaches the rate determined by the delayed term of Eq. (11.85).

It can be readily seen from Eq. (11.85) that if the amount of positive reactivity added equals the value of β_{eff}. In this case, the production of prompt neutrons alone is enough to balance neutron losses and increase the neutron population. The condition where the reactor is critical on prompt neutrons, and the neutron population increases as rapidly as the prompt neutron generation lifetime allows is known as prompt critical, as shown in Fig. 11.22 by the dotted line. The prompt critical condition does not signal a dramatic change in neutron behavior. The reactor period changes in a regular manner between reactivities above and below this reference. Prompt critical is, however, a convenient condition for marking the transition from delayed neutron to prompt neutron time scales. A reactor whose reactivity even approaches prompt critical is likely to suffer damage due to the rapid rise in power to a very high level. For example, a reactor that has gone prompt critical could experience a several thousand percent power increase in less than one second.

Because the prompt critical condition is so important, a specific unit of reactivity has been defined that relates to it. The unit of reactivity is the dollar ($), where one dollar of reactivity is equivalent to the effective delayed neutron fraction. A reactivity unit related to the dollar is the cent, where one cent is one-hundredth of a dollar. If the reactivity of the core is one dollar, the reactor is prompt critical. Because the effective delayed neutron fraction is dependent upon the nuclides used as fuel, the value of the dollar is also dependent on the nuclides used as fuel.

11.11 Nuclear Power Plant Operation

Basic principles of operation of nuclear power plants are almost same as coal-fired power plants. Based on the load demand of the power grid to regulate the reactor power, so the output of nuclear power plant and power grid load demand is balanced. The energy comes from the controllable fission chain reaction within nuclear fueling a nuclear power plant reactor. Nuclear fission not only releases energy but also producing strong radioactivity and neutron activation of fission products. Fission chain reaction control and handling of radioactive products are characteristic of nuclear power plants. Nuclear power plant operation characteristics at run time to ensure that safety, no escape of harmful radioactive material outside the factory to the environment, while improving economic performance, making it competitive in energy market. Different reactor types have essentially the same operating characteristics of nuclear power plants, but there are differences. Here are basic running characteristic of light water reactor nuclear power plant.

Suitable for Base Load Operation
Because of the high safety requirements for nuclear power plants, complex, long construction period, so the cost is much higher than coal-fired power plant, but its relatively low fuel costs. In order to improve the economics, nuclear power plants should continuous run at a base load power level or as close as possible to the rated power. This will make a higher capacity factor for nuclear power plants. In addition, the nuclear power plant base load operation can also reduce the amount of radioactive water produced by fluctuations in power.

Regular Replacement of Fuel
After filled with fuel for light water reactor plants, according to fuel cycle program, regularly shut down to replace part of the nuclear fuel. General fuel cycle is once a year, every change of 1/3 (pressurized water reactor) or 1/4 (boiling water reactor). In parallel with the refueling, to carry out the necessary maintenance, inspection and testing. Refueling takes about a month or two.

Load Change Restrictions
In order to ensure the safety of fuel elements, to strictly limit the load variation speed of nuclear power plants. For example, power increase of pressurized water reactor shall not exceed 5%FP per minute; step change shall not exceed 10%FP. After refueling or extended shutdown, the reboot power growth limit is 3%FP per hour. At the end of the nuclear fuel cycle, because of the low coolant boron concentration, boron dilution capacity decrease, it will take a longer time to raise power to full power. For a base load power plant, according to the power requirement of grid, it can participate in a certain level of frequency adjusting. Now there are some nuclear power plants to meet the needs of load following, its power is controlled by grey rods.

Absolute Guarantee of Residual Heat Removal
After the shutdown of nuclear power plants, reactor core fuel still generate a lot of decay heat, which lasted for a long time. If not removed, has the potential to make the nuclear fuel from overheating and damage. Therefore, in the nuclear power plant has a residual heat removal systems, safety injection system, uninterruptible power

supply (including emergency diesel generator) and important circulating cooling water system. In any case, even with site black out (SBO) and design basis earthquake, also make sure that the cooling of the reactor core is available.

Strict Water Quality Management

Water quality management of the primary loop and secondary loop is key issue related to the safe operation of fuel assembly and equipment within the life period. The water in the primary loop will be poor because: ① dirt deposition of fuel components surface; ② corrosion product increased, and corrosion product in neutron irradiation will be activation, generated radioisotope (mainly of ^{60}Co), bring difficult to maintenance and check; ③ excess of chlorine ion, fluoride ion and dissolved oxygen in water, can makes stainless steel equipment and pipeline, nickel heat transfer tube serious of stress corrosion. Reliability of secondary side water quality is directly related to the steam generator. Secondary side water impurities generates hydroxide (free caustic substances). The excessive concentration and assembly of these compounds led to steam generator tube generated intergranular stress corrosion. For seawater-cooled nuclear power plants, the need to prevent water leakage into the secondary loop. The primary loop and secondary loop have strict water quality standards, with online monitoring instruments and sampling and analysis.

Special Security Installations, Radiation Protection, Environmental Monitoring and Emergency Response Measures

Nuclear fission produces large quantities of radioactive material; the uncontrolled release of radioactive material may result in severe consequences. Nuclear power plant prevents radiation escaping by three barrier: fuel cladding, pressure boundaries of the primary loop and containment. It also has engineered safety features to protect these three barriers. Closely monitoring three barrier integrity in the operation and that the engineered safety features and the availability of safety-related systems. In addition, with many kind of radiation shielding for reactor and components of primary loop to protect staff safety. BWR steam containing radioactive material, so the turbine and steam line also needs to be screened. During normal operation of nuclear power plants, they still have to deal with a certain amount of radioactive waste. Nuclear power plants using a variety of measurement methods for radiation monitoring both inside and outside the plant, including process monitoring, site radiation monitoring and radiation outside the plant environmental monitoring. Before the operation of a nuclear power plant, must also work out a feasible off-site emergency rescue program in case of severe accidents and the implementation of the program.

Strict Requirements on Operator

Operators, in particular, reactor operators and senior reactor operators, directly responsible for the safe operation of nuclear power plants. Nuclear security regulations, the reactor operators and senior operators, through systematic training for a long time, pass the examination of chaired by national authorities, including in the nuclear power plant training simulator for a variety of operational examination. Candidates from the state nuclear safety regulatory bodies to license. After two years of its issue, but also to review once again, make sure, the holder have run straight through and completed the necessary periodic retraining, extension of validity of the license.

Final Decommissioning Disposal Requirements
Nuclear power plant after the termination, disposal must be retired, eventually bringing the site to unrestricted use or compatible with the ecological environment. This is a very long time process, according to retired technology policies; it normally takes years to decades and even centuries.

11.11.1   Startup of Reactor

When a reactor is started up with irradiated fuel, or on those occasions when the reactor is restarted following a long shutdown period, the source neutron population will be very low. In some reactors, the neutron population is frequently low enough that it cannot be detected by the nuclear instrumentation during the approach to criticality. Installed neutron sources are frequently used to provide a safe, easily monitored reactor startup.

The neutron source, together with the subcritical multiplication process, provides a sufficiently large neutron population to allow monitoring by the nuclear instruments throughout the startup procedure. Without the installed source, it may be possible to withdraw the control rods to the point of criticality, and then continue withdrawal without detecting criticality because the reactor goes critical below the indicating range. Continued withdrawal of control rods at this point could cause reactor power to rise at an uncontrollable rate before neutron level first becomes visible on the nuclear instruments.

An alternative to using a startup source is to limit the rate of rod withdrawal, or require waiting periods between rod withdrawal increments. By waiting between rod withdrawals increments, the neutron population is allowed to increase through subcritical multiplication. Subcritical multiplication is the process where source neutrons are used to sustain the chain reaction in a reactor with a multiplication factor (k_{eff}) of less than one. The chain reaction is not "self-sustaining," but if the neutron source is of sufficient magnitude, it compensates for the neutrons lost through absorption and leakage. This process can result in a constant, or increasing, neutron population even though k_{eff} is less than one.

Estimated Critical Position
$1/M$ plots were useful for monitoring the approach to criticality and predicting when criticality will occur based on indications received while the startup is actually in progress. Before the reactor startup is initiated, the operator calculates an estimate of the amount of rod withdrawal that will be necessary to achieve criticality. This process provides an added margin of safety because a large discrepancy between actual and estimated critical rod positions would indicate that the core was not performing as

designed. Depending upon a reactor's design or age, the buildup of xenon within the first several hours following a reactor shutdown may introduce enough negative reactivity to cause the reactor to remain shutdown even with the control rods fully withdrawn. In this situation, it is important to be able to predict whether criticality can be achieved, and if criticality cannot be achieved, the startup should not be attempted.

For a given set of conditions (such as time since shutdown, temperature, pressure, fuel burnup, samarium and xenon poisoning) there is only one position of the control rods (and boron concentrations for a reactor with chemical shim) that results in criticality, using the normal rod withdrawal sequence. Identification of these conditions allows accurate calculation of control rod position at criticality. The calculation of an estimated critical position (ECP) is simply a mathematical procedure that takes into account all of the changes in factors that significantly affect reactivity that have occurred between the time of reactor shutdown and the time that the reactor is brought critical again.

For most reactor designs, the only factors that change significantly after the reactor is shut down are the average reactor temperature and the concentration of fission product poisons. The reactivity normally considered when calculating an ECP include the following.

Firstly, basic reactivity of the core. The reactivity associated with the critical control rod position for a xenon-free core at normal operating temperature. This reactivity varies with the age of the core (amount of fuel burnup). This reactivity is the reactivity of the reactor core at the critical position of control rods, and does not consider the effect of xenon, only consider the impact of fuel burnup.

Secondary, direct xenon reactivity. The reactivity related to the xenon that was actually present in the core at the time it was shutdown. This reactivity is corrected to allow for xenon decay.

Thirdly, indirect xenon reactivity. The reactivity related to the xenon produced by the decay of iodine that was present in the core at the time of shutdown.

Finally, temperature reactivity. The reactivity related to the difference between the actual reactor temperature during startup and the normal operating temperature.

11.11.2 Startup of Nuclear Power Plant

Nuclear power plant startup starts from cold subcritical state to on-grid electricity output. Shutdown is disconnection from the grid back to cold subcritical. This process includes several stages: ① starting from cold subcritical; ② starting from hot subcritical; ③ low power operation; ④ grid-connected power generation. Hot start-up is part stage of the cold start-up in the process. Low power operation is a transition condition; including islanding operation and power transmission prepare status. Different reactor-type of nuclear power plant start-up and stop modes each have their own characteristics. Here is with pressurized water reactor nuclear power plant as an example of how normal startup and normal shutdown of the main stages and characteristics.

Normal startup. Normal starting with cold start and hot start points. Hot start-up including iodine dead-time in the process of starting. Reactor coolant temperature at 60 °C below the start as cold start. Short outages and keep coolant temperatures above 280 °C is hot start. Starting after the first loading fuel for nuclear power plants called starting for the first time.

Cold starting program. After shut down reloading or maintenance, full with approximately 2100 μg/g boron-containing cooling water, all control rods are in the lowest position, the coolant temperature is below 60 °C, the reactor core is in subcritical state. Starting steps are as follows:

1. Fill the primary loop with coolant and exhaust: after the primary loop is filled with water, lower the water level in secondary side of steam generator to zero power level setpoint.

2. Starting coolant pump and the electric heater in pressurizer: First, rise coolant pressure to 2.5 MPa, extract the shutdown control rods and temperature control rods to the top of the reactor core, so that when the unexpected boron dilution accident, to ensure adequate shutdown margin. Starting the main pump and the electric heater in pressurizer, the coolant heats up. When the coolant temperature reaches 90 °C, add Lithium hydroxide (LiOH), to control the pH value of the water and hydrazine, the oxygen dissolved in the coolant reaches the specified value. When the temperature rises to 100 ~ 130 °C, by manually controlling the volume control tank and replacement for N_2 to H_2, establish hydrogen space, set the volume control tank water level control valves to automatic. When internal temperature reaches saturation temperature corresponding to the 2.5–3.0 MPa (221–232 °C), the manual control to reduce water level of pressurizer, in the pressurizer establish upper steam chamber, and adjust the water level of pressurizer, when the water level reaches zero power level setpoint, the coolant system pressure is controlled by pressurizer.

3. Elevate temperature and pressure of the primary loop: by the heat from the electric heater in pressurizer and the mechanical energy from pump rotation, making coolant temperature and pressure gradually raised. The rate of temperature rise should be required.

   When the pressure and temperature of the primary loop reaches hot shutdown state, the pressure control can be changed from manual to automatic. In the heating process, excess heat is removed by the steam generator, secondary side steam released to the atmosphere or bypass to the condenser.

4. Startup of reactor: the reactivity of pressurized water reactor is changed with the nuclear fuel burnup and change of coolant temperature. In the critical startup, the coolant temperature must guarantee at negative temperature coefficient. In the process of startup, the coolant temperature should be kept constant as possible. The coolant boron concentration can be gradually diluted to estimate the critical concentration, then raise control rod slowly, the in-core neutron number increases until critical, bringing the unit to hot standby, due to the secondary water-supply capacity constraints, reactor power at ≤2% FP level.

5. The secondary loop start: after the reactor critical and requires secondary loop systems starting preparations. First starting steam-driven feed-water pump, auxiliary water supply switches to the main water supply, switch atmosphere discharge to the turbine condenser. Then, the reactor power up to 5% ~ 10%FP (according to the different types of units, the power level is different), use the steam from the steam generator to warm up the main steam piping and moisture separator reheater. Turbine starts low speed warming. After warm-up finished, speed up the turbine, up to the rated speed.

6. grid-connected and raise power: after the power increased to about 10% ~ 15%FP, grid-connected generators, and running with minimum load (about 5% of the generator power ratings). Gradually close the bypass valve to the condenser, power balance between the reactor and turbine. Continue to raise the load, when the power exceeds 15%FP, reactor control switch from manual to automatic.

7. Power running: for pressurized water reactor nuclear power plant with power running, generally use the average temperature of the inlet and outlet of reactor core to regulate the system. When the load changes, changing the depth of control rods in the reactor core to change neutron power of the reactor. A new balance makes between neutron power and thermal power. At the same time adjust the boron concentration of coolant; to withdraw control rods in a best position to ensure that core power distribution do not exceed the specified values.

In power running, due to the accumulation of burnup and fission products of the fuel (that is, poison and slagging), the reactivity will go down by adjusting the primary coolant boron concentration to compensate for this loss of reactivity.

In order to ensure the safety of fuel rods, load transient and the rate of it should be less than the specified value.

Hot start-up program. Hot shutdown of nuclear power plants, after the temperature and pressure of the primary loop close to zero power setpoint, directly start the reactor to achieve criticality. Then follow the step (5) and (6) of the "cold start" program. Put steam turbine into operation.

Iodine starting program

Iodine dead-time start belongs to the hot start-up process. Generally refers to starting from hot shutdown after full power, with a balance of xenon poison. After the shutdown, the change of the fission product poison ^{135}Xe is divided into three main stages. The first stage, due to the disappearance of ^{135}Xe slows, xenon poison gradually increased, the reactivity loss increases, known as the accumulation stage of poison. The second stage is about 11 h after the shutdown; Xenon poison reaches the maximum value, maximum reactivity loss, also known as the maximum value of iodine. The third stage, ^{135}Xe decays faster than generating of ^{135}Xe, xenon poison decreases, known as the detoxication stage. About 24 h after shutdown, xenon poison disappeared. Iodine pit in the process of starting to bring a certain degree of operational complexity.

1. Accumulate toxic phase starting: directly according to the order of drawback control rods to making the reactor reached criticality. When you are approaching critical, the average coolant temperature should be avoided catastrophe or coolant boron dilution in action.

2. Starting maximum iodine pit: late in the reactor's fuel cycle, pit depth if iodine may be larger than the excess reactivity after shutdown. At this time, even all control rods are out of the core, it may also be impossible to reach the reactor critical. Only through appropriate boron dilution operation of reactor coolant, it is possible to start the reactor. Soon after starting the reactor, as the power raised, xenon due to absorb neutrons and rapidly reduced, reactivity to rise accordingly. Need to add boron into coolant again. So starting from maximum iodine pit, the operation is very complex, and produces large quantities of wastewater, so you should avoid this situation. At the end of life a reactor, the largest pit in iodine might not able to start.

3. Starting at the detoxication stage: due to the spontaneous decay of xenon will introduce positive reactivity, no need of a coolant boron dilution operation. Instead, in order to avoid critical below the zero power critical estimates rod position, may also need to add appropriate boron. Action must be very careful to prevent the reactivity insertion accident rate is too large and a short period accident.

11.11.3 Nuclear Power Plant Shutdown

Normal shutdown of nuclear power plants is from disconnection from the grid to subcritical state, according to shut down aim to maintain the reactor hot shutdown or back to cold shutdown.

Hot shutdown is a short-term shutdown. Insert the compensate rod group and temperature control rod group into the reactor core to reactor subcritical, shut down rod group remained at top of the reactor core. At this time, the coolant system at or near zero power operating temperature and pressure. Primary loop temperature is maintained by controlling the emission of steam into the atmosphere or the condenser to remove its heat energy from the reactor core and the main pump work, feed water from auxiliary water supply system of steam generator. The pressure of coolant is maintained by the pressurizer. During hot shutdown, at least turn on one primary pump.

When the reactor shutdown time longer than the largest iodine pit value, for a longer period of time the xenon poison will gradually reduce, and the reactor may return to critical. Therefore, hot shutdown time must be based on the estimated boron, so that coolant boron concentration in safety shutdown margin.

After decreasing of the temperature and pressure, the reactor goes from hot shutdown to cold shutdown. To counteract the positive reactivity caused by negative temperature effects in the cooling process, as well as to ensure the sufficient shutdown margin, boron must be added before the cooling of reactor core. The process is: raise temperature control rod to the top of the reactor core, steam by-passed to condenser (such as the condenser is not available, then into the atmosphere), primary loop coolant temperature to 180 °C, pressure to 2.5 MPa. At this point, start the shutdown core cooling system, continues to cool the reactor, annihilate the void space in pressurizer until the temperature is below 90 °C, led the reactor to cold shutdown state.

11.11.4 Status of Nuclear Power Plant

Status of nuclear power plants can be classified into two major categories, operation status and accident status. The operation status are divided into normal operation and anticipated operational occurrences; accidents status are divided into accident conditions (including accidents and limit accidents) and severe accidents. Status of nuclear power plants classified by frequency from high to low, respectively for normal operation, expected transient operation, accidents and severe accident conditions.

In nuclear power plant design, must follow the principle that lead to high doses of radiation or radioactive release state of the lower frequencies, and high frequency radiation to be of little consequence. Status of classification is that of nuclear power plants to different states set different restrictions on the system response, which corresponds to the acceptable limits. So that the design can meet nuclear safety requirements.

Plant status classification is based on engineering judgment, design and operating experience as determined on the basis of. Status classification of nuclear power plants in the United States early in the final safety analysis report of nuclear power plant has been adopted, and more or less stereotyped. For about 30 years, mainly in the analysis of incidents and accidents on the list was perfect. This classification has been more widely used in countries with nuclear power plants, but uses different acceptable limits. Status classification of Chinese nuclear power plant is similar.

Normal operation of nuclear power plants. Running within the operating limits and conditions, including the shutdown state, power operation, shutdown procedures, startup procedures, maintenance, testing and refueling.

Anticipated operational occurrences, or medium-frequency event, in the nuclear power plant's life cycle may occur one or more times. In such cases, when the nuclear power plant operation parameters meet the limit setpoint value, the protection of the system should be able to shut down the reactor, but after you have made the necessary corrective action, the reactor can be put into operation again. The acceptable limits for anticipated operational occurrences follows. Firstly, coolant system pressure is less than 110% of design value. Secondly, without departure from nucleate boiling on the surface of the fuel element cladding. Thirdly, radioactive release below normal operating limits.

For a single nuclear power plant, accidents are not likely to happen. However, for the whole nuclear power plants in the world, is likely to occur once or more. If the occurrence of such accidents, small amounts of fuel may be damaged and need to rely on engineered safety features to mitigate its consequences. Limits for accident are as follows. Firstly, keep the geometry shape of the core and keep cooling. Secondary, pressure of reactor cooling system is less than 120% of design value. Thirdly, in the separation region (2 h within) and the low population district (8 h within), personal radiation dose limited values (by United States standard) are: thyroid dose 300 mSv, whole body dose 25 mSv (for some very low frequency accidents, its limited value are: thyroid dose 750 mSv, whole body dose 60 mSv).

Limit accidents, very unlikely accidents, its frequency is less than 10^{-4}/(core-year). Accidents of this kind may result in fuel elements have a major injury, but without of loss of engineering safety systems limiting the consequences, the reactor coolant system and reactor building will not be affected by the injury. Limits of acceptable limit values are as follows. Firstly, maintain their geometry and cooling of the reactor core. Secondary, reactor coolant system pressure is less than 120% of the design value. Thirdly, in the separation region (2 h within) and the low population district (8 h within), personal radiation dose limited values (by United States standard) are: thyroid dose 3000 mSv, whole body dose 250 mSv.

Severe accidents, resulting in serious damaged fuel elements, the reactor core melted down, containment integrity may be destroyed, release a lot of radioactive material to environment. Frequency of such accidents in nuclear power plants is expected to be limited by the national nuclear security administration security goals.

11.12   Isotope Separation

Isotopic abundance in nature, also known as naturally ratio of the isotope, is the proportion of isotope in all naturally exist isotope of the element. The value is generally expressed as a percentage of the abundance, artificial isotope abundances are zero.

Natural uranium is a mixture of ^{238}U and ^{235}U and ^{234}U, abundance of ^{238}U is 99.27%, ^{235}U abundance is 0.714%, and a trace of remaining ^{234}U. Nuclear power plants generally use ^{235}U enrichment of 3.5% as nuclear fuel. Enrichment is the weight ratio; abundance is the ratio of the number of atoms. The transform relation between abundance and enrichment is Eq. (11.86).

$$C_{235} = \frac{\frac{x_{235}}{M_{235}} A_{00}}{\frac{x_{235}}{M_{235}} A_{00} + \frac{1-x_{235}}{M_{238}} A_{00}} = \frac{1}{1 + 0.9874\left(\frac{1}{x_{235}} - 1\right)} \tag{11.86}$$

where, M_{235} is the ^{235}U molar mass, that is, 235 g/mol; M_{238} is the molar mass of ^{238}U, 238 g/mol; C_{235} is the abundance of ^{235}U (ratio of the number of nucleons), x_{235} is the enrichment of ^{235}U (weight ratio), A_{00} is the Avogadro constant, 6.022×10^{23}.

According to isotope's abundance, you can calculate the nuclear number density. The so-called nuclear number density means the number of nuclei of per unit volume.

For example, the nuclear number density of ^{235}U in UO$_2$ is N_{235}

$$N_{235} = \frac{\rho_u}{M_u} A_{00} C_{235} \qquad (11.87)$$

In view of the natural uranium content of ^{234}U in only 0.008%, so the essence of uranium enrichment is the separation of ^{235}U and ^{238}U. In addition, because essentially the same physical and chemical properties of ^{235}U and ^{238}U, which brings great difficulties to their separation. United States Manhattan project developed four enrichment methods, namely thermal diffusion, the electromagnetic method, gaseous diffusion and gaseous centrifuge method [6]. The first kg amount of ^{235}U in the world were separated in the United States at Oak Ridge laboratory using electromagnetic separation method in 1944. Its former stage of enrichment used thermal diffusion to enrich the natural abundance of ^{235}U to enrichment of 0.86%, and then entered into magnetic separator for further enrichment to weapons-grade. In 1945–1946, it was proven that the gaseous diffusion method has advantage over the other three methods. After that, the other three methods stop to industrial development. From then on until in the 1980, gaseous diffusion has been dominate in the enrichment of uranium. To date most of the world's enriched uranium is still the production of gaseous diffusion. However, gaseous diffusion method has many disadvantages, mainly large amount of electricity power, 70% of the cost, in addition to factories of capital construction investment is also very large. In the separation of the various methods of research, and is capable of industrial application there are three, namely, gaseous diffusion, high-speed centrifugation and separation nozzles. Laser and chemical exchange method are also hope to get the practical application.

11.12.1 SWU and Value Function

In a system of two or more isotope gas components of mixtures, a spontaneous process tends to increase the entropy of system. The process of some isotopes are extracted from a mixture of gas, or separating the various isotopic composition in the mixture, is a non-spontaneous process with entropy decrease [7]. To make the non-spontaneous process happen, you must invest a certain amount of compensation, which is called the SWU (separate work unit). The number of separative work required should be proportional to the decrease of entropy.

Here we introduce the binary separation of ^{235}UF$_6$ and ^{238}UF$_6$ mixture.

Let us look at the material balance of the isotope separation process. Suppose feed (natural uranium) of F kg/s at abundance of C_f, concentrated product (enriched uranium) P kg/s at abundance of C_p and tailings is W kg/s enrichment of C_w of depleted uranium (see Fig. 11.24). Based on the total mass conservation, we have $F = P + W$, according to the conservation of ^{235}U, we have

$$F \cdot C_f = P \cdot C_p + W \cdot C_w \qquad (11.88)$$

Fig. 11.24 Material balance diagram of isotope separation

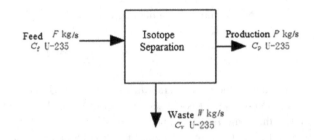

Or

$$\frac{F}{P} = \frac{C_p - C_w}{C_f - C_w} \tag{11.89}$$

Then, to get 1 kg/s of product of C_p, how much separate power needed? The separation work has important significance in practical engineering. Separation of a separator and separation ability of a factory is represented by the amount of separation work. Enriched uranium cost is usually expressed as cost of separate work units.

The value function for calculating separation work is introduced based on the concept of entropy. The relationships between entropy and isotopic abundances for N isotope gaseous mixture is [8]

$$S_i = -R \ln C_i \tag{11.90}$$

where, R is the gas constant, C_i is the ith isotope abundance, the total entropy of the mixture is

$$S = \sum_{i=1}^{N} C_i S_i = -R \sum_{i=1}^{N} C_i \ln C_i \tag{11.91}$$

For a system, it has maximum entropy when all components have the same C_i, which is the most chaotic state. Status of any C_i is not equal to C_j means some kind of order. For a binary system involving two compositions i and j, the total degree of order between the two compositions equals to the degree of order of i composition relative to j composition plus the degree of order of j composition relative to the i composition [9]. Because the entropy reflects the degree of disorder, so the degree of disorder per mole of i composition relative to j composition is $(S_i - S_j)$, its degree of order is the minus of degree of disorder, that is $(S_j - S_i)$. In one mole mixture gas, has C_i mole composition i. So in one mole gas mixture, the degree of order of i composition relative to j composition is $C_i (S_j - S_i)$. Similarly in one mole mixture gas, has C_j mole composition j and the degree of order of j composition relative to i composition is $C_j (S_i - S_j)$. Thus, the total degree of order in one mole of gas mixture is

$$C_i(S_j - S_i) + C_j(S_i - S_j) = (C_i - C_j)(S_j - S_i) \tag{11.92}$$

For a binary system, there are $C_2 = 1 - C_1$, so the total degree of order is

$$\psi = R(2C - 1) \ln \frac{C}{1 - C} \tag{11.93}$$

We define the portion of the independent of gas constant as the value function (also called the separation potential function) is

$$V(C) = (2C - 1) \ln \frac{C}{1 - C} \tag{11.94}$$

Therefore, the value function is proportional to the degree of order of abundance distribution in isotope gas mixture. The image of the value function is shown in Fig. 11.25.

We can see that the value function is zero when the enrichment is 50%. When the enrichment is 0 or 100%, it is infinite. Its physical meaning is, for two-composition mixture, zero when uniform mixed, and infinite if completely separated. In this way, you can use the value function to calculate a separate work or separate power. The value of feed flow of F kg/s at enrichment of C_f is

$$\Phi_f = V(C_f)F \tag{11.95}$$

After separation, the value of waste and product is

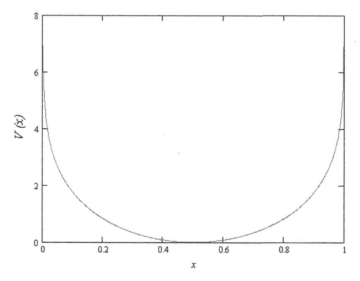

Fig. 11.25 The value Ffnction of binary system

$$\Phi_p + \Phi_w = V(C_p)P + V(C_w)(F - P) \tag{11.96}$$

We define the difference value between Eq. (11.95) and Eq. (11.96) as the separation work, which is

$$\Psi = V(C_p)P + V(C_w)(F - P) - V(C_f)F \tag{11.97}$$

As the value function is dimensionless, and therefore the unit of separation work is same as P and F, that is kg/s. If P and F are not mass flow rate, but the mass of the separation, then is it the separation work unit (SWU).

Example 11.11 Assumed feed is natural uranium, enrichment of waste is 0.2%, calculate the natural uranium required and separation work required.

Solution: the natural uranium required is

$$m = \frac{C_p - C_w}{C_f - C_w} = \frac{0.033 - 0.002}{0.00712 - 0.002} = 6.055 \, \text{kg}$$

Calculate the value function of feed, product and waste, is

$$V(C_f) = 5.009$$
$$V(C_w) = 6.238$$
$$V(C_p) = 3.616$$

According to Eq. (11.97), the separation work required is

$$V(C_p)P + V(C_w)(F - P) - V(C_f)F$$
$$= 3.616 \times 1 + 6.238 \times (6.055 - 1) - 5.009 \times 6.005$$
$$= 4.82 \, \text{kg SWU}$$

11.12.2   Diffusion Method of Isotope Separation

Here introduces the basics of diffusion. Particles such as atoms, molecules or ions in a gas, at any time in disarray for linear motion in all directions, in the way of the movement and collision will change the direction of movement of any objects, but does not change the size of the average velocity. All particles within the same space have the same kinetic energy, if the particles have a different mass, they have a different speed. Therefore, the speed of small mass particle is less than large mass particle.

Based on the above principles, the isotopic composition of two different mass molecules, filled in one side of a membrane partition container (high-pressure side), large speed particles (small mass) spread more easily through the membrane to the other side (low-pressure side). Thus in the low-pressure side, small mass isotope in concentrating. This is the gas diffusion method of uranium enrichment. The component with numerous tiny holes is called membrane. The separation process takes place in the process of gas separation membrane. Usually requires enough tiny holes in the membrane, and the radius of the hole as small as possible (to the extent of no working medium condensation), approximately 10–15 nm. Furthermore, the requirements of membrane are: can work under a mechanical strength by pressure difference, corrosion resistance from working medium, cheap as possible and can be mass-produced.

Here we come to the isotope gas molecules is $^{235}UF_6$ and $^{238}UF_6$, using uranium hexafluoride (UF_6) there are two main reasons: Firstly, UF_6 is gas under working temperature and pressure; Secondary, F has no isotopic.

This is the main method used in the uranium enrichment industrially. The principle is based on two kinds of thermal motion of gas mixtures with different molecular weight in balance, have the same average kinetic energy and different speed, that is,

$$\frac{1}{2}m_1V_1^2 = \frac{1}{2}m_2V_2^2 \tag{11.98}$$

$$V_1/V_2 = \sqrt{m_2/m_1} \tag{11.99}$$

where, V_1 and V_2 are the average speed of molecules $^{235}UF_6$ and $^{238}UF_6$ respectively; m_1, m_2 are the mass number of molecules respectively.

Apparently, from Eq. (11.99), lighter molecules has larger average speed, average speed of the heavier molecules smaller. Therefore, the number of collision of lighter molecules with the container walls, relatively a little bit higher than the heavier molecules. The diffusion membrane allows molecules to pass through the porous of tiny holes; therefore, $^{235}UF_6$ and $^{238}UF_6$ will spread through diffusion membrane at various speeds.

Figure 11.26 shows a single-stage gas diffusion process. When the UF_6 gas mixture flows through this stage, some gas from the high pressure side through diffusion membrane into the low pressure side, be small concentrated at the low

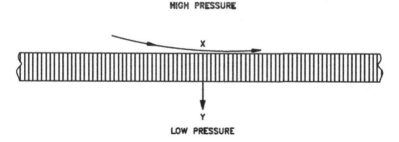

Fig. 11.26 Single stage of gas diffusion process

pressure side, and ^{238}U is concentrated on the high pressure side (^{235}U is depleted), thus realizing the separation of the two isotopes.

Separation coefficient of a single-stage (α) is usually defined as the ratio of relative concentrations of product and waste. Theoretical maximum separation coefficient (α_0) is equal to the square root of the ratio of molecular weight of two components. That is,

$$\alpha_0 = \sqrt{m_2/m_1} = \sqrt{352/349} = 1.0043 \qquad (11.100)$$

Actual separation coefficient α is much lower than the theoretical value; specific value depends on the structure and characteristics of membrane and other conditions. Description of membrane separation factor is defined as

$$\alpha = \frac{y}{1-y} \Big/ \frac{x}{1-x} \qquad (11.101)$$

where, y is the molar fraction of ^{235}U in the low-pressure side, referred to as concentration or concentration of light gases ^{235}UF$_6$; x is the concentration of ^{235}U in the high-pressure side.

A separator with its subsidiary equipment constitute a minimum separation unit of the gaseous diffusion plant, known as a diffusion-stage. Because there is little single-stage separation, in order to be 5% enriched uranium would require nearly hundreds of series diffusion-stage. If you want to produce highly enriched uranium would need thousands of series diffusion-stage. This stage is known as cascading, as shown in Fig. 11.27b.

With the gaseous diffusion method of uranium isotope separation, to process gases through the diffusion membrane, you must recompress it constantly; therefore gaseous diffusion plant consumes large amounts of electricity power. A production capacity of 17,000 tons SWU/a diffusion plant, requires the total electric power

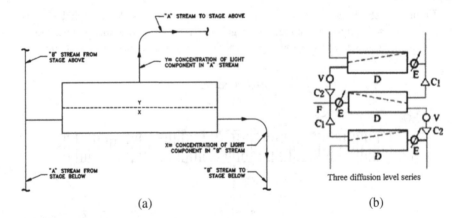

(a)   (b)

Fig. 11.27 A cascade of separation stage

of about 6 million kW. The work to compress gases finally becomes waste heat. Therefore, when you select a diffusion plant site, you must consider a number of cooling water consumption.

Despite the separation of uranium isotopes by gaseous diffusion has these problems, but because of its process is relatively simple, stable and reliable operation of equipment, the advantages of easy to implement projects, so widely used in this method of uranium enrichment for all countries.

11.12.3 High-Speed Centrifugation Method

In the centrifuge is rotating at high speed, with centrifugal force to achieve separation between light and heavy isotopes. In the centrifuge tube, the heavier molecules near the drum peripheral concentrations. The lighter molecules close to the shaft concentration, as shown in Fig. 11.28.

Drawn from the drum peripheral and central gas stream, you can get slightly depleted and slightly concentrated two-stream.

For the enrichment of uranium, high-speed centrifuge method more effective than gaseous diffusion, centrifuge separation factor is not depends on the square root of the mass ratio, but mass number difference. The mass number difference of $^{235}UF_6$ and

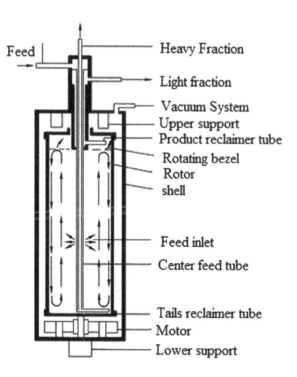

Fig. 11.28 High-speed centrifugation isotope separation

$^{238}UF_6$ elements is 3, if using a drum peripheral speed: 300 m/s centrifuges, separation factor up to 1.058. Thus, as needed to reach a certain degree of uranium enrichment being connected in series will be much less than the diffusion method. As when 5% enriched uranium from natural uranium production, the gas diffusion method takes about 900 diffusion stages, if drum peripheral speed 400 m/s centrifuges, as long as the 27 stages is enough. If you choose the high-speed centrifuges, desired series will be further reduced. However, because production capacity is too small for a single centrifuge, to reach a certain scale of industrial production at all stages must be parallel in many centrifuges.

The advantageous of high-speed centrifugation method are: (1) low power consumption, only 1/10–1/7 of gas diffusion method, (2) construct smaller plant in economically.

At present has built experimental plant of considerable size, centrifuge production capacity, theoretically with the drum peripheral speed of fourth power and is proportional to the length of the drum. The key technology of the development of high-speed centrifugal method is the materials of drum with high specific strength, in order to further improve the speed of the machine.

11.12.4   Laser Method

Based on isotope particle (Atom or molecule) small differences in the absorption spectra, application of excellent color laser selectively particle excitations of one isotope to a given excited state, then by physical or chemical means of stimulating isotope particle and other isotopes not inspire particles apart.

Laser method can be used to separate much kind of isotopes. The most important of which is uranium isotope separation, developed rapidly. Laser method of uranium isotope separation has two ways: atomic process and molecular process. Especially the atomic approach develops fast in recent years, has entered the phase of industrialization.

Atomic approach called atomic vapor laser isotope separation (AVLIS), whose principle is shown in Fig. 11.29, the whole device system including lasers and separator systems. With electrical gun to heat metal uranium, producing high temperature vapour of uranium atomic beam. Then use of copper vapor laser to irradiate uranium vapour atomic beam, selective excitation and ionization of ^{235}U atom, and use magnetic field to deflect ^{235}U ions, separate from ^{238}U atoms in atomic beam, in order to achieve separation.

11.12.5   Separation Nozzle

As shown in Fig. 11.30, (a) the process diagram, (b) schematic for nozzle.

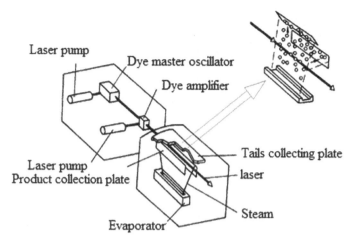

Fig. 11.29 Schematic atom laser method

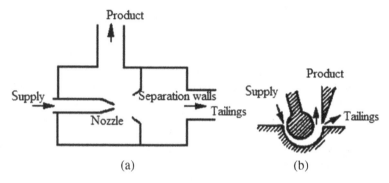

Fig. 11.30 A schematic view of nozzle separation

The principle of jet nozzle process: for feeding fluid, if the apex in the appropriate location, the heavier molecules near the surface enrichment, enrichment the lighter molecules away from the wall. Using isolation wedge tip of nozzle exit to separate $^{235}UF^6$ less air flow and containing more $^{235}UF_6$ product flow, in order to achieve separation of light and heavy isotopes.

Separation nozzles single-stage separation capacity between the gaseous diffusion method and the high-speed centrifugation method (separation coefficient is about 1.015). Because the power consumption is higher than diffusion, and today is still in the pilot stage in the industry.

In addition to the above, basic methods of enriched uranium and some separation of uranium method are still in research and development stage. Table 11.14 lists several methods for uranium enrichment process parameters.

Table 11.14 Method of uranium enrichment process

Method	Separate coefficient	Operation temperature/°C	Operating pressure/MPa	Added material g(U)/SWU·a	Power consumption kWh/kg (SWU)
Gaseous diffusion method	1.0043	70–100	0.071–0.101	100–200	2500
High-speed centrifugation	1.1	40–50	<0.101	0.15	100–400
Separation nozzle	1.015	40	0.025	100	6600
Laser method	大	–	–	少	大
Chemical exchange	1.0013	20	–	500–1000	600

11.13   Nuclear Fuel Cycle

Nuclear fuel cycle includes nuclear fuel production, use, storage or reprocessing, remanufacturing, often referred to as fuel cycle. Because of limited critical mass nuclear fuel cannot be burned out fully in reactor. In reactors using enriched uranium (such as pressurized water reactor), unloading fuel in addition to 0.8–0.9% enrichment of ^{235}U, still contains a certain amount of ^{239}Pu, and fission products. Therefore, spent fuel is not waste, some of the fissile nuclides can reuse. It is estimated that the reuse of plutonium can save natural uranium and uranium isotope separation requirements of 20% and 24% respectively. In the case of fast breeder reactor nuclear power plant, recycles more of the fissile materials for fuels and conversion is an essential part of.

11.13.1   Cyclic Manner

Available for chain reaction in a reactor of the fissile isotopes are ^{235}U, ^{239}Pu and ^{233}U. ^{238}U and ^{232}Th are fertile material, after absorbing neutrons respectively through the following reactions may become fissile nuclides ^{239}Pu, ^{233}U.

Therefore, the nuclear fuel cycle has two systems, one uranium–plutonium fuel cycle, and the other is thorium-uranium fuel cycle. The former starting from natural uranium, ^{235}U fuel, make the ^{238}U absorbs neutrons in a reactor and then converted to ^{239}Pu, ^{239}Pu again as new nuclear-fuel cycle; the latter by the thorium extraction of thorium ore, and placed into a reactor, absorbs a neutron and converted to ^{233}U, ^{233}U as new nuclear-fuel cycle.

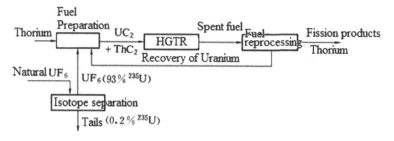

Fig. 11.31 Thorium-uranium cycle

Uranium–plutonium cycle is currently being implemented on an industrial scale, and thorium-uranium is still in the research and testing, to the industrial-scale production is still a big distance, but because of the rich reserves of thorium, ^{233}U has good performance as fissile nuclear fuel, and thorium-uranium cycle will also be developed. General analysis shows that the use of ^{235}U-^{232}Th-^{233}U in high-temperature gas-cooled reactor fuel system works well, but there are still some nuclear fuel processing and reprocessing of special problems. Mainly in the post-processing as well as separating ^{235}U and Th, and extract ^{233}U from Th, also separated fission products. In addition, the spent fuel is mixed with a small amount of a radioactive ^{232}U, to separate ^{232}U from ^{233}U is difficult. Subsequent treatment process is much more complex than uranium–plutonium cycle. Figure 11.31 shows a typical thorium-uranium fuel cycle. The following uranium—plutonium cycle, for example, to discuss specific cyclic manner and process of the fuel cycle.

Process in the nuclear fuel cycle include: uranium exploration, uranium ore mining, uranium extraction and purification, chemical conversion of uranium, uranium enrichment, fuel assemblies manufacture, reactor use, spent fuel storage, spent fuel transport, spent fuel reprocessing, radioactive waste processing and disposal of radioactive waste. Figure 11.32 shows a 1000 MW of pressurized water reactor fuel cycle (load factor 80%) in three ways.

Table 11.15 lists the fuel needed for some kind of thermal neutron reactor.

Once through is a way of the fuel cycle that is not to recycle uranium and plutonium from spent fuel, the utilization of natural uranium is only about 0.5%. Reuse reprocessed uranium, plutonium, can raise the utilization rate of natural uranium to 1–2%. Repeated in a fast breeder reactor using recycled uranium and plutonium, could improve the overall utilization rate of natural uranium to 60–70%. Take once through or recycle mode, depends on the technical feasibility and economic reasonableness. As far as the economy, reprocessing cost and price of natural uranium plays a decisive role. According to a study report [10] of the Working Group of International Nuclear Fuel Cycle Evaluation (INFCE) of the International Atomic Energy Agency (IAEA), when the price of natural uranium is higher than US $88/kg of U_3O_8, fuel economy of recycling to appear in some countries. Fast neutron reactor can save a lot of uranium resources, but the high infrastructure investments, generating costs cannot at present compete with thermal neutron reactors.

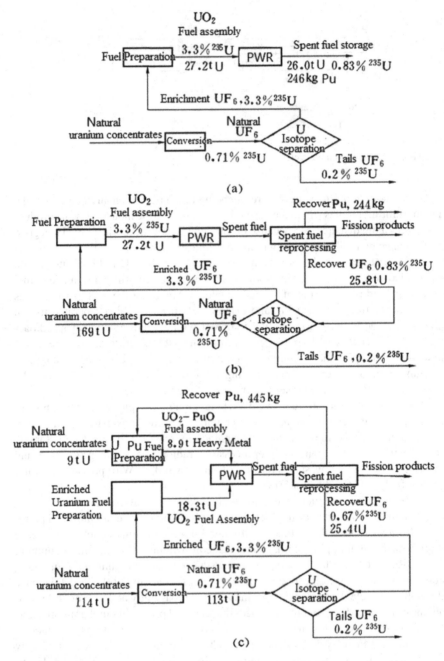

Fig. 11.32 Uranium-plutonium fuel cycle. **a** Once trough, **b** uranium recovery, **c** uranium and plutonium recovery

Table 11.15 Fuel demand for some kind of thermal neutron reactor (electrical power 1000 MW, burnup 33000 MW·d/tU)

Reactor type		PWR	BWR	CANDU
First load	Low enriched uranium/t	75	110	145
	Enrichment/%^{235}U	2.6	2.2	Natural uranium
	Equivalent natural uranium/t	385	466	145
	Separate work/(t SWU)	225	253	–
Refuel per year	Low enriched uranium/t	25	37	140
	Enrichment/%^{235}U	3.3	2.6	Natural uranium
	Equivalent natural uranium/t	165	190	140
	Separate work/(t SWU)	104	111	–

Since in the 1980 of the twentieth century, uranium prices fell and the cost of fuel cycle rising, thus some countries advocated the use of once-through way. Other countries felt once through not conducive to the full use of nuclear resources, recycle uranium and plutonium should be used for thermal neutron reactor. Some claims the temporary storage of spent fuel elements, after uranium price raised to a certain extent then reprocess the uranium, plutonium, and for use in reactors. Development of the nuclear fuel cycle policy must in accordance with the country's economic development, energy supply and demand situation analysis and long-term nuclear power development plan, in order to make the best decision.

11.13.2 Key Aspects of Nuclear Fuel Cycle

Key to the nuclear fuel cycle, including uranium exploration and production, of the ^{235}U isotope enrichment, spent fuel storage and transportation, reprocessing of nuclear fuel.

Natural uranium is the basic raw material for the nuclear industry. Although uranium is widely distributed but limited. Uranium exploration is a means of determining uranium deposits. Ore grade and ore reserves are the main evaluation of uranium deposit indicators. Processing and mining conditions, whether the utilization is synthesized and transport conditions are the references of industrial assessment.

Uranium mining is the first step in production of uranium. At the current level, uranium-containing more than 1 per thousand ore is worth to mine. For open-pit mines or mining, good processing conditions of deposit, grade slightly less is also exploitable. If ore mixed with more rock, the ore traffic and chemical reagent consumption will increase. So chemical treatments should be preceded by the ore dressing. Uranium ore is first processed containing a high concentration of integrated chemical concentrates of uranium U_3O_8, commonly known as yellow cake. Yellow cake containing 40–80% U_3O_8. Wet chemical processing known as hydrometallurgy, in general, it should be carried out in the vicinity of the mine, made after a chemical uranium concentrate, to make an outgoing transport. Chemistry of uranium concentrate still contains a lot of impurities, for further purification, to meet the requirements of nuclear purity. Masterwork of a variety of chemical forms for ease of storage and transportation, often using uranium oxide as the product.

In 1998, the nuclear energy agency of Organization for Economic Co-operation and Development (NEA/OECD) released *the world's uranium resources, production and demand* [11], showed that by the end of 1997, cost less than \$130/kgU reliable and recyclable world resources and class I estimated additional resources of 4.299 million tons of uranium, $\leq$\$80/kgU about 3.085 million tons of uranium, $\leq$\$40/kgU due to lack of data and confidentiality reasons not statistics. In 1997, the world demand for uranium is 60,000 tons. Due to a large number of stores of natural uranium in the 1980 of the twentieth century, annual output of supply only 60% of demand is sufficient. Therefore, although the 1997 capacity has reached 42,000 tons, but only output of 36,000 tons, the gap still filled by inventory. In addition, the military stocks of nuclear-grade uranium removed from nuclear warheads and weapons-grade plutonium into civilian use, so it can be expected that natural uranium supply and demand situation will continue for a long time [12].

The enrichment of isotope ^{235}U. Natural uranium is ^{238}U 99.28%, ^{235}U 0.71% and ^{234}U 0.006%. To raise the level of enrichment of ^{235}U, uranium isotope separation technology is needed to meet the need for nuclear power. Methods for industrial-scale production of enriched uranium are diffusion and centrifugation. Gas diffusion using different isotopes of uranium in uranium hexafluoride by the mass difference of the gas molecules are separated. Diffusion has a history of about 50 years, mature technology and reliable, the disadvantage is that electricity power is too large, to United States of diffusion plant, for example, electricity accounts for about 70% of the separate work unit costs, but diffusion is still the main method of producing enriched uranium. Gas centrifuge method is to use centrifugal force to separate isotopes of uranium hexafluoride gas. Centrifugation is the most recent more than 30 years of development in the 1990 of the twentieth century have been used for industrial production. Main advantages are low power consumption, only about 5% of diffusion, but heavy investment. Laser method of uranium isotope separation is very promising, but still in the testing phase. At present, the total uranium isotope separation in the world of industrial production capacity can meet the needs of future development of nuclear power.

Spent fuel storage and transportation. No matter what kind of the nuclear fuel cycle, spent fuel storage is necessary. High specific activity of spent fuel, and release a large amount of decay heat, must be stored for a period of time to be radioactive and cool down to a certain degree of operation and handling. Spent fuel discharged from reactor core stored in a pool next to reactor a few months to a few years first and then transported to long storage site. It requires spent fuel transport. Because of the highly radioactive, so the transport of spent fuel are not only technically complex, costly, and must be carried out under strict control in order to ensure the safety of transport.

Nuclear fuel reprocessing. Light water reactor nuclear power plant for unloading spent nuclear fuel contains about 0.8% of ^{235}U and 1% industrial plutonium. Industrial plutonium can be used as thermal neutrons or fast reactor fuel [13].

Nuclear fuel reprocessing has about 50 years of history. Solvent extraction processes have been used in industrial production, not only can handle natural metal uranium spent fuel and low enriched uranium oxide fuel, improved potentially fast reactor spent fuel processing. Is now the world's major nuclear countries such as France and the United Kingdom, and Russia, and Japan, have had some nuclear fuel reprocessing capabilities for future recycling of nuclear resources provide the preliminary conditions.

11.13.3 Nuclear Fuel Cycle Cost

Nuclear power plants, the cost relating to the nuclear fuel cycle of 1 kWh electricity production is one of the important components of the nuclear power plant generating cost. In the cost analysis, there are two types of cost calculation methods.

Method A: calculation of nuclear power plant for the production of electricity 1 kWh takes in all aspects of the nuclear fuel cycle costs. It includes both the running cost of fuel consumed in the process, and in order to maintain the reactor critical in long term and the nuclear fuel load in reactor as cost. The former is called variable fuel costs, the latter known as fixed fuel costs. Nuclear fuel cycle costs contains variable fuel costs and fixed fuel costs.

Method B: calculating the nuclear fuel cycle cost for the production of 1 kWh electricity. That is, only the variable fuel costs is considered while the fixed fuel costs is not. The fixed fuel costs is contained into fixed capital investment, depreciation (or reimbursement) the fees into the cost of electricity generation.

Variable fuel cost calculation of equilibrium fuel cycle. In the process of running, often with several fuel cycles, refueling cycle balanced. The cost of all the fuel consumed by, generally a balanced cycle of supplementary fuel cycle cost on behalf of nuclear fuel.

Balance cycle formula of variable fuel costs is

$$C_c = \frac{\tilde{P} \cdot \tilde{M}}{T \cdot E} \qquad (11.102)$$

where, $\tilde{P}$ is nuclear fuel prices and $\tilde{M}$ is replenish nuclear fuel in a balance cycle; T is a balancing cycle period (years); E average annual capacity (kWh) of balance cycle.

Fixed fuel cost calculation, generally use the first cycle fee minus the balance cycle. First cycle fee is loaded for the first time for nuclear fuel enriched ^{235}U fuel costs. Equation for calculation is

11 Reactor Theory

$$Q = \sum_{i=1}^{n} P_i \cdot M_i \qquad (11.103)$$

where, Q is the fuel costs of the first load; P_i, M_i, respectively, for the ith fuel prices and the amount of ^{235}U enrichment of nuclear fuel. The first cycle fuel costs included in that portion of the capital investment costs as

$$I = Q - \tilde{P} \cdot \tilde{M} \qquad (11.104)$$

The fixed cost of the Method A is

$$C_f = \frac{\varepsilon \cdot I}{E} \qquad (11.105)$$

where, ε is an annual depreciation rate (or year recovery rate of fund); E power supply of an average year.

Fuel price P is the nuclear fuel costs per unit weight in nuclear fuel cycle, it covers the direct and indirect costs of all aspects of the nuclear fuel cycle (natural U_3O_8 purchase fee, conversion, enrichment, fuel assembly manufacturing, spent fuel reprocessing, waste disposal, sale of recycle uranium and plutonium, transport, storage and other associated with the above).

Direct costs include:

1. Purchase fee of natural U_3O_8: in calculation of natural U_3O_8 acquisition fee, taking into account the loss of U_3O_8 into UF_6, components manufacturing losses and the uranium enrichment process in the relationship between the amount of input raw materials and production. If the weight of the uranium is U loading nuclear fuel into the reactor, the natural weight of U_3O_8 uranium U need to purchase is

$$U = u \times \frac{1}{1 - f_1} \times \frac{1}{1 - f_2} \times \frac{x_P - x_W}{x_F - x_W} \qquad (11.106)$$

where, x_P, x_F, x_W are the enrichment uranium products, raw materials, tailings of separation process of ^{235}U respectively. f_1, f_2 are the attrition rate of uranium for conversion process and manufacture process respectively.

If calculated in terms of per unit mass of uranium, the price of natural U_3O_8 is c_1, and then the purchase fee of natural U_3O_8 is $c_1 U$.

2. The conversion cost of U_3O_8 into UF_6: U_3O_8 should be all converted into UF_6. Conversion fee per unit weight of uranium into the plant is C_2, the total conversion fees is $C_2 U$.

3. Uranium enrichment fee: SWU needed to produce low enriched uranium per unit weight of separation is

$$S = V(x_P) + \frac{x_P - x_F}{x_F - x_W} V(x_W) - \frac{x_P - x_W}{x_F - x_W} V(x_F) \qquad (11.107)$$

where V is the value function defined by Eq. (11.94). In engineering, generally do not distinguish between enrichment and abundance, since the two are very close.

Provided by the uranium enrichment plants to assembly manufacturing, low enriched uranium of weight $U/(1 - f_2)$, price of SWU is C_3, the uranium enrichment fee is

$$c_3 u \times \frac{S}{1 - f_2} \tag{11.108}$$

4. Assembly manufacturing costs, fuel reprocessing costs, transportation costs, storage and waste disposal costs, according to the amount of processing, handling, transportation, storage and the corresponding price multiplied by calculation.
5. The sale of the recycled uranium and plutonium. Reprocessed uranium and plutonium, only when the user purchases, can be converted into income, this income only as a bookkeeping income.

The revenue of selling recycled plutonium per unit weight of uranium, nuclear fuel and irradiated fissile plutonium contained in the spent fuel (after deduction of the net loss of ^{239}Pu and ^{241}Pu) multiplied by the price of plutonium.

The calculation of revenue of selling recycled plutonium is more complicated. In the assumed recovery of uranium enrichment of ^{235}U for x_E, is generally higher than that of ^{235}U in natural uranium. Sells recycled uranium income should be the same as production of natural uranium for uranium enrichment plants equal to the cost of the same enrichment of uranium. Produced by the uranium enrichment plants concentration per unit weight for x_E low enriched, the uranium natural uranium needed is

$$\frac{1}{1 - f_1} \times \frac{x_E - x_W}{x_F - x_W} \tag{11.109}$$

Production unit weight low enriched uranium required separation work unit is

$$S' = V(x_E) + \frac{x_E - x_F}{x_F - x_W} V(x_W) - \frac{x_E - x_W}{x_F - x_W} V(x_F) \tag{11.110}$$

Sale unit weight of recovered uranium, the income is

$$(c_1 + c_2) \frac{1}{1 - f_1} \frac{x_E - x_W}{x_F - x_W} + c_3 S' \tag{11.111}$$

Indirect costs include spending on the part of the nuclear fuel cycle, interest rates, financial compensation, tax and insurance costs. Indirect costs calculation is more complex, fee of each stage of nuclear fuel cycle, several years before the operation of nuclear power plant, some after a number of years, involving long term, change, and associated with the payment. According to the actual situation in advance, draw up cash flow, and then at a given discount rate, conversion to electricity generation and for comprehensive analysis.

Exercises

1. Calculating fission cross section of ^{235}U at the temperature of 300 °C.
2. At a temperature of 295 K, volume in the 1 cm^3 cube, 10^{18} ^{235}U nuclei uniform random distribution, thermal neutron flux 10^{13} n/(cm^{2}s), calculate the rate of fission.
3. Volume within the cube space of 1 cm^3, 10^5 ^{235}U nuclei with uniform random distribution, comes in one neutron in any direction with energy of 0.0253 eV; calculate the probability of nuclear fission reaction.
4. Why want more neutrons are slowed down, the sooner the better in a thermal reactor?
5. After elastic scattering the neutron kinetic energy E', prove the ratio of E' to E (the kinetic energy before scattering) is

$$\frac{E'}{E} = \frac{1}{2}[(1 + \alpha) + (1 - \alpha) \cos \theta_c]$$

6. Calculate the average generation time of delayed neutrons in the thermal neutron fission of ^{235}U.
7. If the fuel temperature coefficient is -0.2 pcm/°C, the average fuel temperature decreases 50 °C, how much reactivity introduced?
8. If reactivity is 0.001, calculation of k_{eff}.
9. What are the main ways of ^{135}Xe comes from in thermal reactor?
10. Why increase Sm concentration after shutdown?
11. Reactivity of a reactor for -1000 pcm, the count rate is 45 (1/s), when the count rate was increased to 500 (1/s), what is the reactivity?
12. What is the main difference between the two kinds of nuclear fuel cycle costing methods?
13. A refueling of a nuclear power plant needs 27.2 tons of 3.3% enrichment ^{235}U nuclear fuel, assuming that the enrichment of waste is 0.2%, feeding enrichment of 0.71%, calculated how much separate work unit is needed and how much natural uranium is demanded.
14. Enrichment ^{235}U is about 3.5%, calculate the abundance of ^{235}U.
15. Discuss the advantages and disadvantages of recycle of uranium and plutonium.

References

1. Marion JB, Fowler JL. Fast neutron physics. New York: Interscience; 1963.

2. Xie Z, Wu H, Zhang S. Physical analysis of nuclear reactors. Xi'n: Xi'an Jiaotong University Press; 2004.
3. Jacobs AM, Kline DE, Remick FJ. Basic principles of nuclear science and reactors. Van Nostrand Company Inc.; 1960.
4. Lamarsh JR. Introduction to nuclear reactor theory. Addison-Wesley Company; 1972.
5. Lamarsh JR. Introduction to nuclear engineering. Addison-Wesley Company; 1977.
6. World nuclear industry handbook. London: The Special Nuclear Engineering International Publications; 1997.
7. Glasstone S, Sesonske A. Nuclear reactor engineering. 3rd ed. Van Nostrand Reinhold Company; 1981.
8. de Groot SR, Mazur P. Non-equilibrium thermodynamics. Amsterdam: North Holland Publishing Company; 1962.
9. Liu G. The physical meaning of the value function. J. Tsinghua Univ. (Nat. Sci. Ed.). 1994;V34(6).
10. IAEA. International nuclear fuel cycle evaluation. Vienna:IAEA; 1980.
11. IAEA. Nuclear power and fuel cycle: status and trends. Vienna: IAEA; 1988.
12. IAEA. Reprocessing plutonium handling, recycle report of INFCE Working Group 4. Vienna: IAEA; 1980.
13. B. П. Tim lianushin Nuclear power plant fuel reprocessing. Huang Changtai, translation Beijing: Atomic Energy Press; 1996

Chapter 12
Radiation Protection

Radiation is the energy of electromagnetic waves or particles in the form of outward diffusion phenomena. All the objects in the natural world, as long as the temperature is above absolute zero, always radiate heat in the form of electromagnetic waves; this way of transmitting energy is called radiation. Thermal radiation is one of the basic mode of heat transmission, we already introduced in the Chapter 3. In addition to thermal radiation, there are other kind of radiation, depending on their energy level and its ability to ionize material classified as ionizing radiation and non-ionizing radiation [1].

Generally the term used in ionizing radiation. Ionizing radiation has enough energy to the atoms or molecules ionized. Radiation active substance is a substance that emits ionizing radiation. Ionizing radiation are α, β, γ, neutron and X-ray radiation. In this chapter, we discuss is the issue of protection against ionizing radiation.

Radiation protection is an applied discipline that studies the prevention of harmful effects of ionizing radiation. In the United States, Japan and France, also known as health physics, in Russia, Poland, and Hungary and other countries is known as radiation health. Radiation protection relating to the prevention of harmful effects of ionizing radiation effect on all issues, but does not include all aspects of radiation safety, for example, the nuclear reactor safety belongs to the nuclear safety and the radiation effects belongs to the radiation medicine [2].

Radiation protection as an applied discipline, its based subjects: radiation dose, radiation biology, radioecology, radiation shielding and radiation detection, etc.

Radiation protection includes the following five aspects: basic principles of radiation protection and radiation protection standards; the radiation protection methods; radiation monitoring; radiation protection evaluation; emergency of radiation accident.

© Tsinghua University Press 2022
J. Yu, *Fundamental Principles of Nuclear Engineering*,
https://doi.org/10.1007/978-981-16-0839-1_12

12.1   Radiation Quantities and Units

Describe radiation sources or radiation field characteristics and interaction of radiation with matter, referred to as radiation quantities. Some concept, name and unit of radiation quantity experiences a lot of evolution, meaning different of the same names at different times. Early application of radiation quantities and units, because of its ambiguous concept or definition is not precise, imprecise or not easy, being eliminated gradually with applications, such as the biological equivalent of Roentgen, physical equivalent of Roentgen, equivalent grams of radium etc. In1925, international Radiology Conference decided to set up a specialized organization of radiation quantities and units, the International Commission on Radiation Units and later with word of "measurements", changed its name to the International Commission on Radiation Units and Measurements (ICRU)[3]. After years of research, ICRU published a series of reports to promote the harmonization of radiation quantities and units and scientific, made a significant contribution. Current radiation quantities and units used in radiation protection, most of them are defined and recommended by ICRU.

Radiation quantities are mainly classified as the sources of radiation and radiation exposure, radiation dosimetry, description of radiation interactions with matter and radiation protection.

Radiation unit should use SI units, if non-SI units are given, must indicate the proper names and units [4].

12.1.1   Describe the Amount of Radiation Source and Radiation Field

Substance or device capable of emitting ionizing radiation is usually referred to as radiation source, the space is called ionizing radiation field. Describes the amount of radiated field and radiation sources have more than 10 quantities, the most common activity, particle fluence, particle flux, energy fluence and energy fluxes, and so on.

Activity

At a given moment, in a particular energy state of an amount of a radionuclide activity is defined as the quotient of dN divided by dt:

$$A = \frac{dN}{dt} \tag{12.1}$$

where, dN is the number of nuclei of the spontaneous transitions expectations of this species occurs from the excited states within a time interval dt. Activity unit is s^{-1}, its proper name is Becquerel, or Bq. 1 Bq equals 1 s^{-1}.

Particle Fluence

$$\phi = \frac{dN}{da} \tag{12.2}$$

where, dN is injection number of particles in area of da. The unit of ϕ is per square meter (m^{-2}).

Particle Flux

$$\varphi = \frac{d\phi}{dt} \tag{12.3}$$

where, φ is the number of particles per unit time through unit area; the unit of φ is per square meter per second $(m^{-2} s^{-1})$.

Energy Fluence

$$\psi = \frac{dR}{da} \tag{12.4}$$

where, dR is the incident radiant energy on area of da. ψ of joules per square meter (J/m^2).

Energy Flux

$$\dot{\psi} = \frac{d\psi}{dt} \tag{12.5}$$

where, $d\psi$ is energy fluence increment within the interval of time dt; energy flux in watts per square meter (W/m^2).

12.1.2 Usual Quantities of Dosimetry

Radiation dosimetry focused radiation interacts with receptors that occurs when the energy deposition, energy-absorbing characteristics of energy transfer and receptor. In dosimetry, commonly used radiation absorbed dose, dose rate, kerma, and kerma rate, exposure, exposure rate, etc.

Absorbed Dose
Radiation energy absorbed per unit mass of receptor, which is defined as

$$D = \frac{dE}{dm} \tag{12.6}$$

where, dE is the average energy absorbed by the mass of dm (receptor substance); absorbed dose is joules per kilogram (J/kg); special name for the unit is Gy (1 Gy = 1 J/kg); this quantity had used the name Ladd (rad), 1 Gy = 100 rad.

Absorbed Dose Rate

$$\dot{D} = \frac{dD}{dt} \tag{12.7}$$

where, dD is the absorbed dose increment within a certain time interval dt. Absorbed dose rate in joules per kilogram per second (J kg^{-1} s^{-1}), a dedicated unit name Gy per second (Gy/s).

Kerma
When uncharged ionizing particles interaction with matter, transfer its energy to the charged ionized particles firstly, and then re-charged ionized particles by ionization or excitation process to grant the material energy. Kerma is to describe the process of this energy radiation delivery, defined as

$$K = \frac{dE_{tr}}{dm} \tag{12.8}$$

where, dE_{tr} is the initial kinetic energy of all charged ionizing particles released from the inside of dm by uncharged ionizing particles in the total mass of a substance. Kerma in joules per kilogram (J/kg), a dedicated unit name is Gy (Gy).

Kerma Rate

$$\dot{K} = \frac{dK}{dt} \tag{12.9}$$

where, dK is the Kerma increments in the time interval dt. Kerma rate in joules per kilogram per second (J kg^{-1} s^{-1}).

Exposure
To measure air ionization ability of X or γ radiation, defined as the quotient of dQ divided dm

$$x = \frac{dQ}{dm} \tag{12.10}$$

where, dQ is the absolute value of any of the symbols of the total charge (including negative electrons and positrons) in the air with mass of dm when the photon was completely prevented by air. Exposure x has unit of coulomb per kilogram (C/kg), once used unit is Roentgen (R), but is no longer legal units. 1R $= 2.58 \times 10^{-4}$C/kg.

Roentgen was first proposed by a German named Bahnken. In 1928, the International Congress of Radiology defined Roentgen as X-rays or γ "amount" or "dose" of the unit, but the meaning is unclear. In 1956, the ICRU took Roentgen as a "dose" of the unit, but still easy to be confused with the absorbed dose. In 1962, ICRU changed the dose renamed exposure, still with Roentgen as a unit.

Exposure Rate

$$\dot{x} = \frac{\mathrm{d}x}{\mathrm{d}t} \tag{12.11}$$

where, dx is the increment exposure in the time interval dt. Exposure rate unit is coulomb per kilogram per second (C kg^{-1} s^{-1}). With a once used special unit, Roentgen per second (R/s).

12.1.3 Commonly Used Quantities in Radiation Protection

On radiation protection, the quantities used is closely related to radiation hazards, mainly used for radiation protection, radiation protection evaluation, radiation monitoring and radiation protection optimization [5]. Equivalent dose, effective dose, collective dose equivalent, ambient dose equivalent, directional dose equivalent, and personal dose equivalent, and so on.

Radiation Weighting Factors and Dose Equivalent
Probability of stochastic effects of radiation depends on not only the absorbed dose, but also the radiation type and energy absorbed. To take into account this fact, a weighting factor related to absorbed dose and radiation quality, w_R is used. Radiation protection is interested in a tissue or organ absorbed dose average (doses instead of one point) and by a radiation-weighting factor weighted [6], weighted absorbed dose is the dose in the strict sense, known as the equivalent dose in a tissue or organ.

Equivalent dose in tissue T can be expressed as

$$H_T = \sum_R w_R D_{T,R} \tag{12.12}$$

where, $D_{T,R}$ is the averaged absorbed dose from radiation of the tissue or organ T. Equivalent dose in joules per kilogram (J/kg), dedicated to the name Sv.

Tissue Weighting Factors and Effective Dose

The probability of random effects is associated with equivalent doses and the irradiated tissue or organ. Define a quantity derived from the equivalent dose to indicate
when different organizations are at different doses of irradiation a sense of integrated; making it roughly corresponds with the general random effects. On the tissue
or organ T weighted factor of equivalent dose, known as tissue weighting factors w_T,
it reflects overall body evenly under the tissue or organ's relative contribution to total
damage. Effective dose E for all body tissues and organs of the sum of the weighted
equivalent doses, the E given by the following formula

$$E = \sum_T w_T H_T \tag{12.13}$$

where, H_T is the equivalent dose in tissue or organ T; w_T is the organ or tissue
weighting factor. The unit of effective dose of joules per kilogram (J/kg), dedicated
to a name of Sv.

Accumulated Dose Equivalent

External penetrating radiation equivalent dose in tissue exposed to the radiation of
energy deposition is given at the exposure time. However, exposure to radionuclides
entering the body tissue are spread out in time, energy deposition is given gradually
with the decay of radionuclides. Energy deposition of the distribution of physical
and chemical form of the radionuclide and the biological dynamics changes with
the time. Accumulated dose equivalent is dose equivalent after a single intake of
radioactive material, a particular organization to accept integration of the equivalent
dose rate at time τ. When τ is unknown, 50 years for adults, 70 years for children.
Accumulated dose equivalent is

$$H_T(\tau) = \int_{t_0}^{t_0+\tau} \dot{H}_T(t) dt \tag{12.14}$$

For a single intake radiator at time t_0, $\dot{H}_T(t)$ corresponding to the organ or tissue
T at time t equivalent dose rate, τ is integrating period of time, in years.

Accumulated Effective Dose

If the equivalent dose produced by a single intake of radioactive material multiplied
by the corresponding weighting factor w_T, then summed, is the effective dose, which
was expressed as

$$E(\tau) = \sum_T w_T H_T(\tau) \tag{12.15}$$

In determining $E(\tau)$, the integrating time τ in years.

Collective Equivalent Dose

It represents a group designated a tissue or organ suffered total amount of radiation exposure. Collective equivalent dose of tissue T is defined as

$$S_T = \int_0^\infty H_T \frac{dN}{dH_T} dH_T \tag{12.16}$$

where, $(dN/dH_T) \cdot dH_T$ is the people number of accept equivalent dose between H_T and $H_T + dH_T$. Also, be expressed as

$$S_T = \sum_i H_{T,i} N_i \tag{12.17}$$

where N_i is the ith group number with average organ dose equivalent $H_{T,i}$. The unit of S_T is man Sv.

Collective Effective Dose

It represents the total irradiation suffered a population, which is defined as

$$S = \int_0^\infty E \frac{dN}{dE} dE \tag{12.18}$$

or

$$S = \sum_i \overline{E}_i N_i \tag{12.19}$$

where, $\overline{E}_i$ is the ith group, the average effective dose of population accepted; N_i is the ith group number. The definition of collective effective dose does not specify the dosage given elapsed time. Collective effective dose should therefore be specified sum or integration interval and what kind of crowd.

Equivalent Dose Per Capita

In certain irradiation groups according to the average number of equivalent dose, i.e.

$$\overline{H} = S/N \tag{12.20}$$

where, S is the collective equivalent dose with man Sv; N is the number of the group. Literally, appears to be related to the equivalent dose per individual dose, in fact it is characterized by the collective amount according to the situation, because only in occasional cases, it was a truly equal doses to individuals. It is a mean equivalent dose to a group of individuals.

Dose Commitment

Because of a decision or a practice, so that specific groups are sustained per capita dose rate irradiation caused for an unlimited time

$$H_{CT} = \int_0^\infty \dot{H}_T(t)dt \tag{12.21}$$

or

$$E_C = \int_0^\infty \dot{E}(t)dt \tag{12.22}$$

Dose commitment is a computational tool that can estimate the world population, but also on certain key groups estimate.

Ambient Dose Equivalent

A quantity used for external radiation exposure monitoring in strongly penetrating radiation for practical. Radiation field in a point of the ambient dose equivalent $H*(d)$ is the corresponding dose equivalent produced in the field of ICRU sphere at the reverse direction of the radius of the field. This is a quantity with the characteristic of the radiation field. ICRU recommendations for strongly penetrating radiation, d is 10 mm; while weakly penetrating radiation, d is 0.07 mm. ICRU sphere is called a density of 1 g cm^{-3} tissue equivalent material made of 30 cm diameter ball. Material composition (percentage by mass) of 76.2% oxygen, 11.1% carbon, 10.1% hydrogen, 2.61% nitrogen. $H*(10)$ generally reflects the effective dose suffered there. For X and γ radiation, measured $H*(10)$ can only be used to arrange, guide and control of staff and personal, dose measurement is still subject to personal dosimeters. In principle, a response of an isotropic detector, if scaled by $H *(10)$, in the radiation field can be used to measure any uniform ambient dose equivalent.

Directional Dose Equivalent

At a point of radiation field, the directional dose equivalent $H'(d, \Omega)$ is the corresponding extended field specified direction Ω in the ICRU sphere with radius d, dose equivalent produced. ICRU provisions, $d = 0.07$ mm, the direction Ω could be the angle α with the specified radius direction and the reverse incident. For forward irradiation (AP geometry), α = zero, the directional dose equivalent $H'(0.07, 0)$ can be written as $H'(0.07)$. It reflects the specific direction perpendicular to the skin. If a detector capable of determining a direction perpendicular to the specified, by the tissue-equivalent material plate dose equivalent depth at 0.07 mm, the detector can measure directional dose equivalent $H'(0.07)$. $H'(0.07, \alpha)$ is directional, directional dose equivalent at the same point in different directions may be different, so the measurements to indicate the reference point position and the reference direction.

Personal Dose Equivalent

It is a practical quantity used to measure the personal external exposure. Personal dose equivalent $H_P(d)$ refers to the specified point below the soft tissue depth d

at a dose equivalent on the body. For strongly penetrating radiation, d is 10 mm, while weakly penetrating radiation d is 0.07 mm, respectively. It usually reflects the organs and skin dose. $H_P(d)$ can be measured by a hanging personal dosimeter. The dosimeter should cover the appropriate thickness tissue equivalent material. Personal dosimeters must adopt ICRU recommended human body model scale.

12.2 Basic Principles and Standards of Radiation Protection

12.2.1 The Basic Principles of Radiation Protection

This is to protect workers and the public from or less susceptible to radiation damage, the basic principles must be followed. Radiation protection aimed at both people and the environment with the proper safeguards, but also can promote the application and development of nuclear energy and nuclear technology. In order to do that, you must first determine the basic principles of radiation protection, through legislation, to translate these principles into laws and regulations in order to guide people's practical activities. Basic principles of radiation protection consists of three basic components [7].

(1) The legitimacy of the practice: in the execution before any practice associated with radiation exposure, will be subject to justification, confirm that this practice has a valid reason, that is, to obtain positive net interest outweighed the costs.

(2) The optimization of radiation protection: should avoid all unnecessary exposure conditions taking into account economic and social factors, all radiation exposure should be kept to a reasonable level as low as possible.

(3) Personal dosage limits: dose limits for individuals to limit the exposure.

Among the above three principles of radiation protection, the individual dose limits provides an unacceptable dose limit, when the justification and optimization of radiation protection practices results are inconsistent with the principle of individual dose limits, are subject to the principle of individual dose limits.

These basic principles of radiation protection is for controlled radiation source exposure case, principle to say that it does not apply to non-controlled source of radiation protection (such as the case of a nuclear accident), because in these cases, no method for radiation source control to limit or reduce radiation dose suffered by the people. In accident cases, can only follow the basic principle of emergency intervention to control or reduce the radiation exposure.

The International Commission on Radiological Protection (ICRP) 60th publication published in 1991 recommended a set of dose limitation system, and the International Atomic Energy Agency (IAEA) have adopted this system and reflected in it with the International Labour Organization (ILO), the nuclear energy agency of the

Organization for Economic Co-operation and Development (NEA/OECD) and the World Health Organization (WHO) jointly developed in the basic safety standards for radiation protection. This dose-limiting system has also been used in China's radiation protection regulations.

The Legitimacy of Practice

The legitimacy of practice requires the introduction of practice should

$$\text{Interest} > \text{cost} + \text{risk} \tag{12.23}$$

Interest refers to the interest of the society as a whole, which includes economic and social benefits, reduced radiation hazards and so on. Price refers to the sum of all the negative aspects, including economic costs, health risks, and adverse environmental effects, psychological and social problems. Danger is the risk of future damage, is a potential cost.

Despite the justification of the practice is mainly determined by decisions made by authorities, engaged in the practice of management and radiation protection personnel shall provide the necessary information for decision making, enable decision makers to make the right and appropriate decision.

Individual Dose Limits

In order to avoid deterministic effects of radiation and the incidence of stochastic effects to an acceptable level, individual doses must be limited. Basic amount of dose limits is effective dose. Dose limits cannot be directly used in the design and organization of work, the purpose of dose limits do not include medical exposures and natural background radiation contribution.

China's regulations for radiation protection, personal dose limit is divided into the basic limit, export limits, management limits and reference level.

Basic limit values are expressed in effective doses, sometimes referred to as dose limits, for ease of use, according to limit, secondary limit and export limits are used such as internal secondary limit of annual intake limit (ALI). The year intake limits for radiation workers in Appendix E of China national standard GB8703-88, the radiation protection regulations. For members of the public, generally the intake limit is 1/50 of radiation workers. If, during the year external exposure combined with internal radiation, should meet the following formula:

$$\frac{H_E}{50\text{mSv}} + \sum_i \frac{I_i}{(\text{ALI})_i} \leq 1 \tag{12.24}$$

where, H_E is radiation effective dose (mSv), I_i is the radionuclide i intake in a year (Bq a^{-1}), (ALI)$_i$ is the limit of annual intake for radionuclide i (Bq a^{-1}). In order to assess the situation of air pollution in the workplace, set up export limits: derived air concentration (DAC), the values are listed in Appendix E of the China national standard GB8703-88.

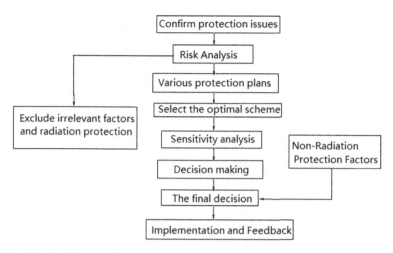

Fig. 12.1 The decision step of radiation protection optimization

The Optimization of Radiation Protection

Optimization of radiation protection is an important principle in radiation protection; you must carry out to associate with radiation exposure in practice. By selecting the best protection level and optimal protection scheme to achieve the goal of maximum benefits with minimum cost. General steps of radiation protection decisions as shown in Fig. 12.1.

First, clear the protection problems, identify protection goals. Then conduct risk analysis, identify sources of danger to assess the size of the various sources of danger, and clear focus of protection, provide the basis for future selection and determination of protection program. Through the risk analysis can also identify factors associated with radiation protection, exclude factors unrelated to radiation protection. Identify sources of risk, analysis of possible sources for each means of implementing protection, which lists all of the available protection schemes. At this stage, but apparently does not reality program, do not be too quick to rule out any option. In addition, by comparing various options choose the best program to determine the best level of protection. For decision on the basis of the quality of the data and assumptions must be evaluated carefully, analyze the data and assumptions of uncertainty and variability impact on results, sensitivity analysis. Through this analysis, learn what factors on the result of the most and when changes in how these factors should change decisions, and clearly the focus of the work. Now that you have available radiation protection optimization decisions. However, some non-radiation protection considerations, such as economic, political or social factors can also affect the outcome of the decision. Therefore, you must consider all these factors in order to make more realistic the final decision. Finally, putting this decision into practice. Gather feedback during the implementation process. It is also very important, because it not only to examine all aspects of decision-making, and be able to identify more effective ways of protection.

Among the various protection schemes and optimized method, qualitative and quantitative of the two categories. Qualitative methods rely on experienced judgment, may also be supported by semi quantitative analysis. In ICRP publication no 37 calls this method for multi-criteria methods. It is for each criterion, on a variety of alternative protection schemes are compared with each other. Because multiple criteria to be considered in the comparison process, so often set relative weights for the standards. By comparison, discard the worst program; keep better programs, until the election of the best program. Generally, the qualitative methods can be preselected some (rather than one) better programs, not fully give the pros and cons of order. In the problems of radiation protection, not all of criterion can be expressed in quantitative, particularly in operational radiation protection problems, quantitative level is usually lower. For example, complete maintenance tasks in high radiation fields, the prior need to make decisions on the following issues: in order to reduce the amount of radiation received by maintenance staff, should be sending people with knowledge and skills to accomplish this task? How many people to send? What tools they should carry? What protective clothing they should wear? What instruments? Etc. While many of these problems is difficult to quantify, you can only use qualitative methods.

Quantitative methods is not compare each other, but each programme is based on the various criteria quantitatively into a single value, then the corresponding values for each programme arranged in order, so as to select the best solution. Call this method as aggregative methods in ICRP publication no 37. Cost–benefit analysis and cost-efficacy of quantitative method of analysis is widely used. Cost–benefit analysis approach is to select a protection level for maximum net benefits to achieve optimization. The net benefits of a certain practice to society associated with radiation exposure is

$$B = V - (P + X + Y) \tag{12.25}$$

where, V is the gross benefit; P is the cost of all production in addition to the cost of radiation protection; X is the cost of protection to achieve the level of protection required to choose to spend; Y is the price of level of protection of the corresponding radiation hazards. In order to obtain the maximum net benefit must have

$$\left.\frac{dB}{dS}\right|_{S=S_0} = 0 \tag{12.26}$$

where, S is the collective dose equivalent; S_0 is the best level of protection corresponding collective dose equivalent. Generally considered V and P is independent of S, so

$$\left.\frac{d(X + Y)}{dS}\right|_{S=S_0} = 0 \tag{12.27}$$

Fig. 12.2 Cost–Benefit
Analysis Method

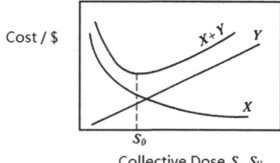

Thus, you can select the best level of protection S_0. Figure 12.2 is a schematic diagram of the cost–benefit analysis method, which shows the relationship between X, Y and S and the selection method of S_0.

Obviously, in order to complete the above calculations must be measured on the same scale X and Y. Now, they are measured with a currency cost. Use currency to measure X seems to be very direct, but the first to determine the measurement of Y is the unit of collective dose equivalent from radiation exposure by monetary cost of the corresponding α values. Alpha value is typically determined by national authorities according to national conditions.

After you decide α value, Y can be represented as

$$Y = \alpha S \tag{12.28}$$

In the calculation of this formula, does not take into account individual dose distribution and the impact of subjective factors on radiation hazards cost. Cost-effectiveness analysis methods by comparing the various protection schemes reduce the protection of collective dose equivalent unit cost to determine the protection scheme. For example, for a practice associated with radiation exposure, if ith protection programme of the collective dose equivalent S_i, then the implementation of the programme i reduce the protection of collective dose equivalent unit cost is

$$\alpha_i = \frac{X_i - X_{i-1}}{S_{i-1} - S_i} \tag{12.29}$$

Smaller α_i, higher performance, which can choose the optimal solution in a variety of protective programs.

Use the cost-effectiveness analysis does not require the establishment of a common measure of the price scale protection and radiation hazards; we do not need a predetermined value α. Thus, relatively easy to use.

12.2.2   Radiation Protection Standards

In order to protect the health and safety of radiation workers and the public, according to the system of dose limitation established basic standards and principles. Used it to normal radiation under control, and prevent the occurrence of deterministic effects and reduce the incidence of stochastic effects to acceptable levels. In the past, radiation protection standard essentially is usually dose limits, and it leads to all kinds of derived levels. Since 1977, the International Commission on Radiological Protection (ICRP) issued 26th publication, extended content involved in radiation protection standards for the entire system of dose limitation: the legitimacy of practices, radiation protection optimization and dose limits. Basic limits for dose limits including dose limits for radiation workers, individuals of the public dose limits and accepted teaching training radiation dose limits for apprentices and students. Also has export limited value (including within irradiation of export limited value with years intake volume limited value ALI, gas contains radioactive concentration of export limited value with derived air concentration DAC), management limited value (to management of purpose, competent sector or enterprise head according to most optimization principles, on radiation protection about of any volume developed of management limited value, they are more strict than basic limited value or derived limited value), reference level (for effective to implementation protection, radiation protection sector prior provides of determine action of level, it including records level, and investigation level and intervention level).

China Atomic Energy industry started in the 1950 of the twentieth century. In order to adapt to the development of atomic energy, was enacted in 1960 the first radioactive radiation protection standards of the provisional regulations on the protection of health at work. In this standard, provides radioactive staff of daily maximum allowed dose (at with of units is Roentgen or biological Roentgen equivalent): gamma ray, and X ray, and beta particles and electronic for 0.05; alpha particles and Proton for 0.005, more charge ion and recoil nuclear for 0.0025; thermal neutron for 0.01; fast neutron for 0.005. Due to work need or other necessary of reasons, in the principle of week dose not over 0.3 biological Roentgen equivalent, day dose can over 0.05 biological Roentgen equivalent. Adjacent to the radiation in the workplace is not within the area of other staff involved in the radiation, external maximum permissible dose of ionizing radiation must not exceed 1/10 of the above standard. Residential and residential area, the external maximum permissible dose of ionizing radiation must not exceed natural background.

In 1974, issued a new standard GBJ8-74 of the radiation protection regulations. In the provisions, the exposed parts of the organ is classified into four categories. Occupational radiation workers in the first category organs (gonads, the whole body, red bone marrow, ocular lens) annual maximum permissible dose equivalent is 5 rem, the 2nd, 3rd and 4th category organs of the annual maximum permissible dose equivalent, respectively 30, 75 and 15 rem. Radioactive adjacent and nearby areas in the workplace and residents of the four category of organ year dose equivalent for radiation less than 1/10 of staff. The first category organ dose equivalent limits for radiation of public less than 1% of the annual maximum permissible dose equivalent of staff, other category of 1/30.

In order to adopt the new concepts and principles of radiological protection of ICRP 26th publication, in 1984 and 1988, respectively, China promulgated the basic standards for radiological protection and radiation protection regulations. Their annual dose of radiation workers incurred during the year for equivalent refers to the sum of external radiation effective dose equivalent and intake of radionuclides and the year accumulated effective dose equivalent, excluding natural background radiation and medical exposure. For radiation workers, in order to prevent harmful deterministic effects, the annual dose equivalent of lens limit is 150 mSv (15 rem); other single organ or tissue is 500 mSv (50 rem). In order to restrict stochastic effects, radiation workers under uniform body exposure annual dose equivalent limit of 50 mSv (5 rem). When long continued exposure to ionizing radiation, and members of the public an effective annual dose equivalent exceeding 1 mSv (0.1 rem). If the lifetime average annual effective dose equivalent dose does not exceed 1 mSv (0.1 rem) under consideration, allowing 5 mSv per year in some years (0.5 rem) as dose limits. Members of the public of the skin and ocular lens of the annual dose limit of 50 mSv (5 rem).

In 1991, the ICRP published its 60th publication, the International Commission on Radiological Protection Recommendations (1990). After that, the United Nations Food and Agricultural Organization (FAO), and International Atomic Energy institutions (IAEA), and International Labour Organization (ILO), and nuclear institutions of Organization for Economic Co-operation and Development (NEA/OECD), and Pan American Health Organization (PAHO) and the World Health Organization (WHO) common initiative developed an *International Basic Safety Standards for Protection against Ionizing Radiation and Radiation Sources* (following referred to BSS). IAEA published 115th security series in 1997, instead of its 9th security series, *Basic Safety Standards for Radiation Protection* (1982). BSS and ICRP60 in the on career irradiation of dose limited value provides: ① within 5 years of regulatory, effective dose not over 100 mSv, average annual not over 20 mSv; ② requirements any a years not over 50 mSv; ③ eye crystal of years equivalent dose for 150 mSv; ④ limbs (hand and feet) or skin of years equivalent dose for 500 mSv. Due to practice on key members of the residents group estimates the average dose limits are: ① annual effective dose of 1 mSv; ② in exceptional circumstances, if the average dose for 5 consecutive years does not exceed 1 mSv/a, the effective doses up to 5 mSv in a single year; ③ annual dose equivalent of lens 15 mSv; ④ skin annual equivalent dose of 50 mSv.

At present, relevant departments have prepared a coalition of groups that are equivalent to BSS modification, Chinese basic standards for radiation protection based on the principles of ICRP-60.

12.3   Radiation Protection Methods

Measures that must be taken in order to achieve protection standards, including technical protection measures and management methods. Within the technical protection methods can be divided into external radiation protection and internal radiation protection. Basic methods of external radiation protection are shortening the exposure time; increased distance with radiation sources; adding shielding between the man and radiation source. Can be summed up as time, distance and shielding. Internal radiation protection in the basic methods are radioactive materials "contain" and "diluted". Various protective devices and equipment is actually the embodiment of the basic method.

Management methods are an important part of protection methods, but it is also one of the aspects. Includes rules and regulations, training of personnel, institutions and funds management.

12.3.1   Human Radiation Effects

Human radiation effects refers to effects after exposure to ionizing radiation. Moreover, any natural ionizing radiation exposure is inevitable. In addition, many people also were in some production, medical and other social practices in the process of being exposed to ionizing radiation from artificial. Exposure to ionizing radiation on human beings may be divided into external and internal exposure in two ways, which is outside the body of sources of ionizing radiation on the human body parts as well as all tissues and organs of the exposure, the latter is the source of radiation into the body (radionuclide) issued by exposure to ionizing radiation on tissues and organs. When the human body subjected to ionizing radiation exposure, ionization of atoms and molecules in the organization change, if these molecules in living cells, cell itself may be directly or indirectly affected by the injury. The International Commission on Radiological Protection (ICRP) according to the modern concepts of radiation protection effect of radiation on people is classified as deterministic effects and random effects, as shown in Fig. 12.3.

12.3.2   Deterministic Effects

When there are enough cells in an organ or tissue killed or does not function properly, it appears clinically observed, reflecting the loss of an organ or tissue function damage. Dose is small, this damage does not occur, zero probability of occurrence; when the dose reaches a certain level (threshold dose), the probability of occurrence will increase rapidly to 1 (100%). Above the threshold dose, the extent of the damage will increase as the dose increases, reflecting more damaged cells, more serious of

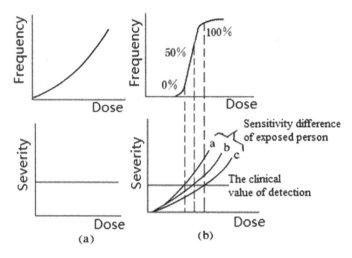

Fig. 12.3 Deterministic effects and random effects

the loss of function. For this effect, although individual cells are killed by radiation exposure is random in nature, but when there are a large number of cells are killed, is the effect of the inevitable. This effect is called deterministic effect, characterized by the severity mentioned above increases with the dosage increases above the threshold dose. Figure 12.3b contains different radio sensitivity in a group of people, a given deterministic effects (Clinical pathological condition can be confirmed) the frequency and severity, dose relationship. In the Figure, curve a, b, and c represent three different levels of radiation sensitivity. In the most sensitive population (curve a), the severity increased with increasing dose quickly, needed to reach the clinical pathology detection threshold dose below the crowds of poor sensitivity (curve b and c). The radiation sensitivity of different organizations are different, but as long as the absorbed dose of radiation is less than a couple of gray (Gy), few organizations appear clinically apparent damage. If the dose is for a number of years have been accepted, in the years when the dose is less than about 0.5 Gy, most organizations are unlikely to have serious effects. However, the gonads, ocular lens of radiation and bone marrow with high sensitivity. Some deterministic effects in these organizations is given in Table 12.1 the threshold dose. It can be seen that in general or prolonged exposure can raise the threshold dose values. Threshold dose here refers to the term in at least 1–5% personnel exposure dose required to cause a particular effect.

Deterministic effect appears to have a process of time; many important and deterministic effects only appears after a long period of latency time. The effects, which may appear within a few weeks after the exposure usually, called the early effect, effects that appear months or years after exposure is called late effects. In total body irradiation (TBI) cases, according to the dose size, there may be varying degrees of early effects, light such as mild blood counts; heavier such as mild discomfort; even acute radiation sickness. In General, 1–8 Gy dose will cause varying degrees of (mild, moderate, severe and very severe) hematopoietic form of acute radiation sickness,

Table 12.1   Threshold dose for an adult testis, ovary, eye lens and bone marrow

Organ and effect	Threshold dose		
	Total dose equivalent suffered from a brief irradiation/Sv	In many separated irradiation or prolonged exposure conditions suffered total dose equivalent/Sv	In many separated irradiation or prolonged exposure conditions for many years in the annual dose equivalent rate/Sv
Testis:			
Temporary infertility	0.5		0.4
Permanent infertility	3.5–6.0		2.0
Ovary:			
Infertility	2.5–6.0	6.0	> 0.2
Eye Lens:			
Detectable cloudy	0.5–2.0	5	> 0.1
Visual impairment (cataracts)	5.0	> 8	> 0.15
Bone marrow:			
Hematopoietic suppression	0.5		> 0.4

when severe acute radiation sickness is reached, if not under active treatment, the mortality rate was very high, cause of death is due to loss of bone marrow stem cells of bone marrow function failure. When the dose of more than about 5 Gy, will have other effects, including severe gastrointestinal damage, bone marrow damage, can cause death in 1–2 weeks. 10 Gy dosage can cause pneumonia and lead to death. When doses larger, nerves and cardiovascular system damage caused by shock and death within a few days. Approximate time of death and doses as shown in Table 12.2. The results in Table 12.2 are of a short time (a few minutes) received a heavy dose of γ-irradiation. If exposure lasts several hours or more, the resulting dose requires more of the effects listed. Human systemic acute lethal irradiation (60d) dose (LD 50/60) is an important parameter of acute radiation effects, but so far there is no recognized

Table 12.2   Sickness caused by acute irradiation, dose range and time of death after γ-ray whole body radiation

Systemic absorption dose/Gy	The main effect of causing death	Time of death/d
3–5	Bone marrow damage (LD 50/60)	30–60
5–15	Gastrointestinal and lung injury	10–20
> 15	Nervous system damage	1–5

value, estimated to be between 3–5 Gy. In the case of parts of body exposed, even if received larger doses in a short time, generally not to cause death, but there will be some other early effects, such as skin redness and dry peeling, the threshold dose of about 3–5 Gy, about 3 weeks after the symptoms appear. Any nuclear facilities under normal operating conditions, with good radiation protection measures, are not normally on the staff (not to mention the public) can lead to the early effects of radiation. Only in case of accidents occurring abnormal exposure of large doses can cause significant effects of early, but the probability is very small, especially can cause lethal effects of large doses of radiation, the probability is extremely small. Late effect of injury also dose-related, doses greater injuries more severe, but generally not fatal, but have the potential to cause the disability. Some organ function may be impaired or is likely to cause changes in other non-malignant, the most familiar examples are cataracts and skin damage. Nuclear facilities under normal operating conditions, as long as the proper protection, nor can also cause late effects of exposure.

12.3.3 Random Effects

If irradiated cells were not killed but is still alive but has changed, the resulting effect will be very different with deterministic effects. The stochastic effects are of two types, one is the cell damage caused. Damage formed by proliferation of cells after a clone, if it had not been removed by the body's defense mechanism, then after a period of a long incubation period, has the potential to develop into a malignant cell proliferation is out of control, commonly referred to as cancer. Radiation carcinogenesis is a major late effect of radiation-induced. Different tissues and organs have different sensitivity to radiation causes cancer. Radiation sensitivity with age, gender and other factors. Other injury due to gonadal germ cells caused by the exposure. Germ cell has the function of the genetic information is passed on to future generations. When the damage (mutations and chromosomal aberrations) occurs, as a possible error of the genetic information is passed on, cause variable degrees of severity in the offspring of irradiated various types of genetic disease, premature death and severe mental retardation, light skin spots.

Stochastic effects are characterized by their probability of occurrence and increases with the dose, but the severity of the dose size is irrelevant. Figure 12.3a illustrates the stochastic effects of this trait. In cancer, for example, is not the size of the dose gives cancer induced by the severity or the light, its severity and the type of cancer is only dependent on the location. The occurrence of cancer and genetic effects may result from damage to individual cells, random nature of its processes, stochastic effects is the name. Stochastic effects may not be a threshold dose, but so far, science is unable to make firm conclusions. For the purposes of radiation protection, usually assumes that there is no threshold dose, meaning that this dose, no matter how small, a dose is always linked to the risk of stochastic effects. In this way, to stochastic effects are impossible to completely prevent it from happening, but only reduce the dosage to limit its probability of occurrence. On radiation protection

Table 12.3 The nominal value of stochastic effects coefficient

Irradiated groups	Nominal value of stochastic effects coefficient/(10^{-2} Sv^{-1})			
	Lethal cancer	Nonfatal cancer	Serious genetic effects	Total
Adult Staff	4.0	0.8	0.8	5.6
The whole crowd	5.0	1.0	1.3	7.3

also assumes that, in daily doses involved in radiation protection equivalent within the entire range of dose equivalent rate is a linear relationship between dose and incidence of stochastic effects. Quantitative representation of the risk of stochastic effects, using the concept of probability coefficients, it refers to a unit of dose equivalent probability of stochastic effects induced by irradiation. Stochastic effects probability coefficient by fatal cancers, non-fatal cancer and severe genetic effects posed by the effect of probability factor, specific values are provided in Table 12.3.

Fatal cancer probability coefficient is based mainly on Japan in Hiroshima, Nagasaki made by survivors of the 1945 atomic bomb attacks after more than 100,000 systems, long-term epidemiological survey results. Surveys show that the incidence of certain cancers in survivors than in the control group, we can estimate the probability of radiation carcinogenesis. However, this is the result of instant exposure to large doses, and people are more concerned with radiation protection involves very small doses of carcinogenic effect of dose rate and exposure conditions. Carcinogenic effect of the latter under lighter than previous conditions, in accordance with Hiroshima, Nagasaki information results need to divide by the appropriate reduction factor can be applied to small dose, low dose rate situations. Results of the current ICRP based on radiation biology research information on Hiroshima, Nagasaki, suggested that the reduction factor is 2. Table 12.3 data are obtained in this way.

ICRP recently discussed the impact of genetic factors on radiogenic cancer risk, put forward that has a strong expression of tumor suppressor gene mutation of autosomal dominant cancer-prone families, increase the probability of radiation carcinogenesis may 5–100 times, certain disorders associated with DNA repair defects in irradiated cancer risk will also increase. Due to familial incidence of cancer in the population is only 1% or lower cancer risk estimates for population is not affected. Have always been very concerned about the genetic effects of radiation, but so far there is no positive evidence due to natural or artificial radiation exposure, one descendant of genetic damage, even descendants of the survivors of Hiroshima and Nagasaki made a large-scale survey found no statistically significant increase in genetic damage. However, the use of animals and plants made by a large number of experimental studies have shown that there are genetic effects of radiation. Therefore, from the radiation protection point of view, it is necessary to assume that humans also has this effect. Table 12.3 data are based mainly on genetic effects on experimental animals (mainly for the mouse) research results derived.

When considering the effects of radiation on genetic diseases, ICRP believe that it is necessary to consider multi-factors disease (such as high blood pressure, coronary heart disease, congenital malformation) and disease-related mutations and environmental factors. The concept of catastrophe share discussed rate increases impact on the incidence of radiation-induced mutations; the basic conclusion is that low levels of radiation-induced mutations on the incidence of multifactorial diseases will not have a significant effect. Different types and energies of radiation-induced risk of stochastic effects is not exactly the same. In radiation protection, on several common of radiation type do has following of divided: gamma and X ray and the electronic for same grade, as assumes that for 1, is neutron for 5–20, specific numerical depending on its energy and set; alpha particles is for 20. In Table 12.3, due to probability coefficient is to unit dose equivalent of probability said, radiation of type and energy on induced randomness effect of effect actually has been considered.

12.4 Radiation Monitoring

For the evaluation and control of radiation or radioactive material exposure, site monitoring of measurement and interpretation of the measurement results are made including environment monitoring and effluent monitoring. Site monitoring can be divided into individual dose monitoring and radiation monitoring in the workplace. Environmental radiation monitoring is to monitor radiation levels in the environment outside the workplace. Environmental monitoring can be divided into investigation before operation and monitoring during operation and after decommissioning, environmental monitoring should pay particular attention to identification and monitoring of key species, critical path and critical population groups. Effluent monitoring object is the connection between site and environment, its main tasks are: ① checking whether the amount of radioactive materials released into the environment meets the requirements of management limits; ② test the effectiveness of radioactive waste disposal facility, discovery could lead to risks of accidents in a timely manner; ③ aim at providing environmental assessment source term [8].

In order to do a monitoring programme, all accompanied by radiation practices and facilities, should be according to the principles of radiation protection optimization, and gives the radiation-monitoring programme. Monitoring plan should include: ① monitoring description; ② major risk factors, ways and may harm people's identification and analysis; ③ monitoring objects and cycle choices; ④the choice of monitoring methods and equipment, including sensitivity and uncertainty analyses; ⑤ monitoring quality assurance; ⑥ recording and reporting system.

12.5 Evaluation of Radiation Protection

According to the principles and standards for radiation protection, made the evaluation to the quality and efficiency of the protection [9]. Specific practices is, according to source items and results of radiation monitoring, select appropriate of mode and parameter, calculation staff and public by personal dose and collective dose; according to optimization principles of radiation protection, integrated analysis protection method and dose data, proposed further improved radiation protection of method and the best distribution of protection resources programme, makes staff and public by dose as reasonable low as possible [10].

Evaluation of radiation protection can be broken down into staff assessment of radiation protection and radiation protection of the public evaluation. Evaluation of radiation protection of the public is the main part of the report on the environmental impact of nuclear facilities. At the time of evaluation for radiation protection of the public, should pay special attention to the model and parameter selection and actual monitoring data to verify the availability of models and parameters.

12.6   Radiation Emergency

Because of the nuclear facility there is a great potential danger, so nuclear contingency has developed an important aspect of radiation protection. Of any nuclear facilities, there should be potential hazard classification, the classification of emergency response, development of emergency response plans accordingly [5].

Emergency response plans should mainly include: ① emergency classification; ② emergency organization; ③ emergency facilities; ④accident consequence assessment; ⑤ the emergency measures.

Emergency response plan is part of the radiation protection programme, in developing platform for radiation protection and radiation monitoring programme, should take into account the requirements of emergency response plans, emergency response in conjunction with normal radiation protection organically [11].

Exercises

1. Briefly describe the time, distance and shielding radiation protection applications.
2. What are the main differences between deterministic effects and random effects?
3. What is currently radiation protection standards used in China?
4. What is the activity? What is the meaning of "There is a source of radioactivity is 0.01 Ci "?
5. What is the absorbed dose? What is a dedicated unit of absorbed dose?
6. What is the difference between equivalent dose and effective dose?
7. What is the difference between personal dose collective doses?
8. In order to restrict stochastic effects, radiation workers under uniform body exposure annual dose limit of 50 mSv. It is equivalent dose or effective dose, why?

Symbol Table

Symbol	Physical name	Value of unit
A	Active of radionuclides	Bq
	Mass number	–
	Area	m^2
	Amplification factor	–
B	Magnetic induction	T
C	Cost	$
	Herald nucleon number density	$1/cm^3$
	Speed of light	2.998×10^8 m/s
	Abundance	–
	Electric charge	C
	Capacitance	F
C_p	Heat capacity	J/(kg°C)
C_R	Count rate	n/s
c	Speed of sound	m/s
D	Absorbed dose	Gy
	Diameter	m
	Neutron diffusion coefficient	cm
d	Inner diameter	m
E	Effective dose	Sv
	The average annual power supply	kW·h
	Energy	J
	Young's modulus	MPa
e	Induced voltage	V
	Charge of the electron	1.602×10^{-19}C
F	Feed mass flow rate	kg/s
	Magnetomotive force	At
	Heat pipe factor	–
	Radiation angle factor	–
	Force	N
f	Thermal neutron utilization factor	Hz
	Frequency	Hz
	Friction coefficient	–
G	Mass flux	$kg/(m^2·s)$
	Shear modulus	MPa
g	Acceleration of gravity	9.81 m/s^2
H	Equivalent dose	Sv
	Magnetic field intensity	At/m
	Height, lift	m

(continued)

(continued)

Symbol	Physical name	Value of unit
	Enthalpy	J
h	Specific enthalpy	J/kg
	Convective heat transfer coefficient	W/(m^2 °C)
I	Electric current	A
	Light intensity	Cd
J	Neutron fluence	n/(cm^{2}s)
K	Kerma	Gy
	Equilibrium constant	–
	Local resistance coefficient	–
	The bulk modulus	MPa
	Relative permittivity	–
k	Multiplication factor	–
	Thermal conductivity	W/(m °C)
	Coulomb constant	9.0×10^9 Nm^2/C^2
L	Length	m
	Probability does not leak	–
	Inductance	H
M	Source multiplication factor	–
	Steam moisture	–
	Molar mass	g/mol
m	Mass	kg
N	The number of collisions	–
	Turns	–
	Nucleon number density	1/m^3
n	Refractive index	–
	Amount of substance	mol
	Rotating speed	rpm
P	Price	\$/kg
	Product flow rate	kg/s
	Power	W
	Wetted diameter	m
p	Pressure	Pa
	Probability	–
Q	Cost of the first fuel load	\$
	Heat	J
	Reactive power	VAR
q	Moderator density	n/(m^{3}s)

(continued)

(continued)

Symbol	Physical name	Value of unit
	Heat flux	W/m^2
	The amount of charge	C
q_m	Mass Flow rate	kg/s
q_V	Volumetric heat release rate	W/m^3
	Volume flow rate	m^3/s
R	Fission rate	$1/(cm^3 \cdot s)$
	Reluctance	At/wb
	Thermal resistance	°C/W
	Resistance	Ω
	Ideal gas constant	$J/(mol\ °K)$
r	Radius	m
S	Collective equivalent dose	man·Sv
	Seperation work unit	kg SWU
	Source strength	$n/(cm^3 s)$
	Intuitive power	VA
	Displacement	m
	Entropy	J
s	Specific entropy	$J/(kg\ °K)$
	Slip ratio	–
T	Half life	s
	Thermodynamic temperature	°K
	Torque	Nm
t	Time	s
	Temperature	°C
U	Internal energy	J
u	Specific internal energy	J/kg
V	Value function	–
	Section average speed	m/s
	Voltage	V
	Volume	m^3
v	Rate of speed	m/s
	Specific volume	m^3/kg
W	Power	J
	Tailings flow	kg/s
w	Specific work	J/kg
X	Impedance	Ω
x	Exposure	C/kg

(continued)

(continued)

Symbol	Physical name	Value of unit
	Steam quality	–
	Enrichment	–
Z	Impedance	Ω
	The number of protons	–
α	Coefficient	–
β	Delayed neutron fraction	–
χ_e	Equilibrium quality	–
ε	Annual depreciation rate	–
	Fast fission factor	–
	Emissivity	–
	Strain	–
	Permittivity	F/m
Φ	Magnetic flux	Wb
ϕ	Particle fluence	m^{-2}
ϕ^2	Multiplication of two-phase pressure drop	–
η	Effectiveness	–
	Proliferation factor	–
φ	Neutron flux, particle flux	$n/(cm^2 \cdot s)$
Λ	Average neutron generation time	s
λ	Decay constant	1/s
μ	Viscosity	Pa s
	Poisson's ratio	–
	Permeability	H/m
θ	Angle	Degree
ρ	Reactivity	$\Delta k/k$
	Density	kg/m^3
	Resistivity	Ω m
$\sum$	Macroscopic cross section	–
σ	Microscopic cross section	b
	Stefan-Boltzmann constant	5.67×10^{-8} W/m^2 K^4
	Stress	Pa
τ	Reactor period	s
	Life expectancy	s
	Fermi age	m^2
ω	Angular frequency	
ψ	Energy fluence	J/m^2
ξ	Logarithmic average loss factor	–

References

1. Cohen BL. Catalog of risks extended and updated. Health Phys Radiat Prot J. 1991;61:317–35.
2. Deping L. Pan Ziqiang radiation protection manual Part III: radiation safety. Beijing: Atomic Energy Press; 1990.
3. ICRU. Determination of dose equivalent resulting from external radiation sources. Maryland: ICRU; 1985.
4. NCRP. Ionizing Radiation Exposure of the Population of the United States, Report No. 93.
5. DOE Order 5480.20A. Personnel Selection, Qualification, and Training Requirements for DOE Nuclear Facilities. U.S. Department of Energy, November, 1994.
6. ICRP. Conversion coefficients for use in radiological protection against external radiation. Oxford: Oxford Pergamon Press; 1995.
7. Ammerich M, Briand A, Laborde JC, Mulcey P, Savornin J. Contamination releases from HEPA filters under high temperature conditions. In: Proceedings of the 20th DOE/NRC nuclear air cleaning conference (M.W. First, Ed), NUREG/CP-0098 (CONF-880822). Boston, MA: Harvard Air Cleaning Laboratory; 1989.
8. U.S. Department of Energy, DOE Radiological Control Standard, [reference Technical Standards Program (TSP) project number SAFT-0039].
9. AIChE. Guidelines for Hazard Evaluation Procedures, 2nd ed. New York, NY: Center for Chemical Process Safety of the AIChE; 1992. P. 10017.
10. U.S. Nuclear Regulatory Commission. Instruction Concerning Risks From Occupational Radiation Exposure. U.S. NRC Regulatory Guide 8.29, Version 1, February 1997.
11. U.S. Department of Energy. Occupational Radiation Protection. 10 CFR 835, November 1998.

Printed in the United States
by Baker & Taylor Publisher Services